Chemistry

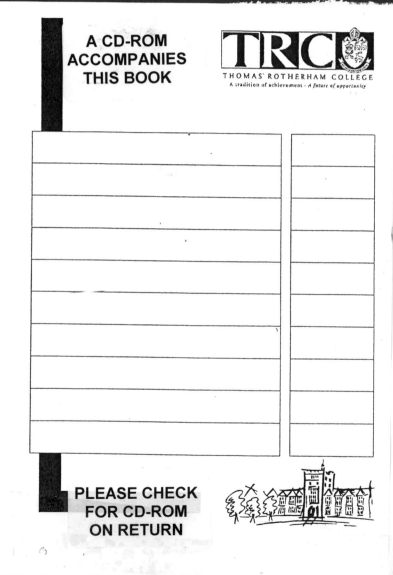

Exclusively endorsed by OCR for GCE Chemistry A

Heinemann is an imprint of Pearson Education Limited, a company incorporated in England and Wales, having its registered office at Edinburgh Gate, Harlow, Essex CM20 2JE. Registered company number: 872828

www.heinemann.co.uk

Heinemann is a registered trademark of Pearson Education Limited

First published 2007

12 11 10 09 08
10 9 8 7 6 5 4 3 2

British Library Cataloguing in Publication Data is available from the British Library on request.

ISBN 978 0 435691 81 3

Edited by Susan Watt, Melissa Wesley
Index compiled by Wendy Simpson
Designed by Kamae Design
Project managed and typeset by Wearset Ltd, Boldon, Tyne and Wear
Original illustrations © Pearson Education Limited 2007
Illustrated by Wearset Ltd, Boldon, Tyne and Wear
Picture research by Q2AMedia
Cover photo © Science Photo Library. The cover image is a scanning electron micrograph of calcium phosphate crystals.
Printed in the UK by Scotprint

Acknowledgements
We would like to thank the following for their invaluable help in the development and trialling of this course: Andrew Gelling, Peter Haigh, Amanda Hawkins, Dave Keble, Maggie Perry, Michael Taylor and Chris Wood.

The authors and publisher would like to thank the following for permission to reproduce photographs:

p3 Andrew Lambert Photography/Science Photo Library; **p4** Science Photo Library; **p7** Mary Evans Picture Library/Alamy; **p10** Rob Ritchie; **p14** The Italian Center for Resistance Philately; **p16 L** Getty Images; **p16 R** Rob Ritchie; **p17 T** Rob Ritchie; **p17 B** Rob Ritchie; **p18 T** Dirk Wiersma/Science Photo Library; **p18 B** Andrew Lambert Photography/Science Photo Library; **p22** Andrew Lambert Photography/Science Photo Library; **p23** Andrew Lambert Photography/Science Photo Library; **p24 T** Photolibrary.com; **p24 B** David Taylor/Science Photo Library; **p25** Holt Studios International Ltd/Alamy; **p26** Andrew Lambert Photography/Science Photo Library; **p27** Charles D. Winters/Science Photo Library; **p28** Martyn F. Chillmaid/Science Photo Library; **p33** Andrew Lambert Photography/Science Photo Library; **p39** Alfred Pasieka/Science Photo Library; **p40** Limber/Istockphoto; **p67 R** Vasyl Helevachuk/Shutterstock; **p49** Harcourt Index/Corbis; **p60** Thomas Hollyman/Science Photo Library; **p62** Martin Harvey/Alamy; **p65 L** Martha Holmes/Nature Picture Library; **p65 R** David M. Denni/Photolibrary.com; **p66** Andrew Lambert Photography/Science Photo Library; **p67 L** Simon Fraser/Science Photo Library; **p70** Andrew Lambert Photography/Science Photo Library; **p71 T** Lawrence Lawry/Science Photo Library; **p71 B** Andrew McClenaghan/Science Photo Library; **p77** TEK Image/Science Photo Library; **p78 T** Bettmann/Corbis; **p78 B** Jean-Loup Charmet/Science Photo Library; **p79** Science Photo Library; **p80 T** CCI Archives/Science Photo Library; **p80 B** Science Photo Library; **p86** Andrew Lambert Photography/Science Photo Library; **p88 T** Andrew Lambert Photography/Science Photo Library; **p88 B** Andrew Lambert Photography/
[…]ience Photo Library; **p90** Ace Stock Limited/Alamy; **p91** Jim Strawser/
L[…]y; **p92** Andrew Lambert Photography/Science Photo Library; **p93** Andrew
Ph[…]t Photography/Science Photo Library; **p94** Mark A. Schneider/Science
Pau[…]rary; **p95** Andrew Lambert Photography/Science Photo Library; **p101**
p102 […]n/Science Photo Library; **p102 T** Dmitry Kosterev/Shutterstock;
GFDL; […]Corporation/Alamy; **p102 B** Photolibrary.com; **p114** Corbis; **p118**
Martyn F[…]nrad Zelazowski/Alamy; **p120 T** Ian Leonard/Alamy; **p120 B**
Photo Lib[…]d/Science Photo Library; **p121** Chemical Design Ltd/Science
Allan/Nature […] Carlos Goldin/Science Photo Library; **p123 TL** Doug
[…]**p123 TR** Andrew B. Singer; **p123 B** Richard Smith/
Cumulus; **p125** Sheila Terry/Science Photo Library; **p129 B** Andrew Lambert Photography/Science Photo Library; **p134** JoLin/Shutterstock; **p135** Maximilian Stock LTD/Photolibrary.com; **p136** Cordelia Molloy/Science Photo Library; **p137** Courtesy of DuPont; **p138 TM** James Holmes/Zedcor/Science Photo Library; **p138 TR** Photolibrary.com; **p138 BR** Kheng Guan Toh/Shutterstock; **p138 L** Moodboard/Corbis; **p139** Bob Krist/Corbis; **p140 T** Robert Brook/Science Photo Library; **p140 M** TEK Image/Science Photo Library; **p140 B** Ferrara PVC Recycling Plant, Italy; **p141 L** Keith M. Law/Alamy; **p141 TR** Caro/Alamy; **p141 MR** NEC Corporation/NTT DoCoMo; **p141 BR** Angela Hampton/Ecoscene; **p147** NASA; **p148** Alex Bartel/Science Photo Library; **p149 T** Maximilian Stock LTD/Science Photo Library; **p149 M** Ron Giling/Still Pictures; **p149 B** Transtock Inc./Alamy; **p152 T** Martyn F. Chillmaid/Science Photo Library; **p152 M** Andrew Lambert Photography/Science Photo Library; **p152 B** Andrew Lambert Photography/Science Photo Library; **p153 L** Rob Ritchie; **p153 R** Rob Ritchie; **p156** Sinclair Stammers/Science Photo Library; **p159** Dave Gent; **p160 T** Leonard Lessin/Science Photo Library; **p160 B** Martyn F. Chillmaid/Science Photo Library; **p161** Science Photo Library; **p164** Robert Brook/Science Photo Library; **p167** TEK Image/Science Photo Library; **p170 T** Library of Congress/Science Photo Library; **p170 B** NASA; **p171** Geoff Tompkinson/Science Photo Library; **p175** James Holmes/Oxford Centre for Molecular Sciences/Science Photo Library; **p183** Biosym Technologies, Inc./Science Photo Library; **p186 T** David Gent; **p186 B** The Boeing Company, 2007. All rights reserved; **p187** Martyn F. Chillmaid/Science Photo Library; **p198** University of Tartu; **p202 T** Charles D. Winters/Science Photo Library; **p202 B** John Mead/Science Photo Library; **p204** Martyn F. Chillmaid/Science Photo Library; **p205** Dave Gent; **p206** David Hay Jones/Science Photo Library; **p207 T** KPT Power Photos/Cumulus; **p207 B** Dave Gent; **p208** Science Photo Library; **p210** Science Photo Library; **p213** Corbis; **p212** Andrew Lambert Photography/Science Photo Library; **p219** Tony Craddock/Science Photo Library; **p220 L** NASA; **p220 M** NASA; **p220 R** Hubble; **p221** Dr Keith Wheeler/Science Photo Library; **p222 T** Joseph Petit – UNEP/Still Pictures; **p222 M** Greenpeace/Beltrà/Archivo Museo Salesiano/De Agostini; **p222 B** Greenpeace/Beltrà/Archivo Museo Salesiano/De Agostini; **p223** Hank Morgan/Science Photo Library; **p224** Richard Folwell/Science Photo Library; **p227** Dave Gent; **p228** NASA; **p229** Jerry Mason/Science Photo Library; **p230** P.G. Adam, Publiphoto Diffusion/Science Photo Library; **p231 L** Dave Gent; **p231 R** London Aerial Photo Library/Alamy; **p232** School of Forestry and Environmental Studies, Yale University; **p233 T** Saturn Stills/Science Photo Library; **p233 B** Stanford University; **p234 T** Adrienne-Hart-Davis/Science Photo Library; **p234 B** TEK Image/Science Photo Library; **p235** Phototake inc./Alamy

The authors and publisher would like to thank the following for permission to use copyright material:

Figure 1, p122 reproduced with permission from *Time for Kids*, Vol. 8, No. 18, February 21, 2003.
Figure 4, p123 reproduced with permission from Andy Singer.
Figure 3, p185 reproduced with permission from *New Scientist Magazine*, 25 April 2004.
Table 1 (data only), p231 reproduced with permission from Chemical Industry Education Centre, University of York, York.

Every effort has been made to contact copyright holders of material reproduced in this book. Any omissions will be rectified in subsequent printings if notice is given to the publisher.

Websites
There are links to websites relevant to this book. In order to ensure that the links are up-to-date, that the links work, and that the links are not inadvertently linked to sites that could be considered offensive, we have made the links available on the Heinemann website at www.heinemann.co.uk/hotlinks. When you access the site, the express code is 1813P.

Exam café student CD-ROM

Original illustrations, screen designs and animation by Michael Heald
Photographs © iStock Ltd
Developed by Elektra Media Ltd

Technical problems
If you encounter technical problems whilst running this software, please contact the Customer Support team on 01865 888108 or email software.enquiries@heinemann.co.uk

OCR

Chemistry

AS

Exclusively endorsed by OCR for GCE Chemistry A

Dave Gent and Rob Ritchie
Series Editor: Rob Ritchie

www.heinemann.co.uk
✓ Free online support
✓ Useful weblinks
✓ 24 hour online ordering

01865 888080

In Exclusive Partnership

Contents

Introduction

How to use this book

In this book you will find a number of features planned to help you.

- **Module openers** – these introductions set the context for the topics covered in the module. They also have a short set of questions that you should already be able to answer from your previous science courses.
- **Double-page spreads** are filled with information and questions about each topic.
- **End-of-module summary and practice** pages help you link together all the topics within each module.
- **End-of-module practice examination questions** have been selected to show you the types of question that may appear in your examination.

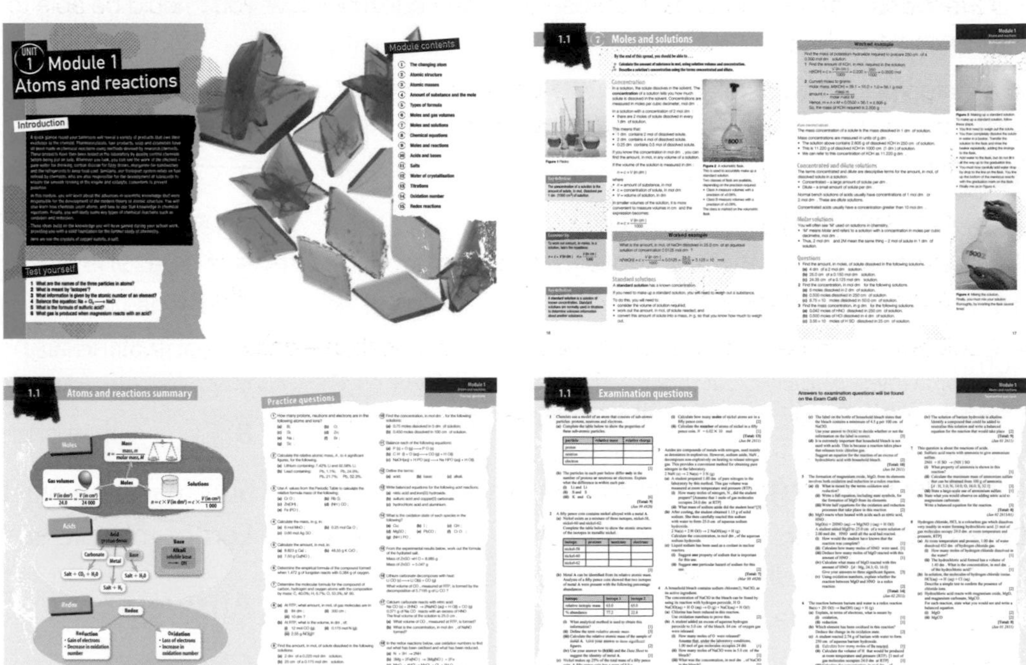

Within each double-page spread you will find other features to highlight important points.

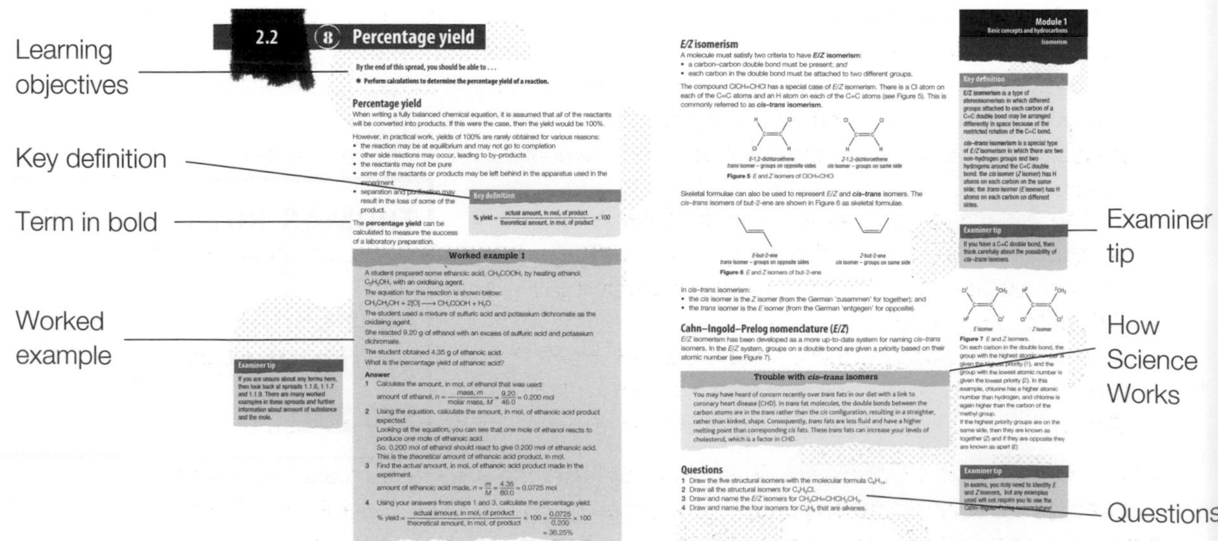

- **Learning objectives** – these are taken from the Chemistry AS specification to highlight what you need to know and to understand.
- **Key definitions** – these are terms in the specification. You must know the definitions and how to use them.
- **Terms in bold** are important terms, used by chemists, that you are expected to know. You will find each term in bold listed in the glossary at the end of the book.
- **Examiner tips** are selected points to help you avoid common errors in the examinations.
- **Worked examples** show you how calculations should be set out.
- **How Science Works** – this book has been written to reflect the way that scientists work. Certain sections have been highlighted as good examples of how science works.
- **Questions** – at the end of each spread are a few questions that you should be able to answer after studying that spread.

In addition, you'll find an Exam Café CD-ROM in the back of the book, with more questions, revision flashcards, study tips, answers to the examination questions in the book and more.

NewScientist

Reinforce your learning and keep up to date with recent developments in science by taking advantage of Heinemann's unique partnership with New Scientist. Visit www.heinemann.co.uk/newscientistmagazine for guidance and discounts on subscriptions.

The examination

It is useful to know some of the language used by examiners. Often this means little more than simply looking closely at the wording used in each question on the paper.

- Look at the number of **marks allocated** to a part question – ensure you write enough points down to gain these marks. The number of marks allocated for any part of a question is a guide to the depth of treatment required for the answer.
- Look for words in **bold**. These are meant to draw your attention.
- Look for words in *italics*. These are often used for emphasis.

Diagrams, tables and equations often communicate your answer better than trying to explain everything in sentences.

Each question will have an **action word**. Some action words are explained below.

- *Define*: only a formal statement of a definition is required.
- *Explain*: a supporting argument is required using chemical background. The depth of treatment should be judged from the marks allocated for the question.
- *State*: a concise answer is expected, with little or no supporting argument.
- *List*: give a number of points with no elaboration. If you are asked for **two** points, then give only two! A third incorrect point will cost you a mark.
- *Describe*: state in words, with diagrams if appropriate, the main points of the topic.
- *Discuss*: give a detailed account of the points involved in the topic.
- *Deduce/Predict*: make a logical connection between pieces of information given in the question.
- *Outline*: restrict the answer to essential detail only.
- *Suggest*: you are expected to apply your chemical knowledge and understanding to a novel situation, which you may not have covered in the specification.
- *Calculate*: a numerical answer is required. Working should be shown.
- *Determine*: the quantity cannot be measured directly but is obtained by calculation, substituting values of other quantities into a standard formula.
- *Sketch*: in graphs, the shape of the curve need only be qualitatively correct.

When you first read the question, do so very carefully. Once you have read something incorrectly, it is very difficult to get the incorrect wording out of your head! If you have time at the end of the examination, read through your answers again to check that you have answered the questions that have been set.

UNIT 1

Module 1 Atoms and reactions

Introduction

A quick glance round your bathroom will reveal a variety of products that owe their existence to the chemist. Pharmaceuticals, hair products, soap and cosmetics have all been made in chemical reactions using methods devised by research chemists. These products have then been tested in the laboratory by quality control chemists before being put on sale. Wherever you look, you can see the work of the chemist – pure water for drinking, carbon dioxide for fizzy drinks, margarine for sandwiches and the refrigerants to keep food cool. Similarly, our transport system relies on fuel refined by chemists, who are also responsible for the development of lubricants to ensure the smooth running of the engine and catalytic converters to prevent pollution.

In this module, you will learn about the advances in scientific knowledge that were responsible for the development of the modern theory of *atomic structure*. You will also learn how chemists count atoms, and how to use that knowledge in chemical equations. Finally, you will study some key types of chemical reactions such as oxidation and reduction.

These ideas build on the knowledge you will have gained during your school work, providing you with a solid foundation for the further study of chemistry.

Here we see crystals of copper sulfate, a salt.

Test yourself

1 What are the names of the three particles in atoms?
2 What is meant by 'isotopes'?
3 What information is given by the atomic number of an element?
4 Balance the equation: $Na + Cl_2 \longrightarrow NaCl$
5 What is the formula of sulfuric acid?
6 What gas is produced when magnesium reacts with an acid?

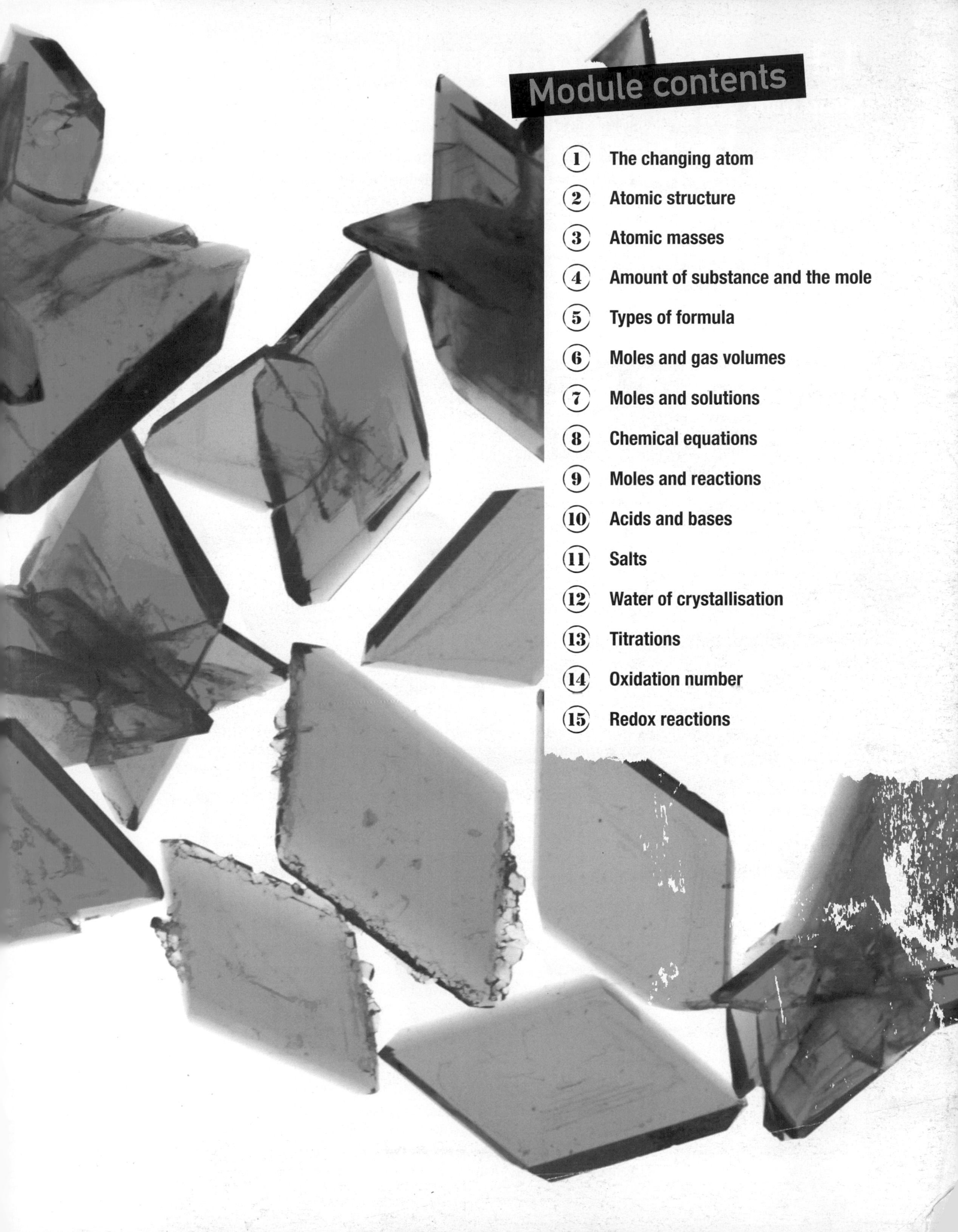

Module contents

1.1 How Science Works

(1) The changing atom

By the end of this spread, you should be able to . . .

- ✱ **Describe how the model of the atom has changed over the years, and how it continues to do so.**
- ✱ **Understand that scientific knowledge is always evolving.**
- ✱ **Describe how new theories are accepted by scientists.**

Figure 1 Dalton's symbols and atomic masses

The Greek atom

In the fifth century BC, the Greek philosopher Democritus developed the first idea of the *atom*. He thought that matter was made out of particles. He suggested that you could divide a sample of matter only so many times. Eventually, he believed, you would end up with a particle that could not be split any further. Democritus called this particle 'átomos' – Greek for indivisible. He also imagined that atoms were held together by tiny hooks.

John Dalton

In the early 1800s, John Dalton developed his atomic theory. This included the following predictions about atoms:

- Atoms are tiny particles that make up elements.
- Atoms cannot be divided.
- All atoms of a given element are the same.
- Atoms of one element are different from those of every other element.

Dalton used his own symbols to represent atoms of different elements. He also developed the first table of atomic masses. Some of these are shown in Figure 1.

The strength of Dalton's atomic theory is that most of his predictions still hold true today. You can apply Dalton's predictions to much of modern-day chemistry.

Thomson's corpuscles

Thomson used the term 'corpuscles' for the negative charges. It was only later that the name *electron* became accepted. By measuring their speed and specific charge, he deduced that these corpuscles must be nearly 2000 times smaller in mass than the hydrogen ion – then the lightest known atomic particle.

Joseph John (J J) Thomson

Between 1897 and 1906, through a series of experiments using cathode rays, Thompson discovered the electron. Using this discovery, he was able to show that atoms were made up of smaller particles.

Scientists had recently discovered cathode rays, but they had no idea what they were. Cathode rays are emitted from the negative electrode inside cathode-ray tubes (CRTs), the fore-runners of CRT televisions.

Thomson discovered that cathode rays were in fact a stream of particles. Each particle had the following properties:

- It had a negative charge.
- It could be deflected by both a magnet and an electric field.
- It had a very, very small mass.

Cathode rays were, in fact, electrons. Thomson concluded that they must have come from within the atoms of the electrodes themselves. The idea that an atom could not be split any further, proposed by the ancient Greeks and Dalton, had been disproved.

Thomson proposed that an atom should be thought of as being made up of negative electrons moving around in a 'sea' of positive charge. In Thomson's atom, the overall negative charge is the same as the overall positive charge. This means that the atom is neutral with no overall charge. This model is commonly called the **plum-pudding atom**, as shown in Figure 2.

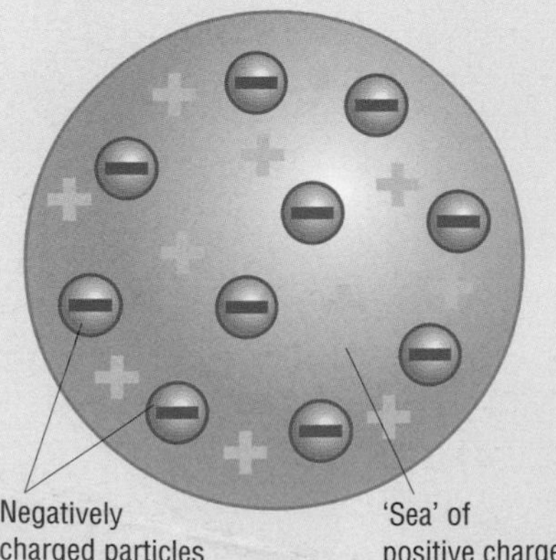

Figure 2 J J Thomson's plum-pudding atom

Ernest Rutherford

In 1909, Rutherford began work that led to the downfall of the plum-pudding model.

Two of Rutherford's students, Hans Geiger and Ernest Marsden, carried out an experiment at the University of Manchester. They directed α-particles (alpha particles) towards a sheet of very thin gold foil and measured any deflection (change in direction) of the particles. Rutherford calculated that a plum-pudding atom would hardly deflect α-particles at all.

The results were astonishing, as shown in Figure 3:

- most of the particles were not deflected at all;
- a very small percentage of particles were deflected through large angles;
- a very few particles were actually deflected back towards the source.

Rutherford commented: 'It was almost as incredible as if you fired a fifteen-inch shell at a piece of tissue paper and it came back and hit you.'

From these observations, Rutherford calculated that an atom's positive charge must be concentrated into a volume much smaller than the size of the atom itself.

In 1911, he proposed the following model for the atom.

- The positive charge of an atom and most of its mass are concentrated in a nucleus, at the centre.
- Negative electrons orbit this nucleus, just like the planets orbit around the Sun.
- Most of an atom's volume would be in the space between the tiny nucleus and the orbiting electrons.
- The overall positive and negative charges must balance.

Rutherford had proposed the *nuclear atom*.

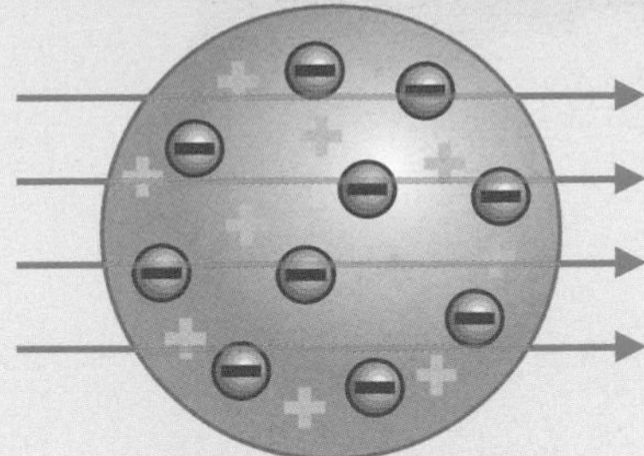

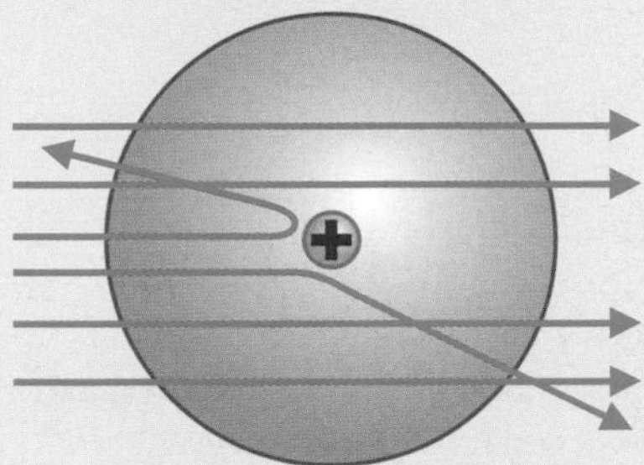

Figure 3 Rutherford's gold-leaf experiment. Top: Expected results: α-particles passing through the plum-pudding atom undisturbed. Bottom: Observed results: a few particles were deflected, indicating a small, concentrated positive charge

Neils Bohr and others

In 1913, the Danish physicist Neils Bohr altered Rutherford's model to allow electrons to follow only certain paths. Otherwise, electrons would spiral into the nucleus. This was the *planetary atom*, in which electrons orbited a central nuclear 'sun' in 'shells'. In the same year, Henry Moseley discovered a link between X-ray frequencies and an element's atomic number (i.e. its order in the Periodic Table). At the time, Moseley couldn't explain this. In 1918, Rutherford discovered the proton. We now know that the atomic number tells us the number of protons in an element's atom.

In 1923, the French physicist Louis de Broglie suggested that particles could have the nature of both a wave and a particle. In 1926, the Austrian physicist Erwin Schrödinger suggested that an electron has wave properties in an atom. He also introduced the idea of atomic orbitals. You will learn about atomic orbitals in spread 1.2.2.

In 1932, the English physicist James Chadwick bombarded light elements with high-energy α-particles. He found that a new type of radiation was emitted, much more penetrating than any that had been detected before. He showed that this new type of radiation was made up of uncharged particles. Each particle had approximately the same mass as a proton. These uncharged particles became known as neutrons, because they have no charge.

Further theories have since been proposed for the structure of the atom. It is now thought that protons and neutrons themselves are made up of even smaller particles called quarks.

Examiner tip

You can find out more about quarks and other very small particles on the Internet. For your A-level course, examiners will not ask you about quarks or even about the development of atomic theory. But examiners will ask you about protons, neutrons and electrons. You will find out more about these in the forthcoming spreads.

So what is an atom, anyway?

The theories and ideas involved in the development of an atom's structure give a fascinating insight into how science works. The scientific method is often represented as the sequence of experiment, observation, analysis and conclusion. The model of the nuclear atom – consisting of protons, neutrons and electrons – is extremely useful and allows us to explain chemical bonding, as well as to understand the pattern of the Periodic Table. The nuclear atom is well accepted by chemists and is firmly based on sound experimental evidence. It is amazing to think that these experiments all began with investigating particles that cannot even be seen.

However, science does advance. The nuclear model of the atom has now been replaced, and future models will also be replaced as science advances further. We have come a long way from the four elements of the ancient Greeks.

Questions

1 What model of the atom did Dalton develop in the early 1800s?

2 What model of the atom was developed following the series of experiments that took place between 1897 and 1911?

1.1 (2) Atomic structure

By the end of this spread, you should be able to . . .

* **Describe protons, neutrons and electrons.**
* **Describe the distribution of mass and charge in an atom**
* **Describe the contribution of protons and neutrons to the nucleus in terms of atomic number and mass number.**
* **Explain the term isotopes.**
* **Deduce the atomic structure in atoms and ions.**

Figure 1 An isotope of beryllium. The beryllium nucleus has four protons (blue) and five neutrons (grey). Four electrons orbit the nucleus. The number of electrically positive protons balances the number of negatively charged electrons. Scientists now believe that electrons exist as clouds of electrical charge in *atomic orbitals*. We will discuss atomic orbitals in spread 1.2.2

Particle	Relative mass	Relative charge
proton	1.0	1+
neutron	1.0	0
electron	$\frac{1}{2000}$	1–

Table 1

Key definition

Isotopes are atoms of the same element with different numbers of neutrons.

Key definitions

Atomic (proton) number is the number of protons in the nucleus of an atom.

Mass (nucleon) number is the number of particles (protons and neutrons) in the nucleus.

Protons, neutrons and electrons

Chemists usually use a model of a nuclear atom first put forward around 100 years ago. Each atom is made up of protons, neutrons and electrons.

The essential features of this model are listed below.

- Protons and neutrons make up the nucleus, which is at the centre of the atom.
- Electrons orbit the nucleus in shells.
- The nucleus is tiny compared with the total volume of an atom.
- The nucleus is extremely dense and makes up almost all of the atom's mass.
- Most of an atom consists of empty space between the tiny nucleus and the electron shells.

Although the nucleus contains almost all the atom's mass, it makes up only a tiny portion of the atom's volume. If the nucleus were the size of a grape, the outer electrons would be about 1.5 km away.

Table 1 shows the relative mass and charge of a proton, neutron and electron.

- Protons have a positive charge and electrons have a negative charge.
- An atom has the same number of protons as electrons, so an atom is electrically neutral.
- A proton has virtually the same mass as a neutron.
- Protons and neutrons make up almost all of the atom's mass.

Isotopes

Most elements are made up of a mixture of *isotopes*.

Isotopes of the same element have:

- different masses;
- the same number of protons and electrons;
- different numbers of neutrons in the nucleus.

In Greek, isotope means 'at the same place'. The word isotope is used for different atoms of the same element that occupy the same position in the Periodic Table.

We describe an isotope using two numbers:

- the **atomic number** (proton number) – the number of protons in the nucleus.
- the **mass number** (nucleon number) – the number of particles (protons and neutrons) in the nucleus.

How to write the symbol for an isotope using these numbers is shown in Figure 2.

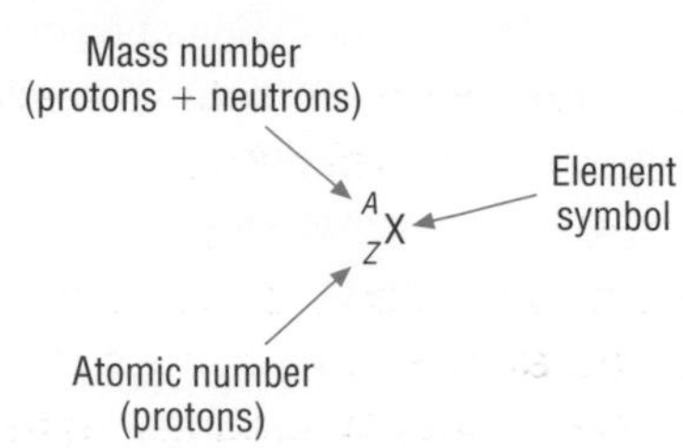

Figure 2 Symbol for an isotope

The isotopes of carbon

Carbon exists as a mixture of the isotopes: $^{12}_{6}C$, $^{13}_{6}C$ and $^{14}_{6}C$.

- The atomic number of each isotope is six.
  So each isotope has six protons.
- The overall charge in the atom must be zero.
  So each isotope must also have six electrons, to balance the charge from the six protons.
- The mass numbers are different because each isotope has a different number of neutrons.
  You can work out the number of neutrons by subtracting the atomic number from the mass number. This is shown in Table 2.

Isotope	$^{12}_{6}C$	$^{13}_{6}C$	$^{14}_{6}C$
Mass number	12	13	14
Atomic number	6	6	6
Number of neutrons	6	7	8

Table 2 Working out the number of neutrons

Different isotopes of the same element react in the same way.

This is because:

- chemical reactions involve electrons
- neutrons make no difference to chemical reactivity.

Figure 3 Frederick Soddy. In 1921 he was awarded the Nobel Prize for investigations into the origin and nature of isotopes

Atomic structures of ions

Many atoms react by losing or gaining electrons to form charged particles called **ions**. Ions are charged because they have different numbers of protons and electrons.

Table 3 shows the atomic structures of the two ions: $^{23}_{11}Na^{+}$ and $^{35}_{17}Cl^{-}$.

Note that:

- Na^{+} has one electron fewer than its number of protons.
- Cl^{-} has one electron more than its number of protons.

Ion	$^{23}_{11}Na^{+}$	$^{35}_{17}Cl^{-}$
Atomic number, *Z*	11	17
Mass number, *A*	23	35
Protons	11	17
Neutrons	12	18
Electrons	10	18
Overall charge	11 – 10 = 1+	17 – 18 = 1–

Table 3 Atomic structures for $^{23}_{11}Na^{+}$ and $^{35}_{17}Cl^{-}$

Key definition

An **ion** is a positively or negatively charged atom or (covalently bonded) group of atoms (a molecular ion).

Examiner tip

In exams, most candidates can easily write down the atomic structures of isotopes.

Ions can pose problems though. You must first be careful to change **only** the number of electrons. e.g. for a 3+ ion, simply take away 3 electrons. For a 2– ion, just add 2 electrons.

Questions

1 How many protons, neutrons and electrons are in the following atoms and ions?
  (a) $^{7}_{3}Li$; **(b)** $^{24}_{11}Na$; **(c)** $^{19}_{9}F$; **(d)** $^{55}_{26}Fe$;
  (e) $^{39}_{19}K^{+}$; **(f)** $^{19}_{9}F^{-}$; **(g)** $^{39}_{20}Ca^{2+}$; **(h)** $^{17}_{8}O^{2-}$.

2 Write down symbols, similar to the format used above, for the following atoms and ions.
  (a) The atom with 13 protons, 14 neutrons and 13 electrons.
  (b) The ion with 16 protons, 18 neutrons and 18 electrons.

1.1 (3) Atomic masses

By the end of this spread, you should be able to . . .

* **Explain why ^{12}C is used as the standard measurement of relative mass.**
* **Define the terms relative isotopic mass and relative atomic mass.**
* **Calculate relative atomic masses.**
* **Work out relative molecular masses and relative formula masses.**

$$^{12}_{6}C$$

Figure 1 Carbon-12: the standard for measuring atomic masses. An atom of the carbon-12 isotope is made up of 6 protons, 6 neutrons and 6 electrons. The mass of an atom of carbon-12 is 12 atomic mass units, 12 u. All atomic masses are measured compared with the mass of a carbon-12 atom

Measurement of relative masses

How would you go about weighing something that you cannot see? This is the situation with atoms.

Instead of finding the mass of atoms directly, we compare the masses of different atoms, using the idea of relative mass. The carbon-12 isotope (also written ^{12}C) has been chosen as the international standard for the measurement of relative mass.

Atomic masses are measured using a unit called the *unified atomic mass unit*, u. 1 u is a tiny mass: $1.660\,538\,782 \times 10^{-27}$ kg.

- The mass of an atom of carbon-12 is defined as 12 u.
- So the mass of one-twelfth of an atom of carbon-12 is 1 u.

Why carbon-12?

Before 1961, oxygen was used as the standard for atomic masses. Oxygen was used rather than the lightest atom, hydrogen, because oxygen combines with many substances to form oxides.

Unfortunately, chemists had based their scale on naturally occurring oxygen, which has a mixture of the isotopes oxygen-16, oxygen-17 and oxygen-18. However, physicists had chosen the single isotope oxygen-16 on which to base atomic masses. This gave two different sets of atomic masses!

In 1961, chemists and physicists finally agreed on a compromise – all atomic masses were to be compared with the carbon-12 isotope.

Relative isotopic mass

All the atoms in a single isotope are identical. They all have the same mass, measured against carbon-12.

For an isotope, the **relative isotopic mass** is the same as the mass number.

- So, for oxygen-16 (also written as ^{16}O), the relative isotopic mass = 16.0; and
- for sodium-23 (also written as ^{23}Na), the relative isotopic mass = 23.0.

Note that we have made two important assumptions.

- We have neglected the tiny contribution that electrons make to the mass of an atom.
- We have taken the masses of both a proton and a neutron as 1.0 u.

Key definition

Relative isotopic mass is the mass of an atom of an isotope compared with one-twelfth of the mass of an atom of carbon-12.

Relative atomic mass, A_r

Most elements contain a mixture of isotopes, each in a different amount and with a different mass. We use the term 'weighted mean mass' to account for the contribution made by each isotope to the overall mass of an element.

The contribution made by an isotope to the overall mass depends on:

- the percentage abundance of the isotope
- the relative mass of the isotope.

Key definition

Relative atomic mass, A_r, is the weighted mean mass of an atom of an element compared with one-twelfth of the mass of an atom of carbon-12.

Chemists then combine together the contribution from each isotope to arrive at the **relative atomic mass** of an element. All masses are relative to the mass of carbon-12.

Worked example

A sample of bromine contains 53.00% of bromine-79 and 47.00% of bromine-81. Determine the relative atomic mass of bromine.

Answer

$$A_r(\text{Br}) = \underbrace{\frac{53.00}{100} \times 79.00}_{\text{contribution from } ^{79}\text{Br}} + \underbrace{\frac{47.00}{100} \times 81.00}_{\text{contribution from } ^{81}\text{Br}} = 41.87 + 38.07 = 79.94$$

Relative molecular mass, M_r

Many elements and compounds are made up of simple molecules. Common examples are gases found in the air: N_2, O_2 and CO_2. The mass of a molecule is measured as the **relative molecular mass** by comparison with carbon-12.

You can find the relative molecular mass, M_r, by adding together the relative atomic masses of each atom making up a molecule.

- $M_r(Cl_2) = 35.5 \times 2 = 71.0$.
- $M_r(H_2O) = (1.0 \times 2) + 16.0 = 18.0$.

Key definition

Relative molecular mass, M_r, is the weighted mean mass of a molecule compared with one-twelfth of the mass of an atom of carbon-12.

Relative formula mass

Compounds with giant structures do not exist as simple molecules. These include ionic compounds, such as NaCl, and covalent compounds, such as SiO_2.
Although the term relative *molecular* mass is often used for these compounds, a better term is **relative formula mass**.

You can find the relative formula mass by adding together the relative atomic masses of each atom making up a formula unit.

- $CaBr_2$: relative formula mass = $40.1 + (79.9 \times 2) = 199.9$.
- Na_3PO_4: relative formula mass = $(23.0 \times 3) + 31.0 + (16.0 \times 4) = 164.0$.

Key definition

Relative formula mass is the weighted mean mass of a formula unit compared with one-twelfth of the mass of an atom of carbon-12.

Examiner tip

Definitions are often asked for in exams. You may be asked to recall definitions for:

- relative isotopic mass
- relative atomic mass.

These really are easy marks but only if you have been bothered to learn them!

Questions

1 Calculate the relative atomic mass, A_r, of the following. Give your answers to 4 significant figures.
 (a) Boron contains: 19.77% ^{10}B; and 80.23% ^{11}B.
 (b) Silicon contains: 92.18% ^{28}Si; 4.70% ^{29}Si; and 3.12% ^{30}Si.
 (c) Chromium contains: 4.31% ^{50}Cr; 83.76% ^{52}Cr; 9.55% ^{53}Cr; and 2.38% ^{54}Cr.

2 Use A_r values from the Periodic Table to calculate the relative molecular mass of the following.
 (a) HCl; (b) CO_2; (c) H_2S; (d) NH_3; (e) H_2SO_4.

3 Use A_r values from the Periodic Table to calculate the relative formula mass of the following.
 (a) Fe_2O_3; (b) Na_2O; (c) $Pb(NO_3)_2$; (d) $(NH_4)_2SO_4$; (e) $Ca_3(PO_4)_2$.

1.1 (4) Amount of substance and the mole

By the end of this spread, you should be able to . . .

- ✱ **Explain the terms amount of substance, mole and the Avogadro constant.**
- ✱ **Define and use the term molar mass.**
- ✱ **Carry out calculations involving masses using the amount of substance in moles.**

Key definition

Amount of substance is the quantity whose unit is the mole. Chemists use 'amount of substance' as a means of counting atoms.

Key definition

The **Avogadro constant, N_A** is the number of atoms per mole of the carbon-12 isotope (6.02×10^{23} mol^{-1}).

Key definition

A **mole** is the amount of any substance containing as many particles as there are carbon atoms in exactly 12 g of the carbon-12 isotope.

Year	Value
1931	6.061×10^{23}
1958	6.02×10^{23}
1981	6.022045×10^{23}
1993	6.0221367×10^{23}
2002	6.0221415×10^{23}
2006	$6.02214179 \times 10^{23}$

Table 1 How the value for the Avogadro constant has altered as measurements have improved

Amount of substance

Chemists use a quantity called **amount of substance** for counting atoms.

Amount of substance is:

- given the symbol *n*
- measured using a unit called the *mole* (abbreviated to *mol*).

Amount of substance is based on a standard count of atoms called the **Avogadro constant**, N_A.

The Avogadro constant, N_A, is the number of atoms per mole of the carbon-12 isotope. (See also spread 1.1.6.)

$$N_A = 6.022\,141\,79 \times 10^{23}\ \text{mol}^{-1}$$
$$= 6.02 \times 10^{23}\ \text{mol}^{-1} \text{ to three significant figures.}$$

A **mole** is the amount of any substance containing as many particles as there are carbon atoms in exactly 12 g of the carbon-12 isotope.

The quest for the Avogadro constant

The number of particles per mole is called the Avogadro constant, after the Italian Amedeo Avogadro (see spread 1.1.6). Table 1 shows how the accepted numerical value for the Avogadro constant has changed over the years as our ability to measure particles accurately has improved.

Austrian chemist Johann Loschmidt was the first person to calculate the value of Avogadro's constant, which is known as the Loschmidt number in some German-speaking countries (hence the occasional use of the symbol *L* rather than N_A).

The changing value of the Avogadro constant illustrates how emerging scientific knowledge is seldom absolute. As Table 1 shows, while our measurement of the constant is becoming more precise, we have not yet reached a perfectly accurate answer – and we never will!

Counting and weighing atoms

Using the definition of one mole, you now know that 12 g of carbon-12 contains 6.02×10^{23} atoms.

Figure 1 shows one mole of atoms of different elements.

Figure 1 One mole of atoms of different elements. Each sample on the right contains 6.02×10^{23} atoms or 1 mol of atoms

The mass of one mole is easy to work out as the relative atomic mass in grams.

You can find the mass of one mole (1 mol) of atoms of any element – it is the relative atomic mass in grams.

- One mole of H atoms has a mass of 1.0 g.
- One mole of C atoms has a mass of 12.0 g.
- One mole of Pb atoms has a mass of 207.2 g.

You can also work in multiples of 1 mol.

For carbon:

- 1.00 mol of C atoms has a mass of 12.0 g.
- 2.00 mol of C atoms has a mass of 24.0 g.
- 0.50 mol of C atoms has a mass of 6.0 g.

Moles of anything!

Amount of substance is not restricted to atoms.

You can have an amount of any chemical species, e.g.:

- amount of atoms
- amount of molecules
- amount of ions

It is important that you always state the chemical formula to which an amount of substance refers.

- *1 mole of oxygen* could either refer to oxygen atoms, O, or to oxygen molecules, O_2.
- 1 mole of O or 1 mole of O_2 is clearer.

How big is a mole?

How do you judge the real size of 6.02×10^{23}?

6.02×10^{23} is a huge number. It is so large that it is almost impossible to imagine its true size.

Let's look at some comparisons.

- 6.02×10^{23} (1 mol of) drink cans would cover the surface of the earth to a depth of over 300 km.
- Even if you were able to count atoms at the rate of 10 million per second, it would still take you about two billion years to count the atoms in one mole of atoms.

Figure 2 One mole of coarse sand – 1 mol of grains would cover the surface of Great Britain to a depth of 1500 km

Molar mass

Molar mass, M, is an extremely useful term as it can be applied to any chemical substance, whether an element, molecule or ion.

You can find the molar mass by adding together all the relative atomic masses for each atom that makes up a formula unit.

- For carbon: $M(C) = 12.0 = 12.0\ g\ mol^{-1}$.
- For carbon dioxide: $M(CO_2) = 12.0 + (16.0 \times 2) = 44.0\ g\ mol^{-1}$.

The amount of substance n, mass m, and molar mass M, are all linked:

$$n = \frac{m}{M}$$

In 48.0 g of C, $n(C) = \frac{48.0}{12.0} = 4.00$ mol

In 11 g of CO_2, $n(CO_2) = \frac{11.0}{44.0} = 0.250$ mol

Key definition

Molar mass, M, is the mass per mole of a substance. The units of molar mass are $g\ mol^{-1}$.

Examiner tip

To tackle exam questions on the mole, you *must* remember the equation for converting between amount of substance and mass:

$n = \frac{m}{M}$, where

n = amount of substance, in mol
m = mass, in g
M = molar mass, in $g\ mol^{-1}$

Questions

1 Calculate the mass, in g, for the following:
(a) 3.00 mol SiO_2; (b) 0.500 mol Fe_2O_3; (c) 0.250 mol Na_2CO_3.

2 Calculate the amount, in mol, for the following:
(a) 8.00 g BeF_2; (b) 23.7 g $KMnO_4$; (c) 10.269 g $Al_2(SO_4)_3$.

1.1 (5) Types of formula

By the end of this spread, you should be able to . . .

* **Explain the terms empirical formula and molecular formula.**
* **Calculate empirical and molecular formulae.**

Empirical formula

In chemistry, an **empirical formula** is used as the simplest way of showing a formula.

Figure 1 shows part of the structure of sodium chloride. If you wanted to show the formula for sodium chloride using the actual number of ions bonded together, it would run into millions of millions – remember how big the mole is! Furthermore, the formula would also depend on the size of the salt crystal. Structures like this are called *giant* structures. Giant structures are discussed in more detail in spreads 1.2.14–16.

For this reason, empirical formulae are always used for compounds with giant structures, including:

- ionic compounds, such as NaCl
- covalent compounds, such as SiO_2.

Key definition

The **empirical formula** is the simplest whole-number ratio of atoms of each element present in a compound.

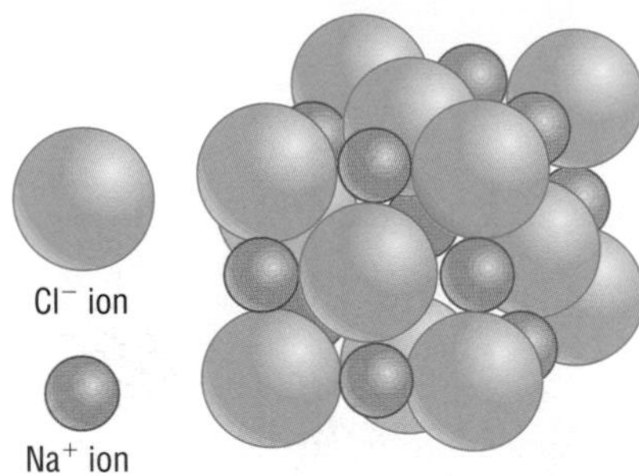

Figure 1 Structure of sodium chloride. The structure is as large as the crystal. We use the empirical formula, NaCl, because we know that there is always one Na^+ ion for every Cl^- ion in a crystal of sodium chloride

Examiner tip

Learn the equations for working out moles,

Remember: *amount*, $n = \dfrac{\textit{mass, m}}{\textit{molar mass, M}}$

Calculating empirical formulae

If you know the mass of each element in a sample of a compound, you can calculate its empirical formula.

Worked example 1

Analysis showed that 0.6075 g of magnesium combines with 3.995 g of bromine to form a compound.

[A_r: Mg, 24.3; Br, 79.9.]

Find the molar ratio of atoms: Mg : Br

$$\frac{0.6075}{24.3} : \frac{3.995}{79.9}$$

0.025 : 0.050

Divide by smallest number (0.025): 1 : 2

Empirical formula is: $MgBr_2$

Worked example 2

Analysis of a compound showed the following percentage composition by mass:

Na: 74.19%; O: 25.81%. [A_r: Na, 23.0; O,16.0.]

100.0 g of the compound contains 74.19 g of Na and 25.81 g of O.

Find the molar ratio of atoms: Na : O

$$\frac{74.19}{23.0} : \frac{25.81}{16.0}$$

3.226 : 1.613

Divide by smallest number (1.613): 2 : 1

Empirical formula is: Na_2O

Molecular formula

Molecular formulae are used for compounds that exist as simple molecules. A **molecular formula** tells you the number of each type of atom that make up a molecule.

Figure 2 shows a propane molecule which contains 3 carbon atoms and 8 hydrogen atoms.

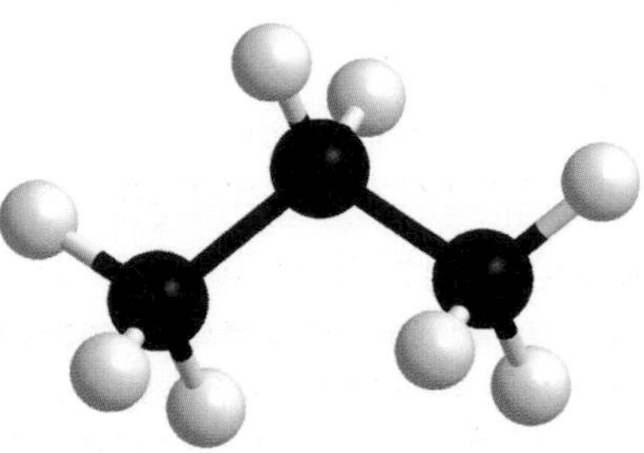

Figure 2 Molecule of propane

Key definition

A **molecule** is a small group of atoms held together by covalent bonds.

Key definition

The **molecular formula** is the actual number of atoms of each element in a molecule.

The molecular formula of propane is C_3H_8.

The simplest ratio of atoms is 3 C to 8 H so the empirical formula of propane is also C_3H_8.

Figure 3 shows a butane molecule which contains 4 carbon atoms and 10 hydrogen atoms. Notice that butane has different molecular and empirical formulae.

For many molecules, the molecular formula is adequate. In organic chemistry, however, the molecular formula does not tell you the order in which the atoms are bonded to each other. You will learn more about the different formulae used in organic chemistry when you study the second unit of the AS course — *Chains, Energy and Resources*.

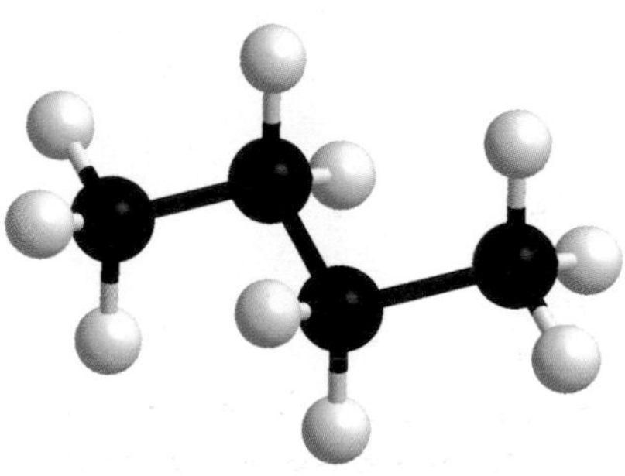

Figure 3 A molecule of butane, C_4H_{10}. The butane molecule has 4 C atoms and 10 H atoms. The simplest ratio of atoms is 2 C to 5 H.
So, the molecular formula = C_4H_{10},
but the empirical formula = C_2H_5

Determination of a molecular formula

We usually determine the empirical formula and relative molecular mass of the compound experimentally.

From this information, you can then find the molecular formula of a compound.

Worked example

A compound has an empirical formula of CH_2 and a relative molecular mass, M_r, of 56.0. What is its molecular formula?

Answer

- empirical formula mass of CH_2: $= 12.0 + (1.0 \times 2) = 14.0$
- number of CH_2 units in a molecule: $= \dfrac{56.0}{14.0} = 4$
- molecular formula: $(4 \times CH_2)$ $=$ C_4H_8

Questions

1 Determine the empirical formula of the following.
   (a) The compound formed when 6.54 g of zinc reacts with 1.60 g of oxygen.
   (b) The compound formed when 5.40 g of aluminium reacts with 4.80 g of oxygen.
   (c) The compound containing: Ag, 69.19%; S, 10.29%; and O, 20.52%.

2 Determine the molecular formula of the following.
   (a) The compound of nitrogen and oxygen containing 0.49 g of nitrogen combined with 1.12 g of oxygen, M_r 92.0.
   (b) The compound consisting of carbon, hydrogen and oxygen atoms containing 1.80 g of carbon combined with 0.30 g of hydrogen and 1.20 g of oxygen, M_r 88.0.
   (c) The compound of carbon, hydrogen and oxygen with the composition by mass: C, 40.0%; H, 6.7%; O, 53.3%; M_r 180.0.

1.1 (6) Moles and gas volumes

By the end of this spread, you should be able to . . .

✱ Calculate the amount of substance in moles, using gas volumes.

Avogadro's hypothesis

In 1811, the Italian Amedeo Avogadro put forward the following hypothesis:

> Equal volumes of gases at the same temperature and pressure contain equal numbers of molecules.

Figure 1 Amedeo Avogadro.
In 1956, Avogadro's original hypothesis was commemorated in an Italian stamp

This idea, now known as Avogadro's Law, is important because it means that we can compare the number of molecules in different gases by simply comparing their volumes. This is just as well. It is very difficult to weigh gases, but comparatively easy to measure gas volumes.

- It does not matter which type of gas is being measured. By measuring the volume of any gas, we are indirectly counting the number of molecules.

At room temperature and pressure (RTP):

- One mole of gas molecules occupies approximately 24.0 dm^3 (24 000 cm^3).
- The volume per mole of gas molecules is 24.0 $dm^3\ mol^{-1}$.

Key definition

Molar volume is the volume per mole of a gas. The units of molar volume are $dm^3\ mol^{-1}$. At room temperature and pressure, the molar volume is approximately 24.0 $dm^3\ mol^{-1}$.

Molar volume is the volume per mole of gas molecules.

At RTP, 1 mol of $H_2(g)$, $O_2(g)$ and $CO_2(g)$ each have a molar volume of 24.0 $dm^3\ mol^{-1}$.

To give an idea of scale, 24.0 dm^3 is equivalent to the amount of air contained in four footballs.

Air consists mainly of N_2 and O_2 gases, which have similar relative molecular masses.

Look at the diagram in Figure 2.

- The same volume of $H_2(g)$ would be much lighter than air.
- The same volume of $CO_2(g)$ would be much heavier than air.

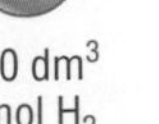

24.0 dm^3	24.0 dm^3	24.0 dm^3
1 mol H_2	1 mol O_2	1 mol CO_2
2.0 g	32.0 g	44.0 g

Figure 2 One mole of different gases. Note that the same volume of each gas has a different mass, so $H_2(g)$, $O_2(g)$ and $CO_2(g)$ have different densities

Calculating amounts using gas volumes

For a gas at room temperature and pressure, you can use the equations below to convert between:

- the amount of gas molecules, n, in mol; and
- the gas volume, V, in dm^3 or cm^3.

You can work out the amount, in mol, by dividing the volume by the molar volume.

At room temperature and pressure,

- if volume V is in dm^3: $n = \frac{V}{24.0}$ mol
- if volume V is in cm^3: $n = \frac{V}{24\,000}$ mol

Worked example 1

What amount, in mol, of gas molecules is in 72 cm^3 of any gas at RTP?

$$n = \frac{V \text{ (in cm}^3)}{24\,000} = \frac{72}{24\,000} = 0.0030 \text{ mol}$$

Worked example 2

What is the volume, in cm^3, of 2.130×10^{-3} mol of a gas at RTP?

$$n = \frac{V \text{ (in cm}^3)}{24\,000}$$

Hence, $V = n \times 24\,000 = 2.130 \times 10^{-3} \times 24\,000 = 51.12$ cm^3.

From hypothesis to law

Most scientific advances start off with a *hypothesis* – a tentative explanation for an observation. It can even be just an educated guess. A hypothesis is an idea that can be tested by further investigation.

Once the hypothesis has been tested repeatedly for many years and is found to hold true by several different scientists, it may become accepted as a *law*.

Avogadro's hypothesis was at first neglected by scientists, partly because it contradicted the views of the respected scientist John Dalton (see spread 1.1.1). Avogadro was the first person to have the idea that atoms grouped together into molecules. Dalton thought that matter was just made of atoms. It was not until after Avogadro's death in 1854 that his brilliant deduction was accepted by the scientific community. Avogadro's *hypothesis* had became a *law*. It had progressed from an *educated guess* to an *accepted explanation*.

In recognition of his contribution to scientific knowledge, the number of fundamental particles per mole of substance is called the Avogadro constant. Avogadro had no idea of the actual number of molecules in equal volumes of gases, and he did nothing to measure it. But his original hypothesis did lead to the eventual determination of this number as 6.02×10^{23} mol^{-1}.

Examiner tip

To work out amount, in moles, in a solution, learn the equations:

$$n = \frac{V \text{ (in dm}^3)}{24.0} \qquad n = \frac{V \text{ (in cm}^3)}{24\,000}$$

Questions

1 (a) What amount, in mol, of gas molecules are in the following gas volumes at RTP?
(i) 36 dm^3; **(ii)** 1080 dm^3; **(iii)** 4.0 dm^3

(b) What is the volume of the following at RTP?
(i) 6 mol SO_2(g); **(ii)** 0.25 mol O_2(g); **(iii)** 20.7 g NO_2(g)

2 What is the mass of the following at RTP?
(a) 0.6 dm^3 N_2; **(b)** 1920 cm^3 C_3H_8(g); **(c)** 84 cm^3 N_2O(g)?

3 What is the volume of the following at RTP?
(a) 1.282 g SO_2(g); **(b)** 1.485 g HCN(g); **(c)** 1.26 g C_3H_6(g)?

1.1 (7) Moles and solutions

By the end of this spread, you should be able to . . .

- ✱ **Calculate the amount of substance in mol, using solution volume and concentration.**
- ✱ **Describe a solution's concentration using the terms concentrated and dilute.**

Figure 1 Flasks

Concentration

In a solution, the solute dissolves in the solvent. The **concentration** of a solution tells you how much solute is dissolved in the solvent. Concentrations are measured in moles per cubic decimeter, mol dm^{-3}.

In a solution with a concentration of 2 mol dm^{-3}:
- there are 2 moles of solute dissolved in every 1 dm^3 of solution.

This means that:
- 1 dm^3 contains 2 mol of dissolved solute.
- 2 dm^3 contains 4 mol of dissolved solute.
- 0.25 dm^3 contains 0.5 mol of dissolved solute.

If you know the concentration in mol dm^{-3}, you can find the amount, in mol, in any volume of a solution.

If the volume of the solution is measured in dm^3:

$$n = c \times V \text{ (in dm}^3\text{)}$$

where
- n = amount of substance, in mol
- c = concentration of solute, in mol dm^{-3}
- V = volume of solution, in dm^3

In smaller volumes of the solution, it is more convenient to measure volumes in cm^3 and the expression becomes:

$$n = c \times \frac{V \text{ (in cm}^3\text{)}}{1000}$$

Figure 2 A volumetric flask.
This is used to accurately make up a standard solution.
Two classes of flask are available, depending on the precision required.
- Class A measure volumes with a precision of ±0.06%.
- Class B measure volumes with a precision of ±0.08%.

The class is marked on the volumetric flask

Key definition

The **concentration** of a solution is the amount of solute, in mol, dissolved per 1 dm^3 (1000 cm^3) of solution.

Examiner tip

To work out amount, in moles, in a solution, learn the equations:

$n = c \times V \text{(in dm}^3\text{)}$ $\quad n = \frac{V \text{(in cm}^3\text{)}}{1000}$

Worked example

What is the amount, in mol, of NaOH dissolved in 25.0 cm^3 of an aqueous solution of concentration 0.0125 mol dm^{-3}?

$$n(\text{NaOH}) = c \times \frac{V \text{ (in cm}^3\text{)}}{1000} = 0.0125 \times \frac{25.0}{1000} = 3.125 \times 10^{-4} \text{ mol}$$

Standard solutions

A **standard solution** has a known concentration.

If you need to make up a standard solution, you will need to weigh out a substance.

To do this, you will need to:
- consider the volume of solution required;
- work out the amount, in mol, of solute needed; and
- convert this amount of solute into a mass, in g, so that you know how much to weigh out.

Key definition

A **standard solution** is a solution of known concentration. Standard solutions are normally used in titrations to determine unknown information about another substance.

Worked example

Find the mass of potassium hydroxide required to prepare 250 cm^3 of a 0.200 mol dm^{-3} solution.

1 Find the amount of KOH, in mol, required in the solution:

$$n(\text{KOH}) = c \times \frac{V\text{ (in cm}^3)}{1000} = 0.200 \times \frac{250}{1000} = 0.0500\text{ mol}$$

2 Convert moles to grams:

molar mass, $M(\text{KOH}) = 39.1 + 16.0 + 1.0 = 56.1\text{ g mol}^{-1}$

$$\text{amount } n = \frac{\text{mass } m}{\text{molar mass } M}$$

Hence, $m = n \times M = 0.0500 \times 56.1 = 2.805$ g

So, the mass of KOH required is 2.805 g.

Figure 3 Making up a standard solution.

To make up a standard solution, follow these steps.

- You first need to weigh out the solute.
- You then completely dissolve the solute in water in a beaker. Transfer the solution to the flask and rinse the beaker repeatedly, adding the rinsings to the flask.
- Add water to the flask, but do not fill it all the way up to the graduation line.
- You must now carefully add water drop by drop to the line on the flask. You line up the *bottom* of the meniscus exactly with the graduation mark on the flask.
- Finally mix as in Figure 4.

Mass concentrations

The mass concentration of a solute is the mass dissolved in 1 dm^3 of solution.

Mass concentrations are measured in units of g dm^{-3}.

- The solution above contains 2.805 g of dissolved KOH in 250 cm^3 of solution.
- This is 11.220 g of dissolved KOH in 1000 cm^3 (1 dm^3) of solution.
- We can refer to this concentration of KOH as 11.220 g dm^{-3}.

Concentrated and dilute solutions

The terms *concentrated* and *dilute* are descriptive terms for the amount, in mol, of dissolved solute in a solution.

- Concentrated – a large amount of solute per dm^3.
- Dilute – a small amount of solute per dm^3.

Normal bench solutions of acids usually have concentrations of 1 mol dm^{-3} or 2 mol dm^{-3}. These are *dilute* solutions.

Concentrated acids usually have a concentration greater than 10 mol dm^{-3}.

Molar solutions

You will often see 'M' used on solutions in chemistry.

- '*M*' means *Molar* and refers to a solution with a concentration in moles per cubic decimetre, mol dm^{-3}.
- Thus, 2 mol dm^{-3} and 2M mean the same thing – 2 mol of solute in 1 dm^3 of solution.

Figure 4 Mixing the solution.

Finally, you must mix your solution thoroughly, by inverting the flask several times

Questions

1 Find the amount, in moles, of solute dissolved in the following solutions.

(a) 4 dm^3 of a 2 mol dm^{-3} solution.

(b) 25.0 cm^3 of a 0.150 mol dm^{-3} solution.

(c) 24.35 cm^3 of a 0.125 mol dm^{-3} solution.

2 Find the concentration, in mol dm^{-3} for the following solutions.

(a) 6 moles dissolved in 2 dm^3 of solution.

(b) 0.500 moles dissolved in 250 cm^3 of solution.

(c) 8.75×10^{-3} moles dissolved in 50.0 cm^3 of solution.

3 Find the mass concentration, in g dm^{-3} for the following solutions.

(a) 0.042 moles of HNO_3 dissolved in 250 cm^3 of solution.

(b) 0.500 moles of HCl dissolved in 4 dm^3 of solution.

(c) 3.56×10^{-3} moles of H_2SO_4 dissolved in 25 cm^3 of solution.

1.1 (8) Chemical equations

By the end of this spread, you should be able to . . .

* **Construct balanced chemical equations for known reactions, and for others given the reactants and products.**

Reactants and products

In a chemical reaction, the number and type of atom is conserved. A chemical reaction does not create or destroy atoms, or change one type of atom into another. A chemical reaction only rearranges the atoms that were originally present into different combinations.

A chemical reaction starts with the **reactants**, which change chemically to form the **products**.

reactants ⟶ products

You will use equations throughout your chemistry course. It is essential that you can construct these equations correctly.

When writing an equation, the first thing you need to do is identify the reactants and products.

Adding species to the equation

A chemical equation is a symbolic representation of the chemical reaction taking place.

In a chemical equation, you can show various chemical **species** in different ways.

- A species is a type of particle that takes part in a reaction.
- A species could be an atom, ion, molecule, empirical formula or even an electron.

Key definition

A species is any type of particle that takes place in a chemical reaction.

Molecules

In equations, a simple molecule is shown by its molecular formula.

- For example: N_2, O_2, H_2, H_2O, CH_4.

Giant structures

A giant structure is formed when many atoms or ions bond together. The number of atoms or ions depends upon the size of the crystalline structure. For this reason, a compound with a giant structure is shown by its empirical formula. This gives the simplest whole number ratio of atoms or ions in the structure. Giant structures are discussed in detail in spreads 1.2.14–1.2.16.

Examples of giant structures include:

- ionic compounds, such as NaCl, $CaCl_2$, $Mg(OH)_2$; and
- giant covalent structures, such as SiO_2.

Some elements have giant structures, for example:

- all metals
- some non-metals, such as C, Si and B.

In equations:

- a compound with a giant structure is shown by its empirical formula
- an element with a giant structure is shown by its symbol – effectively the empirical formula of the element.

Figure 1 The formula of a giant structure. The abundant mineral, quartz, is mainly a compound of silicon and oxygen.

- Each Si atom (black) bonds with four O atoms (red)
- Each O atom bonds with two Si atoms.

The ratio of Si atoms to O atoms is always 1:2, irrespective of the size of the crystal.

We write the formula as this ratio: SiO_2. This is the empirical formula

Adding state symbols to the equation

State symbols are added to provide more information. You will come across these four state symbols:

- (s) solid
- (l) liquid
- (g) gaseous
- (aq) aqueous.

Balancing an equation

Balancing a chemical equation is a convenient way of accounting for all the atoms.

- You balance an equation in order to have the *same number* of atoms for each element on either side of the equation.

With practice, you will find that balancing an equation is quite straightforward. In the following equations, you will be balancing atoms only. Later, we will discuss equations in which electrons must also be balanced.

To balance an equation, first write down the equation using formulae and state symbols.

nitrogen + hydrogen ⟶ ammonia

$N_2(g) + H_2(g) \longrightarrow NH_3(g)$

Then, check on balancing.

- Nitrogen is *not* balanced (there are more nitrogen atoms on the left side of the equation).
- Hydrogen is *not* balanced (there are more hydrogen atoms on the right side of the equation).

	$N_2(g) + H_2(g) \longrightarrow NH_3(g)$		
nitrogen	2	1	✗
oxygen	2	3	✗

Now balance the equation.

You balance an equation to give the *same number* of atoms of each element on each side of the equation.

- Nitrogen can be balanced by placing a '2' in front of NH_3.
- Hydrogen is still unbalanced:

	$N_2(g) + H_2(g) \longrightarrow \mathbf{2}\ NH_3(g)$		
nitrogen	2	2	✓
oxygen	2	6	✗

Hydrogen can be balanced by placing a '3' in front of H_2.

The equation is now balanced – there are the same numbers and types of atoms on both sides of the equation.

	$N_2(g) + \mathbf{3}\ H_2(g) \longrightarrow \mathbf{2}\ NH_3(g)$		
nitrogen	2	2	✓
oxygen	6	6	✓

Question

1 Balance each of the following equations.

(a) $Li(s) + O_2(g) \longrightarrow Li_2O(s)$

(b) $Al(s) + Cl_2(g) \longrightarrow Al_2Cl_6(s)$

(c) $Al(s) + H_2SO_4(aq) \longrightarrow Al_2(SO_4)_3(aq) + H_2(g)$

(d) $C_3H_8(g) + O_2(g) \longrightarrow CO_2(g) + H_2O(l)$

(e) $Zn(s) + HNO_3(aq) \longrightarrow Zn(NO_3)_2(aq) + H_2O(l) + NO_2(g)$

(f) $Cu(s) + HNO_3(aq) \longrightarrow Cu(NO_3)_2(aq) + H_2O(l) + NO(g)$

2 Write balanced equations, with state symbols for the following reactions.

(a) calcium reacts with oxygen to form calcium oxide, CaO.

(b) magnesium reacts with an aqueous solution of silver nitrate, $AgNO_3$, to form silver and a solution of magnesium nitrate, $Mg(NO_3)_2$.

(c) Lead(II) nitrate, $Pb(NO_3)_2$, decomposes when heated to form a solid, lead(II) oxide, PbO, and a mixture of two gases, nitrogen dioxide and oxygen.

(d) copper reacts with concentrated sulfuric acid, $H_2SO_4(l)$, to form water, a blue solution and a gas.

Examiner tip

In exams, candidates often make mistakes by not understanding a chemical formula.

- In a formula, a subscript number applies *only* to the symbol immediately before it, so SO_2 means 1 S and 2 O atoms.
- When balancing an equation, you must *not* change any formula.
- You can only add a balancing number in *front* of a formula.
- Everything in the formula is multiplied by the balancing number:

2 SO_2 means:	3 K_2SO_4 means:
S: $2 \times 1 = 2$	K: $3 \times 2 = 6$
O: $2 \times 2 = 4$	S: $3 \times 1 = 3$
	O: $3 \times 4 = 12$.

- With brackets, the subscript number applies to *everything* in the bracket:

  $Cu(NO_3)_2$ means:
  Cu: 1
  N: $(1 \times 2) = 2$
  O: $(3 \times 2) = 6$.

1.1 (9) Moles and reactions

By the end of this spread, you should be able to . . .

* **Deduce the quantities of reactants and products from balanced equations.**

Stoichiometry and reacting quantities

Stoichiometry studies the amounts of substances that are involved in a chemical reaction. You can use a balanced chemical equation to find the stoichiometry of a reaction.

The balanced equation tells you:

- the reacting quantities that are needed to prepare a required quantity of a product
- the quantities of products formed by reacting together known quantities of reactants.

First, you work out the stoichiometry (molar reacting quantities) from a balanced equation. Then, you can scale these molar quantities up or down to work with any quantity.

Key definition

Stoichiometry is the molar relationship between the relative quantities of substances taking part in a reaction.

Examiner tip

There are three ways to work an amount, in moles.

From mass:

$n = \frac{m}{M}$

From gas volumes:

$n = \frac{V(\text{in dm}^3)}{24.0}$ $\quad n = \frac{V(\text{in cm}^3)}{24\,000}$

From solutions:

$n = c \times V(\text{in dm}^3)$ $\quad n = \frac{V(\text{in cm}^3)}{1000}$

You will need to use these in exams, probably all of them! So learn them.

Worked examples using stoichiometry

The worked examples on these two pages show you how to use a balanced equation using moles. There are three examples:

- reacting masses
- reacting masses and gas volumes
- reacting masses, gas volumes and solution volumes.

Together, these will give you worked examples for working out molar quantities from masses, gas volumes and solution volumes.

Reacting masses

Sodium and chlorine react to form sodium chloride:

$2Na(s) + Cl_2(g) \longrightarrow 2NaCl(s)$

In this example, we will calculate the masses of Na and Cl_2 that would form 2.925 g of NaCl. [A_r: Na, 23.0; Cl, 35.5]

(a) Calculate the amount, in mol, in 2.925 g of NaCl.

$$n = \frac{m}{M} = \frac{2.925}{23.0 + 35.5} = \frac{2.925}{58.5} = 0.0500 \text{ mol}$$

(b) Calculate the amounts, in mol, of Na and Cl_2 that would form 2.925 g NaCl.

equation: $2Na(s) + Cl_2(g) \longrightarrow 2NaCl(s)$

amounts in equation: 2 mol : 1 mol : 2 mol

amounts required: 0.050 mol : 0.025 mol : 0.050 mol

(c) Calculate the masses of Na and Cl_2 that would form 2.925 g NaCl.

Na: $m = n \times M = 0.050 \times 23.0 = 1.15$ g

Cl_2: $m = n \times M = 0.025 \times 71.0 = 1.775$ g

Notes

For (a), we need to use: amount, $n = \frac{\text{mass, } m}{\text{molar mass, } M}$

For (b), we use the equation to work out the reacting quantities.

For (c), we rearrange: $n = \frac{m}{M}$

So, $m = n \times M$

Reacting masses and gas volumes

A student heated 2.55 g of sodium nitrate $NaNO_3$, which decomposes as in the equation below. [A_r: Na, 23.0; N, 14.0; O, 16.0]

$2NaNO_3(s) \longrightarrow 2NaNO_2(s) + O_2(g)$

(a) Calculate the amount, in mol, of $NaNO_3$ that decomposed.

$$n = \frac{m}{M} = \frac{2.55}{23.0 + 14.0 + (16.0 \times 3)} = \frac{2.55}{85.0} = 0.0300 \text{ mol}$$

Notes

For (a), we use: amount, $n = \frac{\text{mass, } m}{\text{molar mass, } M}$

(b) Calculate the amount, in mol, of O_2 that formed.

$2\ NaNO_3(s) \longrightarrow 2\ NaNO_2(s) + O_2(g)$

2 mol 1 mol

0.0300 mol 0.0150 mol

(c) Calculate the volume of O_2 that formed.

$V = n \times 24.0 = 0.0150 \times 24.0 = 0.360\ dm^3$

Notes

For (b), we use the equation to work out the reacting quantities.

For (c), we rearrange:

amount, $n = \frac{V(\text{in } dm^3)}{24.0}$

So, $V = n \times 24.0$

Reacting mass, gas and solution volumes

A chemist reacted 0.23 g of sodium with water to form 250 cm^3 of aqueous sodium hydroxide. Hydrogen gas was also produced. The equation is shown below. [A_r: Na, 23.0; H, 1.0; O, 16.0]

$2Na(s) + 2H_2O(l) \longrightarrow 2NaOH(aq) + H_2(g)$

(a) Calculate the amount, in mol, of Na that reacted.

$$n = \frac{m}{M} = \frac{0.23}{23.0} = 0.010\ mol$$

(b) Calculate the volume of H_2 formed at room temperature and pressure.

$2Na(s) + 2H_2O(l) \longrightarrow 2NaOH(aq) + H_2(g)$

2 mol 1 mol

0.010 mol 0.005 mol

Convert this amount of H_2 to the volume formed.

$V = n \times 24000 = 0.005 \times 24\ 000 = 120\ cm^3$

(c) Calculate the concentration, in mol dm^{-3}, of NaOH(aq) formed.

$2Na(s) + 2H_2O(l) \longrightarrow 2NaOH(aq) + H_2(g)$

2 mol 2 mol

0.010 mol 0.010 mol

Finally, work out the concentration, in mol dm^{-3}, of NaOH(aq).

$$c(NaOH) = \frac{n \times 1000}{V(\text{in } cm^3)} = \frac{0.010 \times 1000}{250} = 0.040\ mol\ dm^{-3}$$

Notes

For (a), use:

amount, $n = \frac{\text{mass, } m}{\text{molar mass, } M}$

For (b), we first use the balanced equation to work out the amount of H_2 that forms.

We then find the volume of this amount of H_2.

We rearrange: amount, $n = \frac{V(\text{in } cm^3)}{24000}$

Hence, $V = n \times 24000$

For (c), we again use the balanced equation, but this time we need to work out the amount of NaOH that forms.

This amount of NaOH is dissolved in 250 cm^3 of solution.

We rearrange: $n = c \times \frac{V(\text{in } cm^3)}{1000}$

Hence, $c = \frac{n \times 1000}{V(\text{in } cm^3)}$

Questions

1 (a) Balance the equation below.

$NaHCO_3(s) \longrightarrow Na_2CO_3(s) + CO_2(g) + H_2O(l)$

(b) What volume of CO_2, measured at RTP, is formed by the decomposition at RTP of 5.04 g of $NaHCO_3$?

2 (a) Balance the equation below.

$Pb(NO_3)_2(s) \longrightarrow PbO(s) + NO_2(g) + O_2(g)$

(b) What is the total volume of gas, measured at RTP, formed by the decomposition of 2.12 g of $Pb(NO_3)_2$?

3 (a) Balance the equation below.

$MgCO_3(s) + HNO_3(aq) \longrightarrow Mg(NO_3)_2(aq) + H_2O(l) + CO_2(g)$

(b) 2.529 g of $MgCO_3$ reacts with an excess of HNO_3.

(i) What volume of CO_2, measured at RTP, is formed?

(ii) The final volume of the solution is 50.0 cm^3.
What is the concentration, in mol dm^{-3}, of $Mg(NO_3)_2$ formed?

1.1 (10) Acids and bases

By the end of this spread, you should be able to ...

- **Know the formulae of common acids and bases.**
- **State that an acid releases H^+ ions in aqueous solution.**
- **State that common bases are metal oxides, metal hydroxides and ammonia.**
- **State that an alkali is a soluble base that releases OH^- ions in aqueous solution.**

Examiner tip

You need to learn the formulae of these three acids.

Figure 1 Everyday acids

Key definition

An **acid** is a species that is a proton donor.

Examiner tip

All acids contain hydrogen. When added to water, acids release this hydrogen as H^+ ions (protons).

Key definition

A **base** is a species that is a proton acceptor.

Acids

When studying chemistry, you will come across acids in both practical and theory work. The word acid comes from the Latin 'acidus' meaning sour. In water, acids give a solution with a pH of less than 7.0.

In the chemistry laboratory, the common acids that you will come across are:

- sulfuric acid H_2SO_4
- hydrochloric acid HCl
- nitric acid HNO_3.

You will also come across weaker acids that occur naturally. These are the acids that you are most likely to have come across in everyday life.

- ethanoic (acetic) acid CH_3COOH in vinegar
- methanoic (formic) acid $HCOOH$ in insect bites
- citric acid $C_6H_8O_7$ in citrus fruits.

You may also be familiar with sulfuric acid and hydrochloric acid without realising it.

- Sulfuric acid is the battery acid used in cars.
- Hydrochloric acid is the acid present in your stomach that helps you digest food.

If you look at the formulae of the acids we have listed above, you will notice they all contain the element hydrogen.

When an **acid** is added to water, the acid releases H^+ ions (protons) into solution:

- hydrochloric acid, HCl
  $HCl(g) + aq \longrightarrow H^+(aq) + Cl^-(aq)$
- sulfuric acid, H_2SO_4
  $H_2SO_4(l) + aq \longrightarrow H^+(aq) + HSO_4^-(aq)$

Notice that 'aq' is used to show water.

The H^+ ion (proton) is the active ingredient in acids.

- A H^+ ion is responsible for all acid reactions.
- An acid is a proton donor

Bases

Bases are also chemicals that you will meet frequently when studying chemistry.

Common bases are metal oxides and hydroxides:

- metal oxides MgO, CuO
- metal hydroxides NaOH, $Mg(OH)_2$.

Ammonia is also a base, as are organic compounds called amines:

- ammonia NH_3
- amines CH_3NH_2.

You may also be familiar with some other bases used in everyday life.

- magnesium oxide MgO 'milk of magnesia' for treating acid indigestion
- calcium hydroxide $Ca(OH)_2$ lime for treating acid soils.

A base is the opposite of an acid:

- A base is a proton, H^+, acceptor.
- Bases neutralise acids.

Alkalis

The word alkali comes from the Arabic 'al kali' meaning the ashes. The ash of burnt plant materials is alkaline and contains potassium compounds (the word potassium comes from 'potash').

In chemistry, an alkali is any chemical compound that gives a solution with a pH of greater than 7.0 when dissolved in water.

Figure 2 Everyday alkalis

In the chemistry laboratory, you will come across the common alkalis:

- sodium hydroxide $NaOH$
- potassium hydroxide KOH
- ammonia NH_3.

Examiner tip

You need to learn the formulae of these three alkalis. $NaOH$, KOH, NH_3.

You may have come across sodium hydroxide without realising it.

- Sodium hydroxide is used in oven cleaners and paint strippers.

Alkalis are very corrosive, even more so than many acids.

- The alkali sodium hydroxide is corrosive even when as dilute as 0.5 mol dm^{-3}.
- Hydrochloric acid needs to be stronger than 6.5 mol dm^{-3} before it is classified as being corrosive.

An **alkali** is a special type of base that dissolves in water forming aqueous hydroxide ions, $OH^-(aq)$, e.g. $NaOH(s) + aq \longrightarrow Na^+(aq) + OH^-(aq)$.

In solution, the hydroxide ions from alkalis *neutralise* the protons from acids, forming water.

- $H^+(aq) + OH^-(aq) \longrightarrow H_2O(l)$

Key definition

An **alkali** is a type of base that dissolves in water forming hydroxide ions, $OH^-(aq)$ ions:

- sodium hydroxide, $NaOH$
  $NaOH(s) + aq \longrightarrow Na^+(aq) + OH^-(aq)$

Ammonia as a weak base

Ammonia (NH_3) is a gas that dissolves in water to form a weak alkaline solution. Dissolved NH_3 reacts with water:

$$NH_3(aq) + H_2O(l) \rightleftharpoons NH_4^+(aq) + OH^-(aq)$$

Ammonia is a weak base because only a small proportion of the dissolved NH_3 reacts with water. This is shown by the equilibrium sign, $\rightleftharpoons$.

You will study equilibrium reactions later in your AS course (Spreads 2.3.14 and 2.3.15) and during A2 chemistry.

Biological acids and bases

In biology, you may have come across other acids:

- fatty acids
- amino acids
- nucleic acids (DNA and RNA)

Amino acids are *amphoteric*, which means they have both acidic and basic properties. An amino acid molecule contains:

- a *carboxyl* acid group, COOH; and
- an *amino* basic group, NH_2.

The amino acid glycine is shown in Figure 3. You will study amino acids in detail during the A2 Chemistry course.

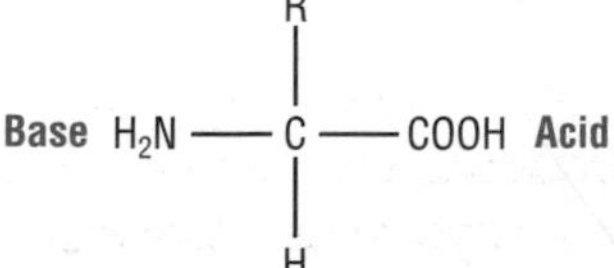

Figure 3 The amino acid glycine. An amino acid has both acid and base parts

Questions

1 Write down the formulae for the following:
(a) sulfuric acid;
(b) nitric acid;
(c) ethanoic acid;
(d) potassium hydroxide;
(e) calcium hydroxide;
(f) ammonia.

2 Define the terms:
(a) acid;
(b) base;
(c) alkali.

1.1 (11) Salts

By the end of this spread, you should be able to . . .

- ✱ **State that a salt is produced when the H^+ ion of an acid is replaced by a metal ion or NH_4^+.**
- ✱ **Describe the reactions of an acid with carbonates, bases and alkalis, to form a salt.**
- ✱ **Understand that a base readily accepts H^+ ions from an acid.**

Key definitions

A salt is any chemical compound formed from an acid when a H^+ ion from the acid has been replaced by a metal ion or another positive ion, such as the ammonium ion, NH_4^+.

A **cation** is a positively charged ion.

An **anion** is a negatively charged ion.

Figure 1 A magnified crystal of common salt, NaCl

Figure 2 Limestone in acid.
Limestone is mainly calcium carbonate ($CaCO_3$). Here, limestone is reacting with hydrochloric acid (HCl). This reaction produces carbon dioxide gas (CO_2) and a solution of the salt, calcium chloride ($CaCl_2$). The equation for this reaction is:
$CaCO_3(s) + 2HCl(aq) \rightarrow CaCl_2(aq) + H_2O(l) + CO_2(g)$
This is one of the reactions by which rainwater (which is slightly acidic) can dissolve limestone buildings and rock formations. If the rain is more strongly acidic, the effect is increased. In areas affected by acid rain, limestone masonry and statues have been badly eroded

Salts

A **salt** is an ionic compound with the following features.

- The positive ion or **cation** in a salt is usually a metal ion or an ammonium ion, NH_4^+.
- The negative ion or **anion** in a salt is derived from an acid.
- The formula of a salt is the same as the parent acid, except that a H^+ ion has been replaced by the positive ion.

Examples

- sulfuric acid $\longrightarrow$ sulfate salts
  H_2SO_4 — K_2SO_4 potassium sulfate
- hydrochloric acid $\longrightarrow$ chloride salts
  HCl — NaCl sodium chloride
- nitric acid $\longrightarrow$ nitrate salts
  HNO_3 — $Ca(NO_3)_2$ calcium nitrate

Acid salts

Sulfuric acid has two replaceable H^+ ions and is an example of a diprotic acid.

If one H^+ ion is replaced, an acid salt is formed, e.g. sodium hydrogensulfate, $NaHSO_4$.

- $H_2SO_4 \longrightarrow NaHSO_4$

The acid salt can itself behave as an acid, because the other H^+ ion can be replaced to form a conventional salt, e.g. sodium sulfate, Na_2SO_4.

- $NaHSO_4 \longrightarrow Na_2SO_4$

Formation of salts

Salts can be produced by neutralising acids with:

- carbonates
- bases
- alkalis.

Examples of neutralisation reactions are shown below. Hydrochloric acid has been used to illustrate these reactions, but any acid will react in a similar way.

In each example, a second equation (called an ionic equation) shows the important role of the H^+ ion in the reaction.

Salts from carbonates

Acids react with carbonates to form a salt, carbon dioxide and water. You will see bubbles of CO_2 (see Figure 2).

$$2HCl(aq) + CaCO_3(s) \longrightarrow CaCl_2(aq) + H_2O(l) + CO_2(g)$$
$$2H^+(aq) + CaCO_3(s) \longrightarrow Ca^{2+}(aq) + H_2O(l) + CO_2(g)$$

Salts from bases

Acids react with bases to form a salt and water. In the example below, you will see solid CaO dissolve.

$$2HCl(aq) + CaO(s) \longrightarrow CaCl_2(aq) + H_2O(l)$$
$$2H^+(aq) + CaO(s) \longrightarrow Ca^{2+}(aq) + H_2O(l)$$

Salts from alkalis

Acids react with alkalis, to form a salt and water.

$$HCl(aq) + NaOH(aq) \longrightarrow NaCl(aq) + H_2O(l)$$
$$H^+(aq) + OH^-(aq) \longrightarrow H_2O(l)$$

Salts from metals

Salts can also be formed from the reaction of reactive metals with acids.

These are known as redox reactions and will be discussed in spread 1.1.15.

Ammonium salts and fertilisers

Ammonium salts are used as artificial fertilisers.

Ammonium salts are formed when acids are neutralised by aqueous ammonia.

$$NH_3(aq) + HNO_3(aq) \longrightarrow NH_4NO_3(aq)$$

Ammonium salts contain the ammonium ion, NH_4^+, in place of the metal ions found in most common salts.

Ammonium nitrate is present in solution as two ions, $NH_4^+(aq)$ and $NO_3^-(aq)$.

Calculating the percentage of nitrogen in a fertiliser

Farmers and gardeners use fertilisers to supply plants with a soluble form of nitrogen. The quantity of a fertiliser required needs to be calculated carefully to provide the best amount of nitrogen for healthy plant growth.

Worked example

Ammonium nitrate, NH_4NO_3, is used as a fertiliser. How much nitrogen does it contain by mass? [A_r: N, 14.0; H, 1.0; O, 16.0]

First, calculate the molar mass of NH_4NO_3

$M(NH_4NO_3) = 14.0 + (1.0 \times 4) + 14.0 + (16.0 \times 3) = 80.0 \text{ g mol}^{-1}$

Then calculate the mass of nitrogen in 1 mole of NH_4NO_3

mass of N = 14.0 + 14.0 = 28.0 g

Finally, calculate the percentage, by mass, of N

$$\text{percentage of N} = \frac{28.0}{80.0} \times 100 = 35.0\%$$

Questions

1 Write balanced equations for the following acid reactions.
 (a) Hydrochloric acid and potassium hydroxide.
 (b) Nitric acid and calcium hydroxide.
 (c) Sulfuric acid and sodium hydroxide.
 (d) Nitric acid and magnesium carbonate, $MgCO_3$.
 (e) Phosphoric acid, H_3PO_4, and sodium carbonate, Na_2CO_3.

2 (a) Write balanced equations for the formation of ammonium salts from ammonia and:
  (i) hydrochloric acid;
  (ii) sulfuric acid;
  (iii) phosphoric acid.
 (b) Each of the ammonium salts above can be used as a fertiliser. Calculate the percentage of nitrogen by mass in each, and hence the 'N' part of its NPK rating.

Examiner tip

Neutralisation reactions are often tested in exams. Learn these general equations

acid + carbonate → salt + CO_2 + H_2O
acid + base → salt + water
acid + alkali → salt + water

and practise writing balanced equations for neutralisation of HCl and H_2SO_4.

NPK rating in fertilisers

Artificial fertilisers contain the three main plant nutrients – nitrogen, N; phosphorus, P; and potassium, K. The proportions of these three elements are shown by an NPK rating.

However, for historical reasons, the amount of phosphorus is measured as phosphorus pentoxide, P_2O_5, and potassium is measured as potash or potassium oxide, K_2O. Only the amount of nitrogen is measured as the element itself.

So, a 20-11-10 fertiliser therefore contains, by mass, 20% nitrogen, N. However the proportions of the elements P and K are actually much less than 11% and 10% – these are the percentages, by mass, of the compounds P_2O_5 and K_2O in the fertiliser.

Figure 3 NPK fertilisers

1.1 (12) Water of crystallisation

By the end of this spread, you should be able to . . .

- ✱ **Explain the terms anhydrous, hydrated and water of crystallisation.**
- ✱ **Calculate the formula of a hydrated salt using percentage composition, mass composition or experimental data.**

Key definition

Hydrated refers to a crystalline compound containing water molecules.

Key definition

Anhydrous refers to a substance that contains no water molecules.

Hydrated crystals

The familiar blue crystals that you see in your chemistry laboratories are the hydrated form of the compound copper sulfate. It seems remarkable that water is needed to give these crystals both their crystalline form and their blue colour.

Without the water, copper sulfate exists as just a white powder.

Chemists use special terms for these two forms:

- **hydrated** for the crystalline form that contains water molecules;
- **anhydrous** for the form without water.

Figure 1 Hydrated copper sulfate, $CuSO_4.5H_2O$

Key definition

Water of crystallisation refers to water molecules that form an essential part of the crystalline structure of a compound.

Examiner tip

Be careful in exams when writing down chemical formulae

Water of crystallisation is one of the rare instances in chemistry in which a number without a subscript is used within a formula, e.g. the '**5**' in $CuSO_4{\cdot}\mathbf{5}H_2O$.

This practice is very unusual and is only used in special cases, such as with hydrated crystals.

A number without a subscript is normally reserved for balancing a chemical equation, when it is always placed before a formula:
$2Na(s) + Cl_2(g) \longrightarrow 2NaCl(s)$.

Compounds crystallised from water frequently contain water molecules within the resulting crystalline structure. This is called **water of crystallisation**. Often, the compound cannot be crystallised if water molecules are not present.

The empirical formula of a hydrated compound is written in a unique way:

- the empirical formula of the compound is separated from the water of crystallisation by a dot
- the relative number of water molecules of crystallisation is shown after the dot.

Thus, for hydrated copper sulfate, which has *five* H_2O molecules for each formula unit of $CuSO_4$, its empirical formula is written as: $CuSO_4{\cdot}\mathbf{5}H_2O$.

Examples

- $CuSO_4{\cdot}5H_2O$ copper sulfate pentahydrate
- $CoCl_2{\cdot}6H_2O$ cobalt chloride hexahydrate
- $Na_2SO_4{\cdot}10H_2O$ sodium sulfate decahydrate

Working out the formula of a hydrated salt

You can work out the formula of a hydrated salt using percentage or mass compositions. You will need to use the same method that you used to work out the empirical formula (see spread 1.1.5). However:

- There is an additional final step required to show the water separately as its *dot formula*.

Table 1 compares some empirical formulae of hydrated salts with their *dot* formulae.

- The secret here is to use the number of hydrogen atoms to work out the number of water molecules of crystallisation.
- The tricky part is that the oxygen may be split between the water molecules and the ions, such as sulfate, nitrate and carbonate ions.

Salt	Empirical formula	Dot formula
hydrated magnesium chloride	$MgCl_2H_{10}O_5$	$MgCl_2{\cdot}5H_2O$
hydrated sodium carbonate	$Na_2CH_{20}O_{13}$	$Na_2CO_3{\cdot}10H_2O$
hydrated calcium nitrate	$CaN_2H_8O_{10}$	$Ca(NO_3)_2{\cdot}4H_2O$

Table 1 Empirical vs dot formulae

Figure 2 Heating hydrated copper sulfate. Although this experimental method is easy to carry out, it will only work if the anhydrous salt is not decomposed by heat. Unfortunately, many salts are decomposed by heat. This is especially true for nitrates

Determination of the formula of a hydrated salt

You can determine the formula of a hydrated salt by heating it so that the water of crystallisation is driven off and evaporates (See Figure 2).

A suitable apparatus for determining water of crystallisation is shown in Figure 3.

From the results, you will need to find:

- the mass of the *hydrated* salt, containing water of crystallisation
- the mass of the *anhydrous* salt, without the water
- the mass of water that was in the hydrated salt.

Worked example

From an experiment to determine the formula of hydrated magnesium sulfate:

mass of the *hydrated* salt, $MgSO_4{\cdot}\mathbf{x}H_2O$ = 4.312 g
mass of the *anhydrous* salt, $MgSO_4$ = 2.107 g
So, mass of H_2O in $MgSO_4{\cdot}\mathbf{x}H_2O$ = 2.205 g

- First, calculate the amount, in mol, of anhydrous $MgSO_4$:
  [A_r: Mg, 24.3; S, 32.1; O, 16.0]
  molar mass, $M(MgSO_4) = 24.3 + 32.1 + (16.0 \times 4) = 120.4\ \text{g mol}^{-1}$
  $$n(MgSO_4) = \frac{2.107}{120.4} = 0.0175\ \text{mol}$$
- Then calculate the amount, in mol, of water ($M = 18.0\ \text{g mol}^{-1}$):
  $$n(H_2O) = \frac{2.205}{18.0} = 0.1225\ \text{mol}$$
- Finally, determine the formula of the hydrated salt:
  Molar ratio $MgSO_4 : H_2O$ = 0.0175: 0.1225 = 1: 7 (divide by the smaller number)
  Hence, formula of the hydrated salt = $MgSO_4{\cdot}\mathbf{7}H_2O$; i.e. **x** = **7**.

Examiner tip

Again, we use:

amount, $n = \frac{m}{M}$

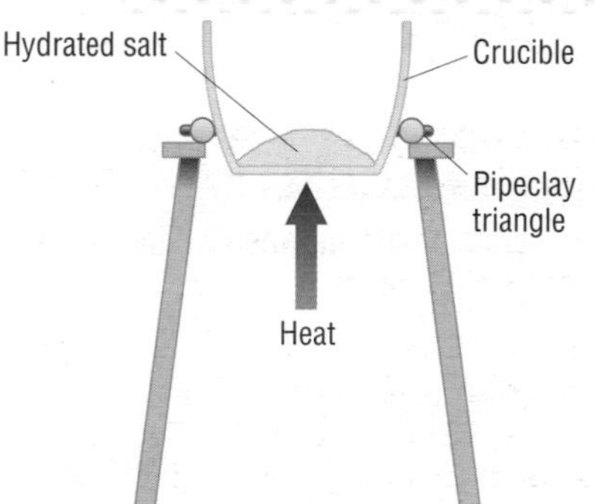

Figure 3 Apparatus for the determination of water of crystallisation

Questions

1 You are supplied with three empirical formulae. Write down the *dot* formula of each salt to show the water of crystallisation.
   (a) $BaCl_2H_4O_2$; **(b)** $ZnSH_{14}O_{11}$; **(c)** $FeN_3H_{12}O_{15}$.

2 From the experimental results shown below, work out the formula of the hydrated salt.
   [A_r: Ca, 40.1; Cl, 35.5; H, 1.0; O, 16.0]
   Mass of $CaCl_2{\cdot}\mathbf{x}H_2O$ = 6.573 g
   Mass of $CaCl_2$ = 3.333 g

Titrations

By the end of this spread, you should be able to . . .

*** Perform acid–base titrations, and carry out structured calculations.**

Acid–base titrations

During volumetric analysis, you measure the volume of one solution that reacts with a measured volume of a second, different solution.

An acid–base titration is a special type of volumetric analysis, in which you react a solution of an acid with a solution of a base.

- You must know the concentration of one of the two solutions. This is usually a *standard solution* (see spread 1.1.7).
- In the analysis, you use this standard solution to find out unknown information about the substance dissolved in the second solution.

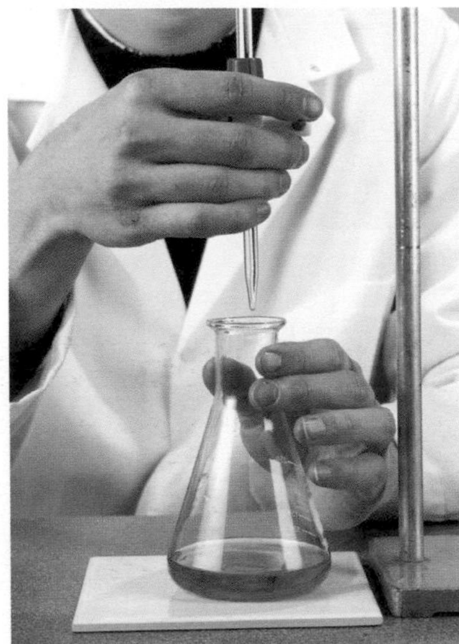

Figure 1 This acid–base titration is using methyl orange as an indicator. Methyl orange is coloured red in acidic solutions and yellow in basic solutions. The end point is the colour in between – orange. The solution in the conical flask above has reached the end point

The unknown information may be:

- the concentration of the solution
- a molar mass
- a formula
- the number of molecules of water of crystallisation.

You carry out a titration as follows.

- Using a *pipette*, you add a measured volume of one solution to a conical flask.
- The other solution is placed in a *burette*.
- The solution in the burette is added to the solution in the conical flask until the reaction has *just* been completed. This is called the *end point* of the titration. The volume of the solution added from the burette is measured.

You now know the volume of one solution that *exactly* reacts with the volume of the other solution.

We identify the end point using an *indicator*.

- The indicator must be a different colour in the acidic solution than in the basic solution.

Table 1 lists the colours of some common acid–base indicators. It also shows the colour at the end point. Notice that this end point colour is in between the colours in the acidic and basic solutions.

Indicator	Colour in acid	Colour in base	End point colour
methyl orange	red	yellow	orange
bromothymol blue	yellow	blue	green
phenolphthalein	colourless	pink	pale pink*

* This assumes that the aqueous base has been added from the burette to the aqueous acid. If acid is added to base, the titration is complete when the solution goes colourless.

Table 1 Colours of some common acid–base indicators

Calculating unknowns from titration results

Analysis of titration results follows a set pattern, as shown in the worked examples:

- the first two steps are always the same;
- the third step may be different, depending on the unknown that you need to work out.

In AS chemistry, any calculations that you carry out will be structured similarly to the examples below.

For A2, you may have to work out these steps yourself.

Worked example 1: Calculating an unknown concentration

In a titration, 25.0 cm^3 of 0.150 mol dm^{-3} sodium hydroxide NaOH(aq) reacted exactly with 23.40 cm^3 of sulfuric acid, H_2SO_4(aq).

$2NaOH(aq) + H_2SO_4(aq) \longrightarrow Na_2SO_4(aq) + 2H_2O(l)$

(a) Calculate the amount, in mol, of NaOH that reacted.

$$n(NaOH) = c \times \frac{V}{1000} = 0.150 \times \frac{25.0}{1000} = 3.75 \times 10^{-3} \text{ mol}$$

(b) Calculate the amount, in mol, of H_2SO_4 that was used.

equation	2NaOH(aq)	+ H_2SO_4(aq)	$\longrightarrow$
moles from equation	2 mol	+ 1 mol	$\longrightarrow$
actual moles	3.75×10^{-3} mol	1.875×10^{-3} mol	

(c) Calculate the concentration, in mol dm^{-3} of the sulfuric acid.

$$c(H_2SO_4) = \frac{n \times 1000}{V} = \frac{1.875 \times 10^{-3} \times 1000}{23.40} = 8.01 \times 10^{-3} \text{ mol dm}^{-3}$$

Notes

For (a), we use:

amount, $n = c \times \frac{V \text{(in cm}^3\text{)}}{1000}$

For (b), we use the balanced equation to work out the reacting quantities of the acid and alkali.
2 mol NaOH reacts with 1 mol H_2SO_4

For (c), we rearrange: $n = c \times \frac{V \text{(in cm}^3\text{)}}{1000}$

Hence, $c = \frac{n \times 1000}{V}$

Worked example 2: Calculating an unknown molar mass

A student dissolved 2.794 g of an acid HX in water and made the solution up to 250 cm^3. The student titrated 25.0 cm^3 of this solution against 0.0614 mol dm^{-3} sodium carbonate Na_2CO_3(aq). 23.45 cm^3 of Na_2CO_3(aq) were needed to reach the end point.
The equation for this reaction is:

$Na_2CO_3(aq) + 2HX(aq) \longrightarrow 2NaX(aq) + CO_2(g) + H_2O(l)$

(a) Calculate the amount, in mol, of Na_2CO_3 that reacted.

$$n(Na_2CO_3) = c \times \frac{V}{1000} = 0.0614 \times \frac{23.45}{1000} = 1.44 \times 10^{-3} \text{ mol}$$

(b) Calculate the amount, in mol, of HX that was used in the titration.

equation	Na_2CO_3 (aq)	+ 2 HX(aq)	$\longrightarrow$
moles from equation	1 mol	+ 2 mol	$\longrightarrow$
actual moles	1.44×10^{-3} mol	2.88×10^{-3} mol	

(c) Calculate the amount, in mol, of HX that was used to make up the 250 cm^3 solution.
25.0 cm^3 HX(aq) contains 2.88×10^{-3} mol
So, the 250 cm^3 solution contains $10 \times 2.88 \times 10^{-3} = 2.88 \times 10^{-2}$ mol

(d) Calculate the concentration, in g mol^{-1}, of the acid HX.

$$n = \frac{m}{M} \quad \text{so, } M = \frac{m}{n} = \frac{2.794}{2.88 \times 10^{-2}} = 97.0 \text{ g mol}^{-1}$$

Notes

For (a) and (b), the steps are the same as in the first worked example.

In this worked example, however, there are two further steps. You would be helpcd through these in an AS exam.

For (c), we scale up by a factor of 10 to find the amount, in mol, of HX in the 250 cm^3 solution that was made up.

For (d), we rearrange:

amount, $n = \frac{\text{mass, } m}{\text{molar mass, } M}$

Hence, $M = \frac{m}{n}$

In this example,

the mass of HX is 2.794 g.

Questions

1 Use the method in *Worked example 1* to calculate the unknown concentration below.
In a titration, 25.0 cm^3 of 0.125 mol dm^{-3} aqueous sodium hydroxide reacted exactly with 22.75 cm^3 of hydrochloric acid.

$HCl(aq) + NaOH(aq) \longrightarrow NaCl(aq) + H_2O(l)$

Find the concentration of the hydrochloric acid.

2 Use the method in *Worked example 2* to calculate the molar mass of the acid H_2X.
A student dissolved 1.571 g of an acid, H_2X, in water and made the solution up to 250 cm^3. She titrated 25.0 cm^3 of this solution against 0.125 mol dm^{-3} sodium hydroxide, NaOH(aq). 21.30 cm^3 of NaOH(aq) were needed to reach the end point.
The equation for this reaction is:

$2NaOH(aq) + H_2X(aq) \longrightarrow Na_2X(aq) + 2H_2O(l)$

1.1 (14) Oxidation number

By the end of this spread, you should be able to . . .

* **Assign oxidation numbers to atoms in elements, compounds and ions.**
* **Use a Roman numeral to indicate oxidation states.**

Key definition

An **oxidation number** is a measure of the number of electrons that an atom uses to bond with atoms of another element. Oxidation numbers are derived from a set of rules.

Oxidation numbers

Oxidation numbers are used extensively in chemistry. You will also come across the term *oxidation state* which essentially means the same thing.

Oxidation numbers cannot be measured directly by experiment. So what is an 'oxidation number'? And why do chemists use them?

Oxidation number rules

In a chemical formula, each atom has an oxidation number.

An atom uses electrons to bond with atoms from other elements.

- The **oxidation number** is the *number* of electrons that an atom uses to bond with atoms of another element.
- Oxidation numbers are worked out by following a set of rules. These are shown in Table 1. You must learn these rules and how to apply them correctly.

Species	Oxidation number	Examples
uncombined element	0	C; Na; O_2; P_4.
combined oxygen	–2	H_2O; CaO.
combined hydrogen	+1	NH_3, H_2S.
simple ion	charge on ion	Na^+, +1; Mg^{2+}, +2; Cl^-, –1
combined fluorine	–1	NaF, CaF_2, AlF_3.

Table 1 Oxidation number rules

As always, there are exceptions to rules:

- When bonded to fluorine, oxygen has an oxidation number of +2. In peroxides, oxygen has an oxidation number of –1.
- When bonded to metals in hydrides, hydrogen has an oxidation number of –1.

Luckily these exceptions are comparatively rare and you will not meet them at AS level.

Oxidation numbers in formulae

Compounds

In sulfur dioxide, SO_2, the overall *charge* on the molecule is zero.

- The sum of the oxidation numbers must equal the overall charge.
- There are 2 oxygen atoms, each with an oxidation number of –2. This gives a total contribution of –4.
- The oxidation number of sulfur must be +4 to balance the overall oxidation number of zero.

	SO_2
O O S	–2 –2 +4
overall	0

Examiner tip

It is essential to apply oxidation numbers to each atom.

You can get yourself into good habits by writing down the oxidation number separately for each atom underneath any formula.

Molecular ions

In a carbonate ion, CO_3^{2-}, the overall *charge* is 2–.

- The sum of the oxidation numbers must equal the overall charge.
- There are 3 oxygen atoms, each with an oxidation number of –2.
  This gives a total contribution of –6.
- The oxidation number of carbon must be +4 to give the overall charge of 2–.

	CO_3^{2-}
O	–2
O	–2
O	–2
C	+4
overall	2–

Oxidation numbers in chemical names

Some elements form compounds and ions in which the element has different oxidation numbers. In this case, the oxidation number of the element is included in the name of the compound or ion as a Roman numeral.

Without the oxidation number assigned in this way, you wouldn't know which oxidation state the element is in.

Compounds of transition elements

Transition elements form ions with different oxidation numbers. Without the oxidation number, the chemical name could apply to more than one compound of a transition element. Some common examples are shown below.

- $FeCl_2$ iron(II) chloride Fe: oxidation number +2
- $FeCl_3$ iron(III) chloride Fe: oxidation number +3
- Cu_2O copper(I) oxide Cu: oxidation number +1
- CuO copper(II) oxide Cu: oxidation number +2

Oxyanions

Oxyanions are negative ions that contain an element along with oxygen.

- Examples: SO_4^{2-}; NO_3^-; CO_3^{2-}.

The names of oxyanions usually end in *-ate*, to indicate oxygen.

As with transition element ions, an element may form oxyanions in which the element has different oxidation numbers. Without the oxidation number, the name would be ambiguous. Some common examples are:

- NO_2^- nitrate(III) N: oxidation number +3
- NO_3^- nitrate(V) N: oxidation number +5

- SO_3^{2-} sulfate(IV) S: oxidation number +4
- SO_4^{2-} sulfate(VI) S: oxidation number +6

Questions

1 What is the oxidation state of each element in the following?
(a) K; **(b)** Br_2; **(c)** NH_4^+; **(d)** CaF_2; **(e)** Al_2O_3;
(f) NO_3^-; **(g)** PO_4^{3-}.

2 Write down the oxidation state of nitrogen in the following.
(a) N_2O; **(b)** N_2O_5; **(c)** N_2H_4; **(d)** NO_3^-; **(e)** HNO_2.

3 Deduce the oxidation state of each element in bold.
(a) Na**Cl**O_3; **(b)** Na_2**S**O_3; **(c)** K**Mn**O_4; **(d)** K_2**Mn**O_4; **(e)** K_2**Cr**$_2O_7$.

4 Use oxidation number rules to work out the formula for the following ions.
(a) Chlorate(III) with an overall charge of 1–.
(b) Chlorate(VII) with an overall charge of 1–.
(c) Phosphate(III) with an overall charge of 3–.
(d) Chromate(VI) with an overall charge of 2–.

Redox reactions

By the end of this spread, you should be able to . . .

* **Describe oxidation and reduction in terms of electron transfer and oxidation number.**
* **Recognise that metals form positive ions by losing electrons, with an increase in oxidation number.**
* **Recognise that non-metals form negative ions by gaining electrons, with a decrease in oxidation number.**
* **Describe redox reactions of metals with either dilute hydrochloric or dilute sulfuric acid.**
* **Interpret and make predictions from redox equations in terms of oxidation numbers and electron loss/gain.**

Key definitions

Oxidation is loss of electrons or an increase in oxidation number.

Reduction is gain of electrons or a decrease in oxidation number.

Key definition

A **redox reaction** is a reaction in which both reduction and oxidation take place.

Examiner tip

No need to get in a muddle:

OILRIG: Oxidation Is Loss
Reduction Is Gain

If one species *gains* electrons (*re*duction), another species *loses* the same number of electrons (*ox*idation).

Key definitions

A **reducing agent** is a reagent that reduces (adds electron to) another species.

An **oxidising agent** is a reagent that oxidises (takes electrons from) another species.

Examiner tip

Always assign the oxidation number to each atom. In exams, candidates may put –2 below chlorine in $MgCl_2$ and then incorrectly claim that the oxidation number of chlorine is –2. The reality is that *each* chlorine atom has an oxidation number of –1.

Oxidation and reduction

The terms **oxidation** and **reduction** were originally used to describe reactions of substances with oxygen.

The definitions were simple.

- Oxidation is the gain of oxygen.
- Reduction is the loss of oxygen.

Oxidation and reduction are now used to describe any reactions in which electrons are transferred.

The definitions are now more general.

- Oxidation is the loss of electrons.
- Reduction is the gain of electrons.

The substance that is *reduced* takes electrons from the substance that is *oxidised*. Reduction must always be accompanied by oxidation. A reaction in which both reduction and oxidation take place is called a **redox reaction**.

Electron transfer in redox reactions

Magnesium reacts with chlorine to form magnesium chloride.

$$Mg + Cl_2 \longrightarrow MgCl_2$$

This is a redox reaction, but the equation for the overall reaction conceals this fact.

At an electronic level, electrons have been transferred from the magnesium atoms to the chlorine atoms. The **half equations** below show this clearly.

$Mg \longrightarrow Mg^{2+} + 2e^-$ oxidation (loss of electrons)
$Cl_2 + 2e^- \longrightarrow 2Cl^-$ reduction (gain of electrons)

Mg is a **reducing agent**: it has reduced the Cl_2 to $2Cl^-$ by donating its electrons.

Cl_2 is an **oxidising agent**: it has oxidised Mg to Mg^{2+} by removing its electrons.

Metals tend to be oxidised – losing electrons to form positive ions.

Non-metals tend to be reduced – gaining electrons to form negative ions.

Oxidation numbers in redox reactions

Oxidation and reduction can also be described in terms of oxidation number.

- Reduction is a decrease in oxidation number.
- Oxidation is an increase in oxidation number.

You can identify the oxidation and reduction processes using oxidation numbers. You assign an oxidation number to each atom in a redox reaction. You can then follow any changes to these oxidation numbers.

The oxidation and reduction processes for the reaction of magnesium and chlorine are shown in Table 1.

$Mg + Cl_2 \longrightarrow MgCl_2$	
$0 \longrightarrow +2$	oxidation (ox. no. increases)
$0 \longrightarrow -1$ $0 \longrightarrow -1$	reduction (ox. no. decreases)

Table 1 Redox reaction for magnesium and chlorine

Redox reactions of metals with acids

Reactive metals react with many acids in redox reactions.

- The metal is oxidised, forming positive metal ions.
- The hydrogen in the acid is reduced, forming the element hydrogen as a gas.

The oxidation and reduction processes for the reactions of magnesium metal with either dilute hydrochloric or dilute sulfuric acid are shown in Tables 2 and 3, respectively. Notice that a salt of the acid is formed together with hydrogen.

$Mg(s) + 2HCl(aq) \longrightarrow MgCl_2(aq) + H_2(g)$	
$0 \longrightarrow +2$	oxidation
$+1 \longrightarrow 0$ $+1 \longrightarrow 0$	reduction

Table 2 Reaction of magnesium metal with dilute hydrochloric acid

$Mg(s) + H_2SO_4(aq) \longrightarrow MgSO_4(aq) + H_2(g)$	
$0 \longrightarrow +2$	oxidation
$+1 \longrightarrow 0$ $+1 \longrightarrow 0$	reduction

Table 3 Reaction of magnesium metal with dilute sulfuric acid

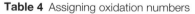

Examiner tip

Exam questions are often set involving a reaction of a metal with an acid.

Remember that the metal is an element and the oxidation number of its atoms is ZERO.

You can get the oxidation number of the metal in the salt from its ionic charge.

Acids react with reactive metals as shown below.

- metal + acid → salt + H_2

The equation can also be written to show the role of the hydrogen ion, H^+ (also called a proton). This equation is the same for each acid (except nitric acid, see Examiner tip).

- $Mg(s) + 2H^+(aq) \longrightarrow Mg^{2+}(aq) + H_2(g)$

Examiner tip

If one species decreases its oxidation number (reduction), another species increases its oxidation number (oxidation)

The total increase in oxidation number ***equals*** the total decrease in oxidation number.

Using oxidation numbers with equations

You can assign oxidation numbers to each atom in any equation. You can then identify whether a redox reaction has taken place. You can also deduce what has been oxidised and what has been reduced.

$MnO_2(s)$	$+ 4HCl(aq) \longrightarrow$	$MnCl_2(aq)$	$+ 2H_2O(l)$	$+ Cl_2(g)$	
+4 −2 −2	+1 −1 +1 −1 +1 −1 +1 −1	+2 −1 −1	+1 −2 +1 −2 +1 +1	0 0	oxidation numbers for each atom
$+4 \longrightarrow +2$					reduction (Mn)
	$-1 \longrightarrow 0$ $-1 \longrightarrow 0$				oxidation (Cl)

Table 4 Assigning oxidation numbers

Figure 1 Magnesium ribbon (Mg) reacting with hydrochloric acid (HCl). This reaction produces hydrogen gas (H_2) and a solution of the salt, magnesium chloride ($MgCl_2$). The equation for this reaction is:

$Mg(s) + 2HCl(aq) \longrightarrow MgCl_2(aq) + H_2(g)$

Questions

1 **(a)** Write equations for the reaction of:

(i) iron and hydrochloric acid forming the salt iron(II) chloride and hydrogen gas

(ii) aluminium with sulfuric acid forming the salt $Al_2(SO_4)_3$ and hydrogen gas.

(b) In each reaction, identify what has been oxidised and what has been reduced.

2 In the redox reactions below, use oxidation numbers to find out what has been oxidised and what has been reduced.

(a) $Cl_2 + 2KBr \rightarrow Br_2 + 2KCl$ **(b)** $2SO_2 + O_2 \rightarrow 2SO_3$

(c) $2HBr + H_2SO_4 \rightarrow SO_2 + Br_2 + 2H_2O$.

1.1 Atoms and reactions summary

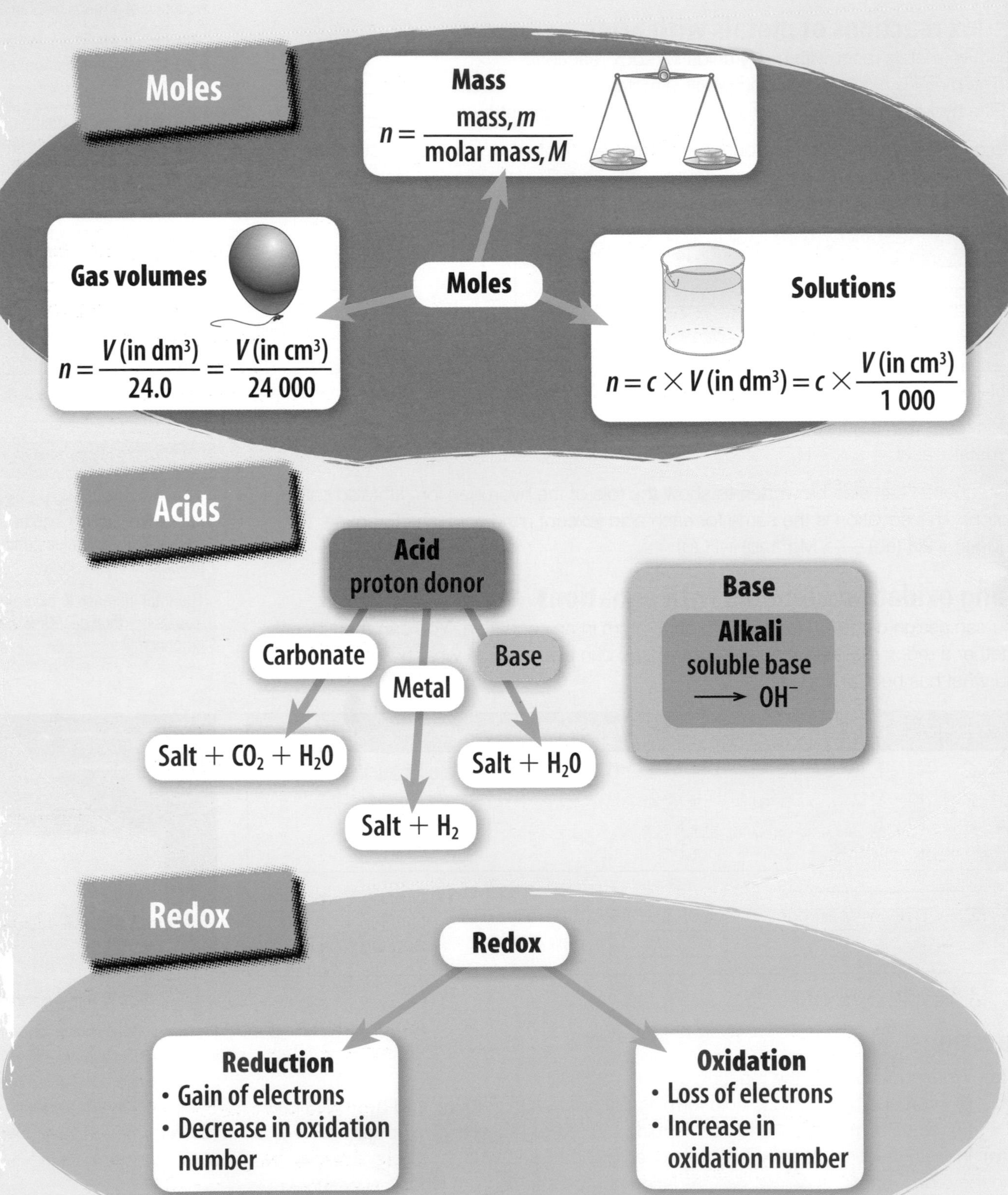

Moles
Mass
$n = \frac{\text{mass}, m}{\text{molar mass}, M}$
Moles
Gas volumes
$n = \frac{V(\text{in dm}^3)}{24.0} = \frac{V(\text{in cm}^3)}{24\,000}$
Solutions
$n = c \times V(\text{in dm}^3) = c \times \frac{V(\text{in cm}^3)}{1\,000}$
Acids
Acid
proton donor
Carbonate
Metal
Base
Salt + CO_2 + H_2O
Salt + H_2
Salt + H_2O
Base
Alkali
soluble base
$\longrightarrow$ OH^-
Redox
Redox
Reduction
• Gain of electrons
• Decrease in oxidation number
Oxidation
• Loss of electrons
• Increase in oxidation number

Practice questions

1 How many protons, neutrons and electrons are in the following atoms and ions?

(a) $^{11}_{5}B$; (b) $^{18}_{8}O$;
(c) $^{30}_{14}Si$; (d) $^{64}_{30}Zn$;
(e) $^{23}_{11}Na^+$; (f) $^{79}_{35}Br^-$;
(g) $^{45}_{21}Sc^{3+}$.

2 Calculate the relative atomic mass, A_r, to 4 significant figures, for the following.

(a) Lithium containing: 7.42% ^6Li and 92.58% ^7Li.
(b) Lead containing: ^{204}Pb, 1.1%; ^{206}Pb, 24.9%; ^{207}Pb, 21.7%; ^{208}Pb, 52.3%.

3 Use A_r values from the Periodic Table to calculate the relative formula mass of the following:

(a) Cr_2O_3; (b) Rb_2S;
(c) $Zn(OH)_2$; (d) $(NH_4)_2CO_3$;
(e) $Fe_3(PO_4)_2$.

4 Calculate the mass, in g, in:

(a) 6 mol MnO_2; (b) 0.25 mol Ga_2O_3;
(c) 0.60 mol Ag_2SO_4.

5 Calculate the amount, in mol, in:

(a) 8.823 g CaI_2; (b) 48.55 g K_2CrO_4;
(c) 7.50 g $Cu(NO_3)_2$.

6 Determine the empirical formula of the compound formed when 1.472 g of tungsten reacts with 0.384 g of oxygen.

7 Determine the molecular formula for the compound of carbon, hydrogen and oxygen atoms with the composition by mass: C, 40.0%; H, 6.7%; O, 53.3%; M_r 90.

8 (a) At RTP, what amount, in mol, of gas molecules are in:

(i) 84 dm^3; (ii) 300 cm^3;
(iii) 10 dm^3?

(b) At RTP, what is the volume, in dm^3, of:

(i) 12 mol $CO_2(g)$; (ii) 0.175 mol $N_2(g)$;
(iii) 2.55 g NO(g)?

9 Find the amount, in mol, of solute dissolved in the following solutions:

(a) 2 dm^3 of a 0.225 mol dm^{-3} solution;
(b) 25 cm^3 of a 0.175 mol dm^{-3} solution.

10 Find the concentration, in mol dm^{-3}, for the following solutions:

(a) 0.75 moles dissolved in 5 dm^3 of solution;
(b) 0.450 moles dissolved in 100 cm^3 of solution.

11 Balance each of the following equations:

(a) $P_4(s) + O_2(g) \longrightarrow P_2O_5(s)$
(b) $C_5H_{12}(l) + O_2(aq) \longrightarrow CO_2(g) + H_2O(l)$
(c) $NaOH(aq) + H_3PO_4(aq) \longrightarrow Na_2HPO_4(aq) + H_2O(l)$.

12 Define the terms:

(a) acid; (b) base; (c) alkali.

13 Write balanced equations for the following acid reactions:

(a) nitric acid and iron(III) hydroxide.
(b) sulfuric acid and copper(II) carbonate.
(c) hydrochloric acid and aluminium.

14 What is the oxidation state of each species in the following?

(a) Cu; (b) I_2; (c) CH_4;
(d) $MgSO_4$; (e) $PbCO_3$; (f) $Cr_2O_7^{2-}$;
(g) $(NH_4)_3PO_4$.

15 From the experimental results below, work out the formula of the hydrated salt.

Mass of $ZnSO_4{\cdot}xH_2O$ = 8.985 g
Mass of $ZnSO_4$ = 5.047 g

16 Lithium carbonate decomposes with heat:
$Li_2CO_3(s) \longrightarrow Li_2O(s) + CO_2(g)$
What volume of CO_2, measured at RTP, is formed by the decomposition of 5.7195 g of Li_2CO_3?

17 Calcium carbonate reacts with nitric acid:
$Na_2CO_3(s) + 2HNO_3 \rightarrow 2NaNO_3(aq) + H_2O(l) + CO_2(g)$
0.371 g of Na_2CO_3 reacts with an excess of HNO_3.
The final volume of the solution is 25.0 cm^3.

(a) What volume of CO_2, measured at RTP, is formed?
(b) What is the concentration, in mol dm^{-3}, of $NaNO_3$ formed?

18 In the redox reactions below, use oxidation numbers to find out what has been oxidised and what has been reduced.

(a) $N_2 + 3H_2 \rightarrow 2NH_3$
(b) $3Mg + 2Fe(NO_3)_3 \rightarrow 3Mg(NO_3)_2 + 2Fe$
(c) $MnO_2 + 4HCl \rightarrow MnCl_2 + Cl_2 + 2H_2O$

1.1 Examination questions

1 Chemists use a model of an atom that consists of sub-atomic particles: protons, neutrons and electrons.

(a) Complete the table below to show the properties of these sub-atomic particles.

particle	relative mass	relative charge
proton		
neutron		
electron		

[3]

(b) The particles in each pair below differ **only** in the number of protons **or** neutrons **or** electrons. Explain what the difference is within each pair.

(i) ^{6}Li and ^{7}Li

(ii) ^{32}S and $^{32}S^{2-}$

(iii) $^{39}K^{+}$ and $^{40}Ca^{2+}$ [6]

[Total: 9]

(Jun 99 4820)

2 A fifty pence coin contains nickel alloyed with a metal **A**.

(a) Nickel exists as a mixture of three isotopes, nickel-58, nickel-60 and nickel-62.
Complete the table below to show the atomic structures of the isotopes in metallic nickel.

isotope	protons	neutrons	electrons
nickel-58			
nickel-60			
nickel-62			

[3]

(b) Metal **A** can be identified from its relative atomic mass. Analysis of a fifty pence coin showed that two isotopes of metal **A** were present with the following percentage abundances.

isotope	isotope 1	isotope 2
relative isotopic mass	63.0	65.0
% abundance	77.2	22.8

(i) What analytical method is used to obtain this information? [1]

(ii) Define the term *relative atomic mass*. [3]

(iii) Calculate the relative atomic mass of the sample of metal **A**. Give your answer to three significant figures. [2]

(iv) Use your answer to **(b)(iii)** and the *Data Sheet* to suggest the identity of metal **A**. [1]

(c) Nickel makes up 25% of the total mass of a fifty pence coin. A fifty pence coin has a mass of 8.0 g.

(i) Calculate how many **moles** of nickel atoms are in a fifty pence coin. [2]

(ii) Calculate the **number** of atoms of nickel in a fifty pence coin. $N_A = 6.02 \times 10^{23}$ mol^{-1} [1]

[Total: 13]

(Jun 04 2811)

3 Azides are compounds of metals with nitrogen, used mainly as detonators in explosives. However, sodium azide, NaN_3, decomposes non-explosively on heating to release nitrogen gas. This provides a convenient method for obtaining pure nitrogen in the laboratory.

$2\ NaN_3(s) \rightarrow 2\ Na(s) + 3\ N_2(g)$

(a) A student prepared 1.80 dm^3 of pure nitrogen in the laboratory by this method. This gas volume was measured at room temperature and pressure (RTP).

(i) How many moles of nitrogen, N_2, did the student prepare? [Assume that 1 mole of gas molecules occupies 24.0 dm^3 at RTP]

(ii) What mass of sodium azide did the student heat? [3]

(b) After cooling, the student obtained 1.15 g of solid sodium. She then carefully reacted this sodium with water to form 25.0 cm^3 of aqueous sodium hydroxide.

$2\ Na(s) + 2\ H_2O(l) \rightarrow 2\ NaOH(aq) + H_2(g)$

Calculate the concentration, in mol dm^{-3}, of the aqueous sodium hydroxide. [2]

(c) Liquid sodium has been used as a coolant in nuclear reactors.

(i) Suggest **one** property of sodium that is important for this use.

(ii) Suggest **one** particular hazard of sodium for this use. [2]

[Total: 7]

(Mar 98 4820)

4 A household bleach contains sodium chlorate(I), NaClO, as its active ingredient.
The concentration of NaClO in the bleach can be found by using its reaction with hydrogen peroxide, H_2O_2.

$NaClO(aq) + H_2O_2(aq) \rightarrow O_2(g) + NaCl(aq) + H_2O(l)$

(a) Chlorine has been reduced in this reaction.
Use oxidation numbers to prove this. [2]

(b) A student added an excess of aqueous hydrogen peroxide to 5.0 cm^3 of the bleach. 84 cm^3 of oxygen gas were released.

(i) How many moles of O_2 were released?
Assume that, under the laboratory conditions, 1.00 mol of gas molecules occupies 24 dm^3. [1]

(ii) How many moles of NaClO were in 5.0 cm^3 of the bleach? [1]

(iii) What was the concentration, in mol dm^{-3}, of NaClO in the bleach? [1]

Answers to examination questions will be found on the Exam Café CD.

(c) The label on the bottle of household bleach states that the bleach contains a minimum of 4.5 g per 100 cm^3 of NaClO.
Use your answer to (b)(iii) to decide whether or not the information on the label is correct. [3]

(d) It is extremely important that household bleach is not used with acids. This is because a reaction takes place that releases toxic chlorine gas.
Suggest an equation for the reaction of an excess of hydrochloric acid with household bleach. [2]

[Total: 10]
(Jun 04 2811)

5 The formation of magnesium oxide, MgO, from its elements involves both oxidation and reduction in a redox reaction.

(a) (i) What is meant by the terms *oxidation* and *reduction*? [2]

(ii) Write a full equation, including state symbols, for the formation of MgO from its elements. [2]

(iii) Write half equations for the oxidation and reduction processes that take place in this reaction. [2]

(b) MgO reacts when heated with acids such as nitric acid, HNO_3.
$MgO(s) + 2HNO_3(aq) \rightarrow Mg(NO_3)_2(aq) + H_2O(l)$
A student added MgO to 25.0 cm^3 of a warm solution of 2.00 mol dm^{-3} HNO_3 until all the acid had reacted.

(i) How would the student have known that the reaction was complete? [1]

(ii) Calculate how many moles of HNO_3 were used. [1]

(iii) Deduce how many moles of MgO reacted with this amount of HNO_3. [1]

(iv) Calculate what mass of MgO reacted with this amount of HNO_3. [A_r: Mg, 24.3; O, 16.0]
Give your answers to three significant figures. [3]

(v) Using oxidation numbers, explain whether the reaction between MgO and HNO_3 is a redox reaction. [2]

[Total: 14]
(Jan 02 2811)

6 The reaction between barium and water is a redox reaction
$Ba(s) + 2H_2O(l) \rightarrow Ba(OH)_2(aq) + H_2(g)$

(a) Explain, in terms of electrons, what is meant by

(i) oxidation, [1]

(ii) reduction. [1]

(b) Which element has been oxidised in this reaction? Deduce the change in its oxidation state. [2]

(c) A student reacted 2.74 g of barium with water to form 250 cm^3 of aqueous barium hydroxide.

(i) Calculate how many moles of Ba reacted. [1]

(ii) Calculate the volume of H_2 that would be produced at room temperature and pressure (RTP). [1 mol of gas molecules occupies 24.0 dm^3 at RTP] [1]

(iii) Calculate the concentration, in mol dm^{-3}, of $Ba(OH)_2$ that was formed. [1]

(iv) The solution of barium hydroxide is alkaline. Identify a compound that could be added to neutralise this solution and write a balanced equation for the reaction that would take place [2]

[Total: 9]
(Jun 01 2811)

7 This question is about the reactions of acids.

(a) Sulfuric acid reacts with ammonia to give ammonium sulfate.
$2NH_3 + H_2SO_4 \rightarrow (NH_4)_2SO_4$

(i) What property of ammonia is shown in this reaction? [1]

(ii) Calculate the maximum mass of ammonium sulfate that can be obtained from 100 g of ammonia.
[A_r: H, 1.0; N, 14.0; O, 16.0; S, 32.1] [3]

(iii) State a large-scale use of ammonium sulfate. [1]

(b) State what you would observe on adding nitric acid to magnesium carbonate.
Write a balanced equation for the reaction. [3]

[Total: 8]
(Jan 02 2813/01)

8 Hydrogen chloride, HCl, is a colourless gas which dissolves very readily in water forming hydrochloric acid. [1 mol of gas molecules occupy 24.0 dm^3 at room temperature and pressure, RTP]

(a) At room temperature and pressure, 1.00 dm^3 of water dissolved 432 dm^3 of hydrogen chloride gas.

(i) How many moles of hydrogen chloride dissolved in the water? [1]

(ii) The hydrochloric acid formed has a volume of 1.40 dm^3. What is the concentration, in mol dm^{-3}, of the hydrochloric acid? [1]

(b) Hydrochloric acid reacts with magnesium oxide, MgO, and magnesium carbonate, $MgCO_3$.
For each reaction, state what you would see and write a balanced equation.

(i) MgO [2]

(ii) $MgCO_3$ [2]

[Total: 6]
(Jun 01 2811)

UNIT 1

Module 2 Electrons, bonding and structure

Introduction

Have you ever wondered how chlorine – a reactive and poisonous gas – can react with sodium – a reactive metal – to produce a substance that you sprinkle on your chips? What is it that makes water a liquid at room temperature, and why do some solids dissolve in water, while others don't?

Most materials we meet on a day-to-day basis are made up of compounds – substances in which two or more elements have bonded together to form a new material with its own special chemical properties.

Electrons play an important role in chemical bonding, being shared or transferred between atoms to hold chemical structures together.

In this module you will study the various types of bonding and find out why different compounds have distinct chemical formulae. You will explore chemical structure and investigate how the properties of materials are related to their bond types. You will also see how the forces between molecules can influence their properties.

The intermolecular forces holding water molecules together are the reason for the intricate patterns we see in snowflakes and frost on a window, shown here.

Test yourself

1 What is the electron structure of an atom with 13 electrons?
2 What is meant by an 'ion'?
3 Name a covalent compound.
4 What are the names of the hard and soft forms of carbon?
5 What is the chemical name for the commonest solid on Earth, sand?
6 What does the word 'alloy' mean?

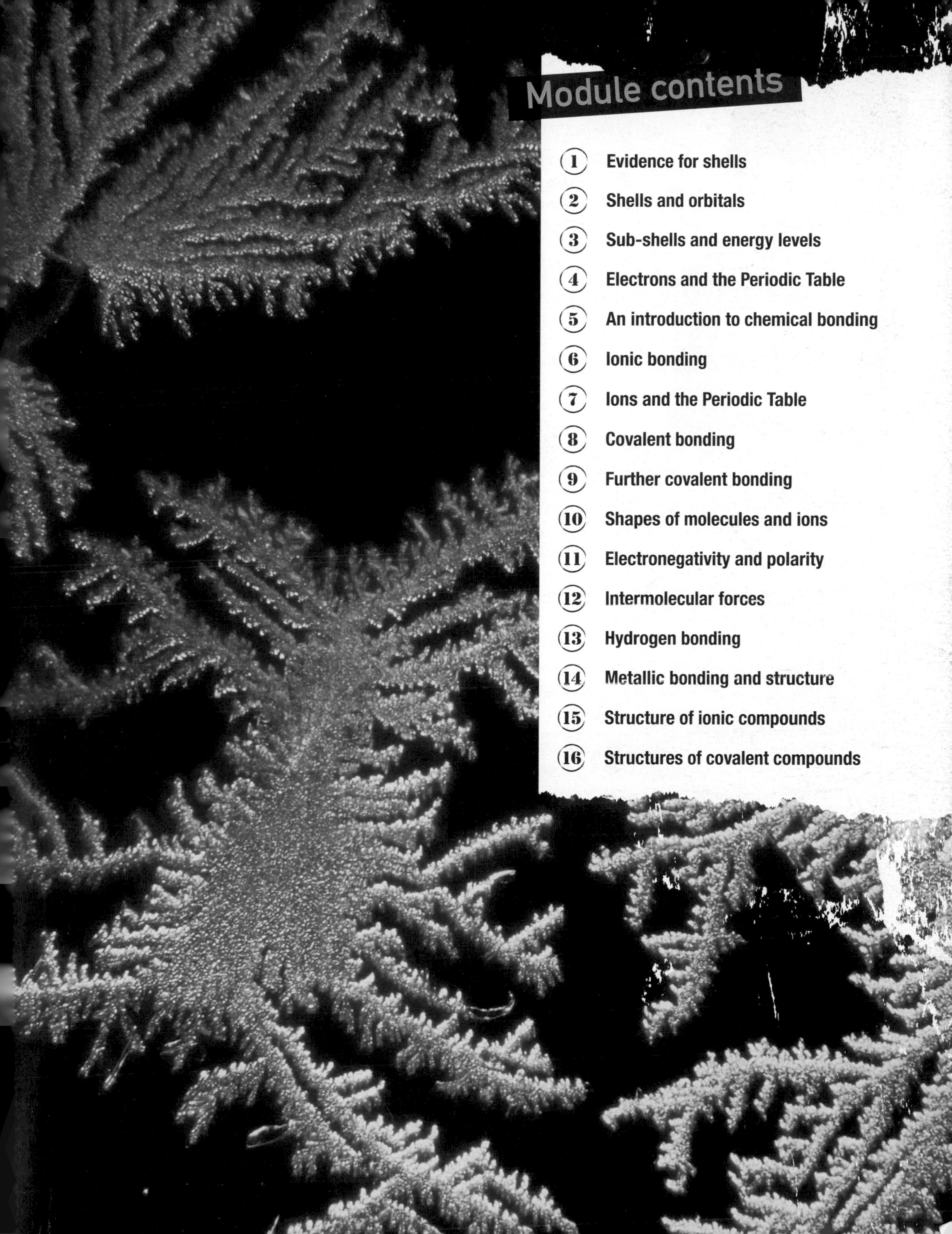

Module contents

1.2 (1) Evidence for shells

By the end of this spread, you should be able to . . .

* **Define first ionisation energy and successive ionisation energy.**
* **Explain the factors that influence ionisation energies.**
* **Predict the number of electrons in each shell as well as the element's group, using successive ionisation energies.**

Figure 1 Ionisation in plasma TVs

Plasma displays

Plasma televisions are large flat-panel displays. The screen consists of hundreds of thousands of tiny cells. Each cell holds an inert mixture of the noble gases neon and xenon between two plates of glass.

The gas in the cells is electrically ionised into a mixture of positive ions and negative electrons.

- $Ne(g) \rightarrow Ne^{+}(g) + e^{-}$
- $Xe(g) \rightarrow Xe^{+}(g) + e^{-}$

The formation of these ions requires energy. The mixture of positive ions and negative electrons is called a *plasma*. The presence of the plasma causes the screen to emit light.

Ionisation energy

The energy needed to form positive ions is known as the ionisation energy. Ionisation energies provide evidence for a model of the atom in which electrons are arranged in shells.

The **first ionisation energy** (1st I.E.) is a measure of how easily an atom loses an electron to form a 1+ ion.

This process is summarised in the equation shown in Table 1.

Key definition

The **first ionisation energy** of an element is the energy required to remove one electron from each atom in one mole of gaseous atoms to form one mole of gaseous 1+ ions.

$Na(g)$	$\longrightarrow$	$Na^{+}(g)$	+	e^{-}	1st I.E. = $+496\ kJ\ mol^{-1}$
1 mol of gaseous atoms		1 mol of gaseous 1+ ions		1 mol of electrons	

Table 1 First ionisation energy (I.E.) for sodium gas

Factors affecting ionisation energy

Negative electrons are held in their shells by their attraction to the positive nucleus.

To form a positive ion, energy must be supplied to an electron to overcome the attraction from the nucleus. Electrons in the outer shell are removed first since they experience the least nuclear attraction. The outer-shell electrons are furthest away from the nucleus and require the least ionisation energy.

The nuclear attraction experienced by an electron depends on three factors.

Atomic radius

- The greater the atomic radius, the smaller the nuclear attraction experienced by the outer electrons.

Nuclear charge

- The greater the nuclear charge, the greater the attractive force on the outer electrons.

Electron shielding or screening

- Inner shells of electrons repel the outer-shell electrons.
- This repelling effect is called **electron shielding** or *screening*.
- The more inner shells there are, the larger the shielding effect and the smaller the nuclear attraction experienced by the outer electrons.

Key definition

Electron shielding is the repulsion between electrons in different inner shells. Shielding reduces the net attractive force from the positive nucleus on the outer-shell electrons.

Successive ionisation energies

Successive ionisation energies are a measure of the energy required to remove each electron in turn. For example, the **second ionisation energy** is a measure of how easily a 1+ ion loses an electron to form a 2+ ion.

An element has as many ionisation energies as it has electrons.

Lithium has three electrons and three successive ionisation energies. Equations to represent the first three ionisation energies of lithium are shown in Table 2.

$Li(g) \longrightarrow Li^+(g) + e^-$	1st I.E. = +520 kJ mol^{-1}
$Li^+(g) \longrightarrow Li^{2+}(g) + e^-$	2nd I.E. = +7298 kJ mol^{-1}
$Li^{2+}(g) \longrightarrow Li^{3+}(g) + e^-$	3rd I.E. = +11 815 kJ mol^{-1}

Table 2 First three ionisation energies for lithium

Each successive ionisation energy is larger than the one before.

- As each electron is removed, there is less repulsion between the electrons and each shell will be drawn in to be slightly closer to the nucleus.
- As the distance of each electron from the nucleus decreases slightly, the nuclear attraction increases. More ionisation energy is needed to remove each successive electron.

Key definition

Successive ionisation energies are a measure of the energy required to remove each electron in turn. For example, the **second ionisation energy** of an element is the energy required to remove one electron from each ion in one mole of gaseous 1+ ions to form one mole of gaseous 2+ ions.

Examiner tip

Notice that the number of the successive ionisation energy is the same as the charge on the ion formed.

The 3rd ionisation energy of Li makes a 3+ ion from a 2+ ion:
$Li^{2+}(g) \longrightarrow Li^{3+}(g) + e^-$.

Evidence for shells

Successive ionisation energies provide experimental evidence for the existence of different shells.

Sometimes you will see a large increase in successive ionisation energies. An example is between the 1st and 2nd ionisation energies of lithium shown in Table 2. This large increase shows that the second electron has been removed from a different shell, closer to the nucleus and with less shielding from inner electrons (see Figure 2).

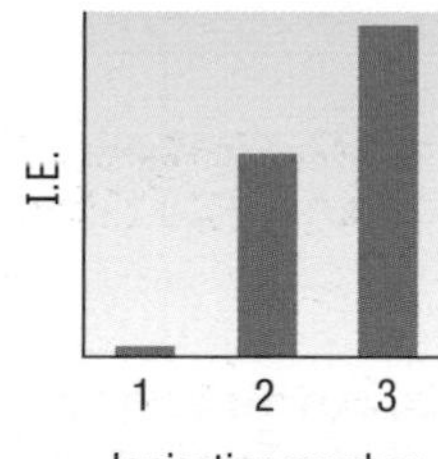

Figure 2 Three successive ionisation energies of lithium

The graph shown in Figure 3 shows how the successive ionisation energies of nitrogen provide evidence for the presence of shells. The shell number is shown by the symbol n. (For more details, see spread 1.2.2.)

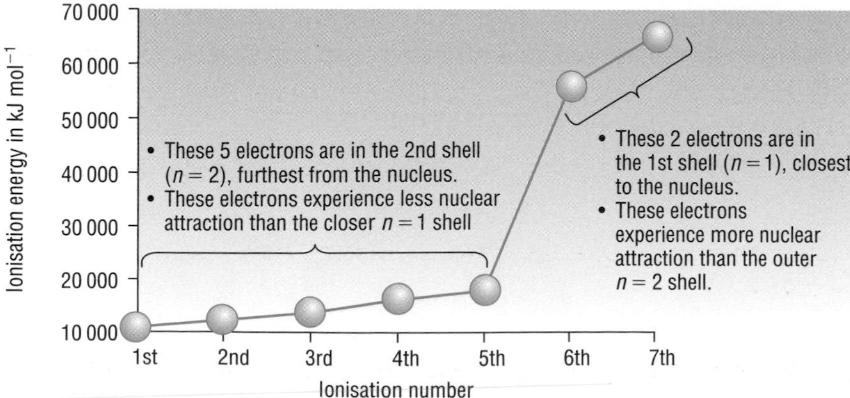

Figure 3 Successive ionisation energies of nitrogen. Notice the large increase in ionisation energy as the 6th electron is removed from the nitrogen atom. The successive ionisation energies are reflecting the change from one shell to another as you get closer to the nucleus. If the electrons were all in the same shell, there would be no sharp increase or jump

Questions

1 Write an equation to represent the 4th ionisation energy of chlorine.

2 Sketch the graph that you would expect for the successive ionisation energies of aluminium.

3 An element in Period 3 (Na–Ar) has the following successive ionisation energies, in kJ mol^{-1}: 789; 1577; 3232; 4356; 16091; 19785; 23787; 29253.
Identify this element, giving your reasons.

1.2 (2) Shells and orbitals

By the end of this spread, you should be able to ...

* **State the number of electrons that can fill the first four shells of an atom.**
* **Define an orbital.**
* **Describe the shapes of s- and p-orbitals.**

Key definitions

A **shell** is a group of atomic orbitals with the same principal quantum number, *n*. Also known as a main energy level.

Principal quantum number, *n*, is a number representing the relative overall energy of each orbital, which increases with distance from the nucleus. The sets of orbitals with the same *n*-value are referred to as electron shells or energy levels.

Energy levels or shells

In spread 1.2.1, we discussed how successive ionisation energies provide evidence for:

- the presence of **shells**;
- the number of electrons in each shell.

Quantum numbers are used to describe the electrons in atoms.

- **The principal quantum number**, *n*, indicates the shell that the electrons occupy.
- Different shells have different principal quantum numbers.
- The larger the value of *n*, the further the shell from the nucleus and the higher the energy level.

Throughout this book, we will use the word *shell*, but its meaning is equivalent to the energy level of the principal quantum number.

The first four shells hold different numbers of electrons, as shown in Table 1.

Each shell holds up to $2n^2$ electrons. Look at Table 1: can you see the pattern?

n	Shell	Electrons
1	1st shell	2
2	2nd shell	8
3	3rd shell	18
4	4th shell	32

Table 1 Numbers of electrons in the first four shells

Atomic orbitals

There are rules governing where the electrons are allowed to be within each shell. We once thought that electrons orbited around the nucleus, much like very fast planets orbiting the Sun. But electrons are not solid particles and this planetary electron idea has now been replaced.

It is now thought that electrons occupy *orbitals*. These are regions around the nucleus in which electrons are found.

Each shell is made up of **atomic orbitals**. Each atomic orbital can hold a maximum of two electrons. Imagine one electron as an electron cloud, with the shape of the orbital. Two electrons would be a cloud with the same shape, but twice as dense.

Key definition

An **atomic orbital** is a region within an atom that can hold up to two electrons, with opposite spins.

There are four different types of orbital – s, p, d and f. Each has a different shape.

s-orbitals

An s-orbital has a spherical shape as illustrated in Figure 1. From $n = 1$ upwards, each shell contains one s-orbital.

- This gives a total of $1 \times 2 = 2$ s electrons in each shell.

Figure 1 An s-orbital is spherical in shape. Each shell contains one s-orbital

p-orbitals

A p-orbital has a 3-dimensional dumb-bell shape as shown in Figure 2.

From $n=2$ upwards, each shell contains three p-orbitals, p_x, p_y and p_z, at right angles to one another.

- This gives a total of $3 \times 2 = 6$ p electrons.

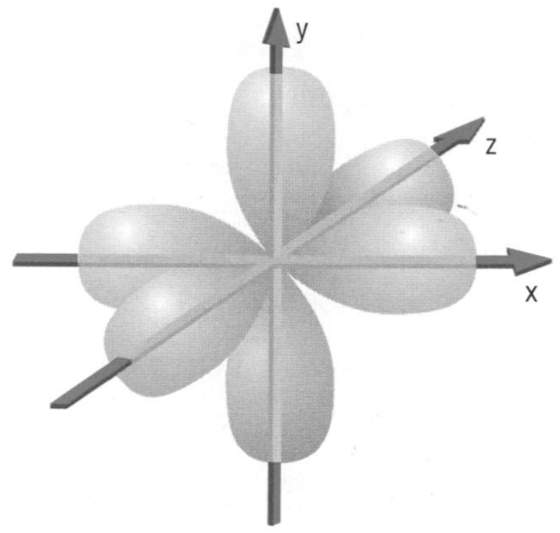

Figure 2 Three p-orbitals. Each p-orbital has a shape rather like a dumb-bell, with its centre at the nucleus

d-orbitals and f-orbitals

The structures of d- and f-orbitals are more complex.

From $n=3$ upwards, each shell contains five d-orbitals.

- This gives a total of $5 \times 2 = 10$ d electrons.

From $n=4$ upwards, each shell contains seven f-orbitals.

- This gives a total of $7 \times 2 = 14$ f electrons.

f 14 electrons

d 10 electrons

p 6 electrons

s 2 electrons

Figure 3 Electrons in boxes. It is convenient to represent an orbital as a box. The diagram above shows boxes for the orbitals in a shell. Each orbital can hold 2 electrons

Representing electrons in orbitals

With orbitals having different types and different shapes, chemists often represent an orbital as a box. Each box can hold two electrons. Chemists call this representation *electrons in boxes*, shown in Figure 3.

However, each electron has a negative charge and we would expect the two electrons in an orbital to repel one another.

An electron has a property called spin. Each of the electrons in an orbital must have opposite spins. We can represent the opposite spins of an electron using arrows, either 'up' or 'down' (see Figure 4).

But what is an electron? The answer to this question is not easy. We don't even know the exact position of the electron, even within an atomic orbital. In some ways, an electron is a particle, but in others it behaves as a wave. And even then, we can't be certain where it is. So it is very difficult, if not impossible, to visualise exactly what an electron is! However, one thing *is* certain – the idea that an electron is like a tiny planet whizzing around the nucleus is not true.

Luckily, chemists are not too concerned with the exact properties of an electron at this level of detail. However, chemists do use information about atomic orbitals and electrons when describing the behaviour of atoms, how elements react and the structure of the Periodic Table.

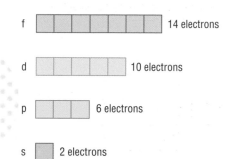

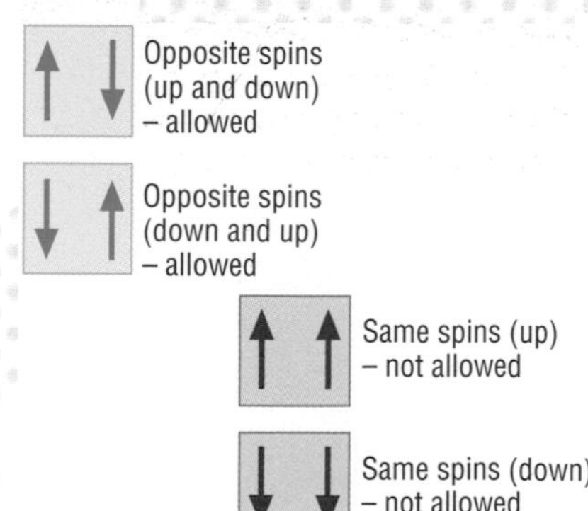

Figure 4 Electrons within an orbital always have opposite spins. This is called Hund's rule

Questions

1 How many electrons are held by the first five shells ($n=1$ to $n=5$)?

2 (a) What is meant by the term *orbital*?

(b) How many electrons can an orbital hold?

(c) Within a shell, state the number of s-, p-, d- and f-orbitals.

1.2 ③

Sub-shells and energy levels

By the end of this spread, you should be able to . . .

* **State the number of orbitals making up s, p and d sub-shells.**
* **State the number of electrons that occupy s, p and d sub-shells.**
* **Describe the relative energies of s-, p- and d-orbitals for shells 1, 2 and 3.**
* **Deduce the electron configurations of atoms for the first two periods.**

Key definition

A **sub-shell** is a group of the same type of atomic orbitals (s, p, d or f) within a shell.

Sub-shells

An electron shell is made up of atomic orbitals with the same principal quantum number, *n*. Within each shell orbitals of the same type are grouped together as a **sub-shell**.

Each sub-shell is made up of one type of atomic orbital only. So there are s, p, d and f sub-shells.

The diagrams below show the orbitals available in the first three shells.

* Each orbital is shown as a box that can hold a maximum of 2 electrons.
* Each shell gains a new type of sub-shell.

***n*=1 shell: 2 electrons**

Sub-shell	1s
Orbital	□
Electrons	2

***n*=2 shell: 8 electrons**

Sub-shell	2s	2p
Orbital	□	□□□
Electrons	2	2 2 2

***n*=3 shell: 18 electrons**

Sub-shell	3s	3p	3d
Orbital	□	□□□	□□□□□
Electrons	2	2 2 2	2 2 2 2 2

Electron energy levels

The sub-shells within a shell have different energy levels. The order of these energy levels is shown in Figure 1. Within a shell, the sub-shell energies increase in the order s, p, d and f.

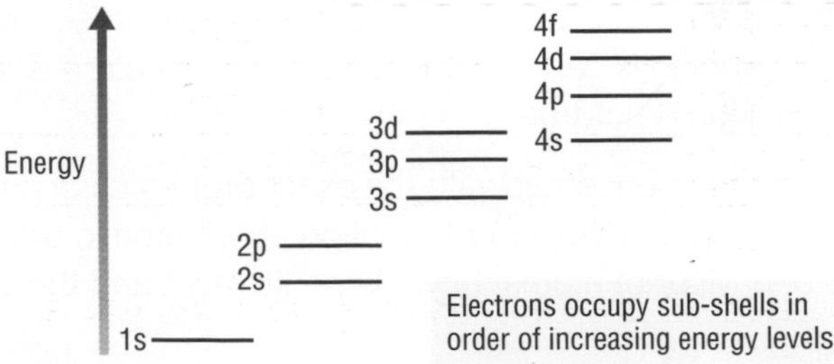

Figure 1 Shells, sub-shells and energy levels

Filling shells and sub-shells

The **electron configuration** is the arrangement of electrons in an atom. You can work out the electron configuration of an atom by following a set of rules. These rules are sometimes called the Aufbau principle, from the German *Aufbauprinzip* – building-up principle.

The electrons in the shells of an atom are arranged as follows:

- Electrons are added, one at a time, to 'build up' the atom.
- The lowest available energy level is filled first.
- Each energy level must be full before the next, higher energy level starts to fill.

Sub-shells are made up of several orbitals, each with the same energy level.

- When a sub-shell is built up with electrons, each orbital is filled singly before pairing starts.
- An orbital can hold a maximum of two electrons, each with an opposite spin.

Filling the orbitals

The diagram in Figure 2 shows how the electron configuration is built up for the elements B, C, N and O.

Notice that:

- orbitals fill from the lowest energy level upwards;
- the 2p-orbitals are filled singly before pairing starts at oxygen
- paired electrons have opposite spins.

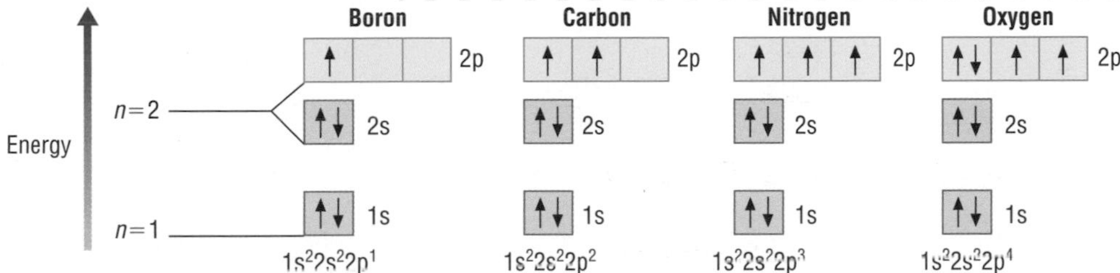

Figure 2 Filling the 2p-orbital

Electron configuration

We use a special shorthand to show **electron configurations**.

Table 1 shows the electron configurations for atoms of B, C, N and O.

Element	Orbitals occupied	Electron configuration
B	$1s^2 2s^2 2p_x^1$	$1s^2 2s^2 2p^1$
C	$1s^2 2s^2 2p_x^1 2p_y^1$	$1s^2 2s^2 2p^2$
N	$1s^2 2s^2 2p_x^1 2p_y^1 2p_z^1$	$1s^2 2s^2 2p^3$
O	$1s^2 2s^2 2p_x^2 2p_y^1 2p_z^1$	$1s^2 2s^2 2p^4$

Table 1 Electron configurations

Notice that each occupied sub-shell is written in the form nx^y, where:

- n is the shell number
- x is the type of orbital
- y is the number of electrons in the orbitals making up the sub-shell.

Key definition

An **electron configuration** is the arrangement of electrons in an atom.

Questions

1 Draw diagrams, similar to those for Figure 2, to show electrons in orbitals for F and Ne.
2 Write the electron configuration in terms of sub-shells for the elements H to Ne.

1.2 (4) Electrons and the Periodic Table

By the end of this spread, you should be able to . . .

- **Describe the relative energies of s-, p- and d-orbitals for the shells 1, 2, 3 and of the 4s- and 4p-orbitals.**
- **Deduce the electron configuration of atoms and ions up to $Z = 36$.**
- **Classify the elements into s-, p- and d-blocks.**

Electron shells overlap

In spreads 1.2.2 and 1.2.3, we discussed that:

- The larger the value of the principal quantum number *n*, the higher the energy level and the further the shell is from the nucleus.
- As we move from one shell to the next, a new type of sub-shell is added.
- Within a shell, the sub-shell energies increase in the order s, p, d and f.

However, you need to take care when you go beyond the 3p sub-shell.

Remember that the orbitals fill in energy level order, but

- the 4s energy level is below the 3d energy level.
- the 4s-orbitals fill before the 3d-orbitals.

Therefore, the 4th shell starts to fill before the 3rd shell has completely filled!

This overlap is shown in Figure 1.

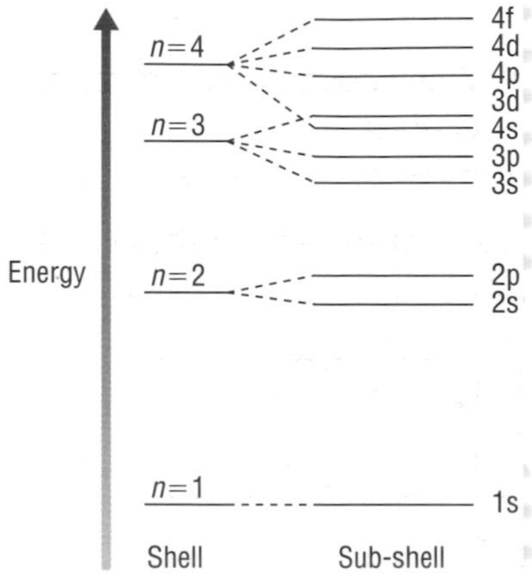

Energy
n=4: 4f, 4d, 4p, 3d, 4s
n=3: 3p, 3s
n=2: 2p, 2s
n=1: 1s
Shell · Sub-shell

Figure 1 Overlap of 4s- and 3d sub-shells

Examiner tip

The 4s sub-shell has a lower energy level than the 3d sub-shell.

The 4s-orbital fills *before* the orbitals in the 3d sub-shell.

The 4p-orbital starts to fill *after* the 3d-orbital is full.

The diagram in Figure 2 shows how sub-shell orbitals are filled for an atom of potassium.

The electron configuration is: $1s^2 2s^2 2p^6 3s^2 3p^6 4s^1$

Electronic configuration of potassium: $1s^2 2s^2 2p^6 3s^2 3p^6 4s^1$

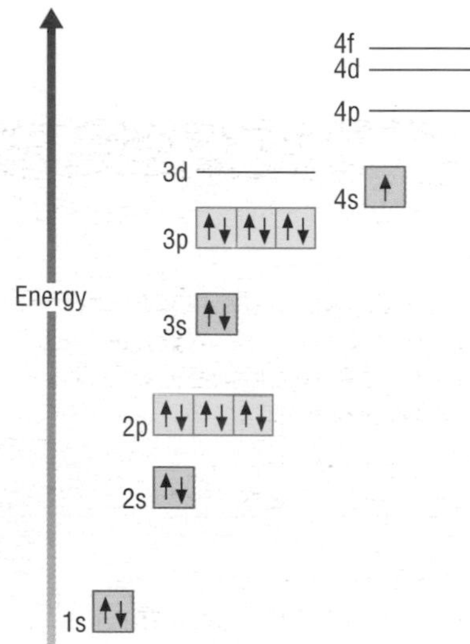

Figure 2 Filling the orbitals in a potassium atom

Sub-shells and the Periodic Table

The Periodic Table is structured in blocks of 2, 6, 10 and 14 elements, linked to sub-shells. The pattern mirrors the sub-shells that are being filled. We can easily see the pattern by dividing the Periodic Table into blocks as shown in Figure 3.

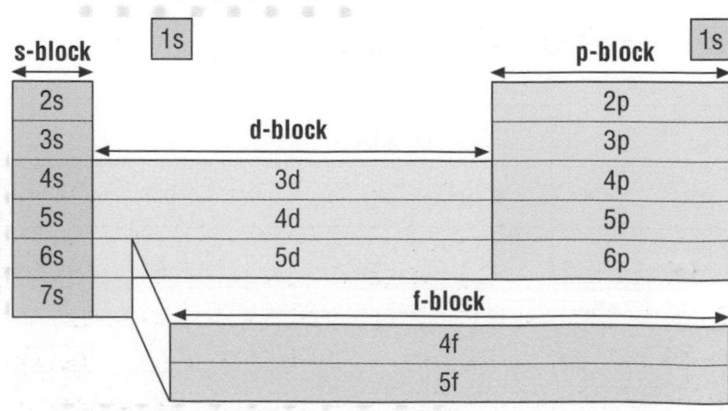

Figure 3 The Periodic Table, sub-shells and blocks

Using the Periodic Table for electron configurations

By following the pattern shown in Figure 3, you can work out the electron configuration for any element using the Periodic Table. Figure 4 shows you how to work out the electron configuration of oxygen.

- Oxygen is the 4th element in the 2p block, i.e. $2p^4$.
- Oxygen has the electron configuration: $1s^2 2s^2 2p^4$.

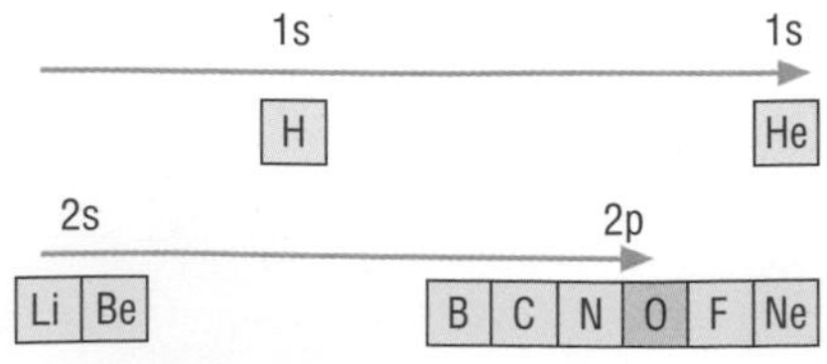

- Oxygen is the 4th element in the 2p block, i.e. $2p^4$
- Oxygen has the electron configuration: $1s^2 2s^2 2p^4$

Figure 4

Shortening an electron configuration

For atoms with many electrons, the electron configuration notation can become lengthy. It is often abbreviated by basing the inner shell configuration on the noble gas that comes before the element in the Periodic Table. This shortening also allows you to concentrate on the important outer shell portion of the electron configuration. It is these electrons that are responsible for the chemical character of the element.

The electronic configurations for the elements in Group I are shown below in their full and abbreviated forms. Note that the noble gas that comes before lithium in the Periodic Table is helium, before sodium is neon, and before potassium is argon.

Li: $1s^22s^1$ or $[He]2s^1$
Na: $1s^22s^22p^63s^1$ or $[Ne]3s^1$
K: $1s^22s^22p^63s^23p^64s^1$ or $[Ar]4s^1$

Electron configuration of ions

- When *positive* ions are formed, electrons are *removed* from the highest energy orbitals.
- When *negative* ions are formed, electrons are *added* to the highest energy orbitals.

Some examples are shown in Table 1.

Element	Number of electrons	Electron configuration
Mg	12	$1s^22s^22p^63s^2$
Mg^{2+}	10	$1s^22s^22p^6$
Cl	17	$1s^22s^22p^63s^23p^5$
Cl^-	18	$1s^22s^22p^63s^23p^6$

Table 1 Electron configuration of elements and their ions

You will meet one exception to this rule – the elements Sc to Zn at the top of the d-block in the Periodic Table. You will remember that the 4s sub-shell is filled before the 3d sub-shell.

The 4s and 3d energy levels are so close together that, after the 4s-orbital has been filled, it is actually at a slightly higher energy than the 3d level.

So, surprisingly, 4s-electrons are lost *before* the 3d-electrons.

An iron atom, Fe, has the electron configuration: $1s^22s^22p^63s^23p^64s^23d^6$

But an Fe^{2+} ion has the electron configuration: $1s^22s^22p^63s^23p^63d^6$

Examiner tip

You need to remember 4s: first in; first out.

The proper way to show the electron configuration is in shell order. But you might find it more logical to write this down in energy level order.

So Fe *should* be shown:
$1s^22s^22p^63s^23p^63d^64s^2$,
rather than:
$1s^22s^22p^63s^23p^64s^23d^6$.
In exams, either representation is perfectly acceptable.

Questions

1 Which block in the Periodic Table contains the following elements? (The atomic number of each element is given in brackets.)
(a) Sr (38); **(b)** Ga (31); **(c)** Re (75); **(d)** Sm (62).

2 Write the electron configuration in terms of sub-shells for the following atoms:
(a) Al; **(b)** Si; **(c)** S; **(d)** Sc; **(e)** Co; **(f)** Zn.

3 Write the electron configuration in terms of sub-shells for the following ions:
(a) Na^+; **(b)** Al^{3+}; **(c)** N^{3-}; **(d)** Mn^{2+}; **(e)** Sc^{3+}; **(f)** Se^{2-}.

4 Write the shorthand electron configuration using the nearest noble gas for the following:
(a) Mg; **(b)** Ba; **(c)** Cl.

An introduction to chemical bonding

By the end of this spread, you should be able to . . .

- **State that noble gases have a stable electron configuration.**
- **Describe the different types of chemical bonding.**
- **Predict the type of bonding from the elements involved.**

The noble gases

An atom is the smallest particle of an element that retains its chemical properties.

Only six elements exist naturally as single unbonded atoms:

- helium, He
- neon, Ne
- argon, Ar
- krypton, Kr
- xenon, Xe
- radon, Rn.

These elements are called the noble gases and are listed in Figure 1. They make up Group 0 of the Periodic Table.

Figure 1 The noble gases

In the atoms of a noble gas:

- all electrons are paired, with opposite spins
- the outer shell contains 2 electrons (for He) or 8 electrons (all other noble gases) (see Figure 2).

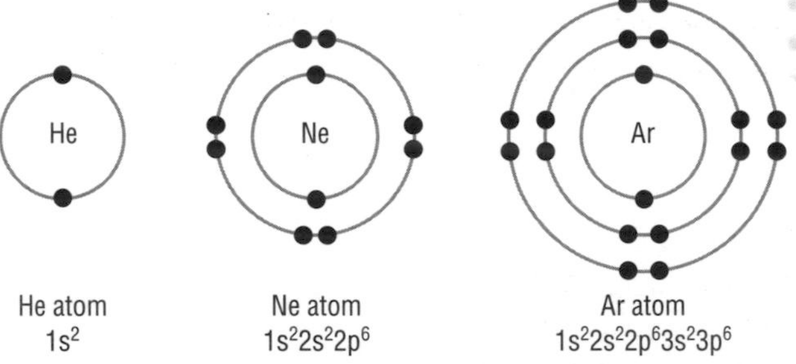

Figure 2 Full outer shells in the noble gases

An outer shell of eight electrons is extremely stable. It is this stability that makes the noble gases so unreactive. The eight electrons in the outer shell of a noble gas are made up of two in the s-orbital and two each in the three p-orbitals. This serves as the model for the Octet Rule (see below). Although other orbitals can be used for bonding, the s- and p-orbitals are perhaps the most important, especially for the formation of compounds involving the first 18 elements.

Bonding

The noble gases are the only elements with atoms containing eight electrons in their outer shell. These atoms are particularly stable and exist on their own. For other elements, atoms combine so that each atom has eight electrons in its outer shell. Unpaired electrons pair up by sharing or transferring electrons to form a chemical bond. This results in each atom having the same electron configuration as a noble gas – eight electrons in its outer shell. This tendency to acquire a noble gas electron configuration is referred to as the *Octet Rule*.

Chemical bonding

Chemical bonds are classified into three main types: ionic; covalent; and metallic.

A **compound** is formed when atoms of different elements are chemically bond together.

In a compound the atoms of the different elements are always in the same proportions.

For example, water, H_2O, always has two hydrogen atoms to one oxygen atom.

Ionic bonding

- In general, ionic bonding occurs in compounds consisting of a *metal* and a *non-metal*.
- If we imagine a bond forming between atoms, electrons are *transferred* from the metal atom to the non-metal atom to form oppositely charged ions that attract.
  e.g. NaCl, MgO, Fe_2O_3.

Covalent bonding

- Covalent bonding occurs in compounds consisting of two *non-metals*.
- If we imagine a bond forming between atoms, electrons are *shared* between the atoms.
  e.g. O_2, H_2, H_2O and C (diamond and graphite).

Metallic bonding

- Metallic bonding occurs in *metals*. Electrons are shared between *all* the atoms.
  e.g. all metals – iron, zinc, aluminium; alloys – brass (copper and zinc), bronze (copper and tin).

Key definition

A compound is a substance formed from two or more chemically bonded elements in a fixed ratio, usually shown by a chemical formula.

Figure 3 City lights. Neon signs and advertisements in Hong Kong at night

Noble gas facts

- The noble gases make up about 1% of air.
- In air, argon is the most abundant noble gas.
- There is 30 times more argon than carbon dioxide in air.
- All the noble gases, except radon, can be separated by the fractional distillation of liquefied air.

Uses

- Helium is used in airships.
- Neon is used in advertising signs.
- Argon is used in filament light bulbs.
- Krypton is used in lasers for eye surgery.
- Xenon is used in car headlights.

Discovery

Helium was discovered in 1868, when scientists noticed an unknown element in the light spectrum from the Sun. Helium was named after the Greek Sun god, Helios. Helium is unique in being the only element not to be discovered on the Earth first. It is the second most abundant gas in the known universe – hydrogen being number one. After helium's discovery, the other noble gases were discovered over the next 32 years.

Unreactivity

- There are no known compounds containing either helium or neon.
- Compounds of argon, xenon and krypton have been prepared, but with extreme difficulty.

Question

1 Predict the type of bonding in the following compounds:
  (a) sodium chloride;
  (b) zinc oxide;
  (c) hydrogen chloride;
  (d) silver bromide;
  (e) nitrogen bromide;
  (f) sulfur dioxide.

1.2 (6)
Ionic bonding

By the end of this spread, you should be able to . . .

- **Describe ionic bonding as the electrostatic attraction between oppositely charged ions.**
- **Draw dot-and-cross diagrams for ionic compounds.**
- **Describe structures with ionic bonding as giant ionic lattices.**

Key definition

An **ionic bond** is the electrostatic attraction between oppositely charged ions.

Ionic bonds

Ionic bonds are present in compounds consisting of a metal and a non-metal.

Imagine an **ionic bond** being formed between two atoms.

- Electrons are *transferred* from the metal atom to the non-metal atom.
- *Oppositely charged ions* are formed which are bonded together by electrostatic attraction.
- The metal ion is positive.
- The non-metal ion is negative.

Ionic bonding in sodium oxide, Na_2O

The compound sodium oxide, Na_2O, is formed from atoms from sodium (the metal) and oxygen (the non-metal).

- A sodium atom has 1 electron in its outer shell.
- An oxygen atom has 6 electrons in its outer shell.

Dot-and-cross diagrams are used to show the origin of electrons in chemical bonding. You use *dots* to label the electrons of one element and *crosses* to label those of the other element. Since reactions involve the outer-shell electrons, you usually only label the electrons in these shells. If there is a third element involved, another symbol is used to represent its electrons.

Figure 1 shows the formation Na_2O from Na and O atoms.

One electron is transferred:

- from each of the two sodium atoms;
- to one oxygen atom;
- with the formation of two Na^+ ions and one O^{2-} ion.

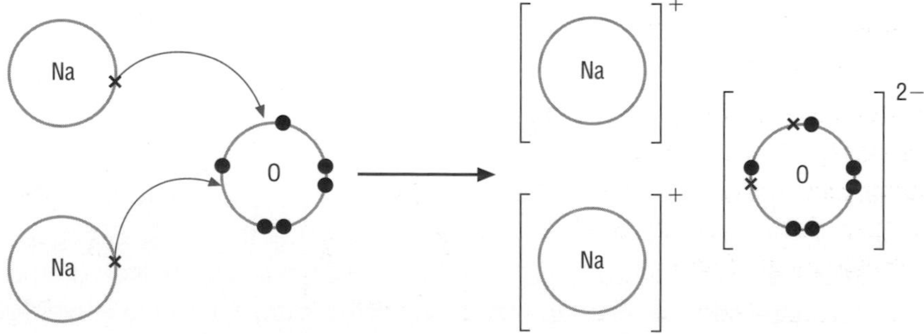

Figure 1 Formation of Na_2O

Each sodium atom has lost one electron to obtain the electron configuration of the noble gas, neon:

$$2Na \longrightarrow 2Na^+ + 2e^-$$
$$1s^22s^22p^63s^1 \longrightarrow 1s^22s^22p^6$$

An oxygen atom has gained two electrons to obtain the electron configuration of the noble gas, neon, too:

$$O + 2e^- \longrightarrow O^{2-}$$
$$1s^22s^22p^4 \longrightarrow 1s^22s^22p^6$$

Examiner tip

In exams, dot-and-cross diagrams are usually drawn for the outer shell only. Note in the sodium oxide example that the sodium shell was shown empty. In exams, you can either show this shell empty, to represent the shell that has lost the electron, or full, to represent the full shell below it.

Giant ionic lattices

In the example above, it looks as if only two Na^+ ions and one O^{2-} ion are involved in the ionic bonding. In reality, the situation is very different.

- Each ion is surrounded by oppositely charged ions.
- These ions attract each other from all directions, forming a **giant ionic lattice** see spread 1.2.15 for more details.

Sodium chloride, NaCl, is also an ionic compound and forms a giant ionic lattice.

- Each Na^+ ion surrounds 6 Cl^- ions.
- Each Cl^- ion surrounds 6 Na^+ ions.

Sodium chloride's giant ionic lattice structure contains many millions of ions – depending upon the size of the crystal.

All ionic compounds exist as a giant ionic lattice in the solid state.

Key definition

A **giant ionic lattice** is a three-dimensional structure of oppositely charged ions, held together by strong ionic bonds.

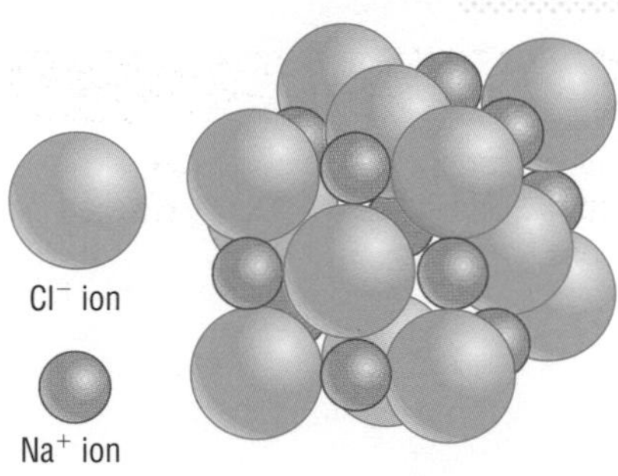

Figure 2 Sodium chloride ionic lattice

Further examples of ionic bonding

Dot-and-cross diagrams are very useful, as they show the number and source of electrons involved in the bonding.

Figures 3 and 4 show dot-and-cross diagrams for calcium oxide, CaO, and aluminium fluoride, AlF_3.

- Look at each example and check to see where the electrons have come from.
- Note the charge on each ion.
- See how the dots and crosses help you to work out the charges on the ions.

Ionic bonding in calcium oxide, CaO

The calcium atom has lost two electrons to form a Ca^{2+} ion, with the electron configuration of argon.

These two electrons, shown as crosses, now form part of the oxide ion, O^{2-}.

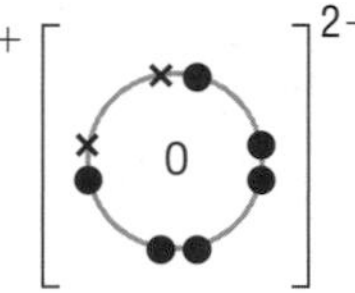

Figure 3 Dot-and-cross diagrams of CaO

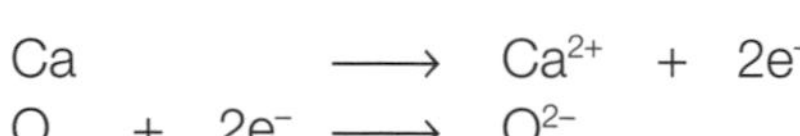

$$Ca \longrightarrow Ca^{2+} + 2e^-$$
$$O + 2e^- \longrightarrow O^{2-}$$

Ionic bonding in aluminium fluoride, AlF_3

The aluminium atom has lost three electrons to form an Al^{3+} ion, with the electron configuration of neon.

Each of these three electrons, shown as crosses, is now part of a fluoride ion, F^-. Three fluoride ions form by taking in these three electrons. Each F^- ion has the electron configuration of neon.

$$Al \longrightarrow Al^{3+} + 3e^-$$
$$3F + 3e^- \longrightarrow 3F^-$$

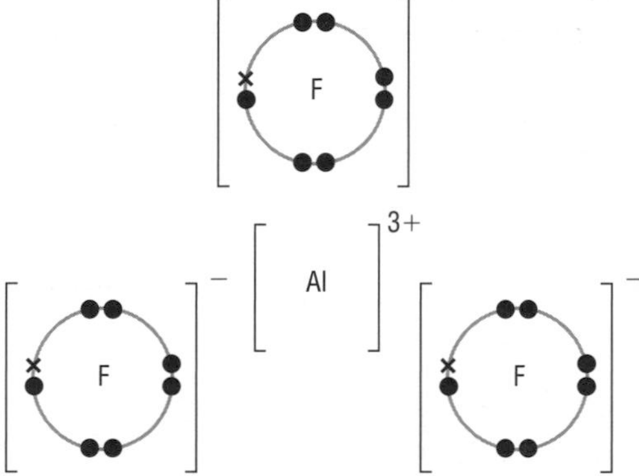

Figure 4 Dot-and-cross diagrams of AlF_3

Examiner tip

In exams, you are often asked to draw a dot-and-cross diagram for an ionic compound. You should only draw out the *final* compound, as in Figure 3 and Figure 4. The examiner is *not* interested in how the ions have been formed from the atoms.

Questions

1 Draw dot-and-cross diagrams for:
 (a) MgO; **(b)** $CaBr_2$; **(c)** Na_3P; **(d)** Al_2O_3.

2 For each atom and ion below, write down the electron configuration, in terms of sub-shells.
 (a) K and K^+; **(b)** Ca and Ca^{2+}; **(c)** S and S^{2-};
 (d) Al and Al^{3+}; **(e)** N and N^{3-}.

1.2 ⑦ Ions and the Periodic Table

By the end of this spread, you should be able to . . .

* **Predict ionic charge from an element's position in the Periodic Table.**
* **State the formulae for the ions NO_3^-; CO_3^{2-}; SO_4^{2-}; and NH_4^+.**

Predicting ionic charges

You can predict the charge on an element's ion from its position in the Periodic Table.

You can find the number of electrons in the outer shell from the element's position in the Periodic Table. It is then a simple step to calculate how many electrons need to be lost or gained in order to reach a noble gas electron configuration. From this, you can predict the likely charge on the resulting ion.

For example:
- lithium, in Group 1, has one electron in its outer shell.

To form the electron configuration of the nearest noble gas, helium:
- a lithium atom must lose one outer electron
- a lithium ion, Li^+, therefore has a charge of 1+.

Elements in the same **group** of the Periodic Table have the same number of outer-shell electrons and react in similar ways.

Key definition

A **group** is a vertical column in the Periodic Table. Elements in the same group have similar chemical properties and their atoms have the same number of outer-shell electrons.

Table 1 shows the charges on ions for the elements in Periods 2 and 3 of the Periodic Table. The same principle can be extended to other members of each group further down the Periodic Table.

Group	1	2	3	4	5	6	7	0
Number of outer-shell electrons	1	2	3	4	5	6	7	8
Element	Li	Be	B	C	N	O	F	Ne
Ion	Li^+				N^{3-}	O^{2-}	F^-	
Element	Na	Mg	Al	Si	P	S	Cl	Ar
Ion	Na^+	Mg^{2+}	Al^{3+}		P^{3-}	S^{2-}	Cl^-	

Table 1 Ion charges for Periods 2 and 3

Atoms of metals in Groups 1–3:
- *lose* electrons
- *form positive ions* with the electron configuration of the *previous* noble gas in the Periodic Table.

Atoms of non-metals in Groups 5–7:
- *gain* electrons
- *form negative ions* with the electron configuration of the *next* noble gas.

Atoms of Be, B, C and Si:
- do not normally form ions
- too much energy is needed to transfer the outer-shell electrons to form ions.

Some elements can form more than one ion, each with a different charge. The oxidation number of the element is written as a Roman numeral.
- iron(II) for Fe^{2+} and iron(III) for Fe^{3+}
- copper(I) for Cu^+ and copper(II) for Cu^{2+}.

You can find more details about oxidation numbers and the use of Roman numerals in spread 1.1.14.

Molecular ions

Groups of covalent-bonded atoms can also lose or gain electrons to form ions. These are called *molecular ions*. Some common molecular ions are shown in Table 2. You will become familiar with most of these during your A-level chemistry course.

1+		1–		2–		3–	
ammonium	NH_4^+	hydroxide	OH^-	carbonate	CO_3^{2-}	phosphate	PO_4^{3-}
		nitrate	NO_3^-	sulfate	SO_4^{2-}		
		nitrite	NO_2^-	sulfite	SO_3^{2-}		
		hydrogencarbonate	HCO_3^-	dichromate	$Cr_2O_7^{2-}$		

Table 2 Molecular ions

Examiner tip

For exams, you must know the charges on common molecular ions:

NO_3^-, CO_3^{2-}, SO_4^{2-} and NH_4^+.

Predicting ionic formulae

Although an ionic compound is made up of oppositely charged ions, its overall charge is zero.

In an ionic compound:

- total number of + charges from positive ions = total number of – charges from negative ions.

Working out an ionic formula from the ionic charges

You can write an ionic formula by following some simple rules.

- In an ionic compound, the overall charge is zero.
- Select as many + ions and – ions as required for the charges to balance.

You may find it useful to list the ions under one another as shown in Table 3.

Calcium chloride:	Ion	Charge	Aluminium sulfate:	Ion	Charge
equalise charges:	Ca^{2+}	2+	*equalise charges:*	Al^{3+}	3+
				Al^{3+}	3+
	Cl^-	1–		SO_4^{2-}	2–
	Cl^-	1–		SO_4^{2-}	2–
				SO_4^{2-}	2–
total charge must be zero:		0	total charge must be zero:		0
formula:	$CaCl_2$		formula:	$Al_2(SO_4)_3$	

Table 3 Ionic formulae

Questions

1 Predict the formula for these ionic compounds:
 (a) calcium iodide; **(b)** lithium nitride;
 (c) aluminium sulfide; **(d)** magnesium phosphide.

2 Predict the formula for these ionic compounds:
 (a) nickel(II) chloride; **(b)** copper(I) oxide;
 (c) iron(III) chloride; **(d)** chromium(III) oxide;
 (e) manganese(VI) oxide; **(f)** titanium(IV) chloride.

3 Predict formulae for these ionic compounds:
 (a) aluminium sulfate; **(b)** calcium hydroxide;
 (c) iron(III) sulfite; **(d)** chromium(III) nitrite;
 (e) ammonium phosphate; **(f)** sodium dichromate.

Covalent bonding

By the end of this spread, you should be able to . . .

- **Describe a covalent bond as a shared pair of electrons.**
- **Describe single and multiple covalent bonding.**
- **Use dot-and-cross diagrams to represent covalent bonding.**

Covalent bonds

Covalent bonding occurs in compounds consisting of non-metals.

Key definition

A **covalent bond** is a bond formed by a *shared* pair of electrons.

In a **covalent bond**, an electron pair occupies the space between the two atoms' nuclei as shown in Figure 1.

- The negatively charged electrons are attracted to the positive charges of *both* nuclei.
- This attraction overcomes the repulsion between the two positively charged nuclei.

positive charge (nucleus) I negative charge (pair of electrons) I positive charge (nucleus)

The resulting attraction is the covalent bond that holds the two atoms together.

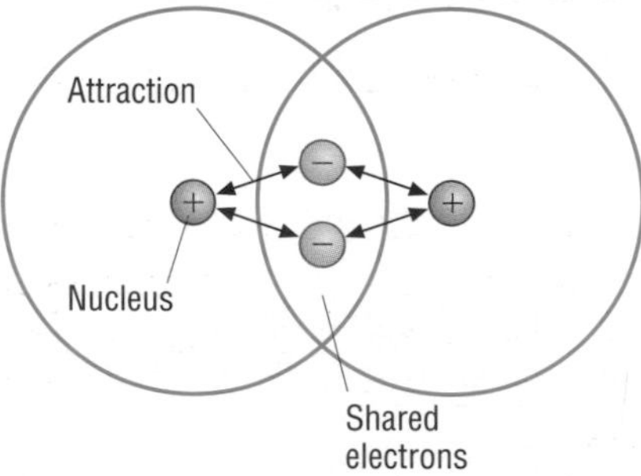

Figure 1 Attraction in a covalent bond

We can imagine a covalent bond formed between two atoms of non-metals.

- Two electrons are *shared*.
- This is in contrast to the *transfer* of electrons that results in an ionic bond.

Single covalent bonds

Hydrogen, H_2, is formed from two hydrogen atoms (see Figure 2).

- Each hydrogen atom has one electron in its outer shell.
- Each hydrogen atom contributes one electron to the covalent bond.
- The covalent bond is often written as a line, e.g. H–H.

Examiner tip

A common exam question is the definition of a covalent bond. The answer is 'a **shared** pair of electrons'. Many candidates omit the word 'shared'.

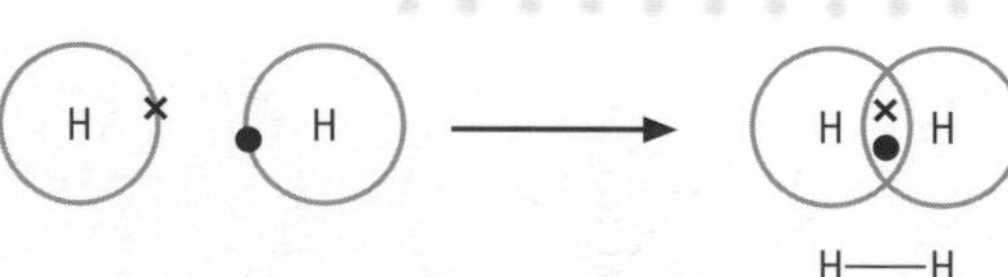

Figure 2 Covalent bond formation in H_2

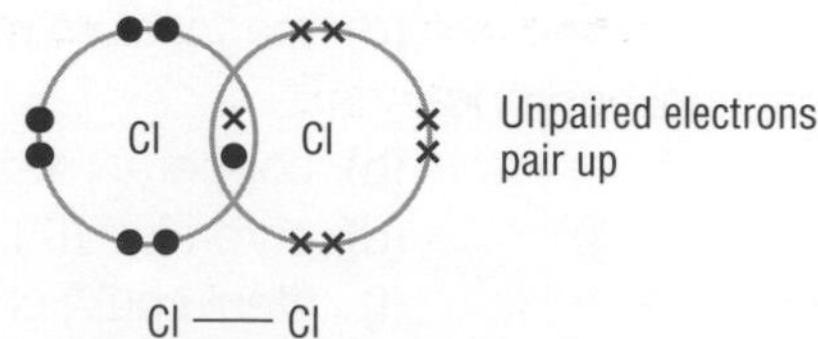

Figure 3 Covalent bonding in Cl_2

A covalent bond is *directional*, acting solely between the two atoms involved in the bond. Remember that an ionic bond attracts in *all* directions.

Further examples of covalent bonding in simple molecules are shown in Figure 4. In all these molecules, H, O, N and C atoms have reached the next noble gas electron configuration – He for the H atom, and Ne for the other three atoms. All unpaired electrons have been paired.

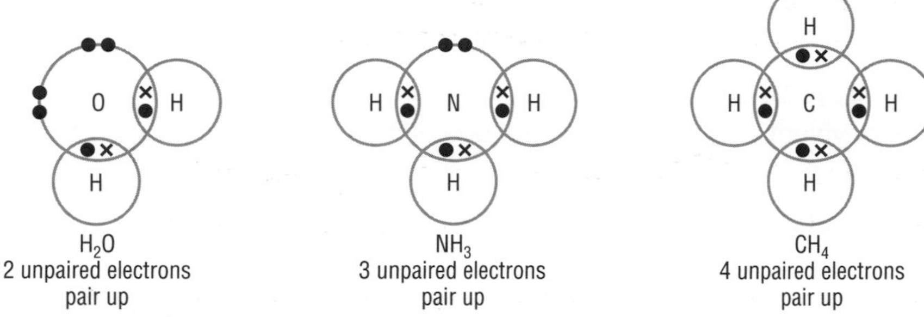

Figure 4 Covalent bonding in H_2O, NH_3 and CH_4

During your AS and A2 chemistry courses, you will become familiar with many covalent compounds involving the four elements: H, O, N, and C. You will need to remember the number of covalent bonds formed by the atoms of these four elements:

C: 4 covalent bonds; N: 3 covalent bonds; O: 2 covalent bonds; H: 1 covalent bond.

Lone pairs

In a covalent bond, an electron pair is shared between two atoms, bonding them together. Sometimes though, the electron pair is not used for bonding. In this instance, the pair is known as a **lone pair** of electrons. In the diagrams of NH_3 and H_2O in Figure 4, an ammonia molecule has *one* lone pair and a water molecule has *two* lone pairs.

A lone pair gives a concentrated region of negative charge around the atom. Lone pairs can influence the chemistry of a molecule in several ways. In the following spreads, we will examine in more detail some of the effects resulting from lone pairs.

Multiple covalent bonds

Some non-metallic atoms can share more than one pair of electrons to form a *multiple bond*.

- Sharing of *two* pairs of electrons forms a *double bond*, e.g. O_2, written as O=O.
- Sharing of *three* pairs of electrons forms a *triple bond*, e.g. N_2, written as N≡N.
- Carbon dioxide, CO_2 has two double bonds, written as O=C=O.

Dot-and-cross diagrams of these structures are shown in Figure 5.

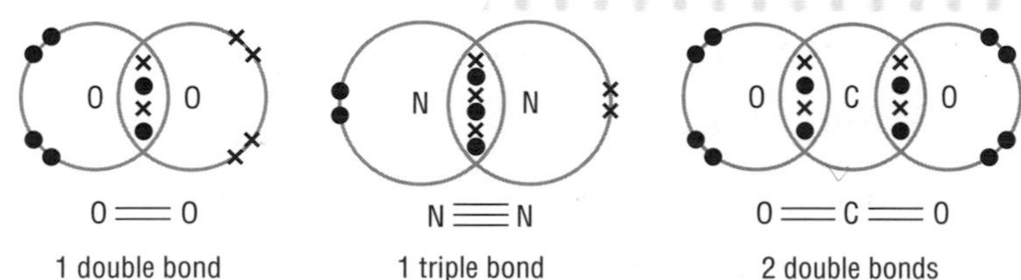

Figure 5 Dot-and-cross diagrams of O_2, N_2 and CO_2

Questions

1 Draw dot-and-cross diagrams for:
(a) F_2; (b) HF; (c) SiF_4; (d) SCl_2.

2 Draw dot-and-cross diagrams for:
(a) CS_2; (b) C_2H_4; (c) HC≡N; (d) H_2C=O.

Key definition

A **lone pair** is an outer-shell pair of electrons that is not involved in chemical bonding.

Examiner tip

When drawing out dot-and-cross diagrams, check to see whether you have included all the electrons in the outer shells. Many candidates will show the correct **bonded** electrons on CO_2, but will forget the **lone pairs** on each O atom.

1.2 (9) Further covalent bonding

By the end of this spread, you should be able to . . .

- **Describe dative covalent (coordinate) bonding.**
- **Use dot-and-cross diagrams to represent covalent bonding.**

Key definition

A dative covalent or coordinate bond is a shared pair of electrons which has been provided by *one* of the bonding atoms only.

Dative covalent bonds

In a **dative covalent** or **coordinate bond**, *one* of the atoms supplies *both* the shared electrons to the covalent bond.

- A dative covalent bond can be written as A→B.
- The direction of the arrow shows the direction in which the electron pair has been donated.

The ammonium ion, NH_4^+

The ammonium ion, NH_4^+, has three covalent bonds and one dative covalent bond.

Figure 1 shows the formation of the NH_4^+ ion from ammonia, NH_3, and H^+. One of the electron pairs around the nitrogen atom in an NH_3 molecule is a *lone pair*. In the formation of an ammonium ion, NH_4^+, this lone pair provides both the bonding electrons when bonding with the H^+ ion, and the resulting NH_4^+ ion has a positive charge of 1+.

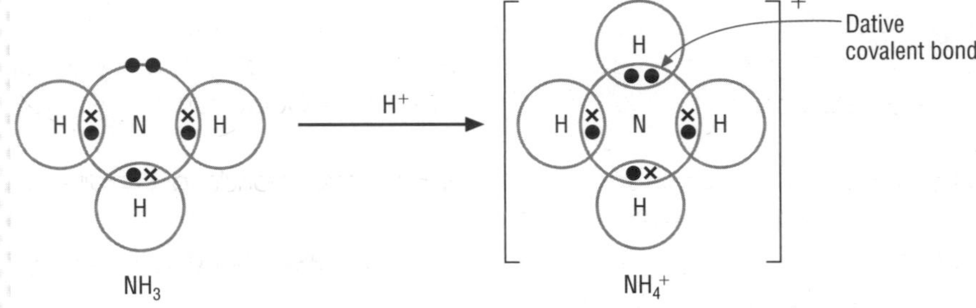

Figure 1 Dative covalent bonding in NH_4^+

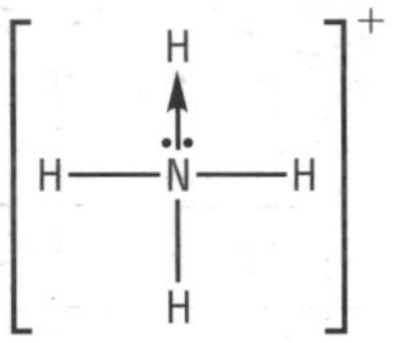

Figure 2 The ammonium ion. Notice the dative covalent bond shown with an arrow. The arrow indicates the origin of the bonded pair

In a dative covalent bond, one atom provides both bonding electrons from a lone pair of electrons. However, once formed, this *dative* covalent bond is equivalent to all the other covalent bonds. For example, in the NH_4^+ ion, you cannot tell which bond was formed from the N lone pair.

The oxonium ion, H_3O^+

When an acid is added to water, water molecules form oxonium ions, H_3O^+. For example, hydrogen chloride gas forms hydrochloric acid when added to water:

$$HCl(g) + H_2O(l) \rightarrow H_3O^+(aq) + Cl^-(aq)$$

H_3O^+ ions are responsible for reactions of acids. In equations, the oxonium ion is often simplified as $H^+(aq)$.

In the oxonium ion, one of the lone pairs around the oxygen atom in an H_2O molecule provides both the bonding electrons to form a dative covalent bond. The bonding in the oxonium ion is shown in Figure 3.

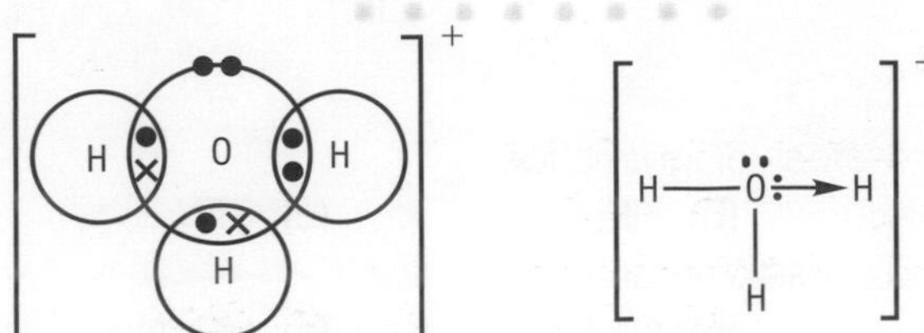

Figure 3 The oxonium ion

How many covalent bonds?

When covalent bonds form, unpaired electrons often pair up so that the bonded atoms obtain a noble gas electron configuration, obeying the Octet Rule (see spread 1.2.5).

This is not always possible:
- there may not be enough electrons to reach an octet
- more than four electrons may pair up in bonding (expansion of the oxtet).

Not enough electrons to reach an octet

Within Period 2, the elements beryllium, Be, and boron, B, both form compounds with covalent bonds. However, Be and B do not have enough unpaired electrons to reach a noble gas electron configuration. But they *can* pair up any unpaired electrons.

Example

Consider the compound boron trifluoride, BF_3 (see Figure 4).

Boron has 3 electrons in its outer shell.

Each fluorine has 7 electrons in its outer shell.
- Three covalent bonds can be formed.
- Each of boron's 3 electrons is paired.
- 6 electrons surround B.
- Each of the three fluorine atoms has 8 electrons in its outer shell, attaining an octet.

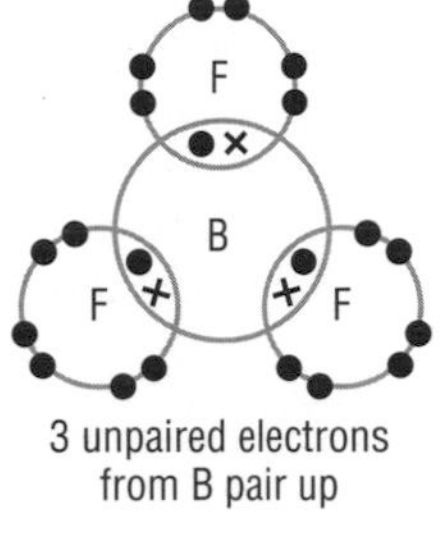

Figure 4 Dot-and-cross diagram of BF_3

Expansion of the octet

For elements in Groups 5–7 of the Periodic Table, something odd happens from Period 3.

As we move down the Periodic Table, more of the outer-shell electrons are able to take part in bonding. In the resulting molecules, one of the bonding atoms may finish up with more than eight electrons in its outer shell. This breaks the Octet Rule and is often called *expansion of the octet*.
- Atoms of non-metals in Group 5 can form 3 or 5 covalent bonds, depending on how many electrons are used in bonding.
- Atoms of non-metals in Group 6 can form 2, 4 or 6 covalent bonds, depending on how many electrons are used in bonding.
- Atoms of non-metals in Group 7 can form 1, 3, 5, or 7 covalent bonds, depending on how many electrons are used in bonding.

Table 1 lists the elements that can expand their octets.

Group 5	Group 6	Group 7
P	S	Cl
As	Se	Br
	Te	I
		At

Table 1 Elements that expand their octet

Example

In the compound sulfur hexafluoride, SF_6 (see Figure 5), sulfur has expanded its octet. Sulfur has 6 electrons in its outer shell.
- Six covalent bonds can be formed.
- Each of sulfur's 6 electrons is paired.
- 12 electrons surround S.
- Each of the six fluorine atoms has 8 electrons in its outer shell, attaining the octet.

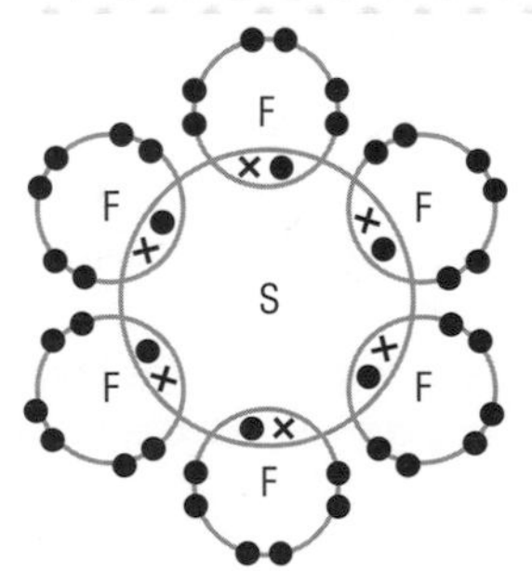

Figure 5 Dot-and-cross diagram of SF_6

A better rule

A better rule than the Octet Rule would be:
- unpaired electrons pair up;
- the maximum number of electrons that can pair up is equivalent to the number of electrons in the outer shell.

Questions

1 Draw dot-and-cross diagrams for:
(a) PCl_4^+; **(b)** H_3O^+; **(c)** H_2F^+.

2 Draw dot-and-cross diagrams for:
(a) BF_3; **(b)** PF_5; **(c)** SO_2; **(d)** SO_3.

Shapes of molecules and ions

By the end of this spread, you should be able to . . .

- ✱ **Explain how the shape of a simple molecule is determined.**
- ✱ **State that lone pairs of electrons repel more strongly than bonded pairs.**
- ✱ **Explain and predict the shapes of, and bond angles in, molecules and ions.**

Examiner tip

In this spread, you will find many different shapes and bond angles – you must learn these. You must also be able to draw the shape of any molecule with up to six pairs of electrons around the central atom. You should also be able to draw them as 3D shapes, using wedges.

Electron-pair repulsion theory

The shape of a molecule or ion is determined by the number of electron pairs in the outer shell surrounding the central atom.

- As electrons all have a negative charge, each electron pair repels other electron pairs.
- The electron pairs push one another as far apart as possible.

Molecules with bonded pairs

A dot-and-cross diagram is used to show the number of electron pairs surrounding the central atom.

- The diagrams in Figure 1 show that molecules have different shapes when there are different numbers of bonded electron pairs around the central atom.
- The shapes and bond angles result from the equal repulsion of the bonded pairs. Each bonded pair is repelled as far away as possible from other bonded pairs of electrons.
- The resulting shapes may be three-dimensional (3D).

Figure 2 shows how chemists represent 3D structures.

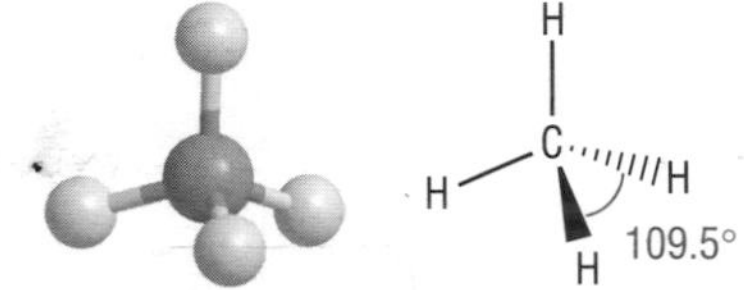

Figure 2 Drawing 3D diagrams.

Many molecules have 3D shapes and we need a simple way of drawing the shape on a flat sheet of paper. Chemists use bold wedges and dotted wedges – as used in the diagrams of methane, CH_4, above.

- A normal line shows a bond in the plane of the paper.
- A bold wedge shows the bond coming out from the plane of the paper towards you.
- A dotted wedge shows the bond going out from the plane of the paper away from you.

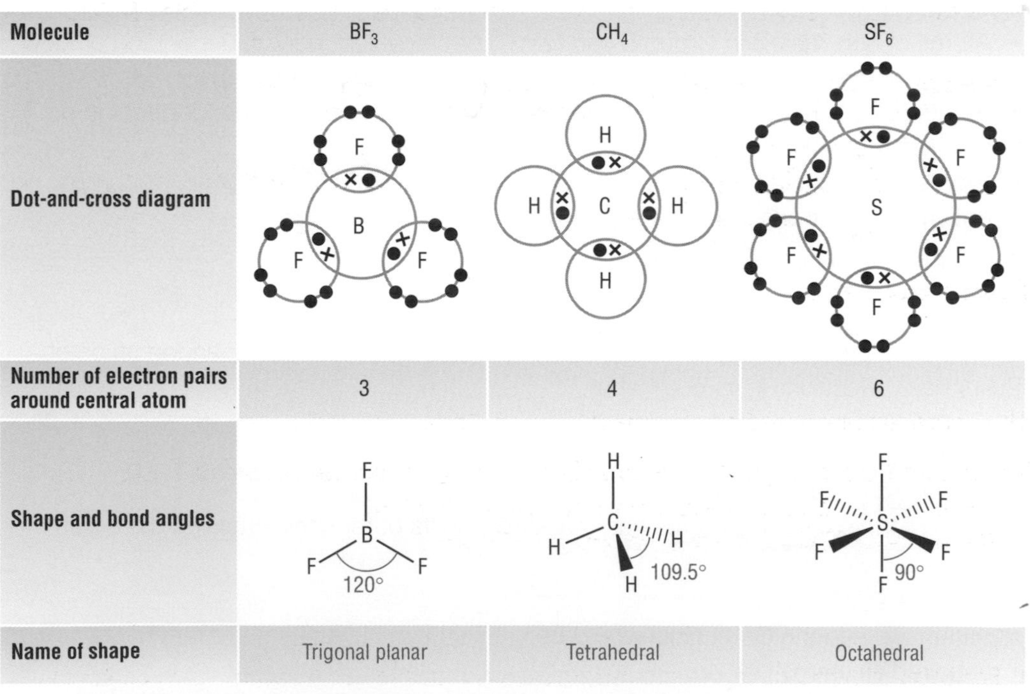

Molecule	BF_3	CH_4	SF_6
Dot-and-cross diagram			
Number of electron pairs around central atom	3	4	6
Shape and bond angles	120°	109.5°	90°
Name of shape	Trigonal planar	Tetrahedral	Octahedral

Figure 1 Shapes of molecules with different numbers of bonded pairs

Molecules with lone pairs

A lone pair of electrons is slightly more electron-dense than a bonded pair. Therefore a lone pair repels more than a bonded pair.

The relative strengths of the repulsion are:

lone pair/**lone** pair > **bonded** pair/**lone** pair > **bonded** pair/**bonded** pair.

- The diagrams in Figure 3 show how the shapes of molecules are affected by the presence of lone pairs of electrons.
- Each lone pair reduces the bond angle by about 2.5°. This is the result of the extra repulsive effect of each lone pair.

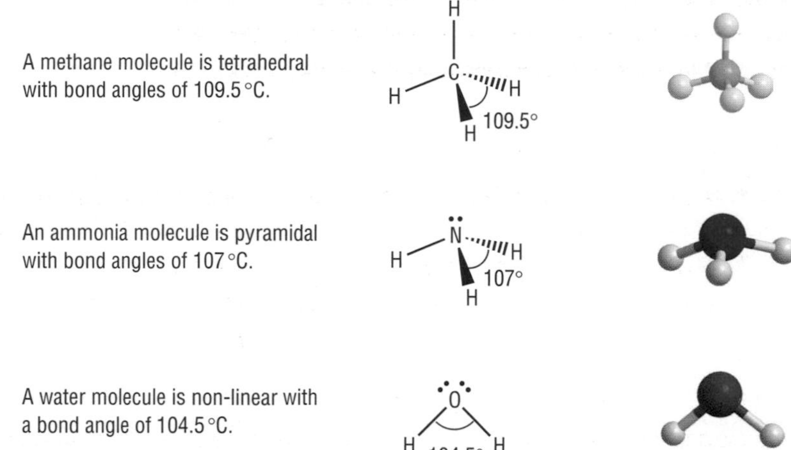

A methane molecule is tetrahedral with bond angles of 109.5 °C.
109.5°
An ammonia molecule is pyramidal with bond angles of 107 °C.
107°
A water molecule is non-linear with a bond angle of 104.5 °C.
104.5°

Figure 3 Shapes of molecules with 0, 1, and 2 lone pairs

Molecules with double bonds

A double bond has four electrons as two bonded pairs. To work out the shape of a molecule with double bonds, each double bond is treated as a *bonded region*, in the same way as a bonded pair.

Carbon dioxide, as shown in Figure 4, has two double bonds. These will repel one another to be as far apart as possible, resulting in a linear molecule with a bond angle of 180°.

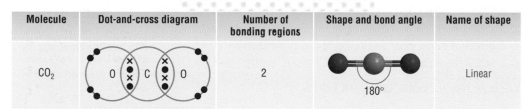

Molecule	Dot-and-cross diagram	Number of bonding regions	Shape and bond angle	Name of shape
CO_2	O C O	2	180°	Linear

Figure 4 Shape of CO_2

Shapes of ions

The principles discussed above can also be applied to any molecular ion.

The dot-and-cross diagram for ammonium ion, NH_4^+, is shown in spread 1.2.9.

- There are four electron pairs around the central N atom.
- The shape will be tetrahedral, as shown in Figure 5.

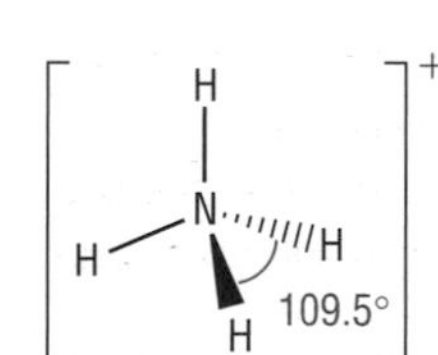

Figure 5 Shape of NH_4^+ ion

Questions

1 For each of the following molecules, draw a dot-and-cross diagram, and predict the shape and bond angles.
(a) H_2S; **(b)** $AlCl_3$; **(c)** SiF_4; **(d)** PH_3.

2 For each of the following ions, draw a dot-and-cross diagram, and predict the shape and bond angles.
(a) NH_4^+; **(b)** H_3O^+; **(c)** NH_2^-.

3 Each of the following molecules has at least one multiple bond. For each molecule, draw a dot-and-cross diagram, and predict the shape and bond angles.
(a) HCN; **(b)** C_2H_4; **(c)** SO_2; **(d)** SO_3.

1.2 (11) Electronegativity and polarity

By the end of this spread, you should be able to . . .

* **Describe electronegativity in terms of an atom attracting bonding electrons in a covalent bond.**
* **Explain how a permanent dipole can result in a polar bond.**

Polar and non-polar bonds

A covalent bond is a shared pair of electrons. In a molecule of hydrogen, H_2, the two bonding atoms are *identical*. Each hydrogen atom has an equal share of the pair of electrons in the bond, resulting in a perfect 100% covalent bond.

The nucleus of each bonded atom is attracted to the bonding electron pair (see spread 1.2.8).

We say that a H–H bond is *non-polar* – the electrons in the bond are evenly distributed between the atoms that make up the bond. The diagrams in Figure 1 show a non-polar covalent bond between two hydrogen atoms and another between two chlorine atoms.

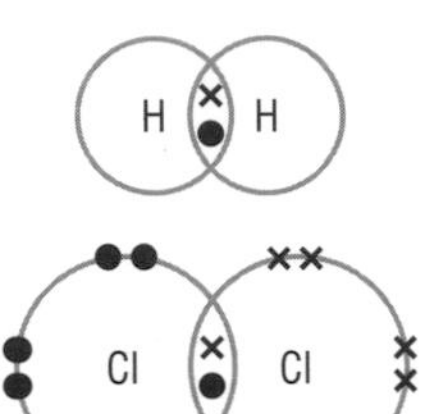

Figure 1 Non-polar bonds

If the bonding atoms are *different*, one of the atoms is likely to attract the bonding electrons more. The bonding atom with a greater attraction for the electron pair is said to be more **electronegative** than the other atom.

Key definition

Electronegativity is a measure of the attraction of a bonded atom for the pair of electrons in a covalent bond.

In a molecule of hydrogen chloride, HCl, the two bonding atoms are *different*.

- The Cl atom is more electronegative than the H atom.
- The Cl atom has a greater attraction for the bonding pair of electrons than the H atom.
- The bonding electrons are closer to the Cl atom than the H atom.

There is now a small charge difference across the H–Cl bond (see Figure 2).

Key definitions

A **permanent dipole** is a small charge difference across a bond that results from a difference in the electronegativities of the bonded atoms.

A **polar covalent bond** has a permanent dipole.

This charge difference is called a **permanent dipole** and is shown by:

- a small positive charge on the H atom, written δ+
- a small negative charge on the Cl atom, written δ–.

$\delta+$ ⟶ $\delta-$
H ×● Cl

The bonded electron pair is attracted towards the Cl atom.

Figure 2 The polar bond in H–Cl

We now have a **polar covalent bond**: $H^{\delta+}$—$Cl^{\delta-}$

Polar molecules

Molecules such as HCl have polar bonds and are **polar molecules**. A hydrogen chloride molecule is non-symmetrical. There is a charge difference across the H–Cl bond and across the whole HCl molecule.

Key definition

A **polar molecule** has an overall dipole, when you take into account any dipoles across the bonds.

Examiner tip

The greater the difference in electronegativity between the bonded atoms, the greater the permanent dipole. The more electronegative atom with take the δ-charge.

For molecules that are symmetrical, the dipoles of any bonds within the molecule may cancel out. Figure 3 shows tetrachloromethane, CCl_4, a non-polar molecule, despite having bonds with permanent dipoles.

- Each C–Cl bond is polar.
- But a CCl_4 molecule is symmetrical.
- So, the dipoles act in different directions and cancel each other out.
- Hence, a CCl_4 molecule is a non-polar molecule with polar bonds!

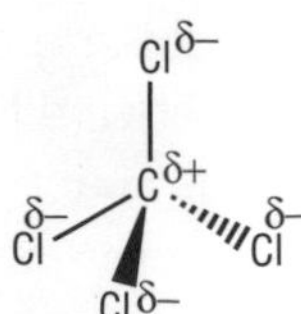

Figure 3 CCl_4 is a non-polar molecule

Figure 4 Linus Pauling

How is electronegativity measured?

In 1932, the US chemist Linus Pauling invented the *Pauling scale* to measure the electronegativity of an atom. Electronegativity measures the attraction of a bonded atom for the pair of electrons in a covalent bond. Electronegativity increases towards the top right of the Periodic Table, with fluorine having the most electronegative atoms.

The Periodic Table in Figure 5 shows the electronegativities of all the elements. The electronegativity depends upon an element's position in the Periodic Table.

- Reactive non-metallic elements (such as O, F and Cl) form compounds with the *most* electronegative atoms.
- Reactive metals (such as Na and K) form compounds with the *least* electronegative atoms.

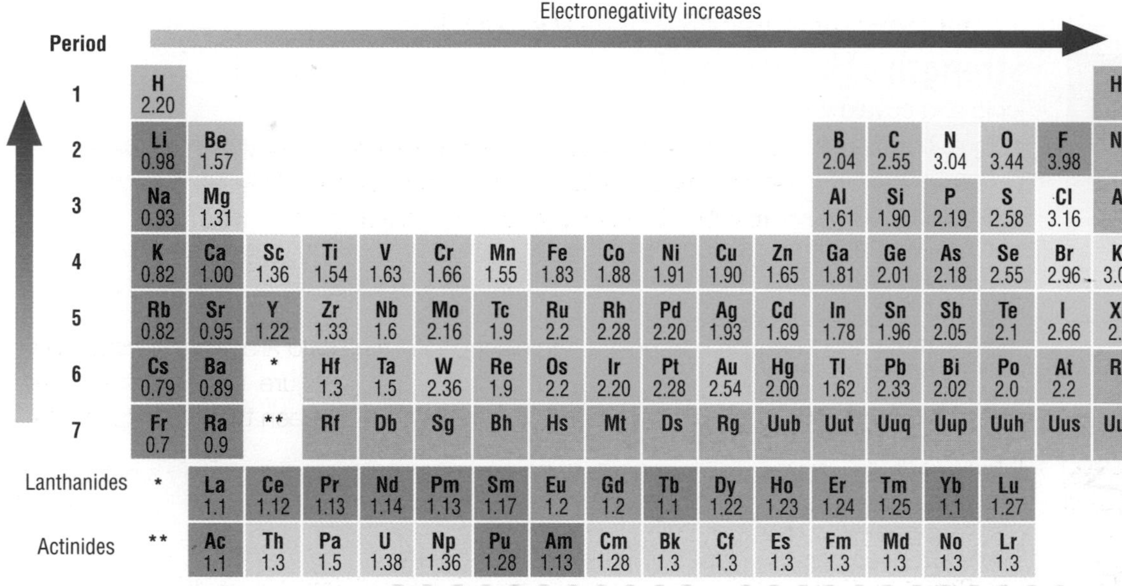

Period																		
1	H 2.20																	He
2	Li 0.98	Be 1.57											B 2.04	C 2.55	N 3.04	O 3.44	F 3.98	Ne
3	Na 0.93	Mg 1.31											Al 1.61	Si 1.90	P 2.19	S 2.58	Cl 3.16	Ar
4	K 0.82	Ca 1.00	Sc 1.36	Ti 1.54	V 1.63	Cr 1.66	Mn 1.55	Fe 1.83	Co 1.88	Ni 1.91	Cu 1.90	Zn 1.65	Ga 1.81	Ge 2.01	As 2.18	Se 2.55	Br 2.96	Kr 3.00
5	Rb 0.82	Sr 0.95	Y 1.22	Zr 1.33	Nb 1.6	Mo 2.16	Tc 1.9	Ru 2.2	Rh 2.28	Pd 2.20	Ag 1.93	Cd 1.69	In 1.78	Sn 1.96	Sb 2.05	Te 2.1	I 2.66	Xe 2.6
6	Cs 0.79	Ba 0.89	*	Hf 1.3	Ta 1.5	W 2.36	Re 1.9	Os 2.2	Ir 2.20	Pt 2.28	Au 2.54	Hg 2.00	Tl 1.62	Pb 2.33	Bi 2.02	Po 2.0	At 2.2	Rn
7	Fr 0.7	Ra 0.9	**	Rf	Db	Sg	Bh	Hs	Mt	Ds	Rg	Uub	Uut	Uuq	Uup	Uuh	Uus	Uuo

Lanthanides *	La 1.1	Ce 1.12	Pr 1.13	Nd 1.14	Pm 1.13	Sm 1.17	Eu 1.2	Gd 1.2	Tb 1.1	Dy 1.22	Ho 1.23	Er 1.24	Tm 1.25	Yb 1.1	Lu 1.27
Actinides **	Ac 1.1	Th 1.3	Pa 1.5	U 1.38	Np 1.36	Pu 1.28	Am 1.13	Cm 1.28	Bk 1.3	Cf 1.3	Es 1.3	Fm 1.3	Md 1.3	No 1.3	Lr 1.3

Figure 5 Pauling electronegativity values across the Periodic Table

Electronegativity and bonding type

Imagine a bond between atoms of two different elements with a *small difference* in electronegativity.

- The more electronegative atom will have slightly more than its fair share of the bonded electrons.
- The result is a polar covalent bond.

Imagine a bond between atoms of two different elements with a *large difference* in electronegativity.

- The more electronegative atom will effectively have captured both the bonding electrons.
- The result is an ionic bond.

Between the extremes of 100% ionic and 100% covalent bonding, we have a whole range of intermediate bonds with both ionic and covalent contributions.

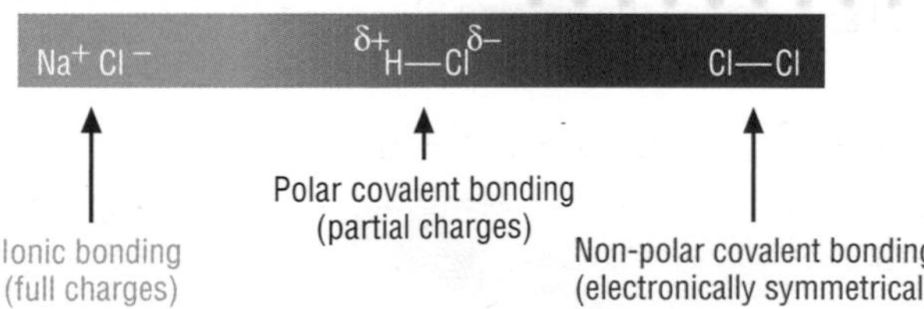

Figure 6 Range of bonding from covalent to ionic

Examiner tip

To predict the type of bonding,

- the greater the difference between the electronegativities of the bonding atoms, the greater the ionic character of the bond
- the greater the similarity in the electronegativities of the bonding atoms, the greater the covalent character of the bond.

Questions

1 For the molecules below, predict any polar covalent bonds. Show dipoles on the diagrams.

(a) Br_2; **(b)** H_2O; **(c)** O_2; **(d)** HBr; **(e)** NH_3; **(f)** CF_4.

2 **(a)** **(i)** Predict the shape of the molecules for BF_3 and PF_3.

(ii) Explain why BF_3 is non-polar, whereas PF_3 is polar.

(b) **(i)** Predict the shape of the molecules for H_2O and CO_2.

(ii) Explain why H_2O is polar, whereas CO_2 is non-polar.

Intermolecular forces

By the end of this spread, you should be able to . . .

* **Describe intermolecular forces in terms of permanent and instantaneous (van der Waals' forces) dipoles.**

Bond type	Relative strength
ionic and covalent bonds	1000
hydrogen bonds	50
dipole–dipole forces	10
van der Waals' forces	1

Table 1 Comparison of approximate strengths of different types of bonds and forces. Ionic and covalent bonds are far stronger than intermolecular forces (hydrogen bonds, permanent dipole–dipole forces and van der Waals' forces)

Key definitions

An **intermolecular force** is an attractive force between neighbouring molecules.

A **permanent dipole–dipole force** is a weak attractive force between *permanent dipoles* in neighbouring polar molecules.

van der Waals' forces are attractive forces between *induced dipoles* in neighbouring molecules.

$\overset{\delta+}{H}$—$\overset{\delta-}{Cl}$ ----------- $\overset{\delta+}{H}$—$\overset{\delta-}{Cl}$

Dipole–dipole interaction between a δ+ atom of one molecule and a δ– atom of another molecule.

Figure 1 Permanent dipole–dipole forces between HCl molecules

Strength of bonds and forces

Ionic and covalent bonds are strong.

- Ionic bonds hold ions together in a giant lattice so that, at room temperature, all ionic compounds are solids. Ionic bonds are strong electrostatic attractions between oppositely charged ions.
- A covalent bond holds atoms together by sharing an electron pair. Many covalent compounds exist as small molecules. The atoms in a molecule are bonded together by strong covalent bonds.

Intermolecular forces are weak.

- Intermolecular forces act *between* different molecules. These forces are very much weaker than ionic or covalent bonds.
- Intermolecular forces are caused by weak attractive forces between very small dipoles in different molecules.

There are three common types of intermolecular forces:

- hydrogen bonds
- permanent dipole–dipole forces
- van der Waals' forces.

Permanent dipole–dipole interactions

Polar molecules have permanent dipoles (see spread 1.2.11)
The permanent dipole of one molecule attracts the permanent dipole in a different polar molecule to form a weak **permanent dipole–dipole force**.

Figure 1 shows a permanent dipole–dipole force between two HCl molecules. Note that the $H^{\delta+}$ on one molecule attracts the $Cl^{\delta-}$ on a neighbouring molecule.

van der Waals' forces (induced dipole–dipole interactions)

van der Waals' forces exist between all molecules, whether polar or non-polar. **van der Waals' forces** are weak intermolecular attractions between very small, temporary dipoles in neighbouring molecules (see Figure 3).

Figure 2 A gecko is a small lizard. A gecko's toes are covered with microscopic hairs that create very weak intermolecular forces known as van der Waals' forces. It is the combined attraction from the great number of hairs that allows the gecko to stick to virtually any surface, even polished glass, or to walk upside down.

What causes van der Waals' forces?

- van der Waals' forces are caused by the movement of electrons in the shells. This movement unbalances the distribution of charge within the electron shells – rather like the electron density in the shells 'wobbling' from side to side.
- At any moment, there will be an *instantaneous* dipole across the molecule.
- The instantaneous dipole *induces* a dipole in neighbouring molecules, which in turn induce further dipoles on their neighbouring molecules.
- The small induced dipoles attract one another causing weak intermolecular forces known as van der Waals' forces.

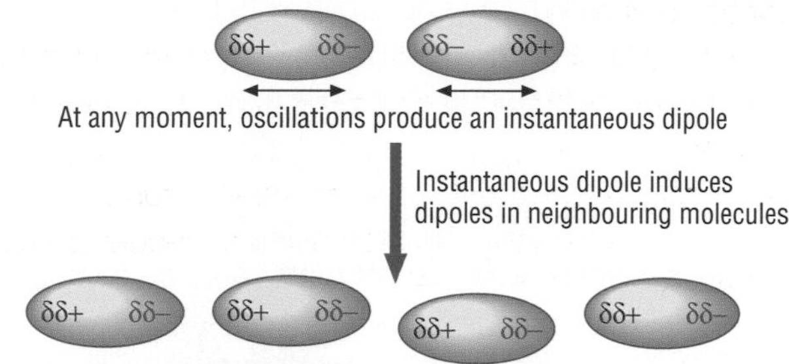
δδ+ δδ−
δδ− δδ+
At any moment, oscillations produce an instantaneous dipole
Instantaneous dipole induces dipoles in neighbouring molecules.
δδ+ δδ−
δδ+ δδ−
δδ+ δδ−
δδ+ δδ−
Induced dipoles attract one another.

Figure 3 van der Waals' forces (induced dipole–dipole interactions)

Examiner tip

In exams, candidates find it difficult to explain the origin of van der Waals' forces. Many have not learnt this! Make sure that you do and test yourself **before** taking the exam.

van der Waals' forces increase with increasing numbers of electrons.

- The greater the number of electrons, the larger the induced dipoles;
- the greater the attractive forces between molecules;
- which are the van der Waals' forces.

Boiling points and van der Waals' forces

van der Waals' forces are the only attractive intermolecular forces acting between non-polar molecules.

Table 2 shows the increasing boiling point of the noble gases as the number of electrons increases.

Noble gas	Boiling point/°C	Number of electrons
He	−269	2
Ne	−246	10
Ar	−186	18
Kr	−153	36
Xe	−108	54
Rn	−62	86

- electrons increase
- van der Waals' forces increase
- boiling point increases

Table 2

As the number of electrons increases, so does the strength of the van der Waals' forces.

The boiling point increases as we move down the noble gas group.

If there were no van der Waals' forces, it would be impossible to liquefy the noble gases.

Questions

1 Describe how van der Waals' forces arise.

2 The boiling point of the Group 7 elements are shown below. Each element exists as diatomic molecules.

(a) F_2, −188 °C; **(b)** Cl_2, −35 °C; **(c)** Br_2, 59 °C; **(d)** I_2, 184 °C.

Explain this trend, in terms of intermolecular forces.

1.2 (13) Hydrogen bonding

By the end of this spread, you should be able to . . .

* **Describe hydrogen bonding between molecules containing –OH and –NH groups.**
* **Describe and explain the anomalous properties of water resulting from hydrogen bonding.**

Examiner tip

It is essential that an H-bond is shown as a dotted line starting at $H^{\delta+}$ and going to a lone pair on $O^{\delta-}$ or $N^{\delta-}$.

For H-bonding, look for molecules with:
O–H
N–H

Common examples: H_2O, NH_3, CH_3OH, C_2H_5OH.

Key definition

A **hydrogen bond** is a strong dipole-dipole attraction between:

- an electron-deficient hydrogen atom ($O–H^{\delta+}$, $N–H^{\delta+}$) on one molecule; and
- a lone pair of electrons on a highly electronegative atom ($H–O^{\delta-}$ or $H–N^{\delta-}$) on a *different* molecule.

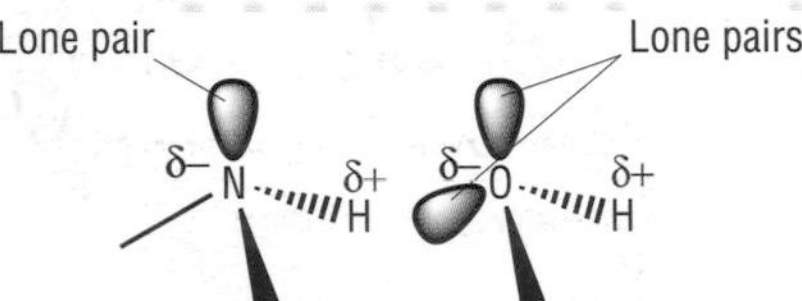

Figure 1 The key features for hydrogen bonding

A hydrogen bond

Molecules containing O–H and N–H bonds are polar with permanent dipoles. These dipoles are particularly strong. The permanent dipole–dipole interaction between molecules containing O–H and N–H bonds is given a special name: a **hydrogen bond**.

In a hydrogen bond the electron deficient $H^{\delta+}$ on one molecule attracts a lone pair of electrons on a $O^{\delta-}$ or $N^{\delta-}$ on a different molecule.

Hydrogen bonding occurs between molecules of water and ice as shown in Figure 2 and Figure 3.

Figure 2 Hydrogen bonding in water. A hydrogen bond is shown between molecules as a dashed line. Notice the role of the lone pair, which is essential in hydrogen bonding

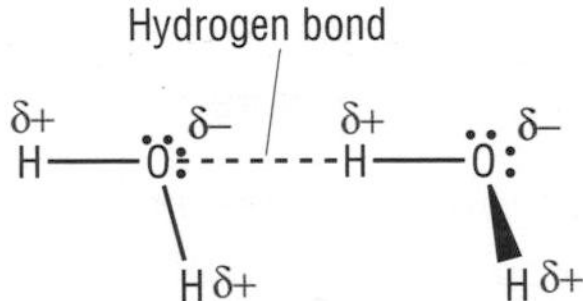

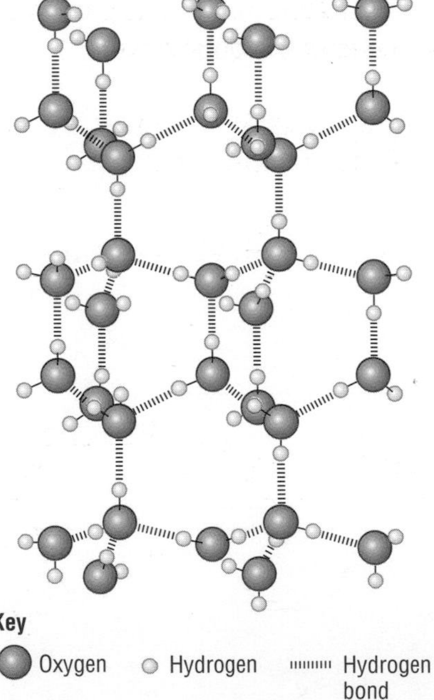

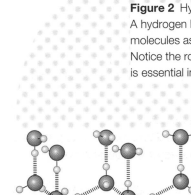

Figure 3 The ice lattice is an open network of H_2O molecules. Notice how the hydrogen bonds hold the water molecules apart. Notice also the shape of the lattice. Each oxygen atom has four bonds: two covalent bonds and two hydrogen bonds. The hydrogen bonds are slightly longer. The open structure is made up of rings of six oxygen atoms. Snowflakes are based on six-sided shapes. There is almost an infinite number of possible six-sided arrangements, so that every snowflake has a unique shape

Special properties of water

A hydrogen bond in water has only about 5% the strength of the O–H covalent bonds. However, hydrogen bonding is strong enough to have significant effects on physical properties. This results in some unexpected properties for water.

Ice is less dense than water

In almost all materials, the solid is denser than the liquid.
Water is the exception, with ice (solid H_2O) being less dense than water (liquid H_2O). This is because:

- ice has an open lattice with hydrogen bonds holding the water molecules apart.

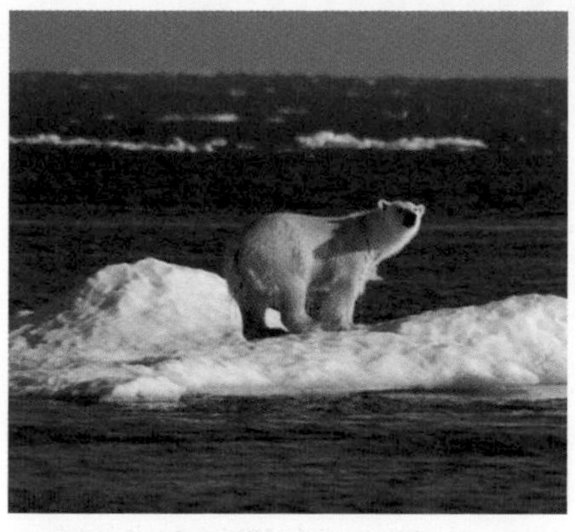

Figure 4 Solids are usually denser than liquids – but ice is less dense than water, and can float on water

- when ice melts, the rigid hydrogen bonds collapse, allowing the H_2O molecules to move closer together.

So, ice is less dense than water (see Figure 4).

Water has relatively high melting and boiling points

There are relatively strong hydrogen bonds between H_2O molecules.

- The hydrogen bonds are extra forces, over and above van der Waals' forces.
- These extra forces have to be overcome in order to melt or boil H_2O. This results in H_2O having higher melting and boiling points than would be expected from just van der Waals' forces.

Other properties

The extra intermolecular bonding from hydrogen bonds also explains the relatively high surface tension and viscosity of water. When small insects walk on water, they are walking across a *raft* of hydrogen bonds – see Figure 5.

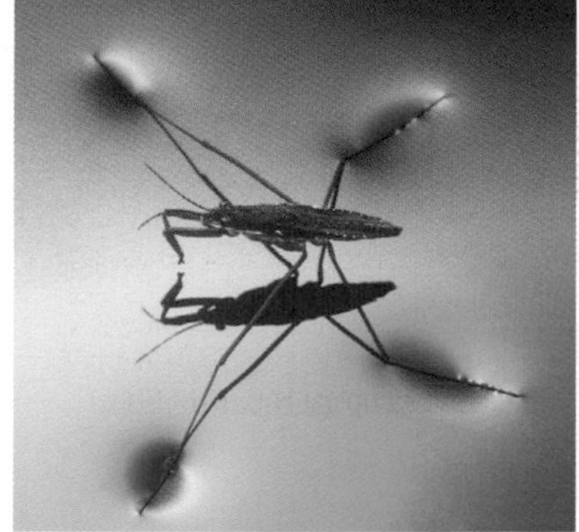

Figure 5 The high surface tension of water caused by the presence of hydrogen bonding keeps this pond skater on the surface

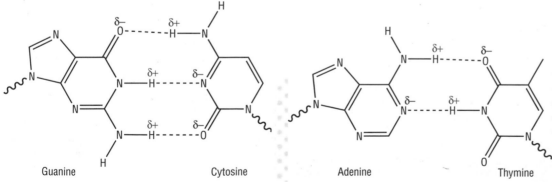

Hydrogen bonding in biological molecules

Hydrogen bonding is important in organic compounds containing O–H or N–H bonds, e.g. alcohols, carboxylic acids, amines and amino acids.

In biological molecules, hydrogen bonds play many essential roles.

- Hydrogen bonds are responsible for the shape of many molecules, such as proteins.
- In DNA, hydrogen bonds are responsible for holding together the double helix structure.
  There are four bases in DNA: guanine (G), cytosine (C), adenine (A) and thymine (T). The bases G and C are bonded together by three hydrogen bonds, and bases A and T by two hydrogen bonds.

The different number of hydrogen bonds ensures that the bases link together correctly (see Figures 6, 7 and 8).

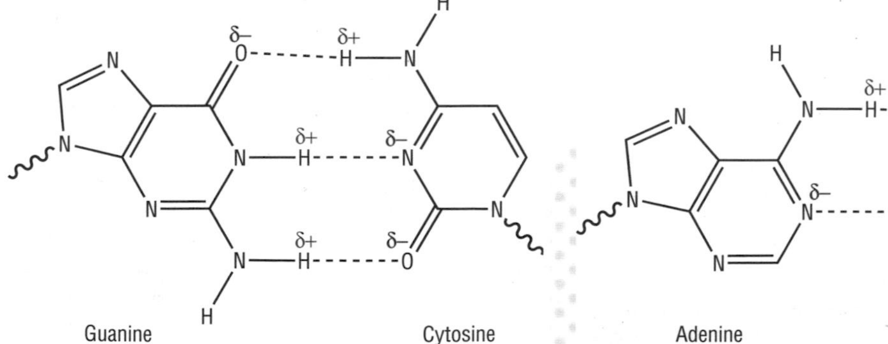

Figure 6 In DNA, three hydrogen bonds link C and G

Figure 7 In DNA, two hydrogen bonds link A and T

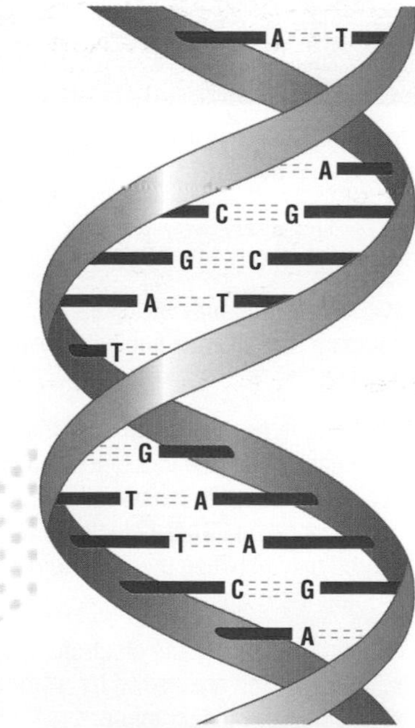

Figure 8 Double helix in DNA

Examiner tip

Remember that when ice melts or water boils, the *hydrogen* bonds *break*.

However, the *covalent* bonds between the H and O atoms in an H_2O molecule are strong and do *not* break.

In exams, many students confuse the bonds and get this wrong.

Questions

1 Which of the following molecules have hydrogen bonding?
H_2S; CH_4; CH_3OH; PH_3; NO_2; CH_3NH_2.

2 Draw diagrams showing hydrogen bonding between:
(a) 2 molecules of water.
(b) 2 molecules of ammonia.
(c) 2 molecules of ethanol, C_2H_5OH.
(d) 1 molecule of water and 1 molecule of ethanol.

1.2 (14) Metallic bonding and structure

By the end of this spread, you should be able to . . .

* **Describe metallic bonding as the attraction of positive ions to delocalised electrons.**
* **Describe giant metallic lattices.**

Key definitions

Metallic bonding is the electrostatic attraction between positive metal ions and delocalised electrons.

Delocalised electrons are shared between more than two atoms.

A giant metallic lattice is a three-dimensional structure of positive ions and delocalised electrons, bonded together by strong metallic bonds.

Figure 1 Copper, mercury and magnesium metals

Examiner tip

In an exam, you may be asked to compare metallic bonding with covalent bonding.

You need to be aware that a covalent bond has a *localised* pair of electrons positioned between the two atoms bonded together. These electrons are shared only between the *two atoms* forming the covalent bond.

In metallic bonding, the electrons are *delocalised* and are shared by *all* the positive ions in the structure.

Examiner tip

A metal conducts electricity because the delocalised electrons move. The *positive ions* do *not* move.

Metallic bonding and structure

The atoms in a solid metal are held together by **metallic bonding**.

In metallic bonding, the atoms are ionised.

- *Positive ions* occupy *fixed* positions in a lattice.
- The outer-shell electrons are **delocalised**. They are shared between all the atoms in the metallic structure.

The metal is held together by the attraction between the positive ions and negative electrons.

In a **giant metallic lattice**:

- the delocalised electrons are spread throughout the metallic structure
- these electrons are able to move within the structure
- it is impossible to tell which electron originated from which particular positive ion
- over the whole structure, the charges must balance.

A giant metallic lattice is often referred to as a lattice of positive ions fixed in position and surrounded by a sea of electrons (see Figure 2). The electrons are delocalised and can move. This model helps us understand the electrical conductivity of metals, which is explained below.

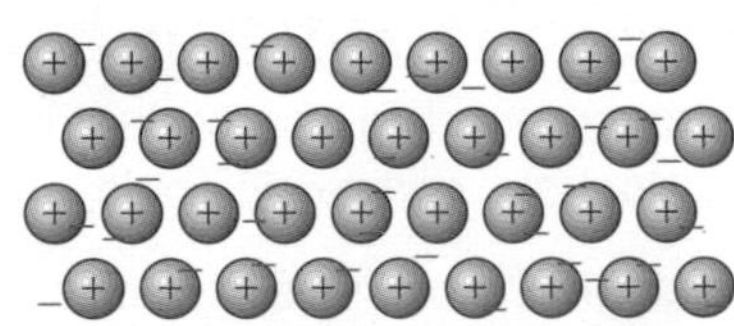

Figure 2 The 'sea of electrons'. Here, there are 36 positive ions balanced by 36 electrons

Properties of giant metallic lattices

A metallic lattice is made up of millions of positive ions and delocalised electrons. The three main properties of this structure are listed below.

High melting and boiling points

Most metals have high melting and boiling points.

- The electrons are free to move throughout the structure, but the positive ions remain where they are.
- The attraction between the positive ions and negative delocalised electrons is strong.
- High temperatures are needed to break the metallic bonds and dislodge the ions from their rigid positions within the lattice.

Good electrical conductivity

Metals are good conductors of electricity.

- The delocalised electrons can move freely anywhere within the metallic lattice.
- This allows the metal to conduct electricity, even in the solid state – see Figure 3.

Figure 3 Electrical conductivity in a metal. The electrons are free to flow between the positive ions. The positive ions do not move

In many ways, metallic bonding is the strangest type of chemical bond.

A metallic lattice is held together by the attraction between oppositely charged particles. But one of these particles – the electron – is free to move anywhere within the structure!

Malleability and ductility

Metals are *ductile* and *malleable*.

Ductile means: can be drawn out or stretched. Ductility permits metals to be drawn into wires.

Malleable means: can be hammered into shape. Many metals can be pressed into shapes or hammered out into thin sheets.

The delocalised electrons are largely responsible for these properties. Because they can move, the metallic structure has a degree of *give*, which allows atoms or layers to slide past each other.

Alloys

Alloys are mixtures of metals. Although the metals mix together freely, the result is not a compound. In a compound, there is always the same proportion of atoms of each element. But metals can mix together in different proportions. In the structure, the positive ions of one element simply replace the ions of another element.

Alloys modify a metal's properties. Often, one ion is a different size from the one that it replaces. This can make a metal harder by creating a barrier, thus preventing the layers of atoms sliding past each other. There are hundreds of different alloys, each with its own particular properties and uses.

Figure 4 Drill bits made from ferrovanadium alloy.
These drill bits are made of an alloy of steel and vanadium. Vanadium has a high resistance to corrosion. Steel is itself an alloy of iron and carbon. Alloying steel with vanadium produces ferrovanadium — a material with greater strength and rust resistance than steel alone. About 80% of the world's vanadium production is used in this way

Figure 5 Angel of the North. This sculpture is situated on reclaimed land in Gateshead, Tyne and Wear. It is 20 metres high, with a wingspan of 54 metres. The sculpture was constructed in 1995 from weather-resistant steel containing copper. The surface of the metal has oxidised (rusted), hence the orange colour

Questions

1 What is meant by *metallic bonding*?
2 How does a metal conduct electricity?
3 Explain how metallic bonding is different from covalent bonding.

Structure of ionic compounds

By the end of this spread, you should be able to . . .

* **Describe structures with ionic bonding as giant ionic lattices.**

Giant ionic lattices

In spread 1.2.6, we discussed how ionic bonds are formed by electron transfer from a metal to a non-metal. We also saw that the resulting giant ionic lattice can be enormous, containing many millions of ions (depending upon the size of the crystal).

Each ion attracts oppositely charged ions from all directions.

- Each ion is surrounded by oppositely charged ions.
- The ions attract each other, forming a *giant ionic lattice.*

The giant ionic lattice structure of sodium chloride, NaCl, is shown in Figure 1.

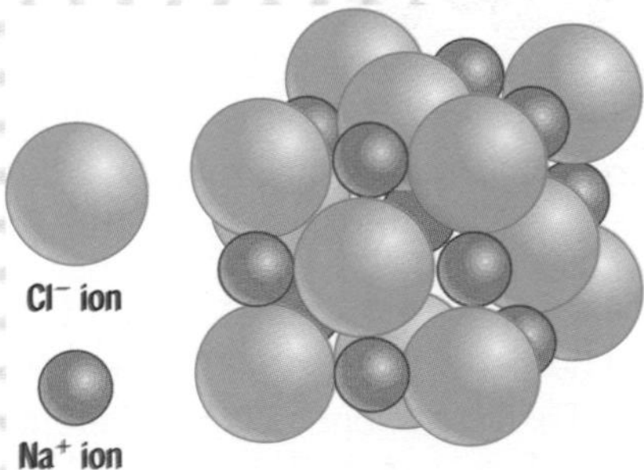

Figure 1 Sodium chloride ionic lattice

- Each Na^+ ion surrounds 6 Cl^- ions.
- Each Cl^- ion surrounds 6 Na^+ ions.

All ionic compounds exist as a giant ionic lattice in the solid state.

Properties of ionic compounds

High melting point and boiling point

- Ionic compounds are solids at room temperature.
- A large amount of energy is needed to break the *strong* electrostatic forces that hold the oppositely charged ions together in the solid lattice. For this reason, ionic compounds have high melting and boiling points.

Table 1 lists the melting point for sodium chloride and for magnesium oxide. The melting point of MgO is higher than that of NaCl. The charges on the Mg^{2+} and O^{2-} ions are greater than Na^+ and Cl^-. The greater the charge, the stronger the electrostatic forces between the ions and the greater the amount of energy required to break up the ionic lattice during melting. The melting point of magnesium oxide is so high that it is used to line furnaces for brick-making.

Compound	Ions	Melting point/°C
NaCl	Na^+ and Cl^-	801
MgO	Mg^{2+} and O^{2-}	2852

Table 1 NaCl compared with MgO

Electrical conductivity

In a solid ionic lattice (Figure 2 top):

- the ions are in a fixed position and no ions can move.
- the ionic compound is a non-conductor of electricity.

When an ionic compound is *melted* or *dissolved* in water (Figure 2 below):

- the solid lattice breaks down and the ions are free to move.
- the ionic compound is now a conductor of electricity.

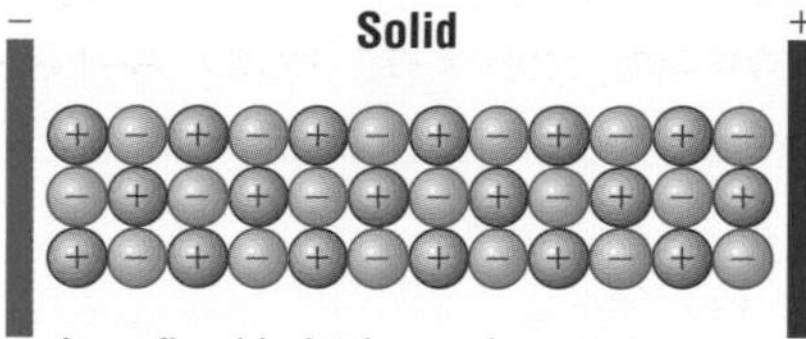

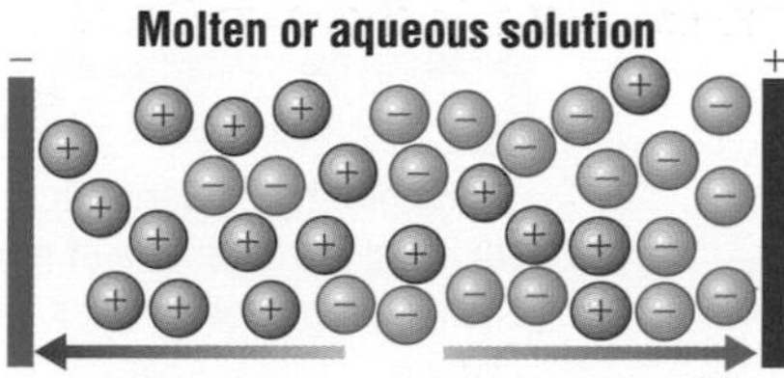

Figure 2 Diagram showing how an ionic compound conducts electricity when molten, but not when solid

Solubility

The ionic lattice dissolves in *polar* solvents, such as water.

The polar water molecules break down the lattice by surrounding each ion to form a solution. This is illustrated in Figure 3 for sodium chloride.

When NaCl is dissolved in water, the giant ionic lattice breaks down.

- Water molecules attract Na^+ and Cl^- ions.
- The ionic lattice breaks down as it dissolves. Water molecules surround the ions.
- Na^+ attracts $\delta-$ charges on the O atoms of the water molecules.
- Cl^- attracts $\delta+$ charges on the H atoms of the water molecules.

Figure 3 Sodium and chloride ions surrounded by water molecules

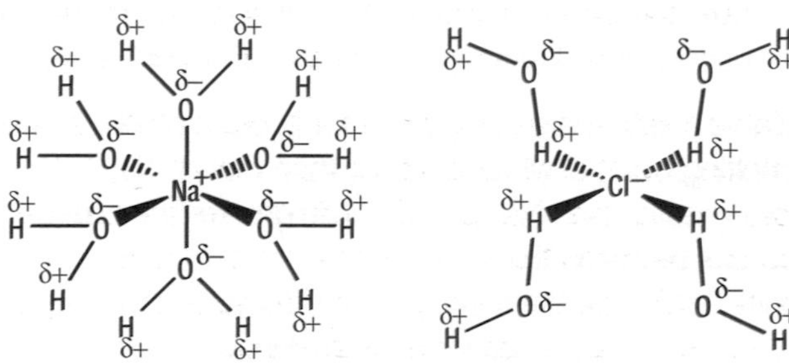

Examiner tip

In exams, you may be asked to explain properties of an ionic compound.

Make sure you get the particles and forces correct – many candidates get in a muddle here.

The *strong* forces between the ions in ionic compounds result in *high* melting and boiling points.

Ionic compounds *conduct* electricity only when ions are free to move – when they are *molten* or in aqueous *solution*.

Ionic compounds tend to dissolve in *polar* solvents, such as water.

Questions

1 Why do ionic compounds have high melting and boiling points?

2 Explain the different electrical conductivities of an ionic compound when solid, molten and aqueous.

3 Explain why water is a good solvent for ionic compounds.

1.2 (16) Structures of covalent compounds

By the end of this spread, you should be able to . . .

* **Describe the structures of simple molecular lattices.**
* **Describe the structures of giant covalent lattices.**
* **Explain physical properties of covalent compounds.**

Types of structure

Elements and compounds with covalent bonds have either of two structures:

- a simple molecular lattice
- a giant covalent lattice.

Simple molecular structures

Simple molecular structures are made up from small, simple molecules, such as Ne, H_2, O_2, N_2 and H_2O.

In a solid **simple molecular lattice** (see Figure 1):

- molecules are held together by weak forces between molecules
- the atoms within each molecule are bonded strongly together by covalent bonds.

Figure 1 Crystal structure of solid iodine. Solid iodine has a regular arrangement of I_2 molecules with weak van der Waals' forces between I_2 molecules

Key definition

A simple molecular lattice is a three-dimensional structure of molecules, bonded together by weak intermolecular forces.

The simple molecular structure of iodine

The different forces within the simple molecular structure of solid I_2 are shown in Figure 2.

- In each I_2 molecule, I atoms are held together by strong covalent bonds.
- When solid I_2 changes state, the weak van der Waals' forces between the I_2 molecules break.

Figure 2 Different forces in iodine

Examiner tip

In exams, you may be asked to describe the changes in bonding that occur when I_2 changes state.

Remember that when the simple molecular structure of I_2 is broken:

- Only the *weak* van der Waals' forces between the I_2 molecules *break*.
- The covalent bonds, I–I, are *strong* and do *not* break.

Properties of simple molecular structures

Low melting point and boiling point

Simple molecular structures have *low* melting and boiling points because:

- the intermolecular forces are weak van der Waals' forces, so a relatively small amount of energy is needed to break them.

Electrical conductivity

- Simple molecular structures are non-conductors of electricity because there are no charged particles free to move.

Solubility

- Simple molecular structures are soluble in *non-polar* solvents, such as hexane. This is because van der Waals' forces form between the simple molecular structure and the non-polar solvent.
- The formation of these van der Waals' forces weakens the lattice structure.

Figure 3 Solubility of iodine. This picture shows a jar of solid iodine next to a beaker containing iodine dissolved in the non-polar solvent hexane (top layer, purple) and iodine dissolved in water (bottom layer). You can see some solid iodine at the bottom of the beaker. Iodine does not dissolve well in water, explaining the faint brown colour observed. It does, however, dissolve well in hexane, producing a strong purple colour

Giant covalent structures

Diamond (Figure 4), graphite (Figure 5) and SiO_2 are examples of a **giant covalent lattice**.

Key definition

A **giant covalent lattice** is a three-dimensional structure of atoms, bonded together by strong covalent bonds.

Properties of giant covalent structures

High melting point and boiling point

Giant covalent structures have *high* melting and boiling points because:

- high temperatures are needed to break the strong covalent bonds in the lattice.

Examiner tip

Covalent bonds are strong. The covalent bonds in a giant covalent structure break when it melts or boils; this only happens at very high temperatures.

Electrical conductivity

Giant covalent structures are non-conductors of electricity because:

- there are *no* free charged particles except in graphite (see Table 1).

Solubility

Giant covalent structures are insoluble in *both* polar and non-polar solvents because:

- the covalent bonds in the lattice are too strong to be broken by either polar or non-polar solvents.

Figure 4 Diamond is the hard, shiny form of the element carbon. Diamond has a giant covalent structure containing many millions of carbon atoms. Each carbon atom is covalently bonded to four other carbon atoms in a tetrahedral shape

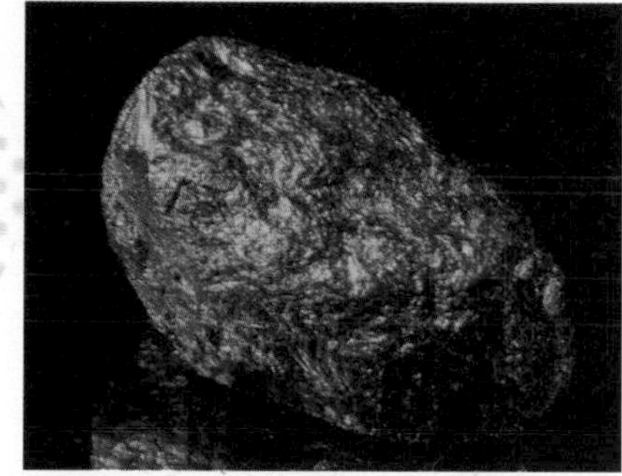

Figure 5 Graphite is the soft, slippery form of carbon. As with diamond, graphite has a giant covalent structure with many millions of carbon atoms. However, in graphite, each carbon atom is covalently bonded to three other carbon atoms in a hexagonal layer structure

Physical property	Diamond	Graphite
structure	tetrahedral structure held together by strong covalent bonds throughout lattice **Figure 6** Tetrahedral structure	strong hexagonal layer structure, but with weak van der Waals' forces between the layers **Figure 7** Hexagonal layer structure
electrical conductivity	*poor conductivity* there are no delocalised electrons as all outer-shell electrons are used for covalent bonds	*good conductivity* there are delocalised electrons between layers electrons are free to move parallel to the layers when a voltage is applied
hardness	*hard* tetrahedral shape allows external forces to be spread throughout the lattice	*soft* bonding within each layer is strong weak forces between layers allow layers to slide easily

Table 1 Structures of diamond and graphite

Questions

1 For each of the following substances:
(i) NaCl; **(ii)** SiO_2 (sand); **(iii)** Br_2; **(iv)** C_2H_5OH.
predict the: **(a)** structure; **(b)** melting point;
(c) electrical conductivity; **(d)** solubility.

2 Explain the different properties of diamond and graphite from their structures.

1.2 Bonding and structure summary

Metallic bonding

Ionic bonding

Giant structures
Strong forces
High melting and boiling point

Giant metallic
e.g. Fe, Cu, Na

Strong metallic bonds between + ions and delocalised electrons

Conduct electricity by movement of delocalised electrons

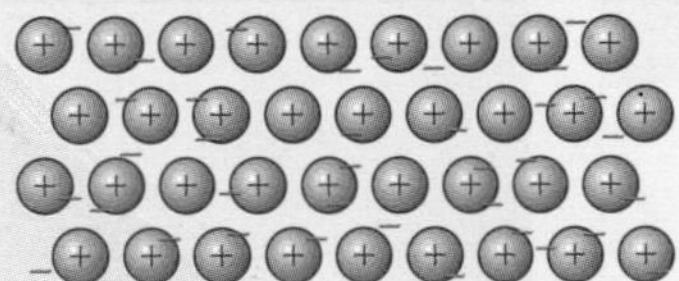

Giant covalent
e.g. diamond, graphite

Strong covalent bonds between atoms

Usually non-conductors as no mobile charges. But graphite does conduct: delocalised electrons between layers

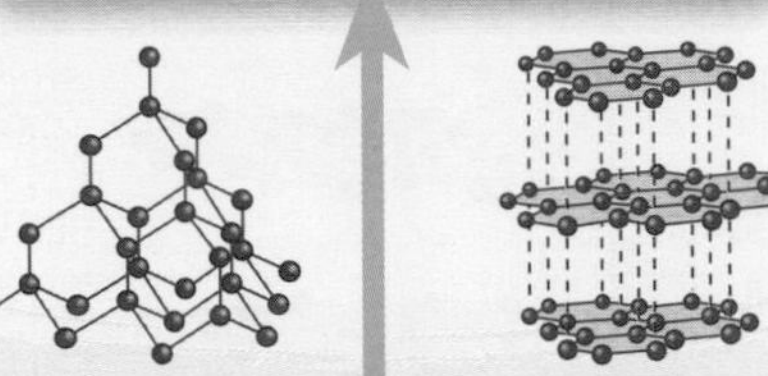

Giant ionic
e.g. NaCl, $MgCl_2$

Strong ionic bonds between oppositely charged ions

Ions fixed in solid lattice but conduct electricity when molten or in solution as + and − ions can then move

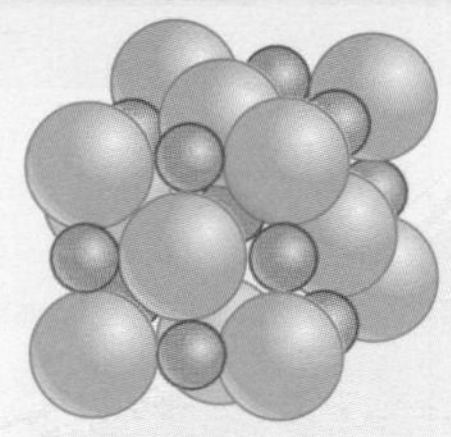

Covalent bonding

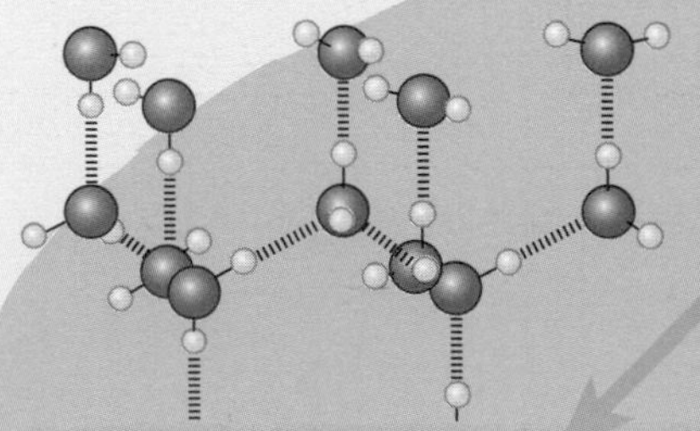

Simple molecular structures
Weak intermolecular forces

Low melting and boiling point

Non-conductors of electricity

Strong covalent bonds within each I_2 molecule

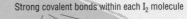

Weak van der Waals' forces between the I_2 molecules

Hydrogen bonded
e.g. H_2O, NH_3

Weak attraction between polar molecules:
H in one molecule
O or N in another molecule
Extra attraction increases melting and boiling points

Dipole–dipole
e.g. HCl

Weak attraction between polar molecules

van der Waals'
e.g. methane, iodine

Weak forces between induced dipoles in neighbouring molecules

Practice questions

1. Write the electron configuration in terms of sub-shells for the following atoms and ions:
   (a) Si; (b) Ca; (c) V;
   (d) Mg^{2+}; (e) S^{2-}; (f) P^{3-};
   (g) Co^{2+}.

2. Write an equation to represent the 4th ionisation energy of silicon.

3. Draw dot-and-cross diagrams for:
   (a) MgO; (b) Na_3N;
   (c) $CaBr_2$; (d) Al_2S_3.

4. Predict the formula of the following ionic compounds:
   (a) barium fluoride; (b) lithium sulfide;
   (c) iron(III) nitrate; (d) calcium phosphate.

5. For the following molecules, draw a dot-and-cross diagram, and predict the shape and bond angles.
   (a) H_2S; (b) BCl_3; (c) $SiCl_4$;
   (d) NI_3; (e) H_2F^+; (f) O_3.

6. Draw diagrams showing hydrogen bonding between:
   (a) 2 molecules of NH_3;
   (b) 2 molecules of CH_3CH_2OH;
   (c) 1 molecule of H_2O and 1 molecule of CH_3OH.

7. For each of the following substances:
   (i) Zn; (ii) C(diamond);
   (iii) $CaCl_2$; (iv) Br_2; (v) CH_3OH
   predict:
   (a) structure; (b) melting point;
   (c) electrical conductivity;
   (d) solubility in water and in hexane (a non-polar solvent).

8. The O–H bonds in water and the N–H bonds in ammonia have dipoles.
   (a) Why do these bonds have dipoles?
   (b) Draw a diagram to show hydrogen bonding between one molecule of H_2O and one molecule of NH_3. Include any relevant lone pairs and dipoles.
   (c) State and explain two anomalous properties of water resulting from hydrogen bonding.

9. Water, ammonia and sulfur dioxide react together to form a compound **A** which has the following percentage composition by mass: N, 24.12%; H, 6.94%; S, 27.61%; O, 41.33%.
   (a) (i) Calculate the empirical formula of compound **A**.
   (ii) Suggest a balanced equation for the formation of compound **A** from the reaction of water, ammonia and sulfur dioxide.
   (b) (i) Draw dot-and-cross diagrams for water, ammonia and sulfur dioxide.
   (ii) Draw the shapes and bond angles of a molecule of H_2O, NH_3 and SO_2.

10. Explain the following physical properties in terms of bonding and structure.
    (a) Solid iron conducts electricity and has a high melting point.
    (b) Solid graphite conducts electricity and has a high melting point.
    (c) Solid nitrogen does not conduct electricity and has a low melting point.

11. Magnesium oxide, MgO, forms a similar ionic lattice to that of sodium chloride.
    (a) Describe the structure of the ionic lattice in MgO.
    (b) Explain why MgO has a high melting point.
    (c) Suggest, with reasons, why magnesium oxide has a higher melting point than sodium chloride.

12. The table below shows some properties of four substances **A**, **B**, **C** and **D**.

compound	A	B	C	D
solubility in water	good	poor	poor	poor
solubility in hexane	poor	good	poor	poor
electrical conductivity	solid: poor liquid: good	solid: poor liquid: poor	solid: poor liquid: poor	solid: good liquid: good
boiling point	high	low	high	high

    (a) Predict the structures of **A**, **B**, **C** and **D**.
    (b) Suggest an identity for each substance.

13. (a) Define the term *first ionisation energy*.
    (b) The table below shows the first six successive ionisation energies of an element that is in **Period 3** of the Periodic Table.

ionisation energy/kJ mol^{-1}					
1st	2nd	3rd	4th	5th	6th
578	1817	2745	11578	14831	18378

    Identify element **X**. Explain how you decided on your answer.

Examination questions

1 Electrons are arranged in energy levels.

(a) An orbital is a region in which an electron may be found. Draw diagrams to show the shape of an s-orbital and of a p-orbital. [2]

(b) Complete the table below to show how many electrons **completely** fill each of the following.

	number of electrons
a d-**orbital**	
a p-**sub-shell**	
the third **shell** ($n = 3$)	

[3]

(c) The energy diagram below is for the eight electrons in an oxygen atom. The diagram is incomplete as it only shows the two electrons in the 1s level.

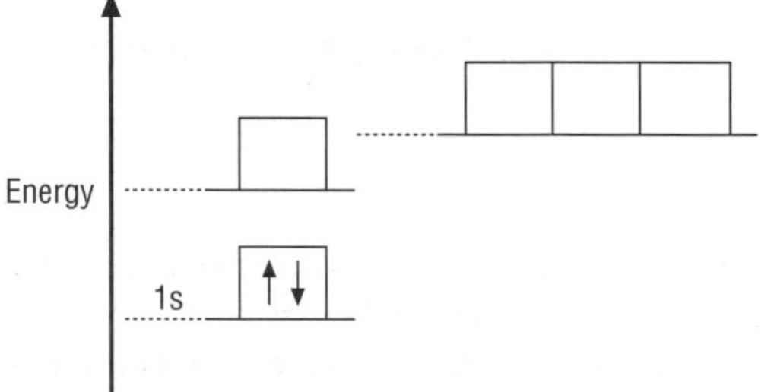

Complete the diagram for the oxygen atom by:

(i) adding labels for the other sub-shell levels [1]

(ii) adding arrows to show how the other electrons are arranged. [1]

(d) Successive ionisation energies provide evidence for the arrangement of electrons in atoms. The table below shows the eight successive ionisation energies of oxygen.

ionisation number	ionisation energy/kJ mol^{-1}
1st	1314
2nd	3388
3rd	5301
4th	7469
5th	10989
6th	13327
7th	71337
8th	84080

(i) Define the term ***first*** ionisation energy. [3]

(ii) Write an equation, with state symbols, to represent the **third** ionisation energy of oxygen. [2]

(iii) Explain how the information in the table above provides evidence for two electron shells in oxygen. [2]

[Total 14]

(Jun 07 2811)

2 Although compounds are usually classified as having ionic or covalent bonding, often the bonding is somewhere in between these two extremes.

(a) State what is meant by the terms

(i) *ionic bond,* [1]

(ii) *covalent bond,* [2]

(b) Compounds with covalent bonding often have polar bonds. Polarity can be explained in terms of electronegativity.

(i) Explain the term *electronegativity.* [2]

(ii) Use a suitable example to show how the presence of a polar bond can be explained in terms of electronegativity.
You may find it useful to draw a diagram in your answer. [2]

(c) Some polar molecules are able to form hydrogen bonds. Draw a diagram to show an example of hydrogen bonding. [2]

[Total 9]

(Jan 07 2811)

3 Limestone contains the ionic compound $CaCO_3$. Limestone decomposes when it is heated strongly, forming an ionic compound, CaO, and a covalent compound, CO_2.

$CaCO_3(s) \rightarrow CaO(s) + CO_2(g)$

(a) State what is meant by *ionic bonding.* [3]

(b) Draw dot-and-cross diagrams to show the bonding in CaO and CO_2. Show outer electron shells only. [3]

(c) Complete the electronic configuration in terms of sub-shells for calcium in CaO.

$1s^2$...... [1]

(d) Calcium oxide neutralises acids such as nitric acid. A student neutralised 1.50 g of CaO with 2.50 mol dm^{-3} nitric acid, HNO_3. The equation for this reaction is shown below.

$CaO(s) + 2HNO_3(aq) \rightarrow Ca(NO_3)_2(aq) + H_2O(l)$

(i) How many moles of CaO reacted? [2]

(ii) Calculate the volume of 2.50 mol dm^{-3} HNO_3 needed to exactly neutralise 1.50 g of CaO. [2]

(e) The nitrate ion, NO_3^-, in $Ca(NO_3)_2$ contains both covalent and dative covalent bonds.

(i) What is the difference between a covalent bond and a dative covalent bond? [1]

(ii) Calcium nitrate decomposes on heating to form calcium oxide, oxygen and nitrogen(IV) oxide, NO_2. Construct a balanced equation for this reaction. [1]

[Total: 13]

(Jun 07 2811)

4 Sodium reacts with chlorine forming sodium chloride.

(a) Describe the bonding in Na, Cl_2 and NaCl. [8]

(b) Relate the **physical** properties of Cl_2 and NaCl to their structure and bonding. [8]

[Total: 16]

(Jan 01 2811)

Answers to examination questions will be found on the Exam Café CD.

5 The compounds NH_3, BF_3 and HI all have covalent bonding and simple molecular structures. The Pauling electronegativity values shown in the table below can be used to predict polarity in these compounds.

H 2.1

Li 1.0	Be 1.5	B 2.0	C 2.5	N 3.0	O 3.5	F 4.0
Na 0.9						Cl 3.0
K 0.8						Br 2.8
						I 2.5

(a) Explain the term *electronegativity*. [2]

(b) The electronegativity values in the table above can be used to predict the polarity of a bond. In the boxes below, show the polarity of each bond by adding δ+ or δ– to each bond. The first box has been completed for you.

$^{\delta-}$O–H$^{\delta+}$	H–N	F–B	H–I

[2]

(c) Using outer electron shells only draw dot-and-cross diagrams for molecules of NH_3 and BF_3. [2]

(d) The diagrams below show the shapes of molecules of NH_3 and BF_3.
In the spaces below each diagram, state the bond angle in each molecule and state the name of each shape.

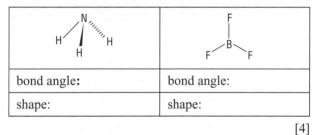

N, H, H, H	F, B, F, F
bond angle:	bond angle:
shape:	shape:

[4]

(e) Explain why NH_3 has polar molecules whereas molecules of BF_3 are non-polar. [2]

(f) Polar molecules of NH_3 form hydrogen bonds. Draw a diagram to show this hydrogen bonding. [1]

(g) NH_3 reacts with HI to form the ionic compound NH_4I, made up of NH_4^+ and I^- ions.
$NH_3 + HI \rightarrow NH_4I$
Explain why the H–N–H bond angle in NH_3 is less than that in NH_4^+. [2]

[Total: 15]
(Jan 02 2811)

6 In this question, one mark is available for the quality of use and organisation of scientific terms.
Nitrogen and oxygen are elements in Period 2 of the Periodic Table. The hydrogen compounds of oxygen and nitrogen, H_2O and NH_3, both form hydrogen bonds.

(a) (i) Draw a diagram containing two H_2O molecules to show what is meant by *hydrogen bonding*. On your diagram, show any lone pairs present and relevant dipoles. [3]

(ii) State and explain **two** anomalous properties of water resulting from hydrogen bonding. [4]

(b) The dot-and-cross diagram of an ammonia molecule is shown below.

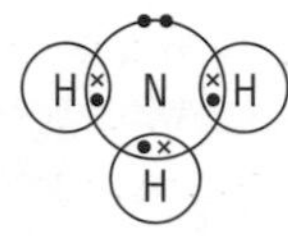

Predict, with reasons, the bond angle in an ammonia molecule. [4]

[Total: 11]
(Jan 05 2811)

7 In this question, one mark is available for the quality of spelling, punctuation and grammar.
Many physical properties can be explained in terms of bonding and structure. The table below shows the structures and some properties of sodium chloride and graphite in the solid state.

substance	sodium chloride	graphite
structure		
electrical conductivity of solid	poor	good
melting and boiling point	high	high
solubility in water	good	insoluble

Explain these properties in terms of bonding and structure. [7]

Quality of Written Communication [1]

[Total: 8]
(Jun 07 2811)

UNIT 1 Module 3 The Periodic Table

Introduction

The Periodic Table is the single most important reference document for the chemist. Scientists have attempted to organise the elements in some sort of logical order for centuries. It was only in the nineteenth century that the first recognisable Periodic Table was pieced together by comparing the properties of different elements. Since then, the Periodic Table has helped generations of chemists – and chemistry students – to understand the complex relationships between the elements.

In this module, you will study the trends and patterns that are revealed by the arrangement of the elements in the Periodic Table. You will also examine the relationships between ionisation energy, electronic configuration and reactivity.

You will study two contrasting groups of elements: those in Group 2, which typify the chemistry of a metallic group; and those in Group 7, a typical non-metallic group.

Test yourself

1 What is meant by an 'element'?
2 State the only two elements that are liquids at room temperature and pressure.
3 What general name is given to a row in the Periodic Table?
4 What do elements in a group have in common?
5 Name three halogens.
6 What is the chemical name of limestone?

Module contents

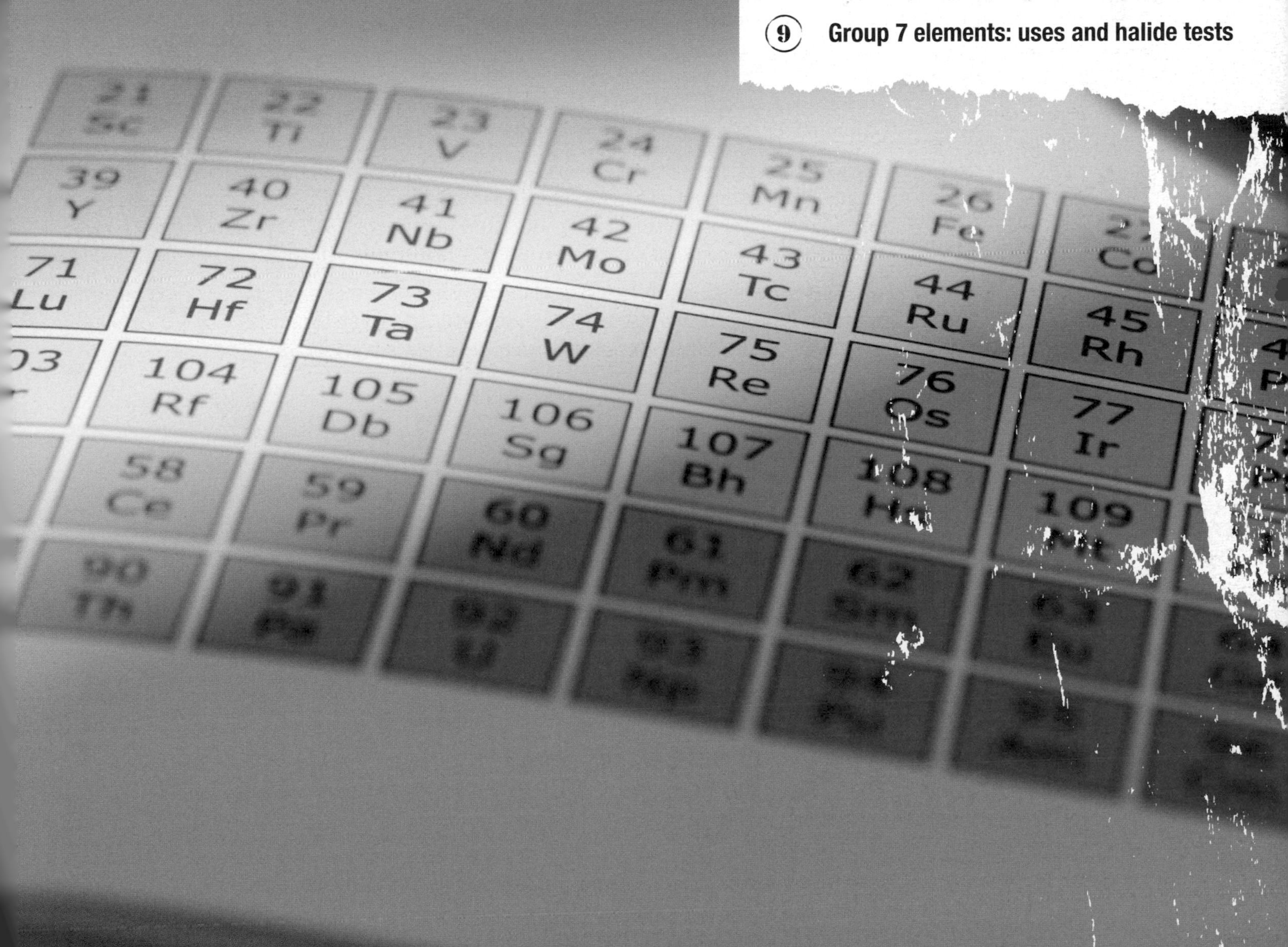

1.3 How Science Works

① The Periodic Table: searching for order

By the end of this spread, you should be able to . . .

* **Describe early attempts to order the elements.**
* **Appreciate that scientific knowledge is not absolute.**
* **Understand how advances in scientific knowledge are accepted.**

Figure 1 Aristotle (384–322 BC) was one of the ancient Greek philosophers who thought that everything was made up of just four elements

Figure 2 Lavoisier, 'the father of modern chemistry'

The distant past

Ancient philosophers believed that our world was made up of four elements: earth, water, air and fire. Three of these are actually examples of what we now call the three states of matter (solid, liquid and gas), while the fourth – fire – is similar to what we now call a plasma (sometimes referred to as the 'fourth state of matter').

Some chemical elements such as gold, silver and sulfur occur naturally, while humans have long known how to extract others – such as copper, tin and iron – from their ores by heating them with carbon. However, our knowledge and understanding of most elements is much more recent.

Antoine-Laurent de Lavoisier

In 1789, the French chemist Antoine-Laurent de Lavoisier, assisted by his wife, produced what is now considered to be the first modern chemical textbook. In his *Traité Élémentaire de Chimie* (Elementary Treatise of Chemistry) he compiled the first extensive list of elements, which he described as 'substances that could not be broken down further'. He also devised a theory for the formation of compounds from elements.

Lavoisier's list of elements included oxygen, nitrogen, hydrogen, phosphorus, mercury, zinc and sulfur. It also distinguished between metals and non-metals. Unfortunately, it also included some compounds and mixtures, along with terms such as 'light' and 'caloric' (heat), which he believed to be material substances.

Branded a traitor during the French revolution, Lavoisier was tried, convicted and executed in a single day, prompting the mathematician Joseph-Louis Lagrange to remark: 'It took them only an instant to cut off that head, and a hundred years may not produce another like it.'

Jöns Jakob Berzelius

In 1828, the Swedish chemist Jöns Jakob Berzelius published a table of atomic weights, and determined the composition by mass of many compounds. Berzelius was also responsible for introducing letter-based symbols for elements. Previously, signs had been used.

Döbereiner's triads

Döbereiner discovered that strontium had similar chemical properties to calcium and barium, and that its atomic weight fell midway between the two. He also discovered the halogen triad of chlorine, bromine and iodine, and the alkali metal triad of lithium, sodium and potassium.

Johann Wolfgang Döbereiner

In 1829 German chemist Johann Wolfgang Döbereiner proposed his 'Law of Triads', stating that:

> *Nature contains triads of elements where the middle element has properties that are an average of the other two members of the triad when ordered by the atomic weight.*

He noted that the atomic weight of the middle element in a triad was the average of the other two members. The densities of some triads followed a similar pattern. These groups became known as 'Döbereiner's triads'.

Alexandre-Emile Béguyer de Chancourtois

The French geologist Alexandre-Emile Béguyer de Chancourtois devised an early form of the Periodic Table. In 1862 he arranged elements in a spiral on a cylinder, ordered by increasing atomic weight. Elements with similar properties were aligned vertically.

John Newlands

The English chemist John Newlands was the first person to devise a Periodic Table of the elements arranged in order of their relative atomic weights (now called relative atomic masses). Building on Döbereiner's work with 'triads', in 1865 he put forward his 'law of octaves', which states that:

> *Any given element will exhibit analogous behaviour to the eighth element following it in the table.*

H 1	F 8	Cl 15	Co & Ni 22	Br 29	Pd 36	I 42	Pt & Ir 50
Li 2	Na 9	K 16	Cu 23	Rb 30	Ag 37	Cs 44	Tl 53
Gl 3	Mg 10	Ca 17	Zn 25	Sr 31	Cd 38	Ba & V 45	Pb 54
Bo 4	Al 11	Cr 18	Y 24	Ce & La 33	U 40	Ta 46	Th 56
C 5	Si 12	Ti 19	In 26	Zr 32	Sn 39	W 47	Hg 52
N 6	P 13	Mn 20	As 27	Di & Mo 34	Sb 41	Nb 48	Bi 55
O 7	S 14	Fe 21	Se 28	Ro & Ru 35	Te 43	Au 49	Os 51

Table 1 Newlands' arrangement of elements in octaves

- Newlands' arrangement showed all known elements arranged in seven 'octaves'.
- The elements are ordered by the atomic weights that were known at the time. They were numbered sequentially to show the order of atomic weights.
- In Newlands' table, periods and groups are shown going down and across the table, respectively – the opposite from our modern Periodic Table.

Newlands also altered the order of some elements, so that those with similar properties appeared in the same row. In Table 1 above, some similarities have been picked out in colour. Unfortunately, he left no gaps for elements that had yet to be discovered. Consequently, the addition of a single new or missing element would throw the entire table out. The table did highlight similarities between certain compounds, but contained many groupings for which there were none. His Di ('didymium') was later found to be a mixture of two elements, praseodymium and neodymium.

Note that none of the noble gases, which had yet to be discovered, appear in Newlands' table. Notice also that some symbols have since been changed (Bo is now B, for example).

Perhaps Newlands' biggest mistake was the use of the musical term 'octave', which prompted ridicule from some other scientists. One scientist suggested that similarities might just as easily be found if the elements were arranged alphabetically.

Dmitri Mendeleev

Four years after Newlands published his law of octaves, the Russian chemist Dmitri Mendeleev independently published a more advanced form of the Periodic Table, also based on atomic masses. This forms the basis of the version we use today, which is arranged by atomic number. While Newlands' law of octaves represented the first application of *periods* to the table, Mendeleev's version is widely regarded as one of the most important developments in chemistry.

Figure 3 John Newlands

Questions

1 Döbereiner proposed the 'Law of Triads'.
Use modern relative atomic mass values to see how well this law works for:
- **(a)** Ca, Sr and Ba;
- **(b)** Cl, Br and I;
- **(c)** Li, Na and K.

2 Look at Newlands' Periodic Table set out in Table 1.
- **(a)** Identify the elements that Newlands placed out of order, according to their atomic weights.
- **(b)** What term is given to the sequence of these elements in the modern Periodic Table?
- **(c)** What are the modern symbols and names for the elements Gl and Bo?
- **(d)** Which names in the modern Periodic Table correspond to the rows and columns in Newlands' Periodic Table?

1.3 (2) The Periodic Table: Mendeleev and beyond

How Science Works

By the end of this spread, you should be able to . . .

*** Understand the work of Mendeleev and others in developing the modern Periodic Table.**

The work of Mendeleev

Dmitri Ivanovich Mendeleev developed the Periodic Table whilst working as a professor of chemistry at St Petersburg University in Russia.

Mendeleev's cards

Mendeleev made a set of 63 cards – one for each of the 63 known elements – on which were written the symbol, atomic weight and properties of an element. It is said that he played 'chemical solitaire' on long train journeys, arranging the cards in order of increasing atomic weight and grouping together elements with similar properties.

Mendeleev's table

Mendeleev's work resulted in the production of a table based on atomic weights, but arranged *periodically*.

- Elements with similar properties were arranged in vertical columns.
- Gaps were left where no element fitted the repeating patterns, and the properties of the missing elements were predicted..
- The order of elements was rearranged where their properties did not fit. For example, while tellurium had a larger atomic weight than iodine, its properties indicated that it should be placed ahead of iodine in the table. Mendeleev was brave enough to reverse the order of these two elements on the basis of their properties, rather than on their established atomic weights. He concluded that the atomic weight of tellurium must actually be less than that of iodine in order to fit properly into his table.

Mendeleev published his table, accompanied by a statement of his 'Periodic Law', in 1869. Crucially, the periodic law made it possible to predict the properties of individual elements.

In 1869, the German chemist Julius Lothar Meyer compiled a Periodic Table containing 56 elements arranged in order of atomic weight, based on the periodicity of properties such as molar volume. Meyer worked independently of Mendeleev, and was unaware of his work. His results were published about a year after those of Mendeleev.

Figure 1 Dmitri Ivanovich Mendeleev (1834–1907), Russian chemist, on a commemorative Soviet stamp issued in 1957

Missing elements

Discovery of each of the missing elements, such as gallium, scandium and germanium, confirmed Mendeleev's predictions and was powerful evidence in favour of Mendeleev's periodic law.

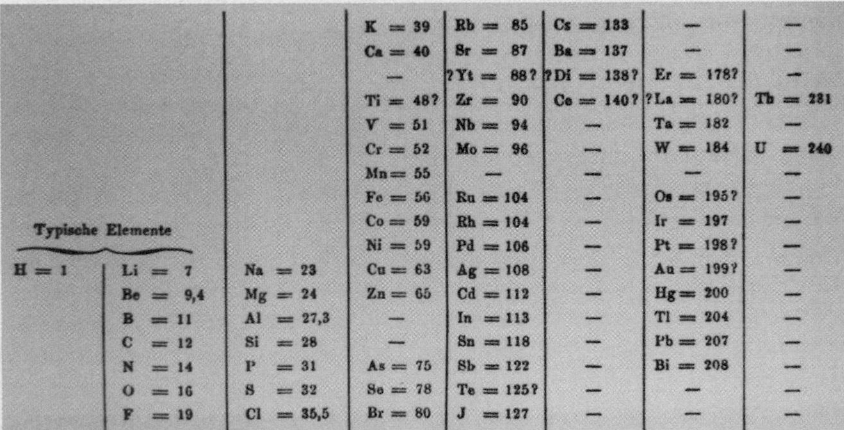

			K = 39	Rb = 85	Cs = 133	—	—
			Ca = 40	Sr = 87	Ba = 137	—	—
			—	?Yt = 88?	?Di = 138?	Er = 178?	—
			Ti = 48?	Zr = 90	Ce = 140?	?La = 180?	Th = 231
			V = 51	Nb = 94	—	Ta = 182	—
			Cr = 52	Mo = 96	—	W = 184	U = 240
			Mn = 55	—	—	—	—
			Fe = 56	Ru = 104	—	Os = 195?	—
			Co = 59	Rh = 104	—	Ir = 197	—
Typische Elemente			Ni = 59	Pd = 106	—	Pt = 198?	—
H = 1	Li = 7	Na = 23	Cu = 63	Ag = 108	—	Au = 199?	—
	Be = 9,4	Mg = 24	Zn = 65	Cd = 112	—	Hg = 200	—
	B = 11	Al = 27,3	—	In = 113	—	Tl = 204	—
	C = 12	Si = 28	—	Sn = 118	—	Pb = 207	—
	N = 14	P = 31	As = 75	Sb = 122	—	Bi = 208	—
	O = 16	S = 32	Se = 78	Te = 125?	—	—	—
	F = 19	Cl = 35,5	Br = 80	J = 127	—	—	—

Figure 2 Mendeleev's Periodic Table of 1869. This early form of the Periodic Table showed similar chemical elements arranged horizontally. Chemical symbols and the atomic weights are used. Mendeleev left gaps in the table for new elements that were indeed later discovered, vindicating his theory. Reading from top to bottom and left to right, the first four gaps awaited scandium (1879), gallium (1875), germanium (1886) and technetium (1937)

Mendeleev's periodic law

Mendeleev's periodic law, which is reproduced below, appears highly complicated, but we must remember that he was searching for patterns among the elements. His version of the Periodic Table is widely regarded as a milestone in the development of chemical science.

- The elements, if arranged according to their atomic weights, exhibit an apparent periodicity of properties.
- Elements which are similar, as regards to their chemical properties, have atomic weights which are either of nearly the same value (e.g., Pt, Ir, Os) or which increase regularly (e.g., K, Rb, Cs).

- The arrangement of the elements, or of groups of elements, in the order of their atomic weights corresponds to their so-called valencies, as well as to their distinctive chemical properties (to some extent). This is apparent in series such as Li, Be, B, C, N, O and F.
- The elements which are the most widely diffused have small atomic weights.
- The magnitude of the atomic weight determines the character of the element, just as the magnitude of the molecule determines the character of a compound body.
- We must expect the discovery of many yet unknown elements – for example, elements analogous to aluminium and silicon, whose atomic weight would be between 65 and 75.
- The atomic weight of an element may sometimes be amended by a knowledge of those of its contiguous elements. Thus the atomic weight of tellurium must lie between 123 and 126, and cannot be 128.
- Certain characteristic properties of elements can be foretold from their atomic weights.

Advantages of Mendeleev's table

Mendeleev predicted the discovery of new elements, and left space for three new elements in his table: 'eka-silicon' (germanium), 'eka-aluminium' (gallium) and 'eka-boron' (scandium). As a result, their eventual discovery did not disturb the balance of his Periodic Table. His work also indicated that some accepted atomic weights were incorrect, while his table provided for variance from atomic weight order.

Disadvantages of Mendeleev's table

Isotopes had yet to be discovered, and there was no place for them in Mendeleev's table. Nor did it include any of the noble gases, which had yet to be discovered. These were added later as Group 0, without any disturbance of the basic concepts underlying the table.

Atomic structure and the Periodic Table

Although Mendeleev's table demonstrated the periodic nature of the elements, it was left for the scientists of the 20th century to explain *why* the properties of the elements are repeated periodically.

In 1913, Henry Moseley determined the **atomic number** for each of the elements (see spread 1.1.1).

- The atomic number is the number of positive charges, or protons, in the atomic nucleus. It is also equivalent to the number of negative charges, or electrons, outside the nucleus in a neutral atom.
- Moseley modified Mendeleev's periodic law to read that the properties of the elements vary periodically with their atomic *numbers*, rather than atomic weight.
- Moseley's modified periodic law puts the elements tellurium and iodine in the right order, as it does for argon and potassium and for cobalt and nickel.

The periodic law could now be explained on the basis of the electronic structure of the atom, the main factor that determines the chemical properties and many physical properties of the elements.

In spite of its great success, the periodic system introduced by Mendeleev did contain some discrepancies. Arranged strictly according to atomic weight, not all elements fell into their proper groups. Based on atomic number, the periodic law now has no exceptions.

In the middle of the twentieth century, the US scientist Glenn Seaborg discovered the transuranic elements from 94 (plutonium) to 102 (nobelium). He also remodelled the Periodic Table by placing the actinide series below the lanthanide series at the bottom of the table. Element 106 has been named Seaborgium (Sg) in honour of his work.

Discovery of the noble gases

In 1868, the noble gas, helium was first detected by studying light emitted from the Sun. Helium gets its name from *Helios*, the Greek word for the Sun. In the 1890s, argon and helium were discovered on Earth as new elements, which were chemically inert. In 1898, Sir William Ramsey suggested that argon be placed into the Periodic Table, with helium, as part of a *zero* group. Argon was placed between chlorine and potassium. This was despite argon having a larger atomic weight than potassium. Ramsey accurately predicted the future discovery and properties of the noble gas neon.

Questions

1 Which elements did Mendeleev leave spaces for in his Periodic Table?

2 **(a)** Look at a modern Periodic Table and find three pairs of elements that have been placed out of order in terms of increasing atomic mass.

(b) What property of an atom is the real order of elements in the Periodic Table based on?

1.3 (3) The modern Periodic Table

By the end of this spread, you should be able to . . .

* **Describe the Periodic Table in terms of the arrangement of elements by increasing atomic number, in periods and in groups.**
* **Explain how periodicity is a repeating pattern across different periods.**
* **Explain how atoms of elements in a group have similar outer-shell electron structures, resulting in similar properties.**
* **Describe the variation in electron structures across Periods 2 and 3.**

Examiner tip

Remember that the atomic number is the number of protons in the nucleus.

Key definition

A **period** is a horizontal row of elements in the Periodic Table. Elements show trends in properties across a period.

Arranging the elements

In the modern Periodic Table, elements are arranged in order of their atomic numbers.

Each horizontal row is called a period.

Elements often show trends (gradual changes) in properties across a **period** in the table.

- These trends are repeated across each period.
- The repeating pattern of trends is called periodicity.

Each vertical column is called a **group**, and contains elements with similar properties.

The Periodic Table is the chemist's way of ordering the elements to show patterns of chemical and physical properties. Figure 1 shows the key areas in the Periodic Table.

The elements in green separate metals (to the left) from non-metals (to the right).

These elements, such as silicon and germanium, are called semi-metals or metalloids. Semi-metals display properties between those of a metal and a non-metal.

Key definition

A **group** is a vertical column in the Periodic Table. Elements in a group have similar chemical properties and their atoms have the same number of outer-shell electrons.

Key: Relative atomic mass / **Atomic symbol** / Name / Atomic (proton number)

Period	Group 1	2											3	4	5	6	7	0
1								1.0 **H** Hydrogen 1										4.0 **He** Helium 2
2	6.9 **Li** Lithium 3	9.0 **Be** Beryllium 4											10.8 **B** Boron 4	12.0 **C** Carbon 6	14.0 **N** Nitrogen 7	16.0 **O** Oxygen 8	19.0 **F** Fluorine 9	20.2 **Ne** Neon 10
3	23.0 **Na** Sodium 11	24.3 **Mg** Magnesium 12											27.0 **Al** Aluminium 13	28.1 **Si** Silicon 14	31.0 **P** Phosphorus 15	32.1 **S** Sulfur 16	35.5 **Cl** Chlorine 17	39.9 **Ar** Argon 18
4	39.1 **K** Potassium 19	40.1 **Ca** Calcium 20	45.0 **Sc** Scandium 21	47.9 **Ti** Titanium 22	50.9 **V** Vanadium 23	52.0 **Cr** Chromium 24	54.9 **Mn** Manganese 25	55.8 **Fe** Iron 26	58.9 **Co** Cobalt 26	58.7 **Ni** Nickel 28	63.5 **Cu** Copper 29	63.5 **Zn** Zinc 30	69.7 **Ga** Gallium 31	72.6 **Ge** Germanium 32	74.9 **As** Arsenic 33	79.0 **Se** Selenium 34	79.9 **Br** Bromine 35	83.8 **Kr** Krypton 36
5	85.5 **Rb** Rubidium 37	87.6 **Sr** Strontium 38	88.9 **Y** Yttrium 39	91.2 **Zr** Zirconium 40	92.9 **Nb** Niobium 41	95.9 **Mo** Molybdenum 42	(98) **Tc** Technetium 43	101.1 **Ru** Ruthenium 44	102.9 **Rh** Rhodium 45	106.4 **Pd** Palladium 46	107.9 **Ag** Silver 47	112.4 **Cd** Cadmium 48	114.8 **In** Indium 49	118.7 **Sn** Tin 50	121.8 **Sb** Antimony 51	127.6 **Te** Tellurium 52	126.9 **I** Iodine 53	131.3 **Xe** Xenon 54
6	132.9 **Cs** Caesium 55	137.3 **Ba** Barium 56	138.9 **La*** Lathanum 57	178.5 **Hf** Hafnium 72	180.9 **Ta** Tantalum 73	183.8 **W** Tungsten 74	186.2 **Re** Rhenium 75	190.2 **Os** Osmium 76	192.2 **Ir** Iridium 77	195.1 **Pt** Platinum 78	197.0 **Au** Gold 79	200.6 **Hg** Mercury 80	204.4 **Ti** Thallium 81	207.2 **Pb** Lead 82	209.0 **Bi** Bismuth 83	(209) **Po** Polonium 84	(210) **At** Astatine 85	(222) **Rn** Radon 86
7	(223) **Fr** Francium 87	(226) **Ra** Radium 88	(227) **Ac*** Actinium 89	(261) **Rf** Rutherfordium 104	(262) **Db** Dubnium 105	(266) **Sg** Seaborgium 106	(264) **Bh** Bohrium 107	(277) **Hs** Hassium 108	(268) **Mt** Meitnerium 109	(271) **Ds** Darmstadtium 110	(272) **Rg** Roentgenium 111	Elements with atomic numbers 112–116 have been reported but not fully authenticated						
				140.1 **Ce** Cerium 58	140.9 **Pr** Praseodymium 59	144.2 **Nd** Neodymium 60	144.9 **Pm** Promethium 61	150.4 **Sm** Samarium 62	152.0 **Eu** Europium 63	157.2 **Gd** Gadolinium 64	158.9 **Tb** Terbium 65	162.5 **Dy** Dysprosium 66	164.9 **Ho** Holium 67	167.3 **Er** Erbium 68	168.9 **Tm** Thulium 69	173.0 **Yb** Ytterbium 70	175.0 **Lu** Lutetium 71	
				232.0 **Th** Thorium 90	(231) **Pa** Protactinium 91	238.1 **U** Uranium 92	(237) **Np** Neptunium 93	(242) **Pu** Plutonium 94	(243) **Am** Americium 95	(247) **Cm** Curium 96	(245) **Bk** Berkelium 97	(251) **Cf** Californium 98	(254) **Es** Einsteinium 99	(253) **Fm** Fermium 100	(256) **Md** Mendeleevium 101	(254) **No** Nobelium 102	(257) **Lr** Lawrencium 103	

Figure 1 The Periodic Table

Periodicity

Periodicity is the trend in properties that is repeated across each period. Using periodicity, predictions can be made about the likely properties of an element and its compounds.

Trends down a group may also affect the periodic trends. For example:

- across each period, elements change from metals to non-metals;
- as you move down the Periodic Table, this change takes place further to the right;
- so, for example, at the top of Group 4, carbon is a non-metal, but at the bottom of Group 4, tin and lead are metals (see Figure 2);
- so, trends in properties can exist vertically down a group as well as horizontally across a period.

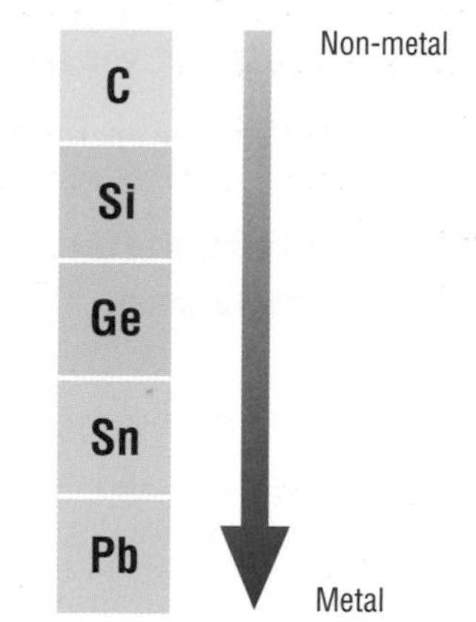

Figure 2 Group 4: Non-metal to metal

Key definition

Periodicity is a regular periodic variation of properties of elements with atomic number and position in the Periodic Table.

Variation in electron structure

Chemical reactions involve electrons in the outer shell. Any similarity in electron configuration will be reflected in the similarity of chemical reactions.

Similar elements are placed in vertical groups – this is a key principle of the Periodic Table.

- Elements within a group have atoms with the same number of electrons in their outer shells. This explains their similar chemical behaviour.
- This repeating pattern of similarity is caused by the underlying repeating pattern of electron configuration.

This is shown in Table 1 for elements in Period 2 and Period 3.

Li	Be	B	C	N	O	F	Ne
[He]$2s^1$	[He]$2s^2$	[He]$2s^22p^1$	[He]$2s^22p^2$	[He]$2s^22p^3$	[He]$2s^22p^4$	[He]$2s^22p^5$	[He]$2s^22p^6$
Na	Mg	Al	Si	P	S	Cl	Ar
[Ne]$3s^1$	[Ne]$3s^2$	[Ne]$3s^23p^1$	[Ne]$3s^23p^2$	[Ne]$3s^23p^3$	[Ne]$3s^23p^4$	[Ne]$3s^23p^5$	[Ne]$3s^23p^6$

Table 1 Repeating pattern of electron configuration across Periods 2 and 3

Each element within a vertical group has:

- the same number of electrons in the outer shell.
- the same type of orbitals.

So, elements in the same group react in a similar way because they have similar electron configurations.

Examiner tip

We often base electron configurations on the previous noble gas.

This means that we can concentrate on the outer-shell electrons – those responsible for reactions.

Na: $1s^22s^22p^63s^1 \rightarrow$ [Ne]$3s^1$

Examiner tip

Similar outer-shell electron configurations give rise to similar reactions.

For example, atoms of Group 1 elements lose one electron to form 1+ ions with a noble gas electron configuration.

Li	Li	$\longrightarrow$	$Li^+ + e^-$
	[He]**$2s^1$**	$\longrightarrow$	[He]
Na	Na	$\longrightarrow$	$Na^+ + e^-$
	[Ne]**$3s^1$**	$\longrightarrow$	[Ne]
K	K	$\longrightarrow$	$K^+ + e^-$
	[Ar]**$4s^1$**	$\longrightarrow$	[Ar]
Rb	Rb	$\longrightarrow$	$Rb^+ + e^-$
	[Kr]**$5s^1$**	$\longrightarrow$	[Kr]
Cs	Cs	$\longrightarrow$	$Cs^+ + e^-$
	[Xe]**$6s^1$**	$\longrightarrow$	[Xe]
Fr	Fr	$\longrightarrow$	$Fr^+ + e^-$
	[Rn]**$7s^1$**	$\longrightarrow$	[Rn]

Questions

1 Describe the periodic trend in terms of electron configurations across the Periodic Table.

2 What is the similarity in the electron configurations for atoms and ions of :
  (a) the Group 2 elements, Be–Ba.
  (b) the Group 7 elements, F–I.

1.3  Periodicity: ionisation energies and atomic radii

By the end of this spread, you should be able to . . .

*** Explain that ionisation energy depends upon: atomic radius; electron shielding; and nuclear charge.**

*** Describe the variation of the first ionisation energies and atomic radii of elements across a period and down a group.**

Variation in first ionisation energies and atomic radii

We first discussed ionisation energies in spread 1.2.1. You may wish to refer back to this spread, especially to revise the factors that affect ionisation energies. These are:

- nuclear charge
- distance from the nucleus
- electron shielding.

The first ionisation energies for the first 20 elements in the Periodic Table, H–Ca, are shown in Figure 1. The ionisation energy for each noble gas, at the end of each period, is shown in green.

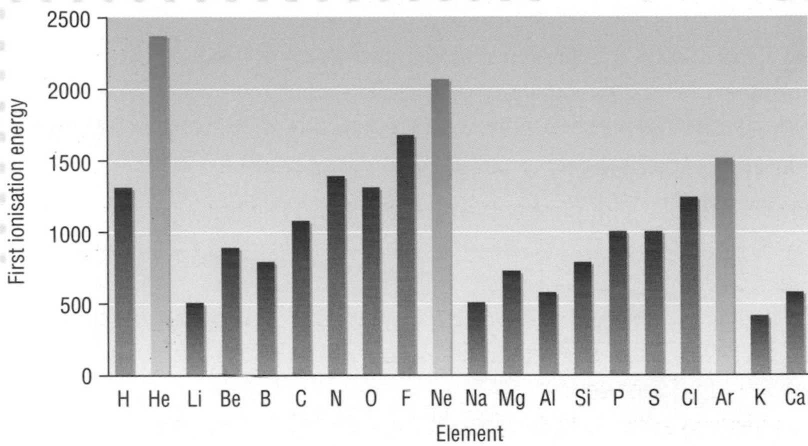

Figure 1 First ionisation energies of the first 20 elements

Trends across a period

Looking at Figure 1, you can see that the ionisation energy shows a general *increase* across each period:

Period 1 H → He
Period 2 Li → Ne
Period 3 Na → Ar.

Across each period:

- The number of protons increases, so there is more attraction acting on the electrons.
- Electrons are added to the same shell, so the outer shell is drawn inwards slightly. There is the same number of inner shells, so *electron shielding* will hardly change.

Across a period, the attraction between the nucleus and outer electrons *increases*, so *more* energy is needed to remove an electron. This means that the first ionisation energy *increases* across a period.

Examiner tip

Across a period, *increased nuclear charge* is the most important factor.

There is also a decrease in atomic radius across a period, because the increased nuclear charge pulls the electrons in towards it (see Figure 2).

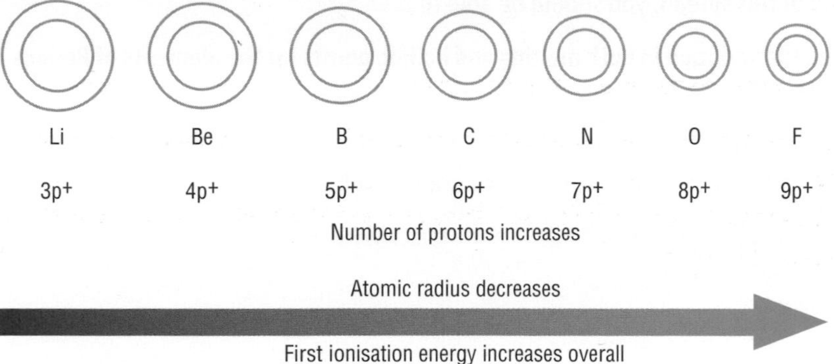

Figure 2 Trend in first ionisation energy and atomic radius across a period

Starting the next period, there is a sharp *decrease* in first ionisation energy between the end of one period and the start of the next period:

- From He at the end of Period 1 to Li at the start of Period 2.
- From Ne at the end of Period 2 to Na at the start of Period 3.
- From Ar at the end of Period 3 to K at the start of Period 4.

This reflects the addition of a new shell, further from the nucleus, which leads to:

- increased distance of the outermost shell from the nucleus; and
- increased electron shielding of the outermost shell by inner shells.

Trends down a group

Down a group, first ionisation energies *decrease*. Again, looking at Figure 1, you can see that the ionisation energy decreases down each group. This is clearest for the three noble gases, He, Ne and Ar, but first ionisation energies decrease down *every* group.

Down each group:

- The number of shells increases,
  so the distance of the outer electrons from the nucleus increases;
  hence, there is a weaker force of attraction on the outer electrons.
- There are more inner shells,
  so the shielding effect on the outer electrons from the nuclear charge increases, hence, again there is less attraction.

The number of protons in the nucleus also increases, but the resulting increased attraction is far outweighed by the increase in distance and shielding.

Taking all these factors into account, the attraction between the nucleus and outer electrons *decreases* down a group, so *less* energy is needed to remove an electron.

- This means that the first ionisation energy *decreases* down a group.
- However, the atomic radius *increases* down a group, because less attraction means that the electrons are not pulled as close to the nucleus.

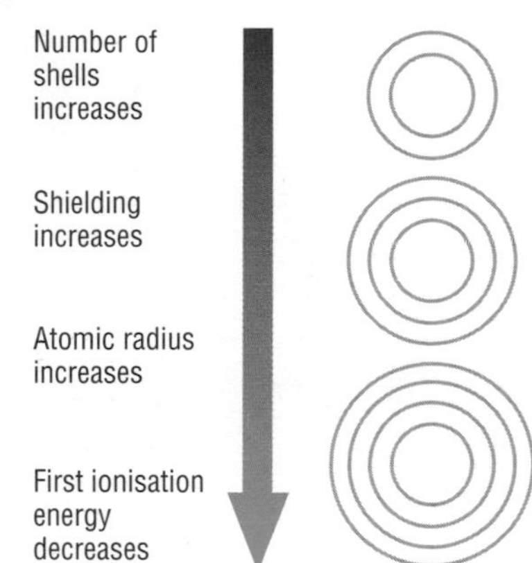

Figure 3 Trend in first ionisation energy and atomic radius down a group

Examiner tip

Down a group, *increased distance* and *shielding* are the most important factors influencing ionisation energy.

Questions

1 Explain the trends in ionisation energy and atomic radius *across* a period. In your answer, use ideas about nuclear charge, distance and shielding.

2 Explain the trends in ionisation energy and atomic radius *down* a group. In your answer, use ideas about nuclear charge, distance and shielding.

Periodicity: boiling points

By the end of this spread, you should be able to . . .

*** Describe the variation in both melting and boiling points for the elements of Periods 2 and 3.**

Metal to non-metal

Samples of the elements in Period 3 are shown in Figure 1.

The most striking trends across the Periodic Table are:
- metals to non-metals
- solid to gas.

Figure 1 Metal to non-metal trend across the Periodic Table

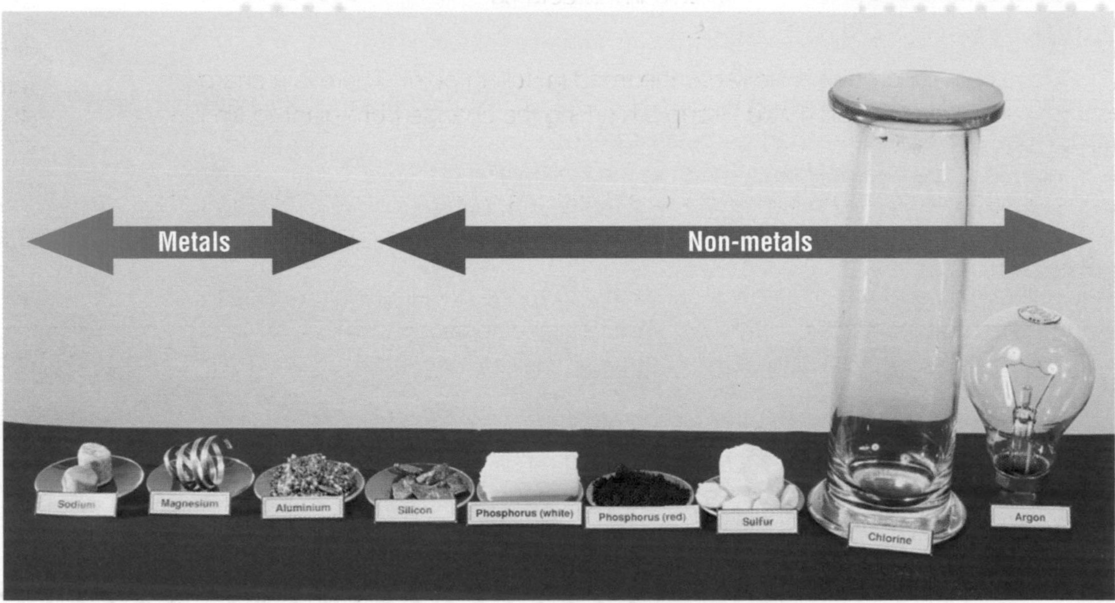

Na, Mg and Al are clearly metals, but Si is much harder to classify:
- Si has the shiny appearance of a metal, but is brittle. It conducts electricity, but very poorly.
- Silicon is an *in-between* element, usually classified as a *semi-metal* or *metalloid*.

Trends in melting and boiling points

Although we can identify broad trends just from the appearance of the elements, we can learn more about the bonding and structure of the elements from their melting and boiling points.

The boiling points of the elements in Period 2 and Period 3 are shown in Figure 2.

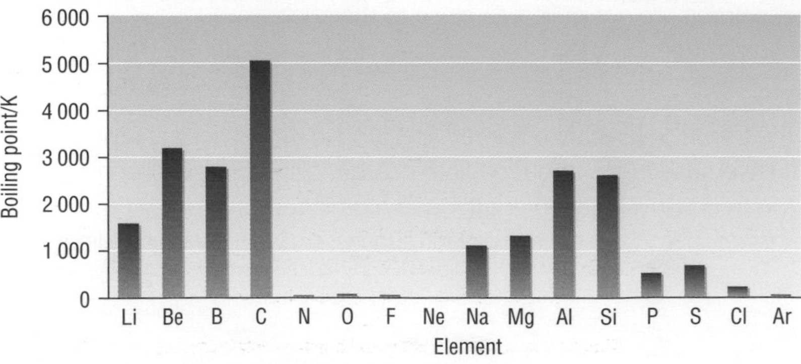

Figure 2 Boiling points of Period 2 and Period 3 elements

The trend in boiling points across each period is shown in Table 1.

Group 1 to Group 4	Group 4 to Group 5	Group 5 to Group 0
Li → C	C → N	N → Ne
Na → Si	Si → P	P → Ar
general increase in boiling points	sharp decrease in boiling point	comparatively low boiling points

Table 1 Boiling points across Periods 2 and 3

There is a distinct change between Group 4 and Group 5 in both the physical structure of the elements and the forces holding the structures together. This change is:

- from giant structures to simple molecular structures
- from strong forces to weak forces.

The trend in melting point is similar to the trend in boiling point. There is a sharp decrease between Group 4 and Group 5 marking the change from giant to simple molecular structure.

The structure and bonding across Periods 2 and 3 is shown in detail in Table 2. The molecules making up the simple molecular structures are shown by their molecular formulae.

Period 2	Li	Be	B	C	N_2	O_2	F_2	Ne
Period 3	Na	Mg	Al	Si	P_4	S_8	Cl_2	Ar
structure	giant metallic			giant covalent	simple molecular structures			
forces	strong forces between positive ions and negative delocalised electrons			strong forces between atoms	weak forces between molecules			
bonding	metallic bonding			covalent bonding	van der Waals' forces			

Table 2 Structure and bonding types across Periods 2 and 3

The melting and boiling points of the metals Na, Mg and Al in Period 3 increase across the period.

Figure 3 illustrates how the attractive forces between the metal ions and delocalised electrons increase from Na → Al.

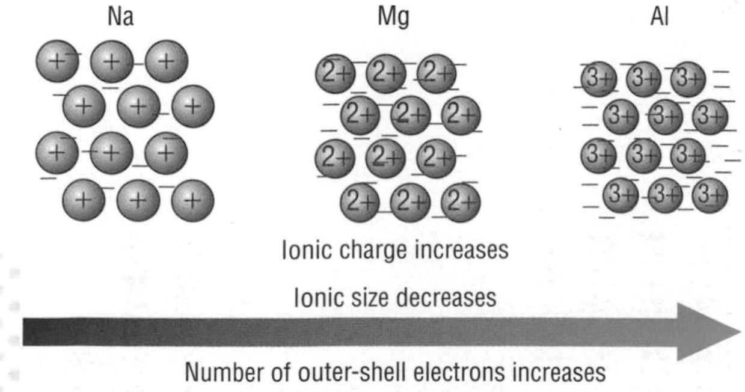

Figure 3 Increase in boiling point from Na to Al

Examiner tip

In exams, candidates find questions on structure and bonding difficult.

You need to concentrate on the *particles* that form the structure and the *forces* between the particles.

If the forces are strong, melting and boiling points will be high.

If the forces are weak, melting and boiling points will be low.

Across the Periodic Table, there is a sharp drop in melting and boiling points between Group 4 and Group 5. This marks the divide between strong forces in giant structures and weak forces in simple molecular structures.

Questions

1 Why is the boiling point of carbon much higher than that of nitrogen?

2 Why is the melting point of aluminium higher than that of magnesium?

Group 2 elements: redox reactions

By the end of this spread, you should be able to . . .

- **Describe the redox reactions of the Group 2 elements Mg → Ba with oxygen and with water.**
- **Explain the trend in reactivity of Group 2 elements down the group.**

Figure 1 Group 2 elements, from left: beryllium; magnesium; calcium; strontium; and barium

- Each element in this group has a tendency to lose two electrons.
- The reactivity of the elements shown in Figure 1 increases from left to right – beryllium is the least reactive, barium the most. In the Periodic Table, this order or tendency goes downwards.

The Group 2 elements

The elements in Group 2 all have hydroxides that are alkaline, which is reflected in the common name for this group: the alkaline earth metals.

Physical properties

The Group 2 elements have the following general properties.

- They have reasonably high melting and boiling points.
- They are light metals with low densities.
- They form colourless compounds.

Electronic configuration

The elements in Group 2 have their highest energy electrons in an s sub-shell. Together with Group 1, they form the s-block of the Periodic Table.

Each Group 2 element has:

- two electrons more than the electronic configuration of a noble gas;
- an outer shell containing two electrons.

Element	Electronic configuration	First ionisation energy /kJ mol^{-1}
Be	[He]$2s^2$	900
Mg	[Ne]$3s^2$	736
Ca	[Ar]$4s^2$	590
Sr	[Kr]$5s^2$	548
Ba	[Xe]$6s^2$	502
Ra	[Rn]$7s^2$	509

Table 1 Group 2 electron configurations and first ionisation energies

- Each Group 2 metal has two electrons in the s sub-shell – the highest-energy sub-shell.
- When the metals react, the two outer electrons are lost.
- The first ionisation energies decrease down the group.
- This is reflected in the increased reactivity down the group.
- Unfortunately, radium does not quite fit in with the trend!

Reactivity of the Group 2 elements

The Group 2 elements are reactive metals and strong reducing agents.

Group 2 elements are oxidised in reactions.

- Each atom loses two electrons from its outer s sub-shell to form a 2+ ion:

$$M \longrightarrow M^{2+} + 2e^- \quad \text{(+2 oxidation state)}$$

- Reactivity increases *down* the group, reflecting the increasing ease of losing electrons.
- The decrease in ionisation energies down Group 2 is an important factor in this process. Details of this trend are explained in spread 1.3.4.

Figure 2 Calcium burns in oxygen with an intense white flame, tinged with red at the edges, to form calcium oxide, CaO. In this reaction, each calcium atom loses two electrons from its outer s sub-shell: $Ca \longrightarrow Ca^{2+} + 2e^-$

Reaction with oxygen

The Group 2 elements react vigorously with oxygen.

- This is a **redox** reaction.
- The product is an ionic oxide with the general formula MO, where M is the Group 2 element.

Calcium reacts with oxygen to form calcium oxide.

$$2Ca(g) + O_2(g) \longrightarrow 2CaO(s)$$

By applying oxidation numbers to this reaction, you can identify the oxidation and reduction processes.

$2Ca(s)$	+	$O_2(g)$	$\longrightarrow$	$2CaO(s)$	
0				+2	oxidation (oxidation number increases)
0				+2	
		0		−2	reduction (oxidation number decreases)
		0		−2	

Table 2 Applying oxidation numbers

Examiner tip

For the reaction of calcium with oxygen, writing separate half equations for each species clearly shows how electrons are *lost* and *gained* in the redox process.

oxidation (loss of electrons)
$Ca \longrightarrow Ca^{2+} + 2e^-$

reduction (gain of electrons)
$O_2 + 4e^- \longrightarrow 2O^{2-}$

Reaction with water

The Group 2 elements react with water to form hydroxides with the general formula $M(OH)_2$. Hydrogen gas is also formed.

Calcium reacts with water to produce calcium hydroxide and hydrogen gas.

$$Ca(s) + 2H_2O(l) \longrightarrow Ca(OH)_2(aq) + H_2(g)$$

By applying oxidation numbers to this reaction, you can identify the oxidation and reduction processes.

$2Ca(s)$	+	$2H_2O(l)$	$\longrightarrow$	$Ca(OH)_2(aq)$	+	$H_2(g)$	
0				+2			oxidation
0				+2			
		+1				0	reduction
		+1				0	

Table 3 Applying oxidation numbers

- Note that only one H atom in each H_2O has been reduced.
- The other H atom does not change its oxidation number.

Mg reacts very slowly with water. As you move further down the group, each metal reacts more vigorously with water.

Examiner tip

The half equations for the reaction of calcium with water are slightly more difficult to write than for those with oxygen.

Note that only one of the H atoms in each H_2O molecule has been reduced.

oxidation (loss of electrons)
$Ca \longrightarrow Ca^{2+} + 2e^-$

reduction (gain of electrons)
$2H_2O + 2e^- \longrightarrow 2OH^- + H_2$

Examiner tip

Sometimes in exams you are asked to state what you would *see* when calcium reacts with water. There is often a clue in the equation.

$Ca(s) + 2H_2O(l) \rightarrow Ca(OH)_2(aq) + H_2(g)$

solid disappears as no '(s)' on right-hand side

H_2 evolved as a 'fizz'

Questions

1 The following reaction is a redox process:

$$Mg + 2HCl \longrightarrow MgCl_2 + H_2$$

(a) Identify the changes in oxidation number.

(b) Which species is being oxidised and which is being reduced?

(c) Identify the oxidising agent and the reducing agent.

2 **(a)** Write down equations for the following reactions:

(i) The reaction of barium with water.

(ii) The reaction of strontium with oxygen.

(b) Use oxidation numbers to identify what has been oxidised and what has been reduced for equations **(a)(i)** and **(a)(ii)**.

1.3 (7) Group 2 compounds: reactions

By the end of this spread, you should be able to . . .

* **Describe the effect of water on Group 2 oxides.**
* **Describe the thermal decomposition of Group 2 carbonates.**
* **Interpret and make predictions from the chemical and physical properties of Group 2 elements/compounds.**
* **Explain the uses of Group 2 hydroxides.**

Group 2 oxides and hydroxides

Group 2 oxides and hydroxides are bases. They are neutralised by acids to form a salt and water.

For example:

- $MgO(s) + 2HCl(aq) \rightarrow MgCl_2(aq) + H_2O(l)$
- $Ca(OH)_2(s) + 2HCl(aq) \rightarrow CaCl_2(aq) + 2H_2O(l)$

As these reactions with hydrochloric acid take place you will see the solid oxides or hydroxides dissolve.

Group 2 oxides

The Group 2 oxides react with water to form a solution of the metal hydroxide.

For example:

$$MgO(s) + H_2O(l) \longrightarrow Mg(OH)_2(aq)$$

The typical pH of these solutions is 10–12.

Group 2 hydroxides

Group 2 hydroxides dissolve in water to form alkaline solutions.

$$Ca(OH)_2(s) + aq \longrightarrow Ca^{2+}(aq) + 2OH^-(aq)$$

The solubility of the hydroxides in water *increases* down the group. The resulting solutions are also *more* alkaline.

- $Mg(OH)_2(s)$ is only slightly soluble in water.
  The resulting solution is dilute with a comparatively low $OH^-(aq)$ concentration.
- $Ba(OH)_2(s)$ is much more soluble in water than $Mg(OH)_2(s)$, with a greater $OH^-(aq)$ concentration. The resulting solution is more alkaline than a solution of $Mg(OH)_2$.

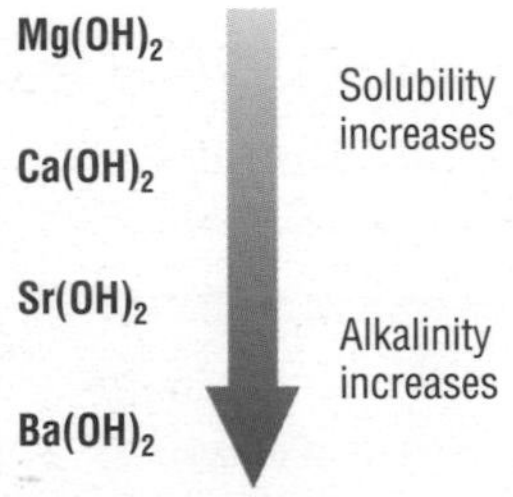

$Mg(OH)_2$
$Ca(OH)_2$
$Sr(OH)_2$
$Ba(OH)_2$
Solubility increases
Alkalinity increases

Figure 1 Alkalinity of Group 2 hydroxides

Group 2 carbonates

The Group 2 carbonates are decomposed by heat, forming the metal oxide and carbon dioxide gas.

- $MgCO_3(s) \longrightarrow MgO(s) + CO_2(g)$

This type of reaction is called **thermal decomposition.**

The carbonates become *more* difficult to decompose with heat as you move *down* the group.

Key definition

Thermal decomposition is the breaking up of a chemical substance with heat into at least two chemical substances.

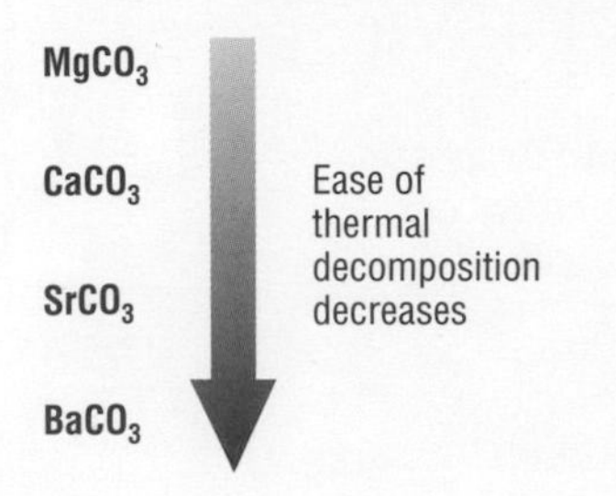

Figure 2 Trend in thermal decomposition of Group 2 carbonates

Figure 3 Limestone quarry in the Derbyshire Peak District, UK. Calcium carbonate is present in limestone and chalk. It is by far the most important calcium compound industrially, as it is used in the building trade and in the manufacture of glass and steel

Properties of Group 2 elements and their compounds

The beauty of Mendeleev's periodic law is that we only need to learn the typical reactions of *one* member of the group.

We can then apply this chemistry to *all* other elements in the group. We do also have to consider trends down a group.

- The Group 2 elements become more reactive down the group.
- The Group 2 carbonates decompose at higher temperatures down the group.
- The hydroxides become more soluble in water, and the resulting solutions become more alkaline.

Table 1 summarises some general properties for any Group 2 element, M.

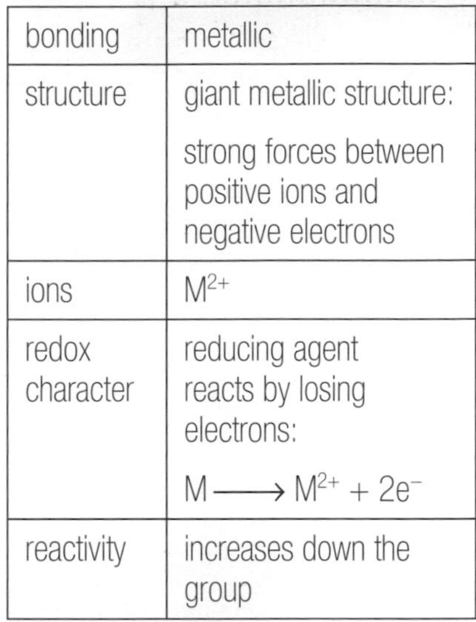

bonding	metallic
structure	giant metallic structure: strong forces between positive ions and negative electrons
ions	M^{2+}
redox character	reducing agent reacts by losing electrons: $M \longrightarrow M^{2+} + 2e^-$
reactivity	increases down the group

Table 1 Properties of Group 2 elements

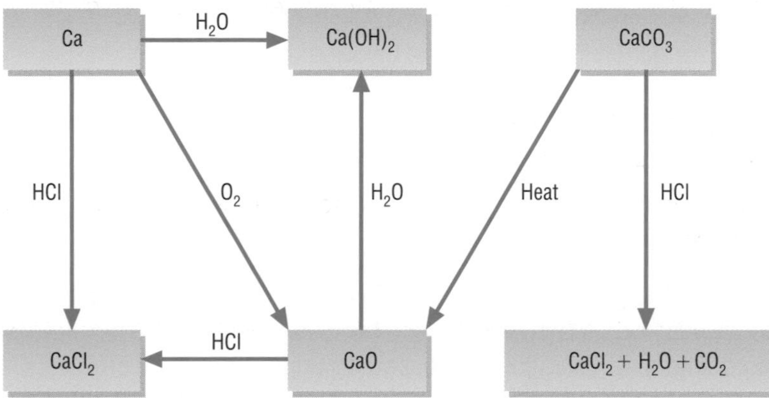

Figure 4 Reactions of calcium and its compounds

Reactions of calcium and its compounds

The flowchart in Figure 4 summarises many of the reactions of calcium and its compounds. The same reactions can be applied to any Group 2 element and its compounds.

Examiner tip

Notice that three of these reactions involve acids. You can find details of acid reactions in spreads 1.1.11 and 1.1.15.

Uses of Group 2 hydroxides

The alkalinity of the Group 2 hydroxides is used to combat acidity.

Two common examples include:

- Calcium hydroxide, $Ca(OH)_2$, is used by farmers and gardeners as 'lime' to neutralise acid soils. See Figure 5.
- Magnesium hydroxide, $Mg(OH)_2$, is used in 'milk of magnesia' to relieve indigestion. It works by neutralising any excess acid in the stomach.

Figure 5 Spreading lime on fields. Lime (the cloud of white powder) consists mostly of calcium hydroxide, and is used to reduce soil acidity and restore pH balance to soils

Questions

1 Write down equations for the following reactions:

(a) Barium oxide with hydrochloric acid.
(b) Radium carbonate with nitric acid.
(c) Strontium with water.
(d) The thermal decomposition of magnesium carbonate.
(e) Calcium oxide with nitric acid.

Group 7 elements: redox reactions

By the end of this spread, you should be able to . . .

* **Explain the trend in boiling points of Cl_2, Br_2 and I_2.**
* **Explain the trend in reactivity for the Group 7 elements.**
* **Describe the redox reactions of the Group 7 elements with other halide ions.**
* **Describe and interpret, using oxidation numbers, the reaction of chlorine with water and aqueous sodium hydroxide.**

The Group 7 elements

The Group 7 elements are fluorine (F), chlorine (Cl), bromine (Br), iodine (I) and astatine (At). These elements are generally known as the halogens (see Figure 1).

Figure 1 Halogen elements chlorine, bromine and iodine have different physical states at room temperature and pressure. Chlorine is a gas, bromine a liquid and iodine a solid

Physical properties

The halogens have the following physical properties.

- They have low melting and boiling points.
- They exist as diatomic molecules, X_2.

As you move down the group the number of electrons increases, leading to an increase in the van der Waals' forces between molecules. See spread 1.2.12 for details.

- The boiling points of the halogens increases down the group.
- The physical states of the halogens at room temperature and pressure show the classic trend of gas ⟶ liquid ⟶ solid as you move down the group (see Figures 1 and 2).

F_2	Gas	
Cl_2	Gas	Boiling point increases down the group
Br_2	Liquid	
I_2	Solid	
At_2	Solid	

Figure 2 The physical states of the halogens

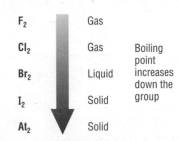

Electronic configuration

The elements in Group 7 have seven electrons in their outer shell. The highest energy electrons are in the p sub-shell and the elements form part of the p-block in the Periodic Table (see Figure 3).

Each Group 7 element has:

- one electron *fewer* than the electronic configuration of a noble gas;
- an outer p sub-shell containing *five* electrons.

F:	...... $2s^22p^5$
Cl:	...... $3s^23p^5$
Br:	.... $4s^24p^5$
I:	...... $5s^25p^5$
At:	.... $6s^26p^5$

Figure 3 Electronic configurations of halogen atoms

Halogens as oxidising agents

The halogens are the most reactive non-metals in the Periodic Table and are strong oxidising agents removing electrons in reactions.

The oxidising power of a halogen is a measure of the strength with which a halogen atom is able to attract and capture an electron to form a halide ion.

In redox reactions each halogen atom gains one electron into a p sub-shell to form a halide ion with a 1– charge.

e.g. $\frac{1}{2}Cl_2(g) + e^- \longrightarrow Cl^-(g)$ (–1 oxidation state)

The halogens become *less* reactive down the group as their oxidising power decreases.

As halogens gain an electron in their reactions, reactivity decreases down the group because:

- the atomic radius increases
- the electron shielding increases
- the ability to gain an electron into the p sub-shell decreases to form a halide ion.

Redox reactions of the halogens

Redox reactions can show that the halogens become less able to form halide ions down the group. We can show the decrease in reactivity using redox reactions of:
aqueous solutions of halide ions: $Cl^-(aq)$, $Br^-(aq)$ and $I^-(aq)$
with aqueous solutions of halogens: $Cl_2(aq)$, $Br_2(aq)$ and $I_2(aq)$.

Each halogen is mixed with aqueous solutions of the different halides. A more-reactive halogen will oxidise and displace a halide of a less-reactive halogen. This is often called a **displacement reaction**.

Halogens form solutions with different colours, so any change in colour will show whether a redox reaction has taken place. The mixture is usually shaken with an organic solvent, such as cyclohexane, to help distinguish between bromine and iodine. Table 1 shows the colours of solutions of Cl_2, Br_2 and I_2 in water and in cyclohexane. The colours in a non-polar solvent are shown in Figure 4.

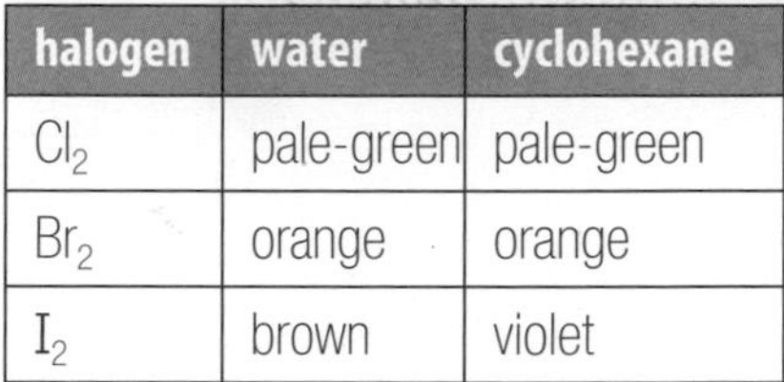

halogen	water	cyclohexane
Cl_2	pale-green	pale-green
Br_2	orange	orange
I_2	brown	violet

Table 1 Colours of halogen solutions in different solvents

Chlorine oxidises both Br^- and I^- ions:

- $Cl_2(aq) + 2Br^-(aq) \longrightarrow 2Cl^-(aq) + Br_2(aq)$ orange in water and in cyclohexane
- $Cl_2(aq) + 2I^-(aq) \longrightarrow 2Cl^-(aq) + I_2(aq)$ brown in water; purple in cyclohexane

The changes in oxidation number for the reaction of chlorine with bromide ions are:

$$Cl_2(aq) + 2Br^-(aq) \longrightarrow 2Cl^-(aq) + Br_2(aq)$$

0 × 2 → 2 × −1 chlorine reduced
2 × −1 → 2 × 0 bromine oxidised

Bromine oxidises I^- ions only:

- $Br_2(aq) + 2I^-(aq) \longrightarrow 2Br^-(aq) + I_2(aq)$ brown in water; purple in cyclohexane

Iodine does *not* oxidise either Cl^- or Br^- ions.

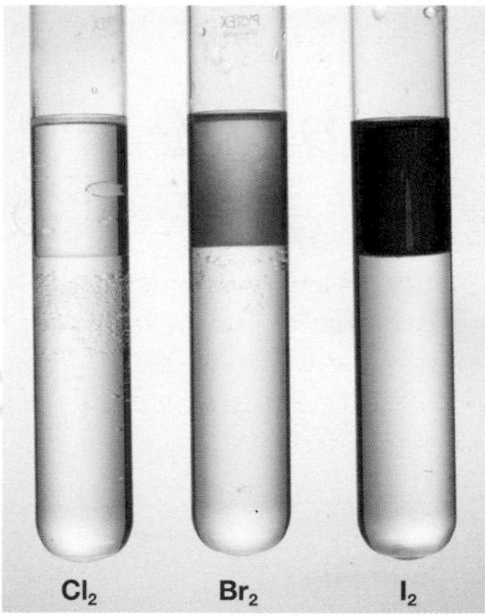

Figure 4 This photograph shows three test tubes containing solutions of halogens.
The lower layer is water and the upper layer is an organic solvent, in which the halogens are far more soluble.
The characteristic colours in the organic solvent make it easy to identify each halogen

Disproportionation

Disproportionation is a reaction in which the same element is both reduced and oxidised.

Disproportionation of chlorine in water

Small amounts of chlorine are added to drinking water to kill bacteria and make the water safer to drink.

Chlorine reacts with water, forming a mixture of two acids, hydrochloric acid, HCl, and chloric(I) acid, HClO.

This is a disproportionation reaction in which chlorine is both reduced and oxidised.

$$Cl_2(aq) + H_2O(l) \longrightarrow HClO(aq) + HCl(aq)$$

0 → −1 chlorine reduced
0 → +1 chlorine oxidised

Disproportionation of chlorine in aqueous sodium hydroxide

Chlorine is only slightly soluble in water and has a mild bleaching action. Household bleach is formed when dilute aqueous sodium hydroxide and chlorine react together at room temperature.

This is also a disproportionation reaction in which chlorine has been both reduced and oxidised.

$$Cl_2(aq) + 2NaOH(aq) \longrightarrow NaCl(aq) + NaClO(aq) + H_2O(l)$$

0 → −1 chlorine reduced
0 → +1 chlorine oxidised

Key definitions

A **displacement reaction** is a reaction in which a more-reactive element displaces a less-reactive element from an aqueous solution of the latter's ions.

Disproportionation is the oxidation and reduction of the same element in a redox reaction.

Questions

1 What are the oxidation number changes in the reaction of bromine with iodide ions?

2 Comment on the changes in the oxidation number of chlorine in the reactions below.
  (a) $Cl_2(aq) + H_2O(l) \longrightarrow HClO(aq) + HCl(aq)$.
  (b) $Cl_2(aq) + 2NaOH(aq) \longrightarrow NaCl(aq) + NaClO(aq) + H_2O(l)$.

Group 7 elements: uses and halide tests

By the end of this spread, you should be able to . . .

* **Interpret and make predictions from the chemical and physical properties of the Group 7 elements/compounds.**
* **Contrast the benefits and risks of chlorine's use as a water treatment.**
* **Describe the precipitation reactions of aqueous anions Cl^-, Br^- and I^- with aqueous silver ions, followed by aqueous ammonia.**
* **Recognise the use of these precipitation reactions as a test for different halide ions.**

Properties of Group 7 elements and compounds

As with Group 2, we can make predictions of chemical and physical properties using Mendeleev's periodic law. We often use the symbol 'X' to represent any of the halogens: F; Cl; Br; I; or At.

For example:

- The Group 7 elements become *less* reactive down the group.
- The Group 7 elements react with metals to form ionic halides with an X^- ion.

The table below summarises some general properties for any Group 7 element, X.

bonding	covalent diatomic molecules, X_2
structure	simple molecular: weak van der Waals' forces between diatomic molecules
redox character	oxidising agent react by gaining electrons to form halide ions: $\frac{1}{2}X_2 + e^- \longrightarrow X^-$ halogen + electron ⟶ halide ion
reactivity	decreases down the group

Table 1 Group 7 properties

Fluorine: the *Tyrannosaurus Rex* of the elements

We have hardly mentioned the first halogen in the group, fluorine. Fluorine is a pale yellow-green gas, rarely seen as the element because it is so reactive. The element fluorine must be the most reactive substance known. Almost anything placed in the path of a stream of fluorine gas will spontaneously burst into flame. It says something that some of the scientists who tried to carry out experiments with fluorine were actually killed by the explosions that followed.

Fluorine reacts with almost everything: glass, steel and even the noble gases krypton, xenon and radon. It is, therefore, extremely difficult to keep fluorine in anything!

Figure 1 This photograph shows crystals of the minerals fluorite, or fluorspar (blue), and quartz (red), fluorescing under ultraviolet radiation.
Fluorite is composed of calcium fluoride, CaF_2, and is sometimes called 'Blue John'

Halides

Ionic halides typically have a halide ion, X^-, with a 1– charge. In contrast to the reactive elements, halide compounds are mainly very stable.

- Sodium chloride, NaCl, is the most familiar halide – common salt.
- Sodium fluoride, NaF, and tin(II) fluoride, SnF_2, are fluoride compounds added to toothpaste, to help prevent tooth decay.
- Crystals of calcium fluoride, CaF_2, also known as fluorite and fluorspar, are used to make lenses to focus infrared light.

Water treatment

Chlorine is a toxic gas. In the UK, small amounts of chlorine have been added to drinking water to kill bacteria since the late 1890s. Some people claim that chlorination of water has done more to improve public health than anything else. Certainly, chlorine is a most effective bacteria killer in our water supplies.

Despite this, the addition of chlorine has not always been welcomed. Some environmentalists are concerned that chlorine reacts with organic matter to form traces of chlorinated hydrocarbons, suspected of causing cancer. In 1991, the government in Peru thought that this was an unacceptable risk and stopped adding chlorine to drinking water. Unfortunately, they had forgotten the very reason for adding chlorine in the first place and an outbreak of cholera followed. This affected a million people and caused 10 000 deaths. The Peruvian government swiftly reversed their decision and resumed adding chlorine to their drinking water.

Testing for halide ions

The presence of halide ions can be detected with a simple test tube test.

- The unknown halide substance is first dissolved in water.
- An aqueous solution of silver nitrate, $AgNO_3(aq)$, is added.
- Silver ions, $Ag^+(aq)$, from the $AgNO_3(aq)$ react with any halide ions, $X^-(aq)$, present, forming a silver halide precipitate, AgX(s).
- The silver halide precipitates are coloured – the colour tells us which halide is present.

Sometimes it is difficult to judge the exact colour.

- If you are unsure, then add aqueous ammonia, $NH_3(aq)$.
- Different halide precipitates have different solubilities in aqueous ammonia – this confirms which halide is present.

Halide test results are:

chloride:	$Ag^+(aq) + Cl^-(aq) \rightarrow AgCl(s)$	white precipitate,	soluble in dilute $NH_3(aq)$
bromide:	$Ag^+(aq) + Br^-(aq) \rightarrow AgBr(s)$	cream precipitate,	soluble in conc $NH_3(aq)$
iodide:	$Ag^+(aq) + I^-(aq) \rightarrow AgI(s)$	yellow precipitate,	insoluble in conc $NH_3(aq)$

This type of reaction is called a **precipitation reaction**. A precipitation reaction takes place in aqueous solution when aqueous ions react together to form a solid precipitate (see Figure 2).

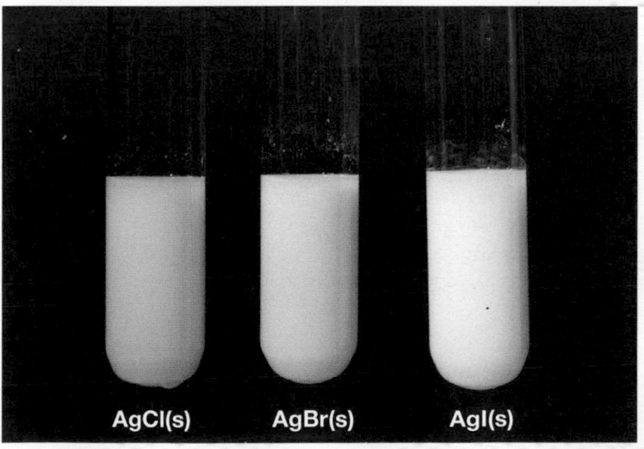

Figure 2 Halide tests with aqueous silver nitrate. Silver halides precipitated by reacting aqueous silver nitrate with aqueous halide solutions

Examiner tip

These equations are easy to write using the ions that react together.

$AgNO_3(aq)$ supplies $Ag^+(aq)$ ions;

The halide supplies $Cl^-(aq)$, $Br^-(aq)$ or $I^-(aq)$ ions.

Then:

$Ag^+(aq) + Cl^-(aq) \longrightarrow AgCl(s)$

You can just ignore the $NO_3^-(aq)$ and metal ions. They don't react anyway.

Key definition

A **precipitation reaction** is the formation of a solid from a solution during a chemical reaction. Precipitates are often formed when two aqueous solutions are mixed together.

Questions

1 Summarise the general properties of a Group 7 element.
2 What is the difference between a halogen and a halide?
3 How could you distinguish between NaCl, NaBr and NaI by using a simple test?

1.3 The Periodic Table summary

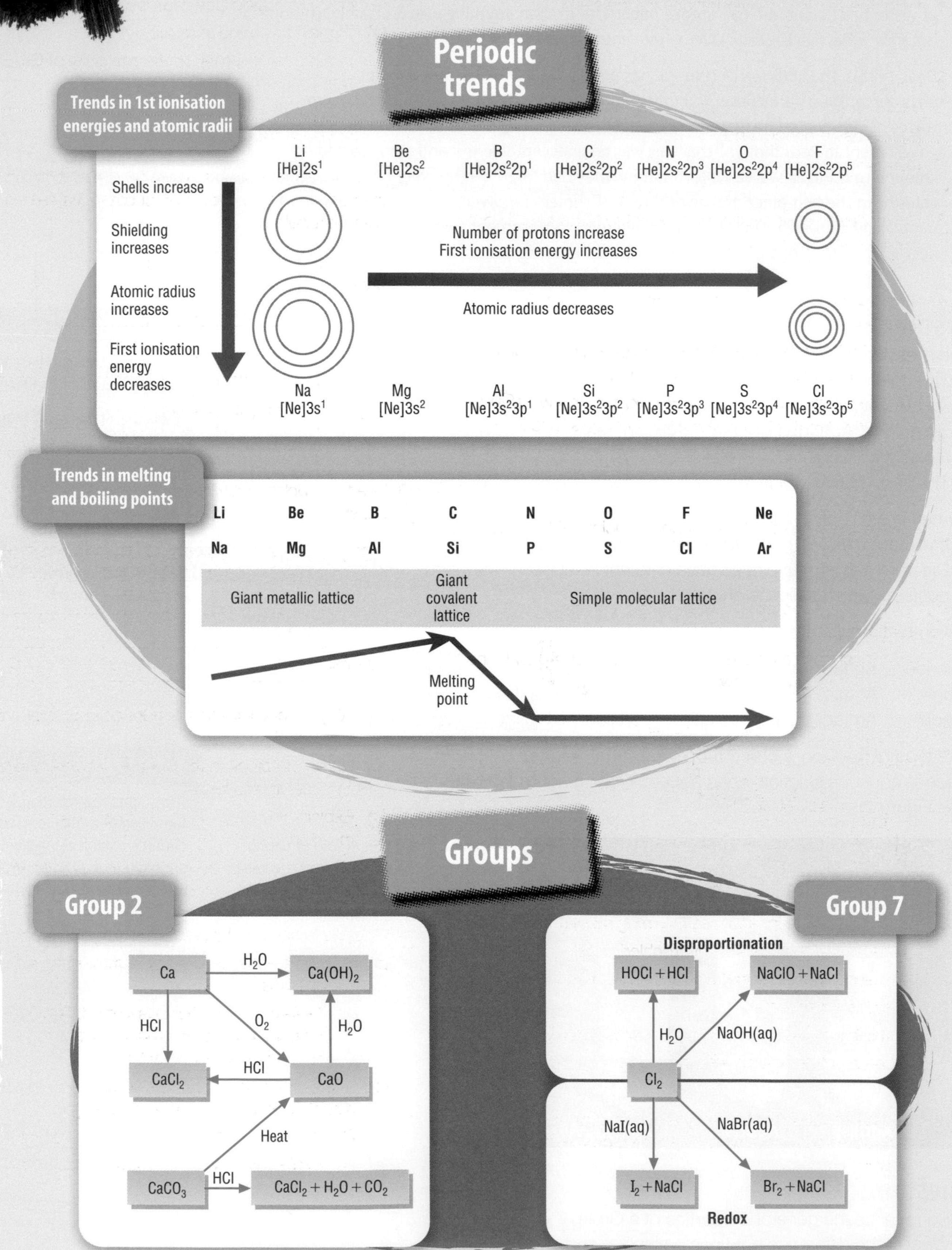

Periodic trends
Trends in 1st ionisation energies and atomic radii
Li [He]2s1
Be [He]2s2
B [He]2s22p1
C [He]2s22p2
N [He]2s22p3
O [He]2s22p4
F [He]2s22p5
Shells increase
Shielding increases
Atomic radius increases
First ionisation energy decreases
Number of protons increase
First ionisation energy increases
Atomic radius decreases
Na [Ne]3s1
Mg [Ne]3s2
Al [Ne]3s23p1
Si [Ne]3s23p2
P [Ne]3s23p3
S [Ne]3s23p4
Cl [Ne]3s23p5
Trends in melting and boiling points
Li Be B C N O F Ne
Na Mg Al Si P S Cl Ar
Giant metallic lattice
Giant covalent lattice
Simple molecular lattice
Melting point
Groups
Group 2
Ca
H2O
Ca(OH)2
HCl
O2
H2O
CaCl2
HCl
CaO
Heat
CaCO3
HCl
CaCl2 + H2O + CO2
Group 7
Disproportionation
HOCl + HCl
NaClO + NaCl
H2O
NaOH(aq)
Cl2
NaI(aq)
NaBr(aq)
I2 + NaCl
Br2 + NaCl
Redox

Practice questions

1. Write the electron configurations for the elements in Period 3, Na → Ar.

2. Using ideas about nuclear charge, attraction and shells explain the trends in ionisation energy and atomic radius:
   (a) across a period; and
   (b) down a group.

3. Using ideas about structure, explain the trend in boiling points across Period 3, Na–Ar.

4. Refer to the flowchart in the summary for Group 2 reactions.
   (a) Write balanced reactions for the seven reactions of calcium and its compounds shown.
   (b) The reactions of calcium with oxygen and water are redox reactions. Use oxidation numbers to identify what has been oxidised and what has been reduced in these reactions.

5. State and explain the trend in boiling points of the halogens fluorine to iodine.

6. Refer to the flowchart in the summary for Group 7 reactions.
   Write balanced reactions for the four reactions of chlorine and its compounds shown.

7. Explain how precipitation reactions can be used to distinguish between different halides.

8. Explain, using oxidation numbers, the displacement reactions of halides with halogens in terms of oxidation and reduction.

9. Write down equations for the following reactions:
   (a) barium hydroxide with hydrochloric acid.
   (b) magnesium carbonate with nitric acid.
   (c) radium with water.
   (d) the thermal decomposition of barium carbonate.
   (e) the reaction of magnesium oxide with sulfuric acid.

10. Write an equation for the likely reaction of bromine with aqueous sodium hydroxide. Comment on the changes in oxidation number.

11. Reactivity *increases* down Group 2.
    (a) Explain this trend in reactivity.
    (b) In contrast to Group 2, the reactivity of Group 7 elements *decreases* down the group.
    Explain why the trends in reactivity for Group 2 and Group 7 are opposite to each other.

12. The reactions of strontium are typical of a Group 2 element. Identify, by formula, substances **A–D** in the flowchart below.

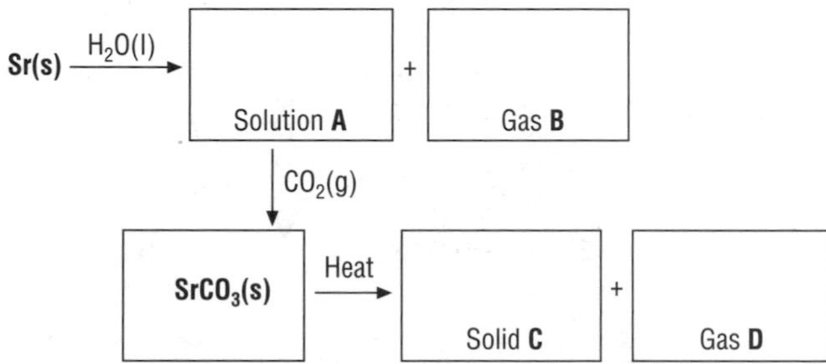

13. The table below shows some properties of barium and caesium.

element	Cs	Ba
group	1	2
atomic number	55	56
atomic radius/nm	0.531	0.435

   (a) Why do caesium and barium have different atomic numbers?
   (b) Name the block in the Periodic Table in which caesium and barium are found.
   (c) Explain why the atomic radius of barium is **less** than the atomic radius of caesium.
   (d) Predict and explain whether a barium **ion** is *larger*, *smaller* or the *same size* as a barium atom.

14. Chlorine is used in the preparation of bleach and iodine.
    (a) (i) Write an equation for the formation of bleach from chlorine.
    (ii) Determine the changes in oxidation number of chlorine during the bleach production and comment on your results.
    (b) Iodine is extracted from seawater by oxidising dissolved iodide ions with chlorine gas.
    (i) Write an ionic equation for this reaction.
    (ii) Explain why chlorine is a stronger oxidising agent than iodine.

1.3 Examination questions

1 The atomic radii of the elements Li to F and Na to Cl are shown in the table below.

element	atomic radius/nm	element	atomic radius/nm
Li	0.134	Na	0.154
Be	0.125	Mg	0.145
B	0.090	Al	0.130
C	0.077	Si	0.118
N	0.075	P	0.110
O	0.073	S	0.102
F	0.071	Cl	0.099

(a) Using **only** the elements in this table, select

(i) an element with **both** metallic and non-metallic properties, [1]

(ii) the element with the largest first ionisation energy, [1]

(iii) an element with a giant covalent structure. [1]

(b) Explain what causes the general **decrease** in atomic radii across each period. [3]

(c) Predict and explain whether a sodium **ion** is *larger* than, *smaller* than or the *same size* as a sodium **atom**. [3]

[Total: 9]

(Jan 01 2811)

2 The diagram below shows the variation in the boiling points of elements across Period 3 of the Periodic Table.

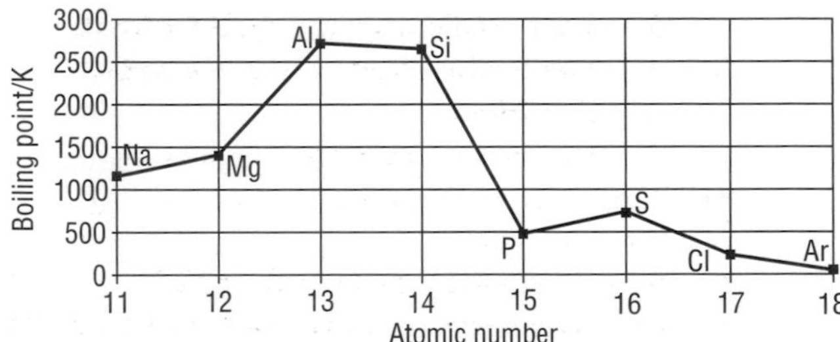

(a) In the table below for the elements Mg, Si and S

- complete the structure column using the word *giant* or *simple*
- complete the bonding column using the word *metallic*, *ionic* or *covalent*.

element	structure	bonding
Mg		
Si		
S		

[3]

(b) Explain why silicon has a much **higher** boiling point than phosphorus. [2]

(c) Explain why the boiling point **increases** from sodium to aluminium. [2]

[Total: 7]

(Jan 07 2811)

3 The table below shows the boiling points of the elements sodium to chlorine in Period 3 of the Periodic Table.

element	Na	Mg	Al	Si	P	S	Cl
boiling point/°C	883	1107	2467	2355	280	445	−35
bonding							
structure							

(a) (i) Complete the *bonding* row of the table using

- **M** for *metallic bonding*,
- **C** for *covalent bonding*. [1]

(ii) Complete the *structure* row of the table using

- **S** for a *simple molecular structure,*
- **G** for a *giant structure*. [1]

(b) State what is meant by *metallic bonding*. You should draw a diagram as part of your answer. [3]

(c) Explain, in terms of their structure and bonding, why the boiling point of

(i) phosphorus is much **lower** than that of silicon, [2]

(ii) aluminium is much **higher** than that of magnesium [2]

[Total: 9]

(Jun 01 2811)

4 The Group 2 element radium, Ra, is used in medicine for the treatment of cancer. Radium was discovered in 1898 by Pierre and Marie Curie by extracting radium chloride from its main ore pitchblende.

(a) Predict the formula of radium chloride. [1]

(b) Pierre and Marie Curie extracted radium from radium chloride by reduction.
Explain what is meant by *reduction*, using this reaction as an example. [2]

(c) Radium reacts vigorously when added to water.
$Ra(s) + 2H_2O(l) \rightarrow Ra(OH)_2(aq) + H_2(g)$

(i) Use the equation to predict **two** observations that you would see during this reaction. [2]

(ii) Predict a pH value for this solution. [1]

(d) Reactions of the Group 2 metals involve removal of electrons. The electrons are removed more easily as the group is descended and this helps to explain the increasing trend in reactivity.

(i) The removal of one electron from each atom in 1 mole of gaseous radium atoms is called the ... [2]
The equation for this process in radium is ... [2]

Answers to examination questions will be found on the Exam Café CD.

(ii) Atoms of radium have a greater nuclear charge than atoms of calcium.
Explain why, despite this, less energy is needed to remove an electron from a radium atom than from a calcium atom. [3]

[Total: 13]

(Jun 05 2811)

5 Aqueous silver nitrate can be used as a test for halide ions. A student decided to carry out this test on a solution of magnesium chloride. The bottle of magnesium chloride that the student used showed the formula $MgCl_2{\cdot}6H_2O$.

(a) The student dissolved a small amount of $MgCl_2{\cdot}6H_2O$ in water and added aqueous silver nitrate to the aqueous solution.

(i) What is the molar mass of $MgCl_2{\cdot}6H_2O$? [1]

(ii) What would the student see after adding the aqueous silver nitrate, $AgNO_3(aq)$? [1]

(iii) Write an ionic equation for this reaction. Include state symbols. [2]

(iv) Using aqueous silver nitrate, it is sometimes difficult to distinguish between chloride, bromide and iodide ions.
How can aqueous ammonia be used to distinguish between these three ions? [3]

(b) Domestic tap water has been chlorinated.
Chlorine reacts with water as shown below.
$Cl_2(g) + H_2O(l) \rightarrow HOCl(aq) + HCl(aq)$

(i) State the oxidation number of chlorine in
Cl_2
HOCl
HCl [3]

(ii) When carrying out halide tests with aqueous silver nitrate, it is important that distilled or deionised water is used for all solutions, rather than tap water.
Suggest why. [1]

[Total: 11]

(Jun 06 2811)

6 Chlorine can be prepared by reacting concentrated hydrochloric acid with manganese(IV) oxide.
$4HCl(aq) + MnO_2(s) \rightarrow Cl_2(g) + MnCl_2(aq) + 2H_2O(l)$

(a) A student reacted 50.0 cm^3 of 12.0 mol dm^{-3} hydrochloric acid with an excess of manganese(IV) oxide.

(i) Calculate how many moles of HCl were reacted. [1]

(ii) Calculate the volume of $Cl_2(g)$ produced, in dm^3.
Under the experimental conditions, one mole of $Cl_2(g)$ occupies 24.0 dm^3. [2]

(b) In this reaction, chlorine is oxidised.
Use oxidation numbers to determine what is reduced. [2]

(c) Sodium reacts with chlorine forming the ionic compound sodium chloride, NaCl.

(i) Write an equation, including state symbols, for this reaction. [2]

(ii) Describe the structure of sodium chloride in the solid state. You may find it useful to draw a diagram. [2]

(d) In this question one mark is available for the quality of spelling, punctuation and grammar.
Chlorine gas was bubbled through an aqueous solution of bromide ions and also through an aqueous solution of iodide ions. An organic solvent was then added and each mixture was shaken.

- State what you would see in each case.
- Write equations for any chemical reactions that take place.
- State and explain the trend in reactivity shown by these observations. [6]

Quality of Written Communication [1]

[Total: 16]

(Jan 07 2811)

7 A student carried out three experiments using chlorine gas, $Cl_2(g)$.

(a) In a first experiment, the student bubbled chlorine through an aqueous solution of potassium bromide, KBr(aq). A reaction took place.

(i) What colour is the solution after the reaction has taken place? [1]

(ii) Write an equation for this reaction. [2]

(iii) This reaction takes place because chlorine has a stronger oxidising power than bromine.
Explain why chlorine has a stronger oxidising power than bromine. [3]

(b) In a second experiment, the student bubbled chlorine through 120 cm^3 of an aqueous solution of 0.275 mol dm^{-3} sodium hydroxide, NaOH(aq).
The equation for this reaction is shown below.
$Cl_2(g) + 2NaOH(aq) \rightarrow$
$NaCl(aq) + NaClO(aq) + H_2O(l)$
Under the reaction conditions, 1 mole of $Cl_2(g)$ occupies 24.0 dm^3.

(i) What is meant by the term *the mole*? [1]

(ii) How many moles of NaOH were in the 120 cm^3 volume of NaOH(aq)? [1]

(iii) Calculate the volume of $Cl_2(g)$ that was needed to react with the NaOH(aq) used. [2]

(iv) What is a common use for the solution that the student prepared? [1]

(c) In a third experiment, the student repeated the procedure in **(b)** but with hot concentrated sodium hydroxide. A different reaction took place in which sodium chlorate(V) was formed instead of NaClO.
Suggest the formula of sodium chlorate(V). [1]

[Total: 12]

(Jan 06 2811)

UNIT 2

Module 1 Basic concepts and hydrocarbons

Introduction

The chemistry of carbon and its compounds is a huge subject, so much so that an entire field of study – known as *organic chemistry* – is devoted to it. Organic chemistry supports a highly creative and profitable industry that is central to the growth of the world's richest economies.

Modern organic chemistry was boosted by the discovery of crude oil and its subsequent conversion into fuels. Synthetic chemists have developed techniques and processes to create new compounds from oil, many of which have improved the quality of life.

In this module, you will learn the language of the organic chemist – how compounds are named and how reactions are carried out.

This module also details the chemistry of the hydrocarbons, alkanes and alkenes. You will look at their reactions, their value to modern society, and the environmental impact of their use. You will discover how scientists hope to solve some of the problems faced by our generation.

Test yourself

1 Which hydrocarbon is used in natural gas for cooking?
2 Name the alkane with four carbon atoms.
3 What does 'unsaturated' mean?
4 Which is healthier, a saturated fat or an unsaturated fat?
5 What is meant by 'cracking'?
6 Name three man-made polymers.

Module contents

Introduction to organic chemistry

By the end of this spread, you should be able to . . .

- ✱ **Use the terms homologous series and functional group.**
- ✱ **Understand the terms hydrocarbon, saturated, unsaturated.**
- ✱ **State that alkanes and cycloalkanes are saturated hydrocarbons.**
- ✱ **Explain the tetrahedral shape around each carbon atom present in an alkane.**
- ✱ **State the names of the first ten members of the alkane homologous series.**

Figure 1 Organic materials are used in the manufacture of sugar, alcohol, petrol, cosmetics, pharmaceuticals and plastics

What is organic chemistry?

Organic chemistry affects all areas of our lives – our clothes, food, cosmetics, as well as many medicines. The common factor for each of these products is the element *carbon*, on which these chemical compounds are based.

All organic compounds contain carbon. The simplest organic compounds, the hydrocarbons, contain carbon and hydrogen only. Almost all our useable supplies of hydrocarbons are obtained from fossil fuels such as coal, oil and natural gas.

The chemistry of carbon is vast; there are in excess of ten million known carbon compounds. It is estimated that 300 000 new carbon compounds are discovered each year.

Carbon can form so many compounds because:

- A carbon atom can form bonds with other carbon atoms to make chains and rings.
- A carbon atom can form a single, double or triple bond to another carbon atom.
- A carbon atom can bond with atoms of other elements, such as oxygen, hydrogen, nitrogen, phosphorus and the halogens.

Bonding in organic compounds

- Carbon is in Group 4 of the Periodic Table.
- Carbon has *four* electrons in its outer shell. These pair up with electrons from other atoms to form *four* covalent bonds.

Looking at Figure 2, you can see that for the alkane methane (CH_4), the carbon atom has four single bonds to four hydrogen atoms.

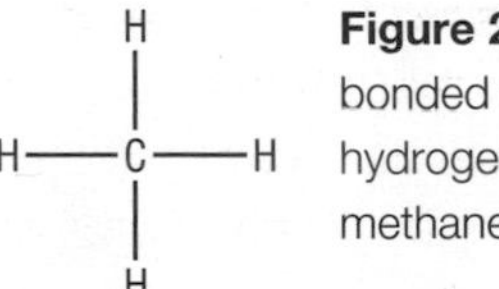

Figure 2 Carbon is bonded to four hydrogen atoms in methane

In Figure 3, you can see that for the alkene ethene (C_2H_4), the two carbon atoms are joined by a double bond. This counts as two bonds, so that each carbon again obeys the rule of having four bonds in total.

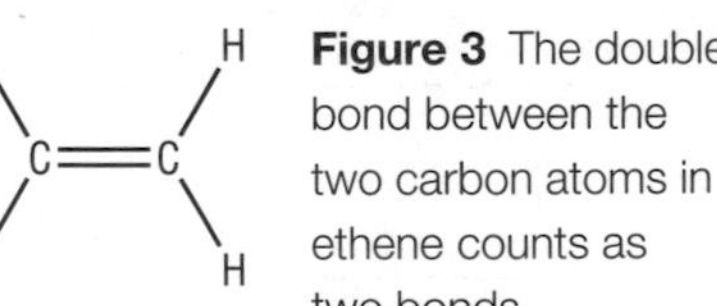

Figure 3 The double bond between the two carbon atoms in ethene counts as two bonds

Saturated and unsaturated hydrocarbons

A **hydrocarbon** is a compound of carbon and hydrogen only.

- Methane, CH_4, and ethane, C_2H_6, are both hydrocarbons.

A **saturated** hydrocarbon has single bonds only.

- Propane, C_3H_8, is a saturated hydrocarbon.

An **unsaturated** hydrocarbon contains carbon-to-carbon multiple bonds.

- Ethene, C_2H_4, is an unsaturated hydrocarbon.

Aliphatic and alicyclic hydrocarbons

Hydrocarbons can be described as:

- **Aliphatic hydrocarbons** – in which the carbon atoms are joined together in straight (unbranched) or branched chains.
- **Alicyclic hydrocarbons** – where the carbon atoms are joined together in a ring structure.

Key definitions

Hydrocarbons are organic compounds that contain carbon and hydrogen only.

A **saturated hydrocarbon** is a hydrocarbon with single bonds only.

An **unsaturated hydrocarbon** is a hydrocarbon containing carbon-to-carbon multiple bonds.

An **aliphatic hydrocarbon** is a hydrocarbon with carbon atoms joined together in straight or branched chains.

An **alicyclic hydrocarbon** is a hydrocarbon with carbon atoms joined together in a ring structure.

Functional groups

A **functional group** is the part of the organic molecule responsible for its chemical properties. The functional group is attached to some point on the carbon backbone of the organic molecule.

A saturated carbon chain has little chemical reactivity. The chemistry of a carbon compound is determined by the functional group.

- Molecules with the same functional group react in a similar way.

Figure 4 The –OH group is the functional group, responsible for the reactivity of the molecule

Homologous series

A **homologous series** is a family of compounds with the following properties:

- They contain the same *functional group*.
- They have similar chemical properties.
- Each successive member of a homologous series differs by one carbon and two hydrogen atoms: a $–CH_2–$ group.

Alkanes

Alkanes are saturated straight-chained hydrocarbons with single C–C bonds only.

- For alkanes each carbon atom is bonded to four other atoms. Each carbon has a tetrahedral shape with a bond angle of 109.5°. The structures for the alkanes methane, ethane, propane and butane are shown in Figure 5.

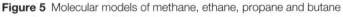

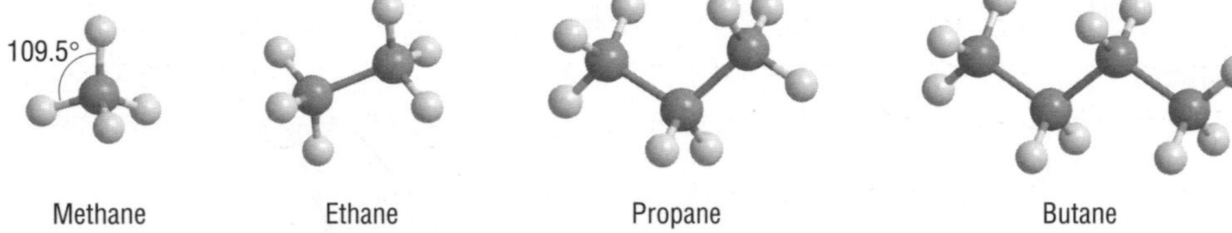

Figure 5 Molecular models of methane, ethane, propane and butane

Table 1 lists the first 10 members of the *alkanes* homologous series. You will need to learn these names. Note that the number of hydrogen atoms is equal to twice the number of carbon atoms *plus* two.

Number of carbons	Name	Formula
1	methane	CH_4
2	ethane	C_2H_6
3	propane	C_3H_8
4	butane	C_4H_{10}
5	pentane	C_5H_{12}
6	hexane	C_6H_{14}
7	heptane	C_7H_{16}
8	octane	C_8H_{18}
9	nonane	C_9H_{20}
10	decane	$C_{10}H_{22}$

Table 1 Homologous series name: alkanes

Examples of homologous series

Eth*ene* is a member of a homologous series known as the alk*enes*.

Butan-1-ol is a member of the homologous series known as the alcoh*ols*.

Key definitions

A **functional group** is the part of the organic molecule responsible for its chemical reactions.

A **homologous series** is a series of organic compounds with the same functional group but with each successive member differing by CH_2.

Alkanes are the homologous series with the general formula: C_nH_{2n+2}

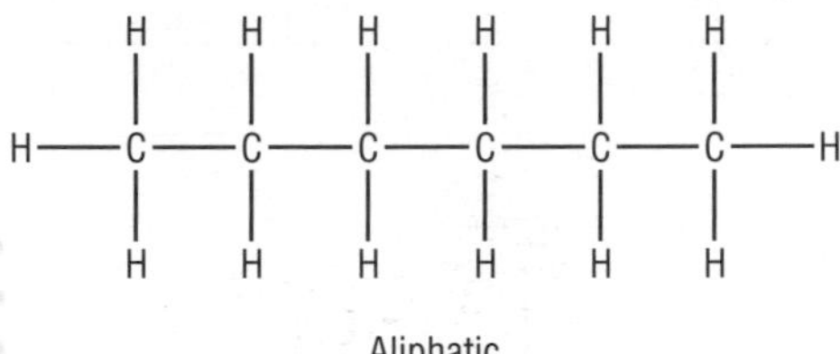

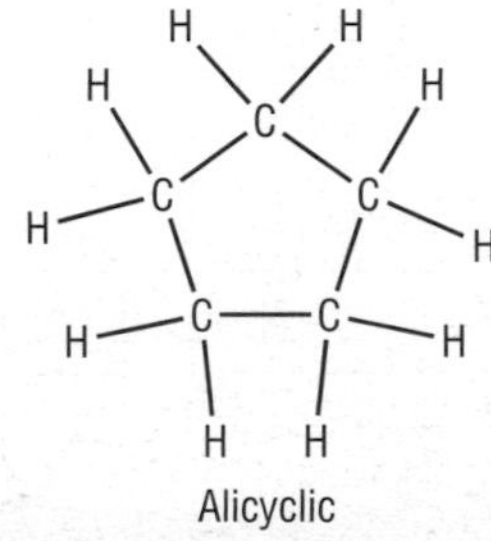

Figure 6 Aliphatic and alicyclic alkanes

Examiner tip

When studying the reactions of a homologous series, remember that the *functional group* reacts. The length of the carbon chain is relatively unimportant.

During your AS level course, you will meet various functional groups, including: single C–C bonds (alkanes); double C=C bonds (alkenes); carbon atoms joined to halogens (halogenoalkanes); and carbon atoms joined to –OH groups (alcohols).

Questions

1 What group of the Periodic Table do you find carbon in?

2 How many bonds does carbon form?

3 Why does carbon form so many compounds?

2.1 (2) Naming hydrocarbons

By the end of this spread, you should be able to . . .

∗ Understand and be able to use the IUPAC rules of nomenclature for naming hydrocarbons.

Key definition

Nomenclature is a system of naming compounds.

Number of carbons	Stem	Alkyl group
1	meth-	methyl
2	eth-	ethyl
3	prop-	propyl
4	but-	butyl
5	pent-	pentyl
6	hex-	hexyl
7	hept-	heptyl
8	oct-	octyl
9	non-	nonyl
10	dec-	decyl

Table 1 Alkyl groups

Key definition

An **alkyl group** is an alkane with a hydrogen atom removed, e.g. CH_3, C_2H_5; any alkyl group is often shown as 'R'.

Examiner tip

When naming compounds, you use the *smallest* numbers possible. So here the compound is 2-methylpentane, rather than 4-methylpentane.

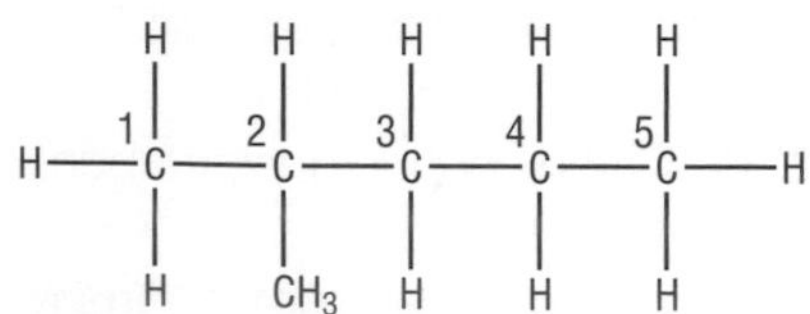

Figure 2 2-Methylpentane

Naming organic compounds

With over ten million organic compounds, it is important that each has its own unique name. Organic **nomenclature** (the system used for naming compounds) was developed by the International Union of Pure and Applied Chemistry and is known as the IUPAC system.

You must be able to:

- name compounds from given structures
- draw molecules from their names.

Organic molecules are typically made up of:

- a carbon chain;
- side chains or alkyl groups;
- one or more functional groups.

Stem, prefix and suffix

The *stem* indicates the number of carbon atoms in the longest carbon chain present in the compound.

A *prefix* is added *before* the stem, as part of the name.

A *suffix* is added *after* the stem, as part of the name.

Alkyl groups

As discussed in spread 2.1.1, methane (CH_4) and ethane (C_2H_6) are members of the homologous series, the *alkanes*.

If you remove a hydrogen atom from an alkane, you will get an **alkyl group**. Stems and alkyl groups for up to ten carbon atoms are shown in Table 1.

Naming alkanes

When naming a molecule from its structure, you just need to follow some simple steps.

Alkanes are the simplest compounds to name. *All* alkanes end their names with the suffix –ane.

Example 1

Identify the *longest unbranched* chain of carbon atoms in the structure. This is called the parent chain.

Name the *stem* of the parent chain using Table 1.

- There are four carbons in the longest unbranched chain, so the stem is *but-*.
- There are no side chains.
- The compound is an alkane, so the suffix is *-ane*.
- The name is *butane*.

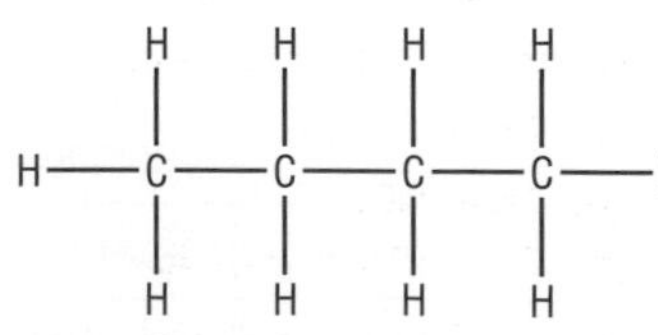

Figure 1 Butane

Example 2

Identify any alkyl groups as side chains. These are added to the stem as a *prefix*. Add a number before each alkyl group to show its position on the parent chain.

- There are five carbons in the longest chain, so the stem is *pent-*.
- There is one alkyl side chain, a *methyl-* group.
- The methyl group is on carbon *2-*.
- The compound is an alkane, so the suffix is *–ane*.
- The name is *2-methylpentane*.

Example 3

If there is more than one alkyl group on the main chain, the groups are named *alphabetically*.

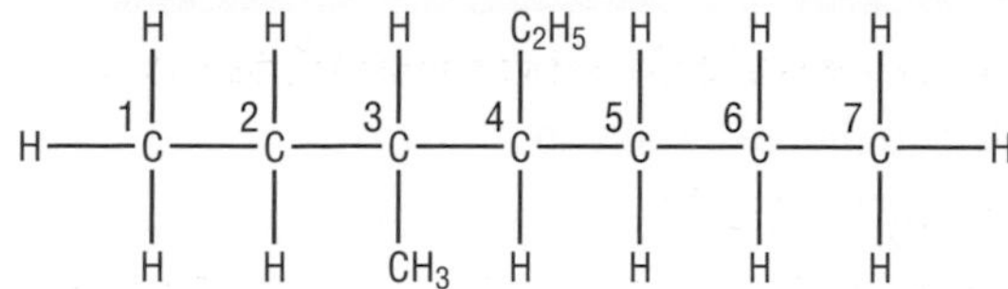

Figure 3 4-Ethyl-3-methylheptane

- There are seven carbons in the longest chain, so the stem is *hept-*.
- There are two alkyl side chains:
  a *methyl-* group on carbon *3-*;
  an *ethyl-* group on carbon *4-*.
- The alkyl groups are listed alphabetically.
- The compound is an alkane, so the suffix is *-ane*.
- The name is *4-ethyl-3-methylheptane*.

Examiner tip

For the molecule in Figure 4, the two alkyl groups are the same and on the same carbon. The name is 2,2-dimethylbutane.

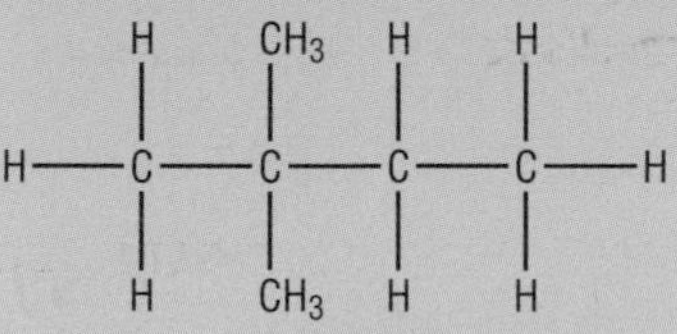

Figure 4 2,2-dimethylbutane

Naming alkenes

An alkene is an unsaturated hydrocarbon with at least one carbon-to-carbon double bond, C=C.

You name alkenes by following similar steps used to name the alkanes.

The main differences are:

- The suffix is *–ene*.
- The position of the double bond has to be stated. This applies to alkenes with *four* or *more* carbons in the longest chain. You only need to use the smaller number. If the double bond starts at carbon-1, it must go to carbon-2.

Examiner tip

When naming compounds, separate numbers by *commas*. Place *hyphens* between numbers and words.

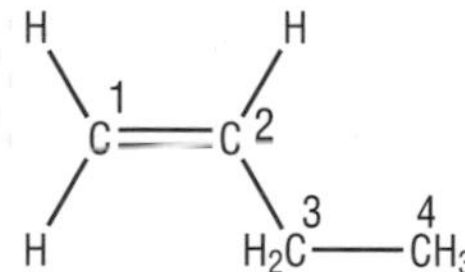

Figure 5 But-1-ene

- There are four carbons in the longest chain, so the stem is *but-*.
- There is a double bond between carbons *-1-* and 2.
- The compound is an alkene, so the suffix is *-ene.*
- The name is *but-1-ene*.

Questions

1 Draw out all the atoms and bonds to display the structures for the following molecules:

(a) pent-2-ene; (b) 2,3-dimethylbutane;
(c) hexane; (d) 2,3,4-trimethylhexane;
(e) 2-methylbut-2-ene.

2 Name each of the following hydrocarbons.

(a)

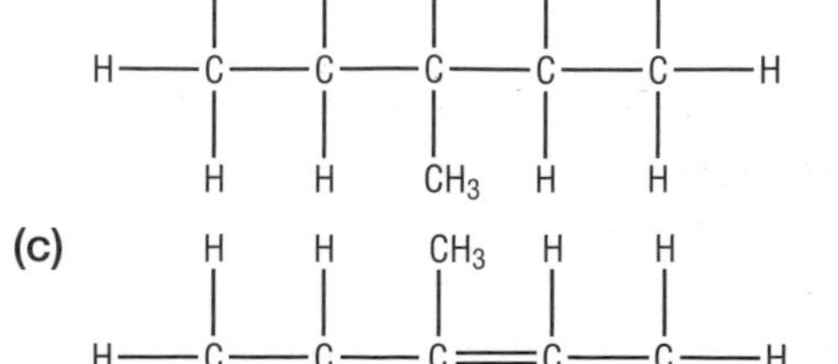

(b)

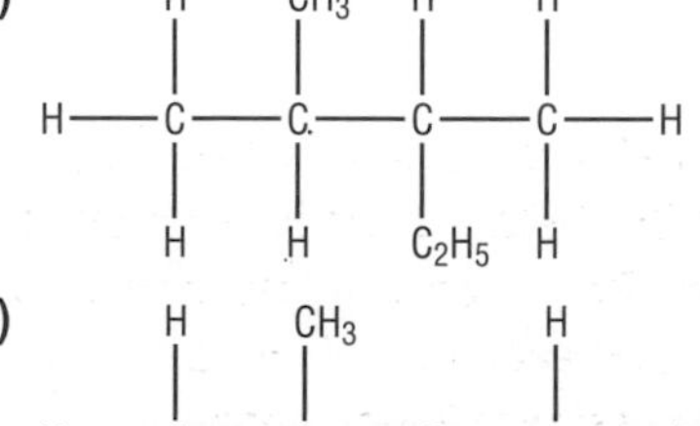

(c)

(d)

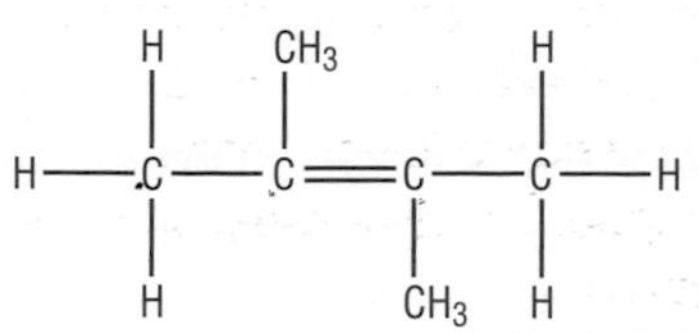

2.1 (3) Naming compounds with functional groups

By the end of this spread, you should be able to . . .

* **Understand and use the IUPAC rules of nomenclature for naming compounds with functional groups.**

Naming molecules with functional groups

You name molecules containing functional groups using similar steps to those when naming branched alkanes (see spread 2.1.2).

- Identify the parent chain – remember that this is the *longest unbranched* chain of carbon atoms in the structure.
- Name the *stem* of the parent chain.
- Identify any alkyl chains or functional groups – remember that alkyl groups are placed *before* the stem name.
- Identify the name of the *prefix* or *suffix* for the functional group – using Table 1.
- Number any alkyl groups and functional groups to show their *positions* on the parent chain.

Type of Compound	Formula	Prefix	Suffix
Alkane	C–C		-ane
Alkene	C=C		-ene
Halogenoalkane	–F –Cl –Br –I	fluoro- chloro- bromo- iodo-	
Alcohol	–OH	hydroxy-	-ol
Aldehyde	–C(=O)H –CHO		-al
Ketone	C–C(=O)–C C–CO–C		-one
Carboxylic acid	–C(=O)OH –COOH		-oic acid

Table 1 Functional groups

If the suffix starts with a *vowel*, you remove the final letter –e from the alkane part of the name: for example, methaneal becomes methanal. This is because it is awkward to pronounce two vowels together to give a chemical meaning, in this case the sound –al for an aldehyde group.

If a suffix starts with a *consonant*, the –e is kept, since the name can be pronounced easily, for example, ethanedioic acid, for a compound containing two carboxylic acid groups.

Example 1

- There are three carbons in the longest chain, so the stem is *prop-*.
- There is –Cl functional group, so the prefix is *chloro-*.
- The chloro group is on carbon *1-*, so the prefix is *1-*.
- There is no suffix on the alkane chain, *–ane*.
- The name is *1-chloropropane*.

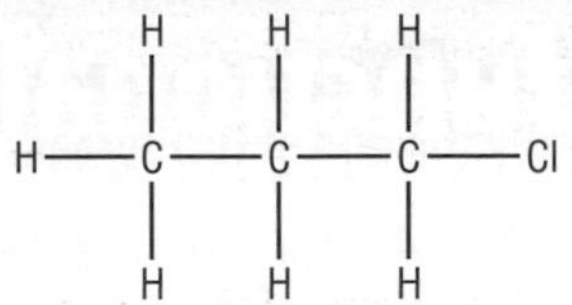

Figure 1 1-Chloropropane

Example 2

- There are five carbons in the longest chain, so the stem is *pent-*.
- There is one –OH functional group, so the suffix is *-ol*.
- The –OH group is on carbon *3-*, so the suffix is *-3-*.
- The suffix begins with a vowel, so the alkane chain ending is shortened to *-an-*.
- The name is *pentan-3-ol*.

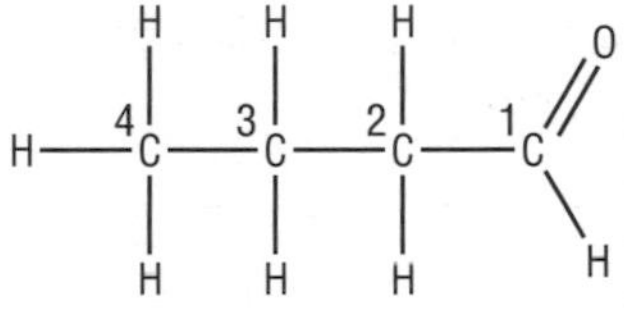

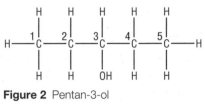

Figure 2 Pentan-3-ol

Example 3

- There are four carbons in the longest chain, so the stem is *but-*.
- There is one –CHO functional group, so the suffix is *-al*.
- There is a suffix beginning with a vowel, so the alkane chain ending is shortened to *–an-*.
- The name is *butanal*.

Figure 3 Butanal

Example 4

- There are four carbons in the longest chain, so the stem is *but-*.
- There is one –Cl functional group, so the prefix is *chloro*. The –Cl group is on carbon *1-*, so the prefix is *1-*.
- There is one –Br functional group, so the prefix is *bromo*. The –Br group is on carbon *2-*, so the prefix is *2-*.
- There is no suffix on the alkane chain, *–ane*.
- The name is *2-bromo-1-chlorobutane* (alphabetical order).

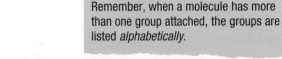

Figure 4 2-Bromo-1-chlorobutane

Example 5

- There are three carbons in the longest chain, so the stem is *prop-*.
- There are two groups attached to the parent chain: one methyl group on carbon-2, giving a prefix: *2-methyl*; one –OH group on carbon-1, giving a suffix: *-1-ol*.
- The suffix begins with a vowel, so the alkane chain ending is shortened to *-an-*.
- The name is *2-methylpropan-1-ol*.

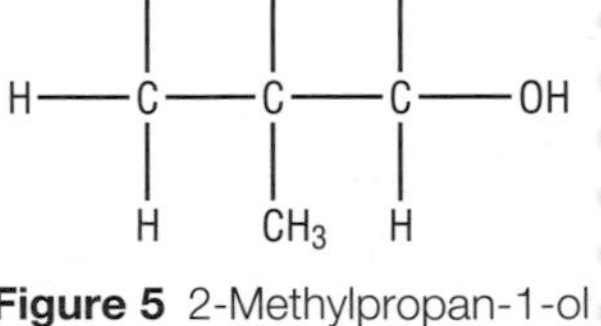

Figure 5 2-Methylpropan-1-ol

Example 6

- There are two carbons in the longest chain, so the stem is *eth-*.
- There are four –Cl functional groups attached to the parent chains, so the name *tetra* is added to indicate four –Cl groups giving the prefix, *tetrachloro-*.
- There is no suffix on the alkane chain, *–ane*.
- The name is *1,1,2,2-tetrachloroethane*.

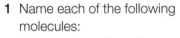

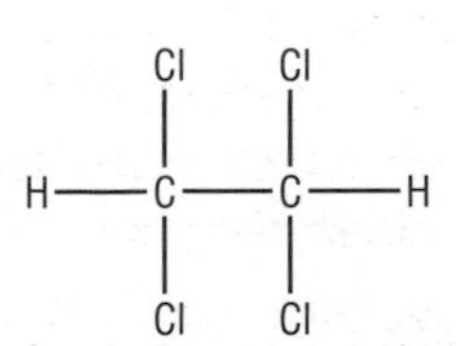

Figure 6 1,1,2,2-Tetrachloroethane

Examiner tip

When naming aldehydes, you don't need to add a number before the suffix –al.

The aldehyde carbon is *always* on carbon-1.

Examiner tip

Remember, when a molecule has more than one group attached, the groups are listed *alphabetically*.

Examiner tip

In names:

- *di-* is added to indicate two identical groups;
- *tri-* is added to indicate three identical groups; and
- *tetra-* is added to indicate four identical groups.

Questions

1 Name each of the following molecules:

(a)

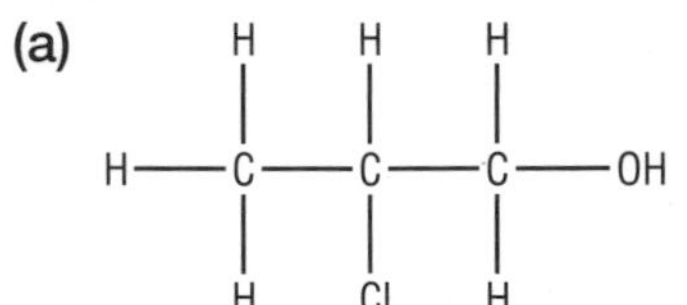

(b)

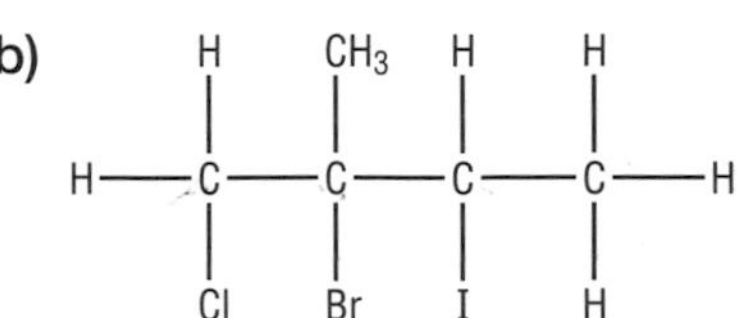

(c)

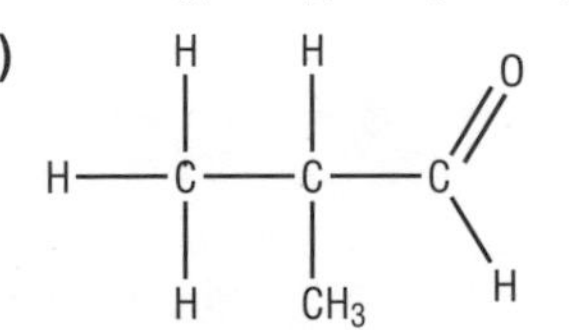

(d)

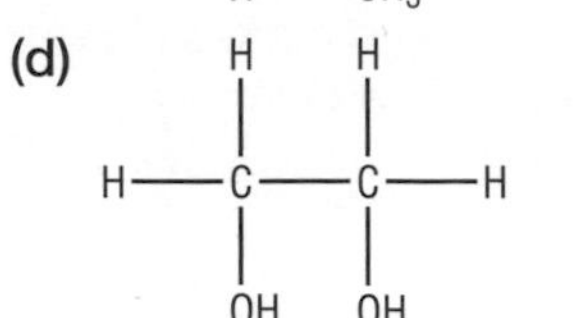

Formulae of organic compounds

By the end of this spread, you should be able to . . .

* **Understand and use the following terms: empirical formula; molecular formula; general formula; displayed formula.**

Empirical formula

An *empirical formula* is the simplest whole-number ratio of atoms of each element present in a compound. (See also spread 1.1.5.)

Examples

- Ethanoic acid (CH_3COOH) has two carbon, four hydrogen and two oxygen atoms. The empirical formula for ethanoic acid is CH_2O as the simplest whole-number ratio.
- Benzene (C_6H_6) and ethyne (C_2H_2) have the same empirical formula, which is CH.

Remember

$$amount, n = \frac{mass, m}{molar\ mass, M}$$

Examiner tip

Make sure that you divide by the smallest number of moles.

Worked example

A hydrocarbon was found to contain 85.63% carbon and 14.37% hydrogen. Calculate its empirical formula.

Answer

100.0 g of the compound contains 85.63 g of C and 14.37 g of H.

Find the molar ratio of atoms: C : H

$$\frac{85.63}{12.0} : \frac{14.37}{1.0}$$

7.14 : 14.37

Divide by smallest number (7.14): 1 : 2

Empirical formula = CH_2

Molecular formula

A *molecular formula* is the actual number of atoms of each element in a molecule. (See also spread 1.1.5.)

Example

The molecular formula of ethene is C_2H_4.

- In a molecule of ethene, there are two atoms of carbon and four atoms of hydrogen.

Worked example

A compound has an empirical formula of CH_2O and a relative molecular mass of 60.0.

Find its molecular formula.

Answer

Find the relative empirical mass of the empirical formula.

$CH_2O = 12.0 + 1.0 + 1.0 + 16.0 = 30.0$

Then, divide the relative molecular mass by the relative empirical mass:

$$\frac{\text{relative molecular mass}}{\text{relative empirical mass}} = \frac{60.0}{30.0} = 2$$

The molecular formula is twice the empirical formula (CH_2O).

This gives the molecular formula as $C_2H_4O_2$.

Examiner tip

Even if the answer seems obvious to you, it is important to show all of your working. You will get marks for it, even if you make a mistake in your calculations.

General formula

A **general formula** is the simplest algebraic formula for a member of a homologous series.

Examples

Alkanes: C_nH_{2n+2}.

n	Molecular formula	Name
1	CH_4	Methane
2	C_2H_6	Ethane
3	C_3H_8	Propane
4	C_4H_{10}	Butane

Table 1 The alkanes

Alkenes: C_nH_{2n}.

n	Molecular formula	Name
1	–	–
2	C_2H_4	Ethene
3	C_3H_6	Propene
4	C_4H_8	Butene

Table 2 The alkenes

Alcohols: $C_nH_{2n+1}OH$.

n	Molecular formula	Name
1	CH_3OH	Methanol
2	C_2H_5OH	Ethanol
3	C_3H_7OH	Propanol
4	C_4H_9OH	Butanol

Table 3 The alcohols

Displayed formula

A **displayed formula** shows the relative positioning of all the atoms in a molecule, and the bonds between them.

Examples

Butane has the molecular formula C_4H_{10}. Its displayed formula is shown in Figure 1.

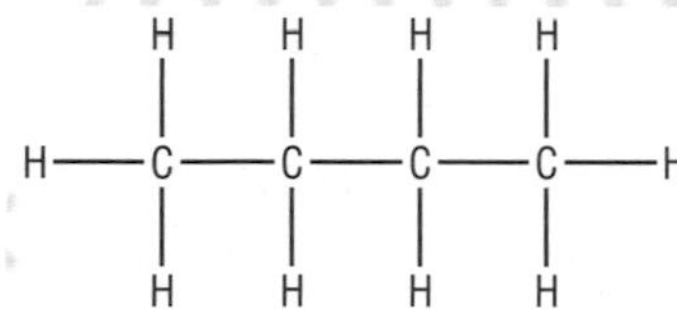

Figure 1 Butane

Propan-2-ol has the molecular formula C_3H_8O. Its displayed formula is shown in Figure 2.

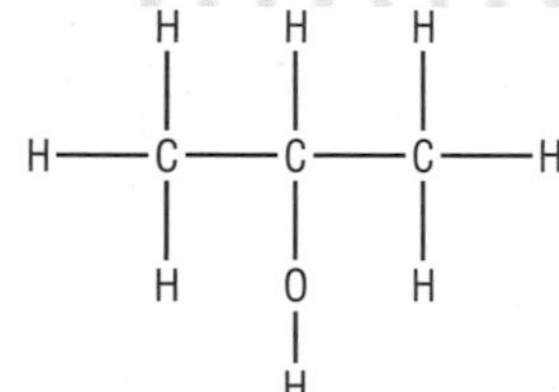

Figure 2 Propan-2-ol

Key definitions

A **general formula** is the simplest algebraic formula of a member of a homologous series. For example, the general formula of the alkanes is C_nH_{2n+2}.

A **displayed formula** shows the relative positioning of all the atoms in a molecule and the bonds between them.

Examiner tip

When drawing the full displayed formula of a compound, you must remember to draw *all* atoms and *all* bonds, even those between the oxygen and hydrogen in an alcohol group.

Questions

1 A molecule contains 12.79% C, 2.15% H and 85.06% Br. The molecule has a relative molecular mass of 187.9.

  (a) Find the empirical and molecular formula.

  (b) Draw a possible displayed formula for the molecule.

2 What is the molecular formula for $n = 5$ to $n = 8$ for

  (a) the alkenes?

  (b) the alcohols?

3 Allicin is a powerful medicinal compound formed from garlic. Allicin has the following composition by mass: C, 40.66%; H, 6.26%; N, 7.90%; O, 27.08%; S, 18.10%. Allicin has a relative molecular mass of 177.1.

  (a) Determine the empirical formula of allicin.

  (b) Show that the molecular formula of allicin is the same as its empirical formula.

2.1 5 Structural and skeletal formula

By the end of this spread, you should be able to . . .

* Understand and use the term structural formula.
* Understand and use the term skeletal formula.

Structural formula

Key definition

A **structural formula** shows the minimal detail for the arrangement of atoms in a molecule.

A **structural formula** shows the minimal detail for the arrangement of atoms in a molecule.

- This is a shorthand form of writing the displayed formula (see spread 2.1.4), with no bonds between atoms.

Example 1

The displayed and structural formulae for propane and 1-bromopropane are shown in Figure 1.

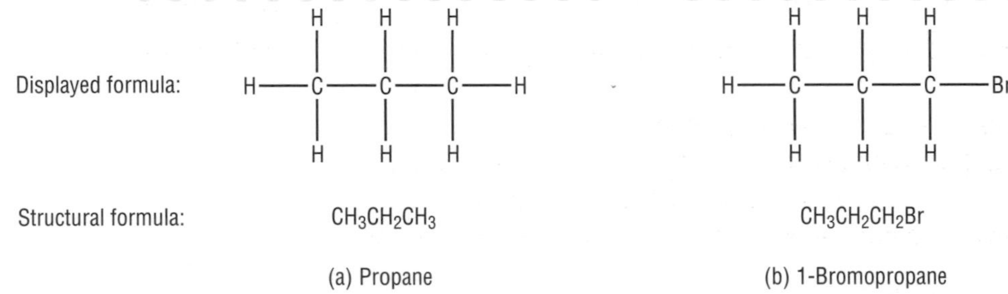

Figure 1 Displayed and structural formulae for (a) propane and (b) 1-bromopropane

Example 2

The displayed and structural formulae for decanoic acid, $C_{10}H_{20}O_2$, are shown in Figure 2.

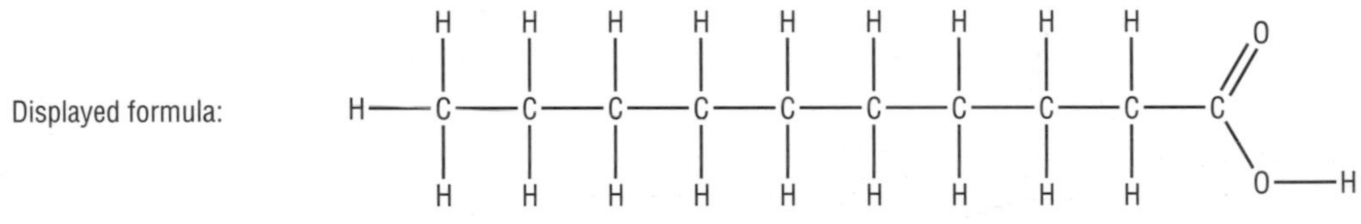

Structural formula: $CH_3CH_2CH_2CH_2CH_2CH_2CH_2CH_2CH_2COOH$

Figure 2 Displayed and structural formulae for decanoic acid

The structural formula can be further simplified by collecting all the CH_2 groups together. The formula can then be represented using brackets.

- The structural formula for decanoic acid then becomes: $CH_3(CH_2)_8COOH$.

Skeletal formula

Key definition

A **skeletal formula** is a simplified organic formula, with hydrogen atoms removed from alkyl chains, leaving just a carbon skeleton and associated functional groups.

A **skeletal formula** is a simplified organic formula, with the hydrogen atoms removed from the alkyl chains. This leaves just a carbon skeleton with associated functional groups. The displayed and skeletal formulae for hexane are shown in Figure 3.

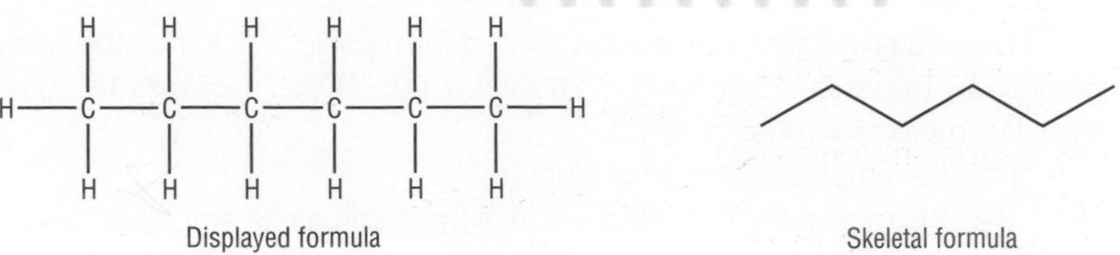

Figure 3 Displayed and skeletal formulae of hexane

Saturated hydrocarbons

The displayed, structural and skeletal formulae for pentane and 3-methylpentane are shown in Figure 4.

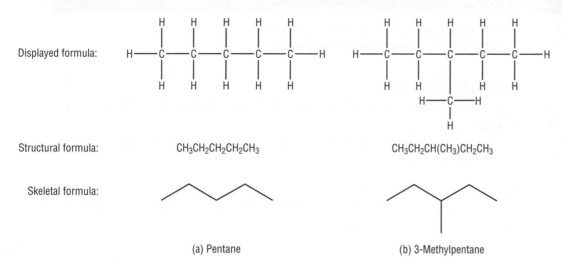

	(a) Pentane	(b) 3-Methylpentane
Displayed formula:		
Structural formula:	$CH_3CH_2CH_2CH_2CH_3$	$CH_3CH_2CH(CH_3)CH_2CH_3$
Skeletal formula:		

Figure 4 Displayed, structural and skeletal formulae for (a) pentane and (b) 3-methylpentane

For a skeletal formula of a hydrocarbon:

- note that *no* carbon or hydrogen atoms are shown.
- there is a carbon atom at *each end* of the chain.
- there is a carbon atom at each point where two lines *meet*.

Cyclic compounds

When you need to draw a cycloalkane or cycloalkene, you usually represent the compound as a *skeletal* formula rather than as a displayed one. The displayed and skeletal formulae for cyclopentane and cyclohexene are shown in Figure 5.

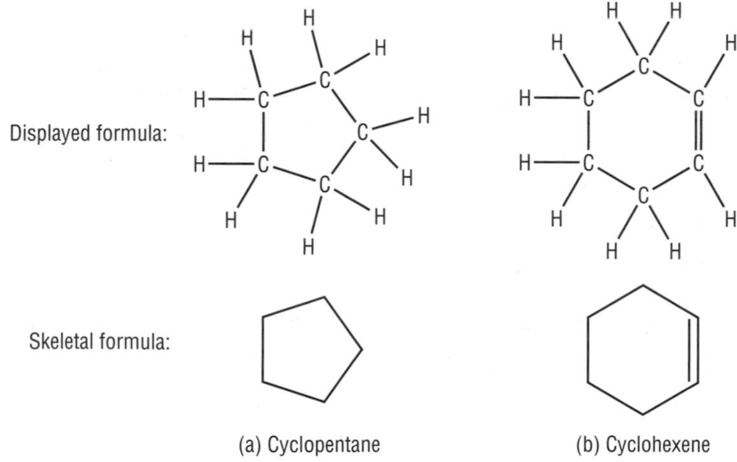

Figure 5 Displayed and skeletal formulae for (a) cyclopentane and (b) cyclohexene

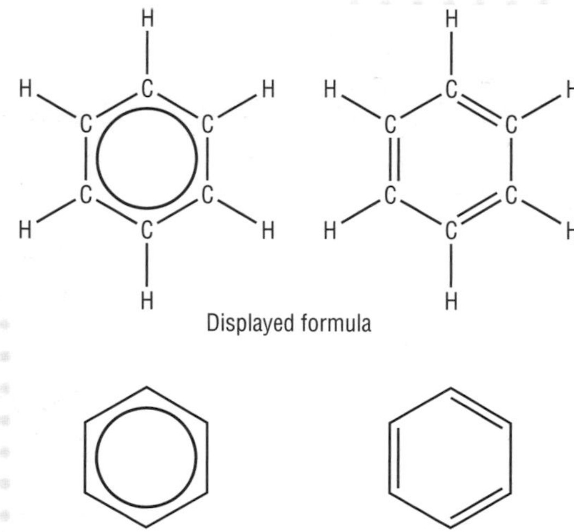

Figure 6 Aromatic hydrocarbons. Benzene, C_6H_6, is an aromatic hydrocarbon. The structure of benzene is shown in different ways, and some of these are shown above. You will learn about aromatic compounds in detail during A2 Chemistry.

Questions

1 Draw the skeletal formula for:
(a) hexane; **(b)** 4-methylnonane; **(c)** cyclobutane.

2 Show the structural formula for:
(a) octane; **(b)** 2,3-dimethylhexane.

3 Draw the skeletal formula for:
(a) methylcyclohexane; **(b)** cyclopentene; **(c)** methylbenzene.

Skeletal formulae and functional groups

By the end of this spread, you should be able to . . .

- **Draw skeletal formulae for simple molecules containing functional groups.**
- **Draw and interpret the skeletal formula for complex molecules.**

Unsaturated hydrocarbons

Pent-2-ene can be represented by the skeletal formula shown in Figure 1.

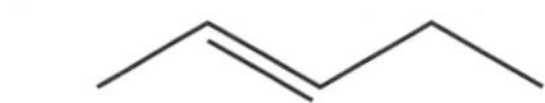

Figure 1 Skeletal formula for pent-2-ene

- The double bond is shown between carbon-2 and carbon-3.
- When drawing skeletal formulae for alkenes, two parallel lines are used to represent the double bond.

Compounds with functional groups

When functional groups are present, they must be included in the skeletal formula. Butan-2-ol has the structural formula: $CH_3CH(OH)CH_2CH_3$.

The skeletal formula for butan-2-ol is shown in Figure 2.

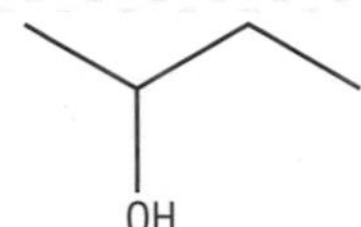

Figure 2 Skeletal formula for butan-2-ol

- There are four carbon atoms in the main chain.
- There is an –OH group on carbon-2.
- The hydrogen on the –OH is shown, as it is part of the functional group.

Pentanoic acid has the structural formula: $CH_3CH_2CH_2CH_2COOH$ or $CH_3(CH_2)_3COOH$. The skeletal formula for pentanoic acid is shown in Figure 3.

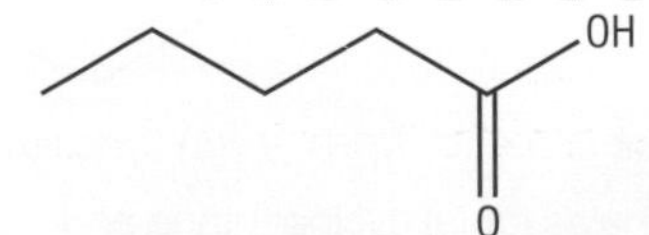

Figure 3 Skeletal formula for pentanoic acid

- There are five carbon atoms in the main chain.
- Carbon-1 is part of a –COOH group.
- The hydrogen on the –OH is shown, as it is part of the functional group.

More complex structures

During your AS and A2 course, you will come across some more complex structures for organic compounds. These are usually drawn in skeletal form. The skeletal formulae for paracetamol, retinol and salbutamol are shown below.

These compounds are all used as medicinal drugs.

- Paracetamol is used for pain relief.
- Retinol is used to treat acne.
- Salbutamol is used to treat asthma.

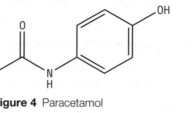

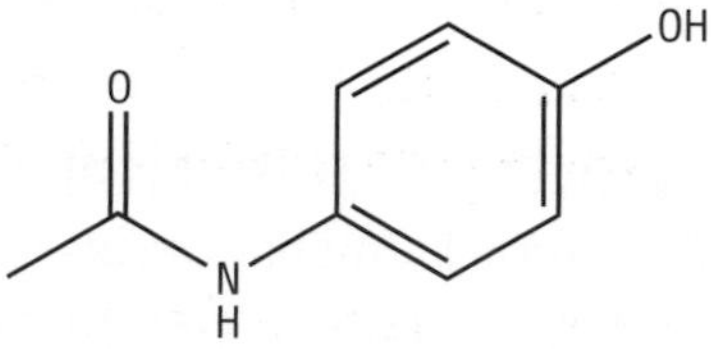

Figure 4 Paracetamol

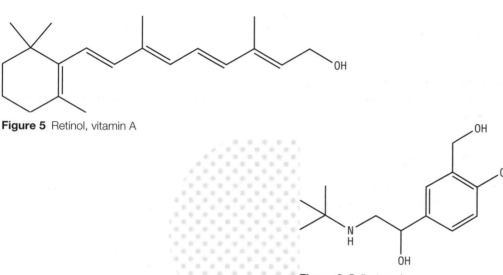

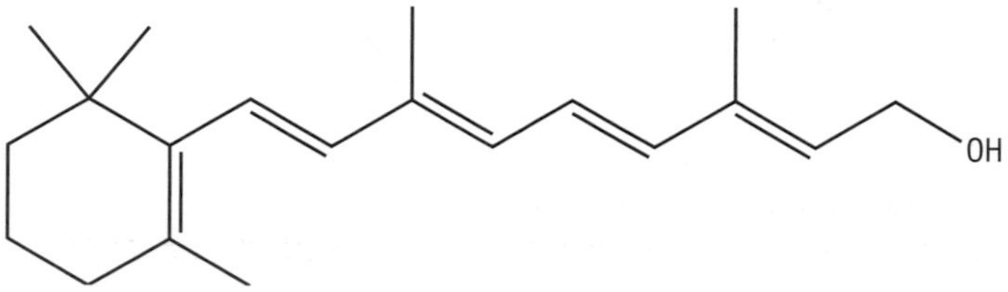

Figure 5 Retinol, vitamin A

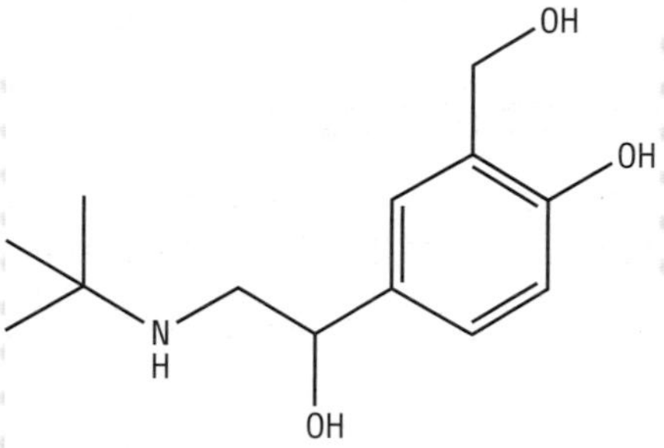

Figure 6 Salbutamol

Notice that all these names end in -ol, used by chemists to indicate an –OH group.

Questions

1 Draw the skeletal formula for $CH_3CH_2CH_2CH_2CH_2CH_2OH$.

2 A molecule has the following skeletal formula.

(a) Deduce the empirical, molecular and structural formulae.

(b) Draw a displayed formula, showing *all* atoms and bonds.

(c) Name the compound.

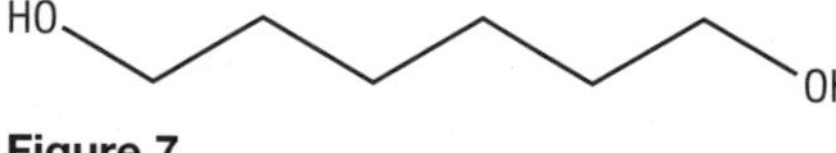

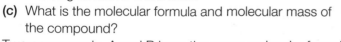

Figure 7

3 A compound has the following skeletal formula.

(a) Name the compound.

(b) Draw the displayed formula of the compound, showing all atoms and bonds.

(c) What is the molecular formula and molecular mass of the compound?

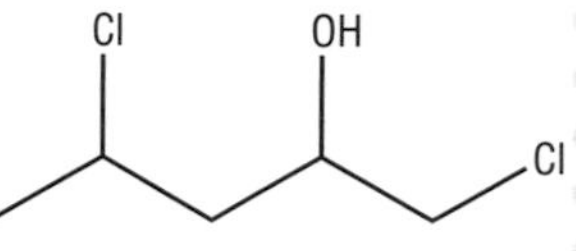

Figure 8

4 Two compounds, A and B have the same molecular formula but different structural formulae. Compound A has the structural formula $CH_3(CH_2)_3OH$ and compound B has the structural formula $CH_3CH(OH)CH_2CH_3$.

(a) Name compounds A and B.

(b) Draw the skeletal formulae for compounds A and B.

2.1 (7) Isomerism

By the end of this spread, you should be able to . . .

* **Describe structural isomers, stereoisomers and *E/Z* isomerism, including *cis–trans* isomerism.**
* **Determine possible structural formulae and/or stereoisomers from a molecular formula.**

Key definition

Structural isomers are molecules with the same molecular formula but with different structural arrangements of atoms.

Figure 1 Friedrich Wöhler (1800–82). The German chemist Friedrich Wöhler prepared cyanic acid in 1827. He noticed that this acid was made from the same elements, and in the same proportions, as another known acid: fulminic acid. However, the chemical properties of these two acids were *different*. We now know that these two acids are structural isomers of CHNO

Structural isomerism

Structural isomerism exists when molecules are composed of the same number and type of atoms, but these atoms are arranged in different ways.

Structural isomerism can occur in three ways:

- The hydrocarbon chain can be unbranched or branched, as in butane and 2-methylpropane shown in Figure 2. Both compounds have the same molecular formula, C_4H_{10}.

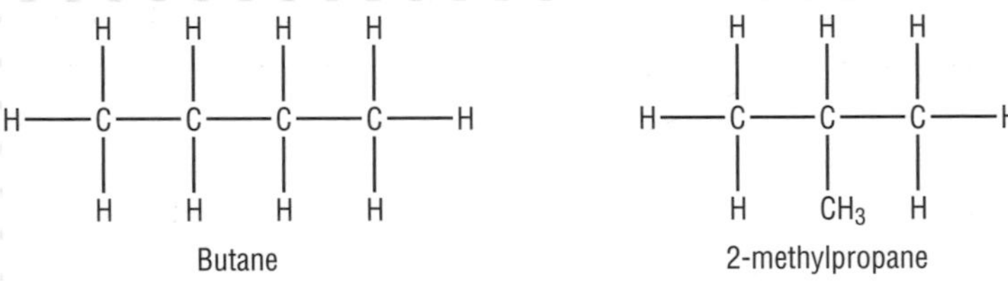

Figure 2 Structural isomers of C_4H_{10}

- A functional group can be in different positions along the main hydrocarbon chain, as in propan-1-ol and propan-2-ol, shown in Figure 3. Both compounds have the same molecular formula, C_3H_8O.

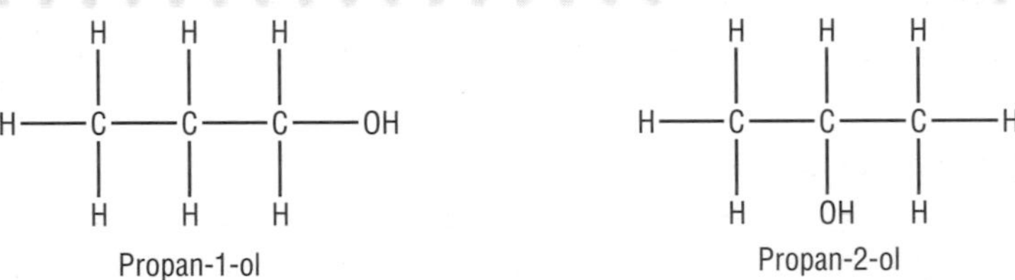

Figure 3 Structural isomers of C_3H_8O

- The functional groups may be different, as in propanal and propanone, shown in Figure 4. Both compounds have the same molecular formula, C_3H_6O.

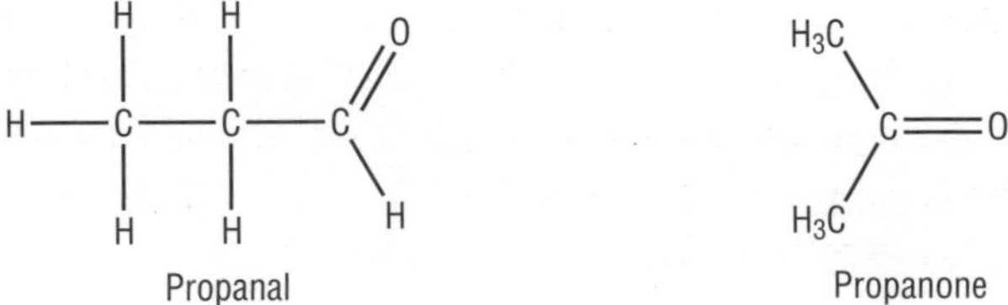

Figure 4 Structural isomers of C_3H_6O

Examiner tip

It's a good idea to draw the isomer with the *longest* chain *first*, and then shorten the chain or move the functional group along the chain.

Key definition

Stereoisomers are compounds with the same structural formula, but with a *different arrangement* of the atoms in space.

Stereoisomerism

In **stereoisomerism**, the atoms making up the isomers are joined up in the same order, but have a different arrangement in space.

During your AS chemistry course, you will study two types of stereoisomerism – *E/Z* isomerism and optical isomerism. (Optical isomerism is not introduced until your A2 course.)

E/Z isomerism

A molecule must satisfy two criteria to have ***E/Z* isomerism**:

- a carbon–carbon double bond must be present; *and*
- each carbon in the double bond must be attached to two different groups.

The compound ClCH=CHCl has a special case of *E/Z* isomerism. There is a Cl atom on each of the C=C atoms and an H atom on each of the C=C atoms (see Figure 5). This is commonly referred to as ***cis–trans* isomerism**.

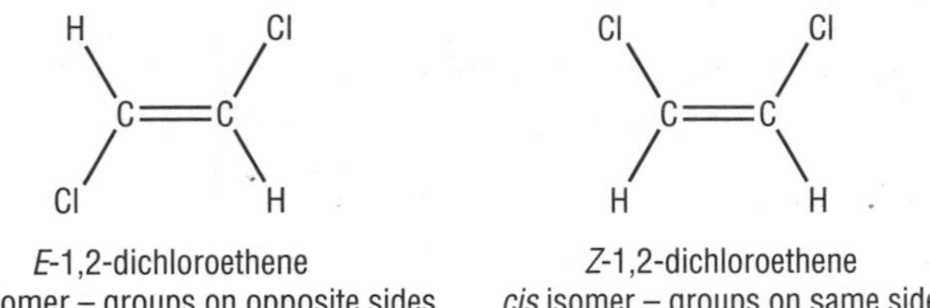

Figure 5 *E* and *Z* isomers of ClCH=CHCl

Skeletal formulae can also be used to represent *E/Z* and ***cis–trans*** isomers. The *cis–trans* isomers of but-2-ene are shown in Figure 6 as skeletal formulae.

Figure 6 *E* and *Z* isomers of but-2-ene

In *cis–trans* isomerism:

- the *cis* isomer is the *Z* isomer (from the German 'zusammen' for together); and
- the *trans* isomer is the *E* isomer (from the German 'entgegen' for opposite).

Cahn–Ingold–Prelog nomenclature (*E/Z*)

E/Z isomerism has been developed as a more up-to-date system for naming *cis–trans* isomers. In the *E/Z* system, groups on a double bond are given a priority based on their atomic number (see Figure 7).

Trouble with *cis–trans* isomers

You may have heard of concern recently over *trans* fats in our diet with a link to coronary heart disease (CHD). In *trans* fat molecules, the double bonds between the carbon atoms are in the *trans* rather than the *cis* configuration, resulting in a straighter, rather than kinked, shape. Consequently, *trans* fats are less fluid and have a higher melting point than corresponding *cis* fats. These *trans* fats can increase your levels of cholesterol, which is a factor in CHD.

Questions

1 Draw the five structural isomers with the molecular formula C_6H_{14}.

2 Draw all the structural isomers for C_4H_9Cl.

3 Draw and name the *E/Z* isomers for $CH_3CH{=}CHCH_2CH_3$.

4 Draw and name the four isomers for C_4H_8 that are alkenes.

Key definition

***E/Z* isomerism** is a type of stereoisomerism in which different groups attached to each carbon of a C=C double bond may be arranged differently in space because of the restricted rotation of the C=C bond.

***cis–trans* isomerism** is a special type of *E/Z* isomerism in which there is a non-hydrogen group and hydrogen on each C of a C=C double bond: the *cis* isomer (*Z* isomer) has the H atoms on each carbon on the same side; the *trans* isomer (*E* isomer) has the H atoms on each carbon on different sides.

Examiner tip

If you have a C=C double bond, then think carefully about the possibility of *cis–trans* isomers.

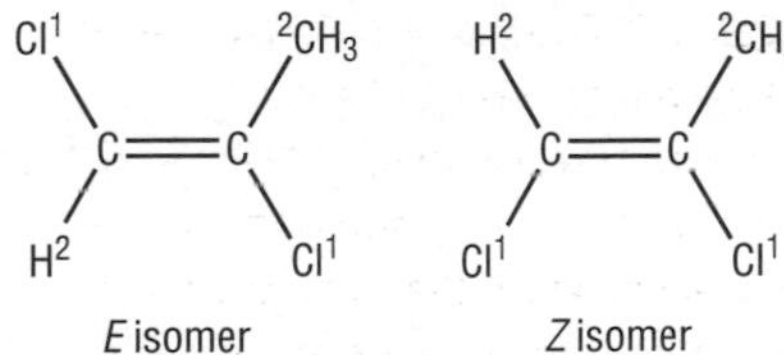

Figure 7 *E* and *Z* isomers.
On each carbon in the double bond, the group with the highest atomic number is given the highest priority (1), and the group with the lowest atomic number is given the lowest priority (2). In this example, chlorine has a higher atomic number than hydrogen, and chlorine is again higher than the carbon of the methyl group.
If the highest priority groups are on the same side, then they are known as together (*Z*) and if they are opposite they are known as apart (*E*)

Examiner tip

In exams, you may need to identify *E* and *Z* isomers, but any examples used will not require you to use the Cahn–Ingold–Prelog nomenclature!

2.1 (8) Organic reagents and their reactions

By the end of this spread, you should be able to . . .

- **Describe the different types of covalent bond fission – homolytic and heterolytic.**
- **Define the terms: nucleophile; electrophile; radical; addition; and substitution.**

Organic reactions

During a chemical reaction, bonds are broken in a process called bond fission.

In organic chemistry, a covalent bond can be broken in one of two ways:
- homolytic fission;
- heterolytic fission.

Homolytic fission

In **homolytic fission**, *each* bonded atom takes *one* of the shared pair of electrons.
- Each atom now has an unpaired electron and is called a **radical**.
- Two species of the same type are produced.

X–Y ⟶ X• + Y•
two radicals

Heterolytic fission

In **heterolytic fission**, *one* of the bonded atoms takes *both* of the shared pair of electrons.

Two ions are produced:
- The atom that *takes* both shared electrons becomes a negatively charged ion (anion).
- The atom that does *not* take the shared electrons becomes a positively charged ion (cation).

X–Y ⟶ X^+ + Y^-
two ions

Nucleophile

A **nucleophile** is a reactant that attacks an electron-deficient carbon atom, donating an electron pair. Nucleophiles are often negative ions with a lone pair of electrons. Nucleophiles often include an electronegative atom with a lone pair of electrons and a δ– partial charge.

Examples
- $:Br^-$, $:OH^-$, $H_2O:$, $:NH_3$.

Electrophile

An **electrophile** is a reactant that attacks an area of high electron density, accepting an electron pair. Electrophiles are often positive ions. Some molecules can act as electrophiles these often include an atom with a δ+ partial charge.

Examples
- Br_2, HBr, NO_2^+.

Addition reactions

In an **addition** reaction, two reactants combine together to make one product.

A molecule is added across the double bond of an unsaturated molecule to make a saturated molecule.

During your AS course, you will become familiar with addition reactions when studying the chemistry of the alkenes.

Key definitions

Homolytic fission is the breaking of a covalent bond, with one of the bonded electrons going to each atom, forming two radicals.

The two species are of the same type: *homo = same.*

A **radical** is a species with an unpaired electron.

A single dot is often written next to the species to represent the unpaired electron.

Heterolytic fission is the breaking of a covalent bond with both of the bonded electrons going to one of the atoms, forming a cation (+ ion) and an anion (– ion).

The two species are different: *hetero = different.*

Key definitions

A **nucleophile** is an atom (or group of atoms) that is attracted to an electron-deficient centre or atom, where it donates a pair of electrons to form a new covalent bond.

An **electrophile** is an atom (or group of atoms) that is attracted to an electron-rich centre or atom, where it accepts a pair of electrons to form a new covalent bond.

An **addition reaction** is a reaction in which a reactant is added to an unsaturated molecule to make a saturated molecule.

Examiner tip

In an addition reaction:

2 reactants ⟶ 1 product.

Example

In the addition reaction of ethene and bromine, the reactants combine together to produce 1,2-dibromoethane (see Figure 1).

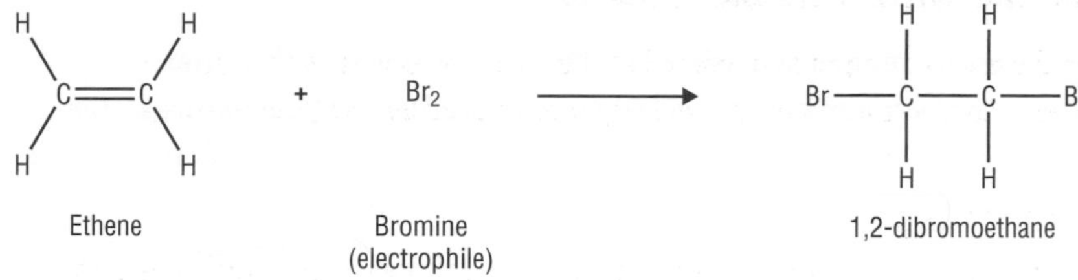

Figure 1 Addition reaction

Substitution reactions

In a **substitution** reaction, an atom or group of atoms is replaced with a different atom or group of atoms.

During your AS course, you will become familiar with substitution reactions in the chemistry of halogenoalkanes.

Example

In the substitution reaction of bromoethane and hydroxide ions, Br is replaced by OH, forming the products ethanol and a Br^- ion (see Figure 2).

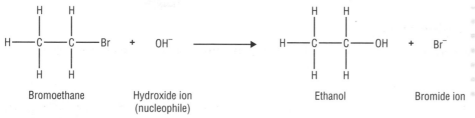

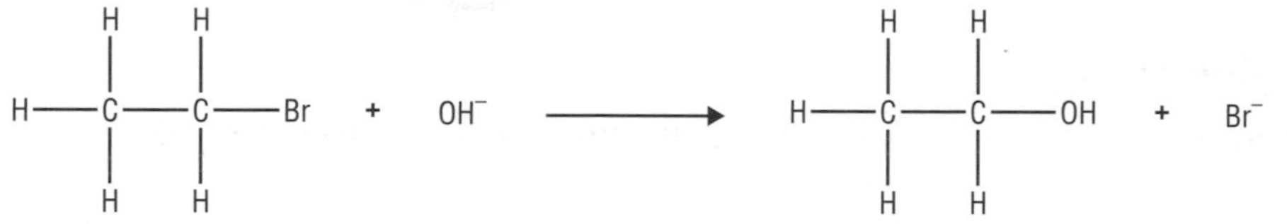

$$CH_3CH_2Br + OH^- \longrightarrow CH_3CH_2OH + Br^-$$

Bromoethane | Hydroxide ion (nucleophile) | Ethanol | Bromide ion

Figure 2 Substitution reaction

> **Key definition**
>
> A **substitution reaction** is a reaction in which an atom or group of atoms is replaced with a different atom or group of atoms.
>
> In a substitution reaction:
>
> 2 reactants ⟶ 2 products.

Elimination reactions

In an **elimination** reaction, one reactant reacts to form two products.

A molecule is removed from a saturated molecule to make an unsaturated molecule.

During your AS course, you will become familiar with elimination reactions in the chemistry of alcohols.

Example

In the elimination reaction of ethanol, using an acid catalyst, water is eliminated from ethanol to form the unsaturated molecule ethene (see Figure 3).

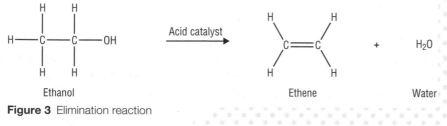

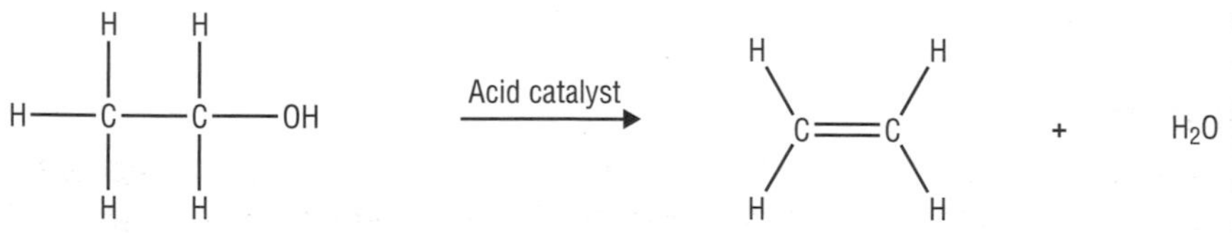

$$CH_3CH_2OH \xrightarrow{\text{Acid catalyst}} CH_2{=}CH_2 + H_2O$$

Ethanol | Ethene | Water

Figure 3 Elimination reaction

> **Key definition**
>
> An **elimination reaction** refers to the removal of a molecule from a saturated molecule to make an unsaturated molecule.
>
> In an elimination reaction:
>
> 1 reactant ⟶ 2 products.

Yields in chemistry

A chemical equation assumes that all the reactants have been converted into products but this is rarely the case. *Percentage yield* measures the proportion of products formed in a reaction.

Chemists have been concerned that some of the products may be waste. *Atom economy* measures the proportion of products that are used.

The important concepts of percentage yield and atom economy are discussed in detail in spreads 2.2.8 and 2.2.9.

Questions

1 Define and give an example of:
 (a) a radical;
 (b) a nucleophile;
 (c) an electrophile.

2 Why are radicals extremely reactive?

3 Write equations using structural formulae for:
 (a) an addition reaction;
 (b) a substitution reaction;
 (c) an elimination reaction.

2.1 (9) Hydrocarbons from crude oil

By the end of this spread, you should be able to . . .

* **Explain the use of crude oil as a source of hydrocarbons, as well as their uses.**
* **Explain that the hydrocarbons in crude oil are separated using fractional distillation.**
* **Explain, in terms of van der Waals' forces, the variations in the boiling points of different alkanes.**

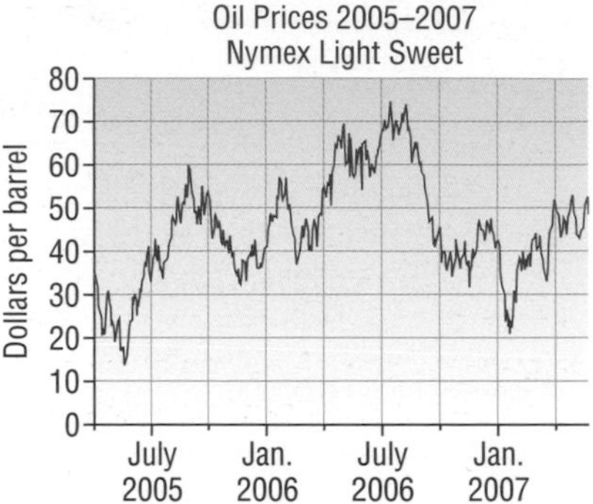

Figure 1 Oil prices hit a record high in July 2006

Hydrocarbons

A hydrocarbon is a compound that contains the elements hydrogen and carbon only.

Crude oil is a fossil fuel. It has been made from naturally decaying plants and animals that once lived in ancient seas, millions of years ago. The composition of crude oil varies greatly from one source to another. For example, crude oil from the Middle East contains a high proportion of unbranched alkanes.

Generally, crude oil is a mixture of over 150 different hydrocarbons, most of which are unbranched (straight-chained) alkanes. Crude oil does not ignite easily, and so is not very useful in its native state. However, some of its components are valuable, and are used in petrol, kerosene, and heating and lubricating oils (see Figure 1).

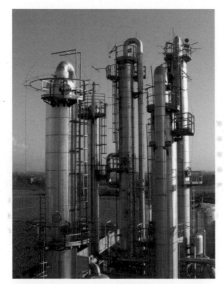

Figure 2 Distillation plant

Key definition

Fractional distillation is the separation of the components in a liquid mixture into fractions which differ in boiling point (and hence chemical composition) by means of distillation, typically using a fractionating column.

Examiner tip

In fractional distillation, the hydrocarbons in crude oil are separated on the basis of their different boiling points.

Fractional distillation of crude oil

Crude oil is refined in a distillation plant (Figure 2).

- The complex mixture of hydrocarbons in crude oil is first separated into *fractions*.
- Each fraction consists of a mixture of hydrocarbons with similar boiling points.
- Pure liquids have a fixed boiling point. A *pure* hydrocarbon can be obtained by further distillation of a crude oil fraction.

Fractional distillation takes place in a fractionating column. The crude oil is first vaporised by heating and is then passed into the fractionating column. The column is hotter at the bottom than at the top, and the gases pass up the column through a series of bubble caps. Eventually, the gases reach a temperature that is *lower* than their boiling points. Here, the vapour condenses to a liquid. The liquid fractions are then tapped off into storage containers.

- Short-chained hydrocarbons with *lower* boiling points condense near the *top* of the column.
- Longer-chained hydrocarbons with *higher* boiling points condense nearer to the *bottom*.
- Gases do not condense and pass through an outlet at the top of the column as 'petroleum gas'.
- The residue from the process is bitumen, which is removed from the bottom of the column.

The fractions obtained from crude oil can be used as fuels or may be further processed to make petrochemicals – chemicals made from natural gas and oil. The fractions can also be distilled further to form pure liquids in other fractionating columns, operating over a narrower range of temperatures. Even the tarry bitumen can be used, in road surfacing or roof coverings.

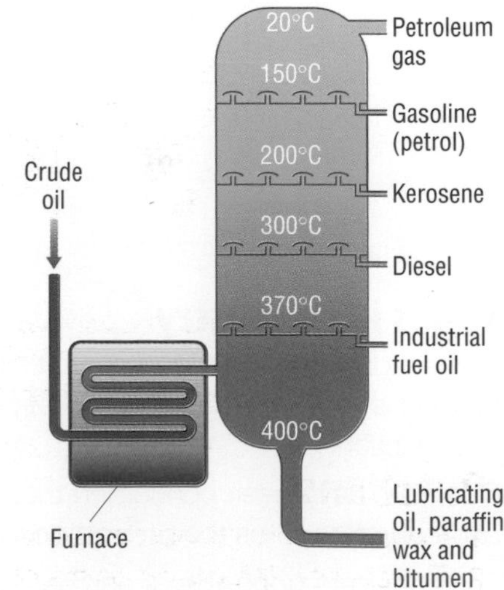

Figure 3 Fractionating column

Boiling points of alkanes

In a crude oil fraction there are many different alkane molecules. There are weak intermolecular forces of attraction between the molecules called van der Waals' forces. In order to boil, these forces have to be broken.

Effect of chain length

As the chain length increases, the boiling point increases because the intermolecular forces between the molecules get stronger. In a longer-chained alkane, there are more points of contact between the molecules leading to more van der Waals' forces between the molecules. It takes more energy, and therefore a higher temperature, to separate the molecules. This allows the different alkanes in crude oil to be separated by their boiling points.

Figure 4 Petrol, viewed as one of life's essentials, is produced from crude oil

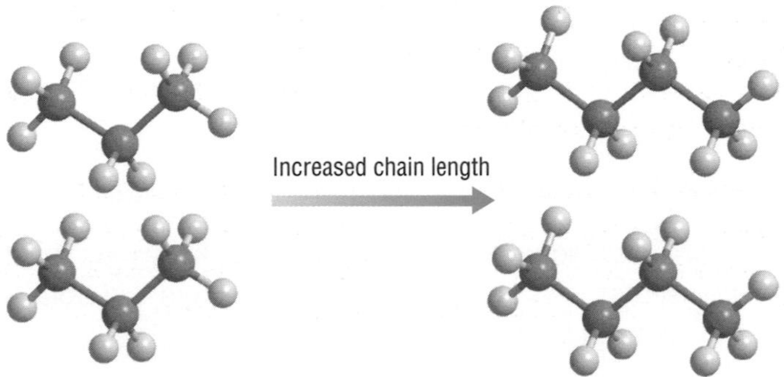

As the chain length increases there are more points of contact leading to more van der Waals' forces between the molecules.

Figure 5 Effect of increasing chain length on the boiling points of alkanes

Effect of branching

Isomers of alkanes have the same molecular mass. A branched isomer has a lower boiling point than the unbranched isomer. In a branched alkane, there are fewer points of contact between molecules, leading to fewer van der Waals' forces between the molecules. Also, branched molecules cannot get as close to each other as unbranched ones, thus decreasing the intermolecular forces between the molecules. Therefore, less energy (heat) is required to separate the molecules.

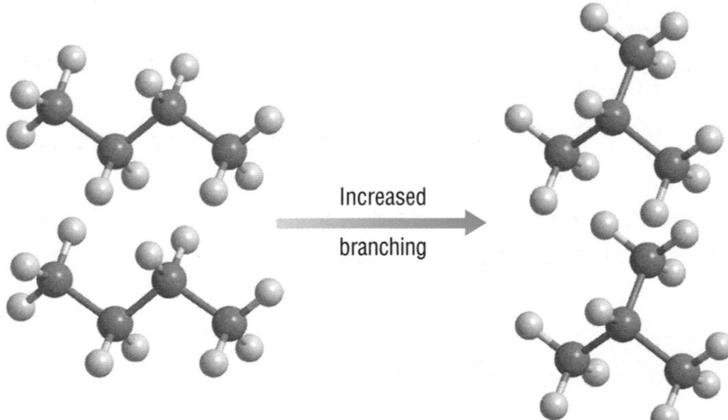

Branched molecules have fewer points of contact and are further apart than unbranched molecules so there are fewer intermolecular forces.

Figure 6 Effect of branching on the boiling points of alkanes

Questions

1 Explain how fractional distillation works.

2 Explain how the boiling points of alkanes are affected by:
 (a) chain length; (b) branching.

2.1 (10) Hydrocarbons as fuels

By the end of this spread, you should be able to . . .

- * **Describe the combustion of alkanes.**
- * **Explain the incomplete combustion of alkanes.**
- * **Describe catalytic cracking to obtain more useful alkanes and alkenes from long-chain hydrocarbons.**
- * **Describe further processing to form branched alkanes and cyclic hydrocarbons.**

Figure 1 In Tokyo LPG powers all taxis

Examiner tip

When balancing combustion equations of hydrocarbons, look at the molecular formula of the hydrocarbon.

After combustion, there will always be:

- the same number of CO_2 molecules as carbon atoms
- half the number of H_2O molecules as hydrogen atoms.

Figure 2 A carbon monoxide detector fitted in the home could save lives

Key definitions

Cracking refers to the breaking down of long-chained saturated hydrocarbons to form a mixture of shorter-chained alkanes and alkenes.

A **catalyst** is a substance that increases the rate of a chemical reaction without being used up in the process.

The combustion of alkanes

Short-chain alkanes are valuable as clean fuels. They burn in a plentiful supply of oxygen to form carbon dioxide and water. Methane is the main constituent of natural gas and is commonly used for domestic heating and cooking.

Methane burns cleanly in oxygen, according to the equation:

$$CH_4(g) + 2O_2(g) \longrightarrow CO_2(g) + 2H_2O(l)$$

Propane and butane are easily liquefied and are commonly known as liquefied petroleum gas or LPG. They are used as fuels in barbecues, patio heaters and portable cooking appliances such as those used in camping.

Octane is present in petrol, used to fuel the internal combustion engine. The equation for the complete combustion of octane is:

$$C_8H_{18}(l) + 12\tfrac{1}{2}O_2(g) \longrightarrow 8CO_2(g) + 9H_2O(l)$$

Incomplete combustion

In practice, the internal combustion engine in most cars has a limited supply of oxygen. Some of the octane will undergo *incomplete* combustion to form carbon monoxide, CO, rather than carbon dioxide.

$$C_8H_{18}(l) + 8\tfrac{1}{2}O_2(g) \longrightarrow 8CO(g) + 9H_2O(l)$$

Carbon monoxide is a colourless, odourless gas. It is produced when any fossil fuel is burned with an insufficient supply of oxygen.

Carbon monoxide is poisonous. It prevents the haemoglobin in red blood cells from binding with oxygen and the body's tissues become starved of oxygen. This can be fatal.

In the home, carbon monoxide can be formed from faulty domestic heating systems, blocked flues/chimneys, or inadequate ventilation. It is advisable to fit carbon monoxide detectors in the home to protect from the toxic effects of carbon monoxide (see Figure 2).

Cracking

After fractional distillation of crude oil, there is a surplus of long-chained hydrocarbons. Short chain hydrocarbons are in high demand:

- short chain alkanes for use as fuels
- short chain alkenes for use in polymer production.

Cracking is used to break down long-chained saturated hydrocarbons to form a mixture of shorter-chained alkanes and alkenes.

Catalytic cracking was first used in the mid-1930s. Initially, the catalysts Al_2O_3 and SiO_2 were used in the cracking process. However, scientists have now developed more efficient catalytic systems. Today, most catalytic cracking uses a zeolite **catalyst** at about 450 °C. See Figure 4.

In cracking long-chained alkanes are broken randomly. There are many possible equations for cracking a particular starting material. One *possible* cracking reaction of dodecane, $C_{12}H_{26}$ is shown in Figure 3 below.

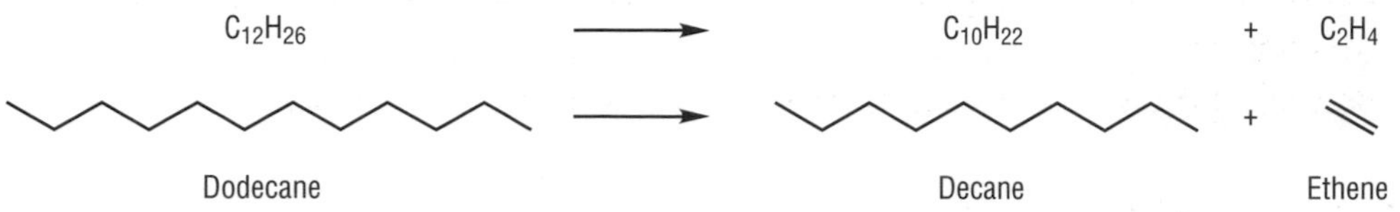

Figure 3

Examiner tip

The equations must be *balanced* and a mixture of alkenes and alkanes is always formed.

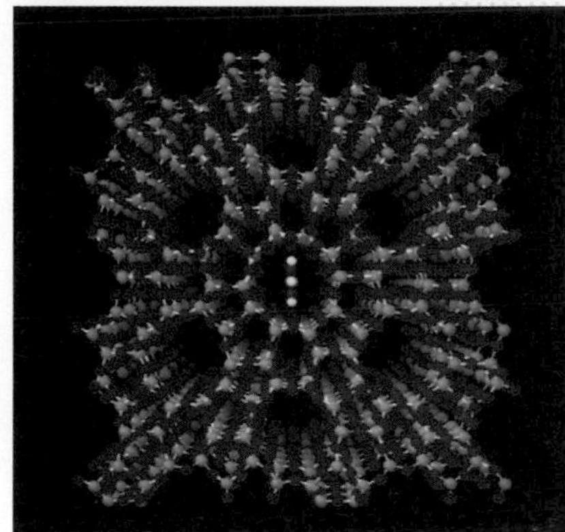

Figure 4 Computer graphics showing a 'zeolite' mineral. Zeolites are used as catalysts in catalytic cracking. Cracking fragments long-chain alkane molecules into shorter-chain alkanes and alkenes. Ths shorter-chain alkanes can be used as fuels such as petrol. The alkenes are used to make polymers

Producing branched alkanes

Unbranched (straight-chained) alkanes can be converted into branched alkanes in a process sometimes referred to as *isomerisation*. Figure 5 shows one *possible* isomerisation reaction of pentane, C_5H_{12}.

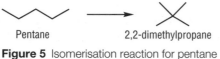

Figure 5 Isomerisation reaction for pentane

Producing cyclic hydrocarbons

Aliphatic hydrocarbons can be converted into cyclic or aromatic hydrocarbons in a process sometimes known as *reforming*. Figure 6 shows pentane being reformed into the cyclic hydrocarbon cyclopentane. Figure 7 shows heptane being reformed into the aromatic hydrocarbon methylbenzene. Hydrogen gas is also produced.

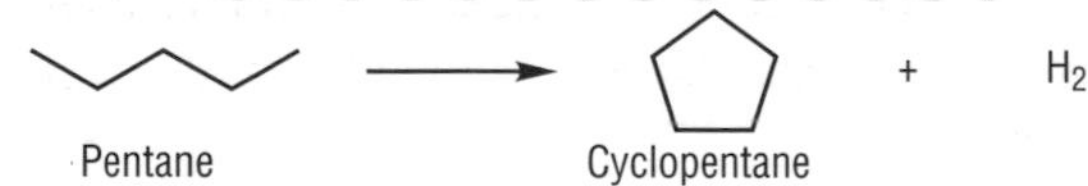

Figure 6 Reforming process for pentane into cyclopentane

Figure 7 Reforming process for pentane into cyclopentane

Improving fuels

The Research Octane Number (RON) rates how well a fuel burns. Fuels with high octane ratings close to 100 burn efficiently. The straight-chained alkane heptane is a poor fuel and has an octane rating of zero.

Branched and cyclic alkanes are important petrol additives as they promote more efficient combustion than straight-chain alkanes. Consequently, branched alkanes are used extensively in fuels for car engines.

The hydrogen that is also produced during reforming is used in other chemical processes such as ammonia production (spread 2.3.15) and margarine production (spread 2.1.17).

This is a good example of using by-products of a process to make other useful products.

Questions

1 Write equations for the combustion of propane in both a *plentiful* and *limited* supply of oxygen.
What volume of oxygen would be required to completely burn one mole of propane at RTP?

2 Decane can be cracked to form ethene and one other important hydrocarbon.
Write an equation for this reaction.

3 Hexane can be reformed to produce a six-membered cyclic compound.
Using displayed formulae, write a balanced equation for this reaction.

4 Give two advantages of cracking long-chained hydrocarbons.

Fossil fuels and fuels of the future

By the end of this spread, you should be able to . . .

* **Contrast the value of fossil fuels with an over-reliance on non-renewable fossil fuel reserves.**
* **Examine the importance of developing renewable plant-based fuels, such as alcohols and biodiesel.**
* **Contrast the value of fossil fuels with increased CO_2 levels, leading to global warming and climate change.**

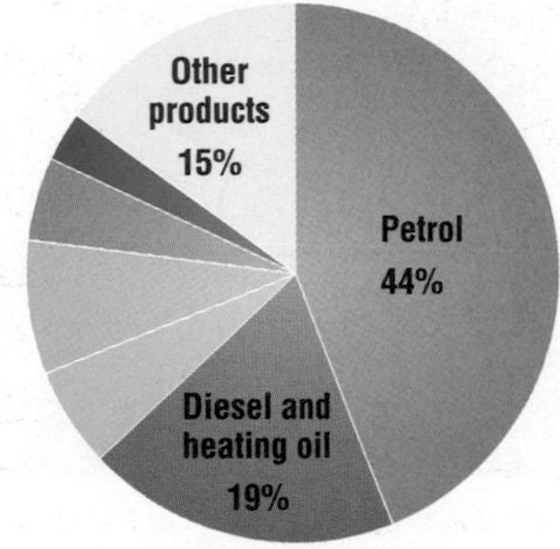

Figure 1 Uses of crude oil. Other products include chemical and polymer feedstock

The crude oil economy

Crude oil and the materials produced from it play an important part in the global economy and have a significant impact on our standard of living. We have come to rely on crude oil as a source of power for electrical generation and for major forms of transport.

Many of the chemicals produced from crude oil have become an important feedstock for the chemical industry.

- Over 90% of crude oil produced in the world is used as a source of fuel.
- Plastics, pharmaceuticals, cosmetics, dyes and inks are just some of the everyday items that can be made indirectly from crude oil. These are called *petrochemicals*.

This reliance on crude oil comes at a price – the Earth's known deposits of crude oil are depleting at an alarming rate!

The use of crude oil for fuels

Many of the fuels produced from crude oil are alkanes. Branched and cyclic alkanes resulting from refining and cracking are amongst the best fuels. A good fuel needs to be readily available, easily transported and inexpensive. Oil certainly fits the bill, but in recent years its price has increased significantly and now scientists are looking into alternatives to fossil fuels.

It is not just the price of oil that has caused scientists to look for alternative forms of energy. At the start of the twenty-first century, there has been a major shift in public opinion and a new concern for the environment in which we live. All hydrocarbons produce pollutants. The over-reliance on crude oil as a source of fuel has led environmentalists to call for urgent action.

Burning hydrocarbons leads to an increase in atmospheric pollutants, such as:

- carbon monoxide – a toxic gas formed by incomplete combustion in the internal combustion engine.
- carbon dioxide – a major contributor to global warming via the greenhouse effect.
- nitrogen oxides – contributors to acid rain and destruction of forests.
- sulfur dioxide – a contributor to acid rain.

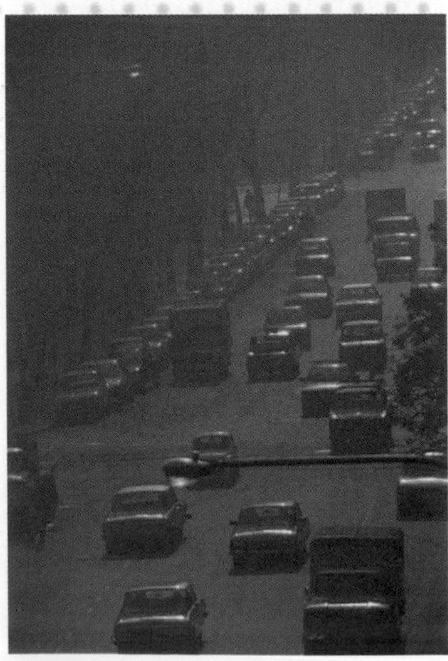

Figure 2 Smog over busy street in city of Buenos Aires, Argentina

Global warming and climate change

Global warming refers to an increase in the Earth's average temperature. There is growing evidence that this increase could be related to human activity such as burning coal, oil and natural gas.

Burning fuels to generate energy releases carbon dioxide and other greenhouse gases into the atmosphere. These greenhouse gases prevent heat from escaping the atmosphere, and hence lead to increased temperatures on Earth. This practice could be one of the causes for our changing climate.

A warmer Earth may mean a change in climate, with heavier rain and more frequent violent storms. Changes in temperature threaten to melt the polar ice caps, with a

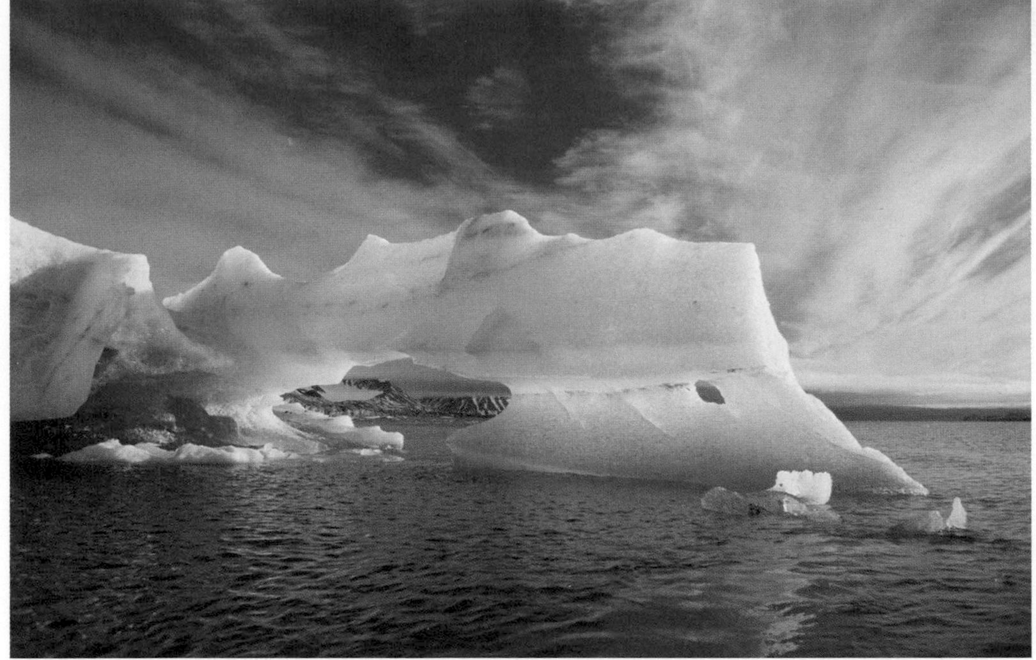

Figure 3 Global warming is blamed for a rise in sea levels as the polar ice caps start to melt. In December 1997, representatives from around 180 countries met in Kyoto, Japan, to sign the Kyoto Protocol. This committed 38 industrialised countries to cut their emissions of greenhouse gases in a five-year period, starting in 2008, to levels that are 5.2% below 1990 levels

Figure 4 Our rising demands to replace fossil fuels and reduce CO_2 emissions may lead to other environmental issues

consequential rise in sea levels and localised flooding. If this happens, it will have an impact on all plant and animal life forms and will certainly lead to a change of lifestyle for many of the world's inhabitants.

Fuels of the future

A *biofuel* is a fuel that is derived from recently living material such as plants, or from animal waste.

Agricultural crops specifically grown for energy use include sugar cane (in such countries as Brazil) and rape – its bright yellow flowers commonly seen in the British countryside during spring.

Ethanol can be made by fermenting sugar and other carbohydrates. It burns efficiently, in a plentiful supply of oxygen, to give carbon dioxide and water. Ethanol can be blended with petroleum, causing the fuel to burn more efficiently. Ethanol blends can be used in engines without the need for any major modifications and it significantly reduces harmful exhaust emissions.

In the UK, *bioethanol* is getting a big boost as the country's first production plants become operational. The first plant, located in Somerset, is capable of producing 105 000 tonnes of bioethanol a year.

Biodiesel is a fuel oil derived from natural sources such as plants. Rapeseed is the most popular source for biodiesel processing in Europe. In compatible engines, biodiesel can be used 100% pure. However, it is usually blended with normal diesel.

Figure 5 Oilseed rape crop. This plant is cultivated for its seeds, which contain an oil that is used in the production of vegetable oil, animal feed and, more recently, biodiesel

Questions

1 Why is the production of motor fuel from cane sugar as a raw material of long-term interest?
2 Why is ethanol manufactured by fermentation of cane sugar in Brazil?
3 What is a fossil fuel? Explain why ethanol can be regarded as both a renewable and non-renewable fuel.

2.1 (12) Substitution reactions of alkanes

By the end of this spread, you should be able to . . .

* **Describe the substitution of alkanes in the presence of ultraviolet radiation.**
* **Describe how homolytic fission leads to the mechanism of radical substitution in alkanes.**
* **Explain the limitations of radical substitution in synthesis.**

Halogenation of alkanes

Alkanes react with halogens in the presence of ultraviolet radiation or at a temperature of about 300 °C.

For example, methane reacts with chlorine to produce chloromethane:

$$\underset{\text{methane}}{CH_4} + Cl_2 \longrightarrow \underset{\text{chloromethane}}{CH_3Cl} + HCl$$

This reaction is **radical substitution**:

- covalent bonds are broken by homolytic fission to form **radicals** with an unpaired electron;
- a hydrogen atom in the alkane is *substituted* by a halogen atom.

The halogens bromine and iodine also react with alkanes in a similar way.

Mechanism for chlorination

Many reactions take place in more than one step.

A **mechanism** is a sequence of steps, showing the path taken by electrons in a reaction.

It outlines how the reaction is thought to proceed in clear diagrams.

The mechanism of radical substitution takes place in *three* stages. The example below outlines the mechanism for the chlorination of methane.

Step 1: initiation

In the **initiation** stage, the Cl–Cl bond in a chlorine molecule is broken by homolytic fission, forming two chlorine radicals. Ultraviolet radiation provides the energy for this bond fission.

$$Cl\text{–}Cl \longrightarrow Cl\cdot + Cl\cdot$$

After initiation, the reaction can continue without the need for further energy or more radicals.

Chlorine radicals attack methane in the second reaction stage.

Step 2: propagation

The **propagation** stage has two steps:

- In the *first* propagation step, methane reacts with a chlorine radical. A single C–H bond is broken by homolytic fission, forming a methyl radical $\cdot CH_3$. Hydrogen chloride, HCl, is also formed.
- In the *second* propagation step, the methyl radical reacts with a chlorine molecule. The organic product chloromethane, CH_3Cl, is formed together with a further chlorine radical. The chlorine radical can be used again in the first propagation step.

First propagation step:

$$CH_4 + Cl\cdot \longrightarrow \cdot CH_3 + HCl$$

Second propagation step:

$$\cdot CH_3 + Cl_2 \longrightarrow CH_3Cl + Cl\cdot$$

Key definition

Propagation is the two repeated steps in radical substitution that build up the products in a chain reaction.

Examiner tip

Often, examination questions ask for a description of a *mechanism*. This is answered best using equations and diagrams, rather than by describing with many words. Mechanisms are good mark-gatherers! So do ensure that you learn them thoroughly.

Key definitions

Radical substitution is a type of substitution reaction in which a radical replaces a different atom or group of atoms.

Mechanism is a sequence of steps showing the path taken by electrons in a reaction.

Initiation is the first step in a radical substitution in which the free radicals are generated by ultraviolet radiation.

Examiner tip

Remember that a radical is a species with an unpaired electron.

Radicals are very reactive intermediates.

A radical is represented by a single dot before or after the chemical symbol.

Examiner tip

The number of radicals in a propagation step is always maintained. If there is one radical on the left of an equation, then there will be one on the right.

Two steps are required to form the products.

The chlorine radical is regenerated in the second step.

- Propagation reactions are rapid, and the reactions continue until no reactants remain.
- Propagation is a *chain* reaction. Although a chlorine radical is used up in the first propagation step, another chlorine radical is generated in the second step. This chlorine radical can now react with another methane molecule. This continues until all the chlorine has been used or the termination stage (step 3) has removed all the radicals.

Step 3: termination

In the *termination* stage, two radicals combine to form a molecule. There are a number of possible termination steps. This is because of the large numbers of radicals in the reaction mixture. In this example, the following molecules can be produced:

$$Cl\cdot + Cl\cdot \longrightarrow Cl_2$$
$$\cdot CH_3 + \cdot CH_3 \longrightarrow C_2H_6$$
$$\cdot CH_3 + \cdot Cl \longrightarrow CH_3Cl$$

The termination stage removes radicals, stopping the reaction. Before this happens, though, the propagation stage will have cycled through many thousands of times.

Key definition

Termination is the step at the end of a radical substitution when two radicals combine to form a molecule.

Examiner tip

In termination steps, *all* of the radicals are removed.

Further reactions of chloromethane

Radical substitution reactions lead to the formation of a mixture of products.

In the reaction of methane and chlorine described above, chloromethane is formed in the propagation stage. However, other organic products can form.

- In the termination stage, chlorine, ethane and chloromethane are produced.
- Chloromethane, made in the propagation stage, may react with further chlorine radicals until all of the hydrogen atoms have been replaced.
  This results in a mixture of chloromethane (CH_3Cl), dichloromethane (CH_2Cl_2), trichloromethane ($CHCl_3$) and tetrachloromethane (CCl_4).

Beware the radicals

Scientists have discovered that every cell in the human body produces tens of thousands of radicals every day. Radicals are a natural by-product of normal cell metabolism, such as from fighting infection or burning glucose for energy. They are important in hormone and enzyme production.

Exposure to radicals and their reactions has been linked to premature aging. Left unchecked, radicals can cause extensive cell damage and contribute to a whole list of chronic illnesses. Radicals damage the cell membranes of virtually all cells and the very source of our genetic material – DNA.

Fortunately, the body has a natural defence mechanism – chemicals called antioxidants – that keep radicals in check and protect the body against the harmful effects of radicals.

Vitamins C and E are antioxidants.

- Vitamin C comes from fresh fruit and vegetables.
- Vitamin E comes from fats, oils, nuts and green leafy vegetables, egg yolk and liver.

We are also exposed to radicals from cigarette smoke (see Figure 1), sunbathing, general pollution and food contaminated by herbicides.

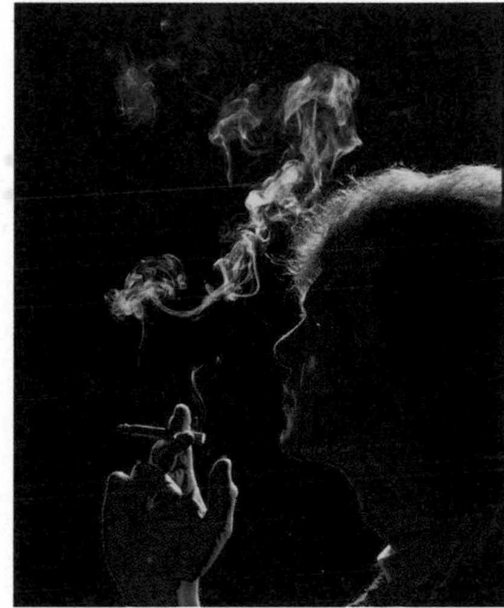

Figure 1 Cigarettes contain cancer-causing chemicals and the addictive drug nicotine. Tobacco smoking is a major cause of lung cancer and chronic bronchitis. It has been estimated that 10^{15} radicals are present in one puff of cigarette smoke. This includes nitrogen monoxide and nitrogen dioxide radicals, which interact with many of the biological molecules in the lungs. It is no wonder that the United Kingdom Government has recently passed legislation to prevent smoking in enclosed public spaces

Questions

1 Bromine reacts with propane in ultraviolet radiation.
  (a) Write an equation to show the initiation step.
  (b) One of the products of the reaction will be 1-bromopropane. Show the two propagation steps of this reaction.

2 In the chlorination of ethane, one of the products is a molecule with four carbon atoms. Explain using equations how this molecule can be formed.

2.1 (13) Alkenes

By the end of this spread, you should be able to . . .

* **Understand that alkenes and cycloalkenes are unsaturated hydrocarbons.**
* **Explain the overlap of adjacent p-orbitals to form a π-bond.**
* **Explain the trigonal planar shape around each carbon making up the C=C bond of alkenes.**

Introduction

Alkenes are *unsaturated hydrocarbons* with at least one C=C double bond.

- Aliphatic alkenes with *one* double bond have the general formula: C_nH_{2n}.
- Alkenes are *more* reactive than alkanes, and typically take part in *addition* reactions.
- Alkenes can form *E/Z* isomers, including *cis–trans* isomers.

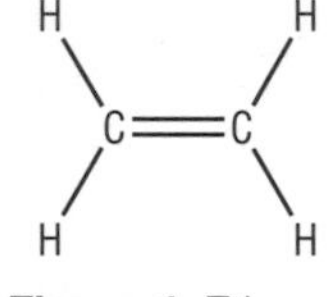

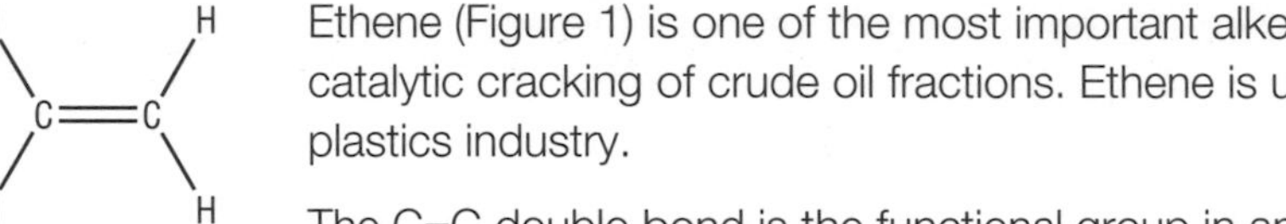

Figure 1 Ethene

Ethene (Figure 1) is one of the most important alkenes. It is obtained from the catalytic cracking of crude oil fractions. Ethene is used to make polymers for the plastics industry.

The C=C double bond is the functional group in an alkene and is responsible for its reactions.

Key definition

A **pi-bond (π-bond)** is the reactive part of a double bond formed above and below the plane of the bonded atoms by sideways overlap of p-orbitals.

The nature of the double bond

The C=C double bond (Figure 2) is made up from two parts: a sigma, σ, bond and a **pi, π, bond**. σ-bonds are also present in alkanes.

In the formation of the double bond:

- A σ-bond is formed directly between two carbon atoms by the overlap of orbitals. Each carbon atom contributes one electron to the electron pair in the σ-bond.
- A π-bond is formed above and below the plane of the carbon atoms by sideways overlap of p-orbitals. Each carbon atom contributes one electron from a p-orbital to the electron pair in the π-bond.

Examiner tip

In exams, do use diagrams in your answers. Annotated diagrams can often communicate far more than trying to explain something in words alone.

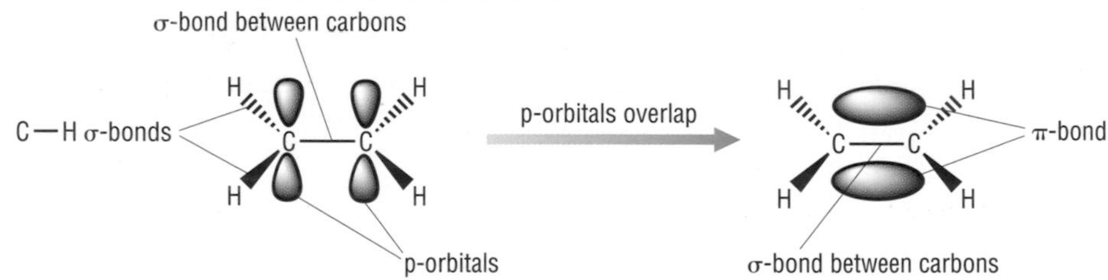

Figure 2 Formation of the C=C double bond

The π-bond *fixes* the carbon atoms in position, at either end of the double bond. This prevents any rotation of the bond.

In alkenes, each carbon involved in the double bond:

- uses three of its electrons in the formation of three σ-bonds; and
- uses one of its electrons in the formation of a π-bond.

The shape of an alkene molecule

- Three regions of electron density surround each carbon atom in the double bond.
- Pairs of electrons repel each other as far apart as possible to minimise repulsions.
- The electron pairs repel each other to give a *trigonal planar* shape, with bond angles of approximately *120°*.
- Ethene is a flat *planar molecule*, with all of the atoms in the same plane (see Figure 3).

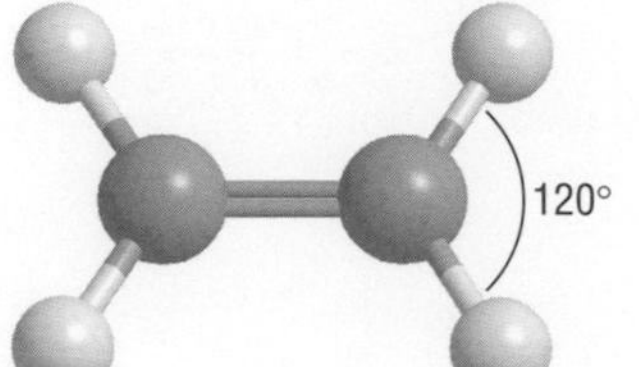

Figure 3 Shape of an ethene molecule

Cyclic alkenes

- Cylic alkenes have closed rings of carbon atoms containing one or more double bonds.
- The most common cyclic alkene is cyclohexene (Figure 4).
- Cyclic alkenes do *not* follow the same general formula as the aliphatic alkenes.

Figure 4 Cyclohexene

Carotene, Figure 5, is the orange pigment in carrots, used by the body to make vitamin A. The carotene molecule has 11 double bonds and both cyclic and straight chain sections. Notice that carotene's name ends in '–ene' to indicate the presence of a double bond (in this case, many!).

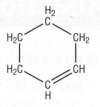

Examiner tip

If you are asked about the shape of an alkene, use the following important terms:
- trigonal planar
- 120° bond angle
- electron pair repulsion.

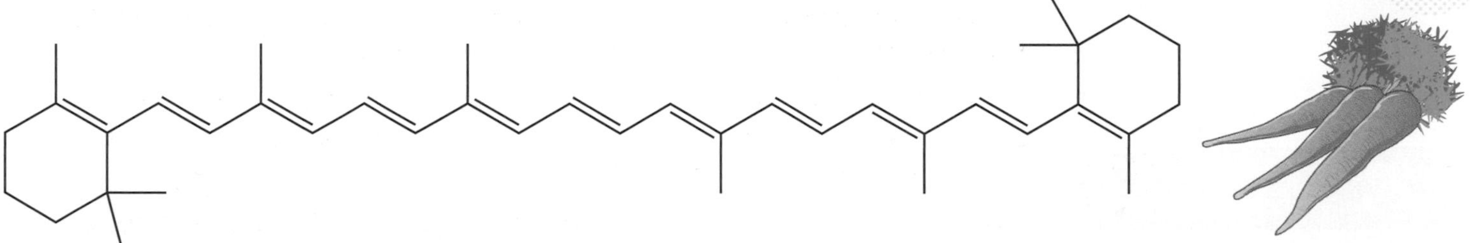

Figure 5 Carotene, $C_{40}H_{56}$

Developing alkene pharmaceuticals

The antiviral drug Tamiflu has been stockpiled by governments for use in the event of an influenza pandemic. It is manufactured from plant extracts, but research into its synthesis from the relatively inexpensive 1,4-cyclohexadiene is under way.

Examiner tip

Cyclic compounds, such as cyclohexene, are sometimes used in examination questions. These can be treated in the same way as straight-chain and branched alkenes as they react in a similar way!

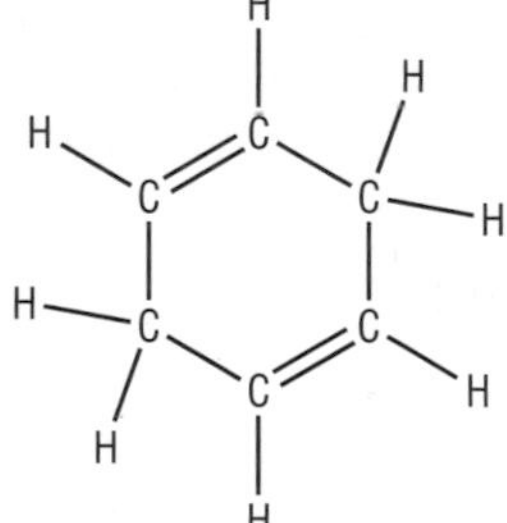

Figure 6 Tamiflu and 1,4-cyclohexadiene

1,4-cyclohexadiene

Questions

1 What is the molecular formula for the aliphatic alkene containing four carbon atoms?

2 Draw the displayed formula for the following alkenes:
  (a) 2-methylpent-2-ene;
  (b) buta-1,3-diene.

3 Draw the *E* and *Z* isomers for pent-2-ene.

4 Name the following alkenes:

(a)

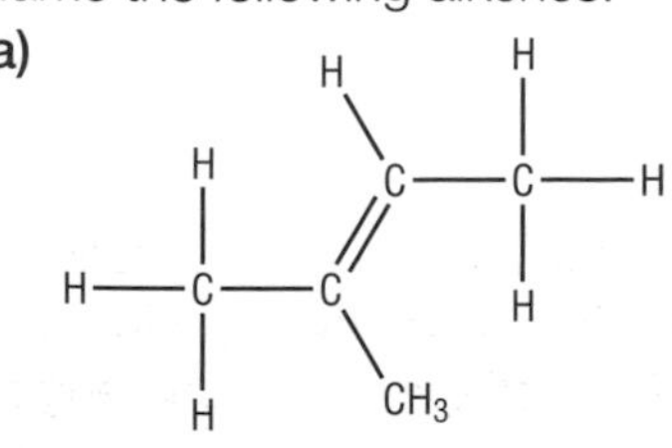

(b)

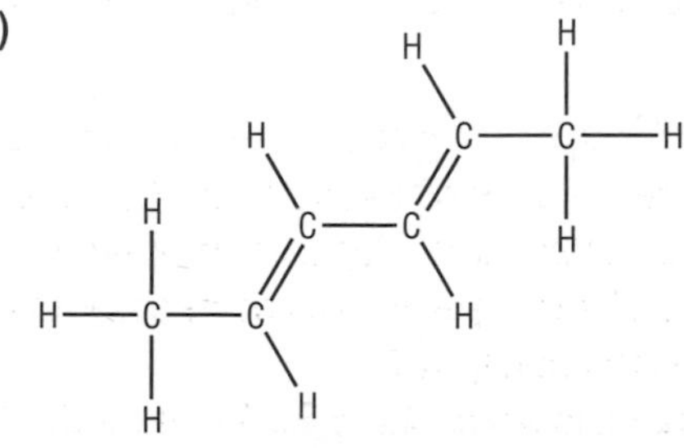

2.1 (14) Reactions of alkenes

By the end of this spread, you should be able to . . .

* **Describe addition reactions of alkenes with both hydrogen and halogens.**
* **Describe the use of bromine as a test for unsaturation.**

The reactivity of the C=C double bond

Alkenes are more reactive than alkanes because of the C=C double bond. The average *bond enthalpies* for a C–C single bond and a C=C double bond are shown in Table 1. Bond enthalpies measure the strengths of covalent bonds. You will study bond enthalpies in detail in spread 2.3.7.

- In an alkane, the single bond between two carbon atoms is a σ-bond.
- In an alkene, the double bond between two carbon atoms is made up of a σ-bond *and* a π-bond.

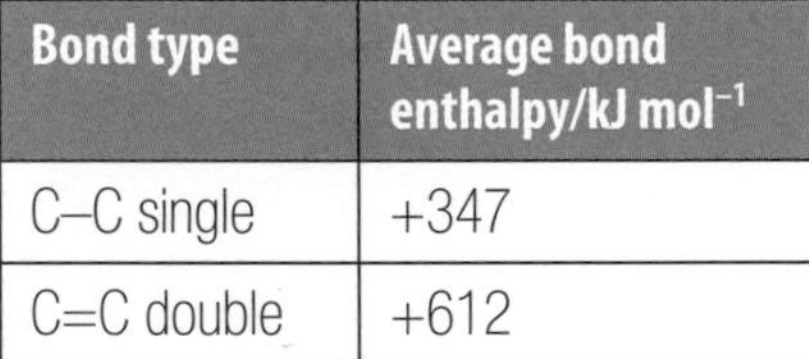

Bond type	Average bond enthalpy/kJ mol^{-1}
C–C single	+347
C=C double	+612

Table 1 Bond enthalpies

Look at Table 1. Note that the difference in the bond enthalpies for the single C–C bond and a double C=C bond is 612 – 347 = 265 kJ mol^{-1}. This reflects the strength of the π-bond.

The bond enthalpies show that:

- a double bond is *stronger* than a single bond; and
- a π-bond is *weaker* than a σ-bond.

When an alkene reacts:

- the π-bond breaks; and
- the σ-bond remains intact.

Examiner tip

Remember that *saturated* means the compound has only *single* bonds between its atoms.

Unsaturated means that there is a C=C *double* bond present.

Reactions of the alkenes

Alkenes typically take part in addition reactions.

- A small molecule is added across the double bond causing the π-bond to be broken.
- Two reactant molecules react together to form one product molecule.
- An unsaturated alkene reacts and forms a saturated product.

A general equation for an alkene reacting with a small molecule in an addition reaction is shown in Figure 1.

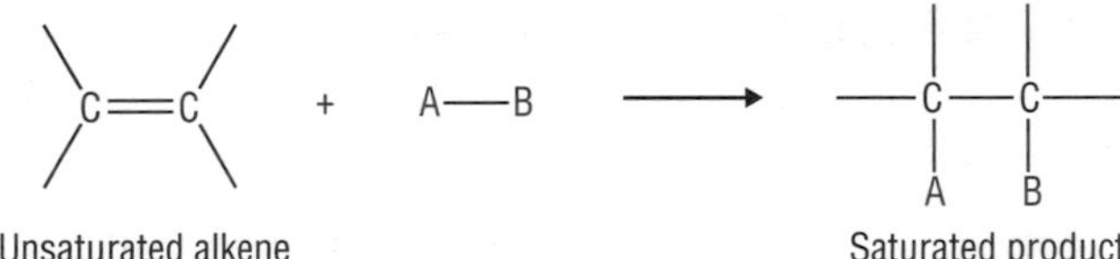

Figure 1 Typical addition reaction for an alkene

Addition of hydrogen

A mixture of hydrogen gas and a gaseous alkene is passed over a catalyst of nickel at a temperature of 150 °C.

- The hydrogen adds across the double bond and an alkane is formed.
- The reaction is sometimes known as *hydrogenation* and is an example of reduction.

The equation in Figure 2 shows the addition of hydrogen to ethene.

- Longer-chained alkenes also take part in addition reactions in exactly the same way.

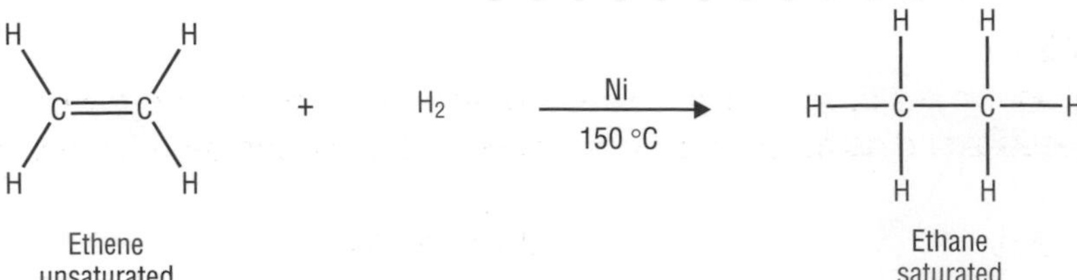

Figure 2 Hydrogenation of ethene

Figure 3 Raney nickel is a nickel–aluminium alloy that can be used as a catalyst in addition reactions. This catalyst has a large surface area, giving a high catalytic activity. (Raney is a trademark of W.R. Grace & Co., a multinational chemical company)

Examiner tip

In exams, you may be asked to predict reactions of unfamiliar organic molecules. If a molecule has a C=C bond, then think '*addition reaction*'.

Addition of halogens

Alkenes react rapidly with halogens such as chlorine, bromine and iodine at room temperature.

- The halogen adds across the double bond to give a di-substituted halogenoalkane.
- The reaction is sometimes known as *halogenation*.

The equation in Figure 4 shows the addition of chlorine to ethene.

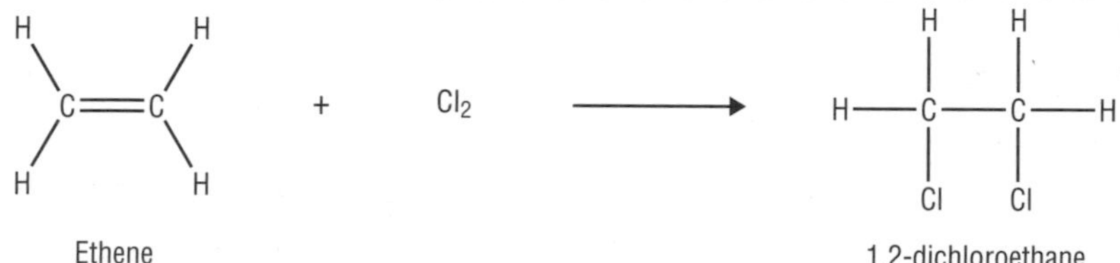

Figure 4 Halogenation of ethene with chlorine

- When bromine is added to a sample containing an alkene, the colour changes from orange to colourless. The colour change indicates that the bromine has reacted with the double bond. This reaction is used as a test for unsaturation and shows the presence of a double C=C bond.

The equation in Figure 5 shows the addition of bromine to propene.

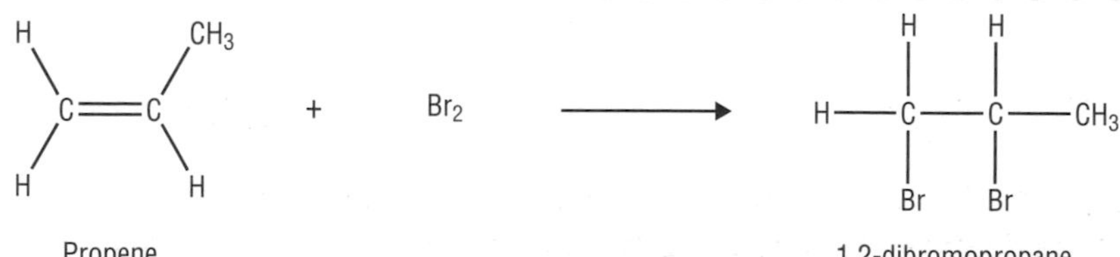

Figure 5 Halogenation of propene with bromine

Figure 6 Bromine test for unsaturation. Two test tubes containing an organic layer floating on top of an aqueous layer are shown in Figure 6. The aqueous layer contains bromine (orange). The organic layer in the test tube on the right is cyclohexene, and the bromine has been decolourised by the C=C bond present. The organic layer on the left is a saturated alkane. With no double bond present, the bromine has not been decolourised

Questions

1 In a C=C double bond, why does the π-bond react rather than the σ-bond?

2 Write equations and name the products formed in the reaction between but-2-ene and:

(a) hydrogen; **(b)** bromine.

2.1  Further addition reactions of alkenes

By the end of this spread, you should be able to . . .

* **Describe addition reactions of alkenes with both hydrogen halides and steam.**
* **Define an electrophile as an electron pair acceptor.**
* **Describe a curly arrow as the movement of an electron pair.**
* **Describe, using curly arrows, the mechanism of electrophilic addition in alkenes.**

Addition of hydrogen halides

A hydrogen halide adds across the double bond of an alkene to produce a halogenoalkane.

- Hydrogen halides include HCl, HBr and HI.
- Hydrogen halides are gases at room temperature and are bubbled into liquid alkanes.

The equation in Figure 1 shows the addition of hydrogen bromide to ethene.

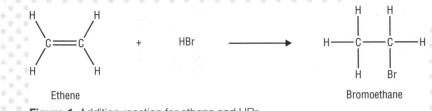

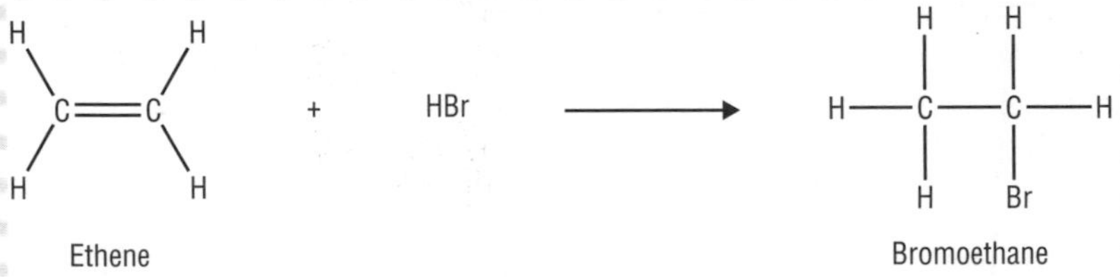

Ethene + HBr → Bromoethane

Figure 1 Addition reaction for ethene and HBr

Addition of steam

The addition of steam to alkenes is one method of preparing alcohols. This reaction is used widely in industry.

- Steam and the gaseous alkene are heated to a high temperature and pressure in the presence of a phosphoric acid catalyst.
- The process is sometimes called the *hydration* of an alkene.

The equation in Figure 2 shows the addition of steam to ethene.

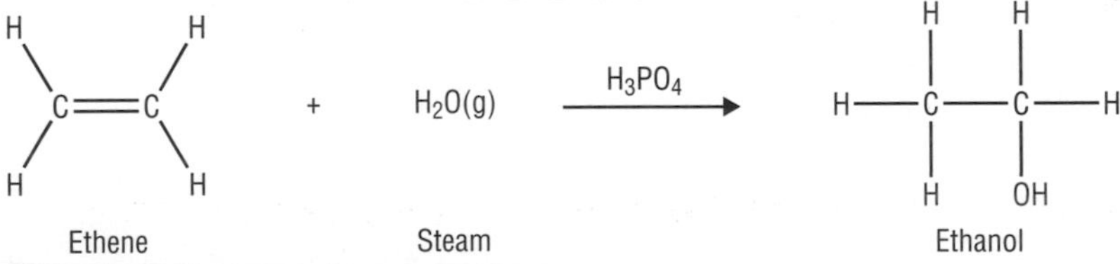

Figure 2 Hydration of ethene

Examiner tip

You need to know that H_2O is present as *steam* and that an acid catalyst is required. Don't worry about the actual conditions, though – they will not be tested in the exams.

Addition to unsymmetrical alkenes

Unsymmetrical alkenes also take part in addition reactions, forming a mixture of organic products. Propene reacts in this way with hydrogen bromide as shown in Figure 3. Note that more of one organic product is formed than the other. In this case, the major product is 2-bromopropane and the minor one 1-bromopropane.

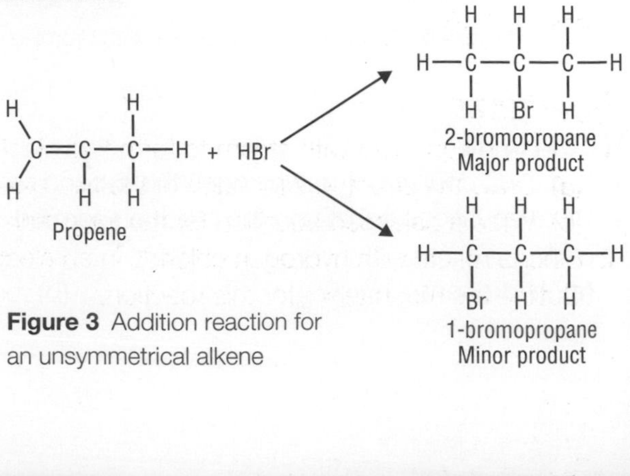

Propene + HBr → 2-bromopropane (Major product) / 1-bromopropane (Minor product)

Figure 3 Addition reaction for an unsymmetrical alkene

Examiner tip

In exams, you may be asked to predict possible products from addition to an unsymmetrical alkene. You will not be asked, though, to predict the major and minor products. You can find out more about this reaction by researching 'Markovnikoff' on the Internet.

Electrophilic addition reactions

The double bond in alkenes represents a region of high electron density. This density is due to the π-electrons in the double bond. Electrophiles are attracted to these π-electrons. The typical reaction of alkenes is an **electrophilic addition**.

Addition of hydrogen bromide

Hydrogen bromide is a polar molecule. Bromine is more electronegative than hydrogen. This causes a dipole: $H^{\delta+}$—$Br^{\delta-}$.

- The electron pair in the π-bond is attracted to the slightly positive hydrogen atom, causing the double bond to break.
- A new bond forms between one of the carbon atoms and the hydrogen atom.
- The H–Br bond breaks by *heterolytic fission*, with the electron pair going to the bromine.
- A bromide ion, Br^-, and a **carbocation** are formed.

The mechanism for this stage of the reaction is shown in Figure 4. The **curly arrows** represent movement of a pair of electrons.

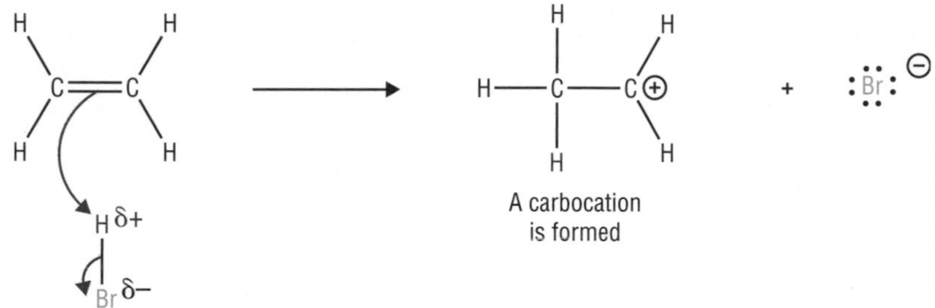
A carbocation is formed

Figure 4 Electrophilic reaction for ethene and hydrogen bromide

- The positively charged carbocation is unstable and quickly reacts with the bromide ion to form the organic product.

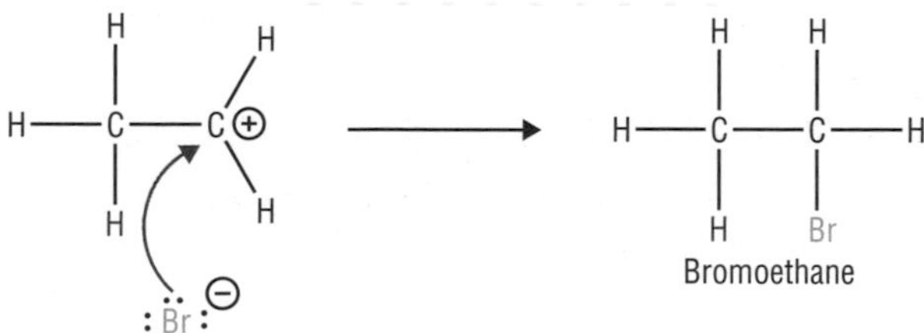
Bromoethane

Figure 6 Carbocation reacts with bromide ion

The product of the reaction is a halogenoalkane, in this case bromoethane.

In the mechanism described above, hydrogen bromide is the *electrophile*. The $\delta+$ end of the dipole accepts the electron pair.

Non-polar molecules can also react with alkenes. This reaction is discussed in detail on spread 2.1.16.

Questions

1 Propene can react with steam to form two alcohols that are structural isomers.
   (a) Draw the structures for each of the two alcohols formed.
   (b) Write a balanced equation for the formation of one of these products.
2 Ethene reacts with hydrogen chloride in an electrophilic addition reaction. Outline the mechanism for this reaction.

Examiner tip

Remember that an electrophile is an electron pair *acceptor*. See spread 2.1.8 for more details.

Key definitions

Electrophilic addition is a type of addition reaction in which an electrophile is attracted to an electron-rich centre or atom, where it accepts a pair of electrons to form a new covalent bond.

Carbocation is an organic ion in which a carbon atom has a positive charge.

A **curly arrow** is a symbol used in reaction mechanisms to show the movement of an electron pair in the breaking or formation of a covalent bond.

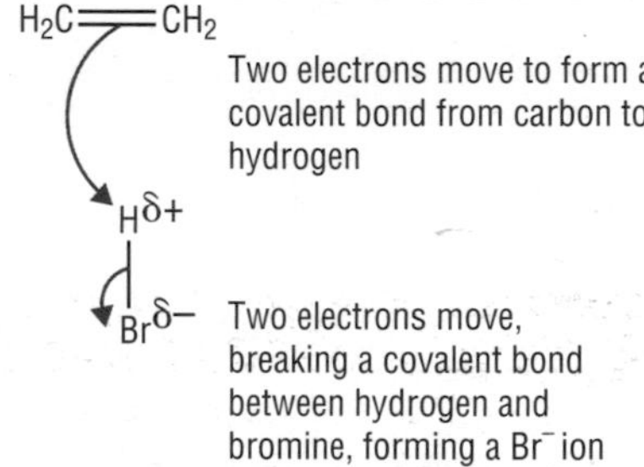
Two electrons move to form a covalent bond from carbon to hydrogen
Two electrons move, breaking a covalent bond between hydrogen and bromine, forming a Br⁻ ion

Figure 5 Curly arrows, as seen in this mechanism, represent the movement of a pair of electrons to form or break a covalent bond

Examiner tip

Take care when drawing curly arrows, especially their starting and finishing points. In an exam, a mechanism could be worth 4 marks, the majority being for correct curly arrows.

Alkenes and bromine

By the end of this spread, you should be able to . . .

- **Describe how heterolytic fission leads to electrophilic addition in alkenes, using bromine as an example.**
- **Recall the reactions of alkenes.**

Addition of bromine

Bromine reacts with alkenes at room temperature, producing a di-substituted halogenoalkane. This reaction is used as a test for unsaturation (see spread 2.1.14).

Mechanism

Bromine is a non-polar molecule. When bromine approaches an alkene, the electrons in the π-bond repel the electrons in the Br–Br bond. This induces a dipole in the Br_2 molecule, $Br^{\delta+}$—$Br^{\delta-}$. The Br_2 molecule is now slightly *polar*.

- The electron pair in the π-bond is attracted to the slightly positive bromine atom, causing the double bond to break.
- A new bond forms between one of the carbon atoms and the bromine atom.
- The Br–Br bond breaks by *heterolytic fission*, with the electron pair going to the bromine.
- A bromide ion, Br^-, and a carbocation are formed.

The mechanism for this reaction is shown in Figure 1.

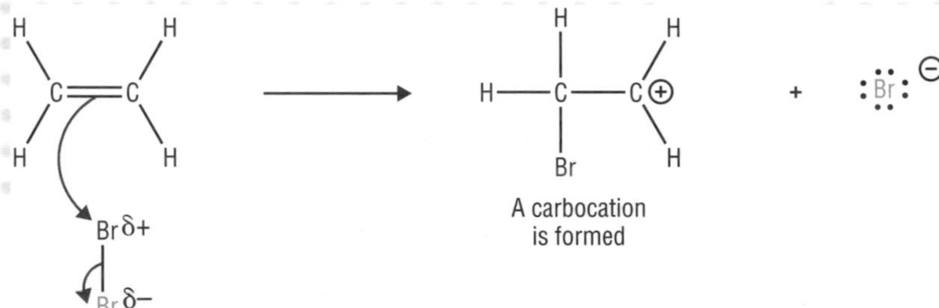

Figure 1 Electrophilic reaction for ethene with Br_2

Examiner tip

Sometimes the examiner will ask you to explain what the curly arrows are for.

Remember, they represent the movement of a pair of electrons to form or break a covalent bond.

The positively charged carbocation is unstable and quickly reacts with the bromide ion to form the organic product as shown in Figure 2.

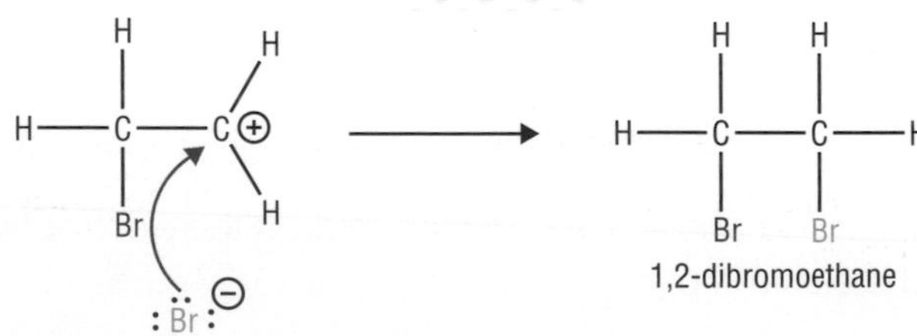

Figure 2 Carbocation reacts with bromide ion

Examiner tip

You will come across this mechanism frequently in examination papers, so make sure you can draw it for yourself!

In this mechanism, the bonds are broken by heterolytic fission and ions are formed as the intermediates.

Heterolytic fission

In this mechanism, notice how the bromine molecule breaks by heterolytic fission.

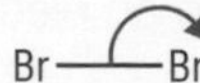

Figure 3 Heterolytic fission of a bromine molecule

Alkenes with more than one double bond.

Myrcene ($C_{10}H_{16}$) is a member of the terpene family of compounds. Terpenes are alkenes built from whole numbers of isoprene (C_5H_8) molecules. The isoprene molecules can either be linked together head to tail forming chains or can be formed into rings.

Terpenes contain carbon-to-carbon double bonds. If reacted with an excess of hydrogen in the presence of a nickel catalyst, all the double bonds will react, forming a saturated molecule.

$$C_{10}H_{16} \text{ (unsaturated)} + 3H_2 \longrightarrow C_{10}H_{22} \text{ (saturated)}$$

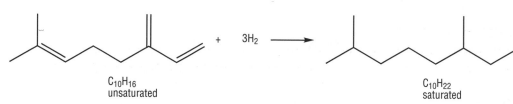

Figure 5 Reaction of myrcene with excess hydrogen to form a saturated product

Revision – the chemistry of alkenes

The flowchart in Figure 4 summarises the reactions of alkenes.

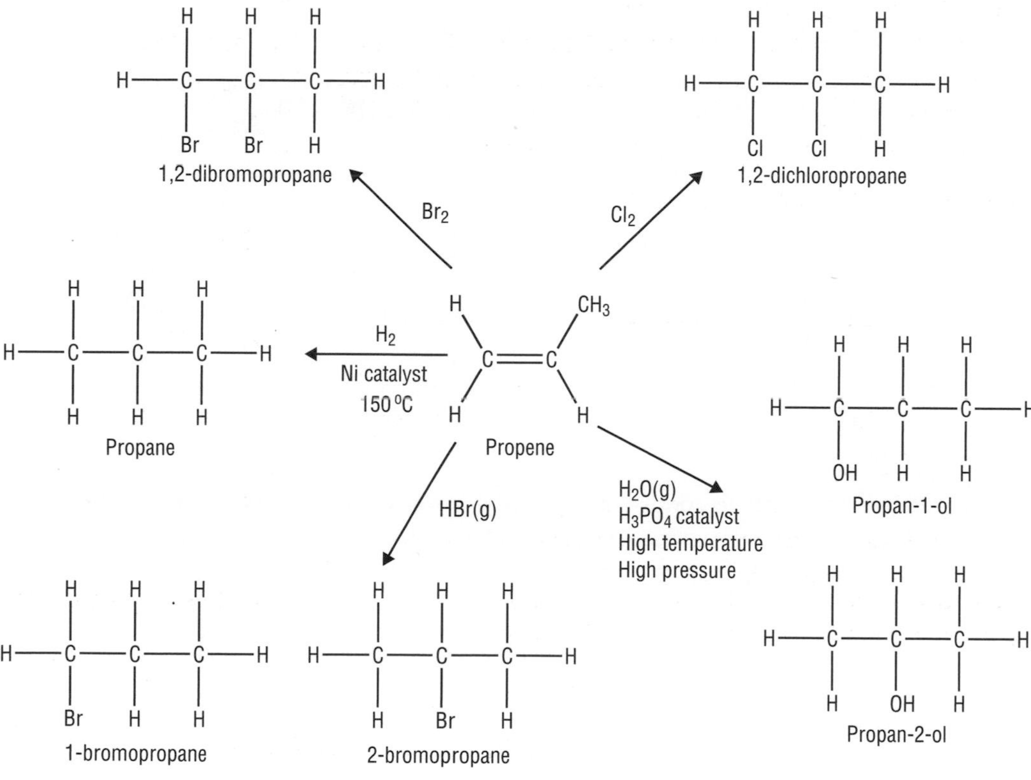

Figure 6 Reactions of alkenes

Questions

1 Propene, C_3H_6, reacts with chlorine by electrophilic addition. Using curly arrows, outline the mechanism for this reaction.

2 But-2-ene takes part in the addition reactions shown in Figure 7.
   (a) Identify the reagents and conditions used in reactions A and B.
   (b) Name product C.

3 Myrcene (see Figure 4) was reacted with an excess of bromine to form a saturated compound D.
   (a) Draw the structure of compound D.
   (b) **(i)** Deduce the molecular formula of compound D.
   (ii) Deduce the empirical formula of compound D.

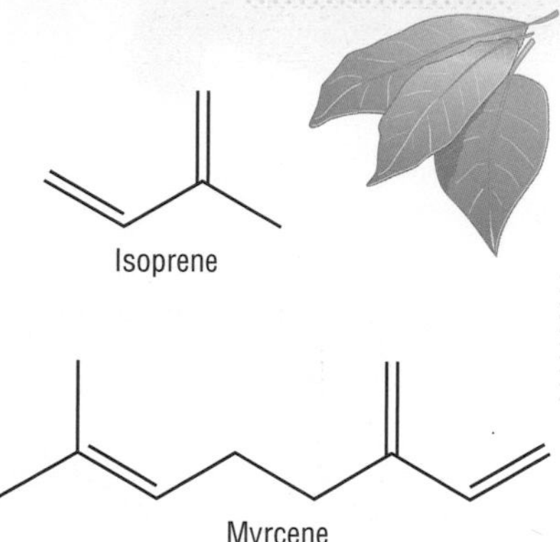

Figure 4 The structures of isoprene and myrcene. Terpenes are important biological molecules, being found in essential oils, insect pheromones and fruit pigments. Myrcene is present in bay leaves, used in cooking to add flavour to sauces, stews and meats

Examiner tip

Parts of the reaction scheme in Figure 6 have featured on a number of examination papers – make sure you know *all* the reactants and products.

This reaction scheme is a good way of revising. Could you write it out from memory? Why not try?

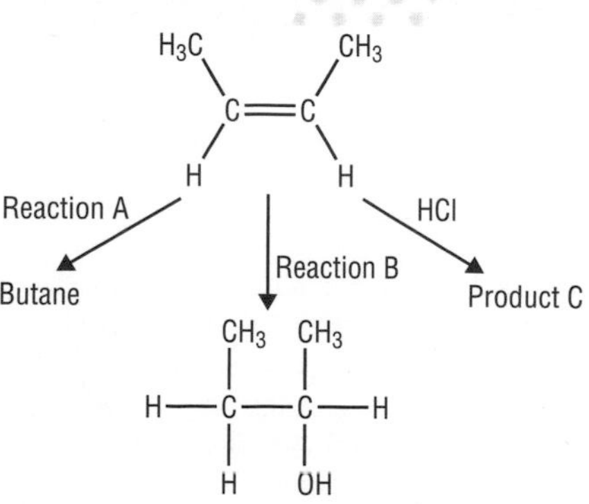

Figure 7

2.1  Industrial importance of alkenes

By the end of this spread, you should be able to . . .

* **Outline the use of alkenes in the manufacture of margarine by catalytic hydrogenation of unsaturated vegetable oils.**
* **Outline the use of alkenes in the formation of a range of polymers using unsaturated monomer units.**

Case study

Figure 1 Margarine

Napoleon III offered a prize in the 1860s for the production of a synthetic edible fat.

French pharmacist and chemist Hippolyte Mège-Mouriés developed margarine in 1869. He carried out research on modifying animal fats, which were partially solid at room temperature. He flavoured beef fats with a small quantity of skimmed milk to make the substance palatable. Unilever began the first commercial margarine plant in 1878.

Unsaturated compounds in industry

Unsaturated compounds are important starting materials for many common industrial processes. A large proportion of all industrially produced ethene is polymerised to make plastics.

Ethene can also be used to make many other important chemicals such as:

- 1,2-dichloroethane, $ClCH_2CH_2Cl$ – used as degreaser and paint remover;
- ethane-1,2-diol, $HOCH_2CH_2OH$ – widely used as an antifreeze and one of the key materials for making polyesters, such as *Terylene*, for clothing;
- ethanoic acid, CH_3COOH – used as vinegar and in organic synthesis.

The manufacture of margarine

Margarine is a general word for any substance which acts as a butter substitute. In many parts of the world, margarine has a bigger market share than butter.

- Margarine is made from animal or vegetable fats, mixed with skimmed milk and salt.
- Margarine made from vegetable oil is becoming popular on health grounds, as people turn away from animal fats in their diet.
- Margarines with a high content of mono- or polyunsaturated fats are said to be healthier than butter and other margarines. These are often made from sunflower or olive oil.

Vegetable oils are liquids containing long hydrocarbon chains, often with many double bonds, and are polyunsaturated. The oil must be hardened, so that it can be spread on bread without soaking into it.

A process called *hydrogenation* hardens these oils:

- Hydrogenation adds hydrogen molecules across double bonds in an *addition* reaction (see spread 2.1.14).
- Hydrogenation alters the individual molecule in the oil in such a way that the oil partially solidifies and hardens.
- By adding hydrogen molecules across different numbers of the many double bonds, we can make margarines with different amounts of hardness. This affects the spreadability of the margarine.
- Partial hydrogenation of unsaturated fats can transform some *cis* double bonds into *trans* double bonds (see spread 2.1.7) as a by-product. These *trans* fats are now thought to be bad for health.

Key definitions

A polymer is a long molecular chain built up from monomer units.

A monomer is a small molecule that combines with many other monomers to form a polymer.

Addition polymerisation is the process in which unsaturated alkene molecules (monomers) add on to a growing polymer chain one at a time, to form a very long saturated molecular chain (the addition polymer).

Making polymers

Polymers are long-chained molecules with large molecular masses. They are made from short-chained molecules called **monomers**. If these monomers are alkenes, and many monomer molecules can be added together to form a long polymer chain. This process is called **addition polymerisation**.

- The monomer units are based on different alkenes.
- Monomers are *unsaturated* and have a double bond.
- **Addition polymers** have *saturated* chains with no double bonds.

Key definition

An **addition polymer** is a very long molecular chain, formed by repeated addition reactions of many unsaturated alkene molecules (monomers).

Addition polymers are made from *one* type of monomer only. For example, poly(ethene) is made from the addition of thousands of ethene molecules as shown in Figure 2.

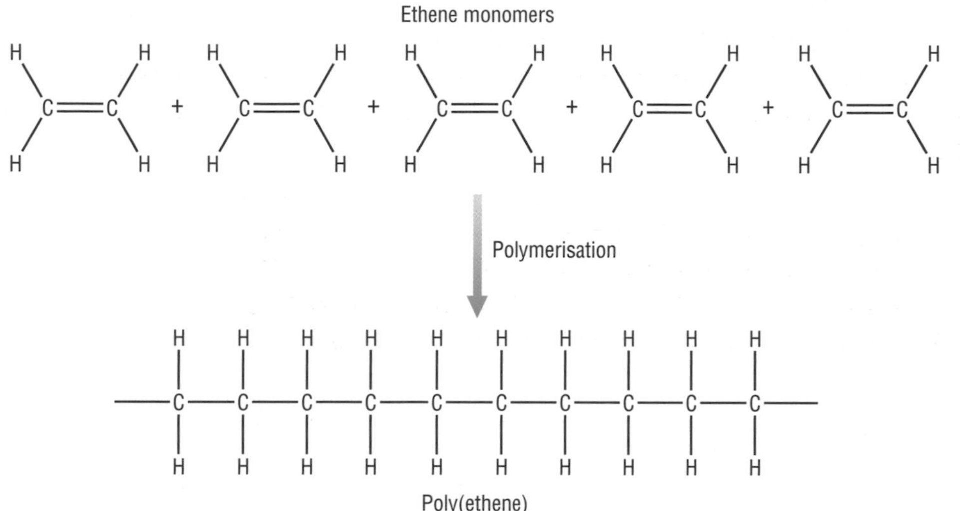

Figure 2 Alkene molecules form addition polymers

Case study

Although natural polymers have been known for many years, the early synthetic polymer industry started in the 1900s. Here are the key milestones in the birth of the polymer:

- 1909 Bakelite developed
- 1930 Poly(styrene)
- 1938 Poly(vinyl chloride) (PVC)
- 1939 Poly(ethene)
- 1940 Polyurethane
- 1942 Polyester
- 1950 Teflon

Polymers and industry

The polymer sector of the chemical industry is made up of about 7500 companies, employing nearly 190 000 people in the UK. Addition polymerisation is carried out in two important industrial processes, as described below.

Radical polymerisation

Radical polymerisation requires temperatures of 200 °C and very high pressures.

This reaction can lead to branching of the polymer chain and the production of polymer mixtures. Typically poly(phenylethene), poly(styrene) and branched poly(ethene) are made in this way.

The Ziegler–Natta process

The Ziegler–Natta process involves the use of specialist catalysts such as $TiCl_3$ and $Al(C_2H_5)_2Cl$ at 60 °C.

The German chemist Karl *Ziegler* discovered this titanium-based catalyst in 1953. Later, an Italian, Giulio *Natta*, modified the catalytic mixture, improving control of the polymer formed.

In this process, the alkene is passed over the catalyst. The conversion is low and any unreacted alkene is recycled and passed over the catalyst repeatedly. This is the most common method for the manufacture of non-branched poly(ethene).

Figure 3 Plastics plant. Plastics are polymers produced from natural gas and oil derivatives

Questions

1 Name three important chemicals made from ethene.
2 State one key advantage of the Ziegler–Natta process over that of radical polymerisation.
3 Draw a section of a polymer made from three molecules of ethene.

2.1 (18) Polymer chemistry

By the end of this spread, you should be able to . . .

- **Describe the addition polymerisation of alkenes.**
- **Deduce the repeat unit of an addition polymer obtained from a given monomer.**
- **Identify the monomer that would produce a given section of an addition polymer.**

Figure 1 Plastics for the home

> **Key definition**
>
> A **repeat unit** is a specific arrangement of atoms that occurs in the structure over and over again. Repeat units are included in brackets, outside of which is the symbol *n*.

> **Examiner tip**
>
> You will sometimes be asked to draw a polymer from a monomer.
>
> This is easy if you draw the 4 groups at right angles. Then simply replace the double bond with a single bond. Draw side links and you have the repeat unit.

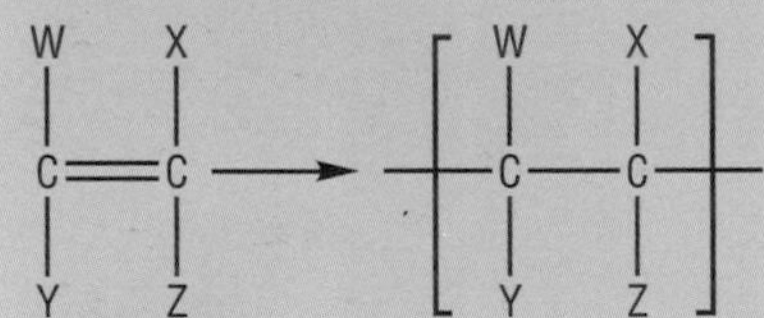

Addition polymerisation

Addition polymerisation is used in the manufacture of a variety of polymers, starting from different alkene monomers. Polymers can be put to various uses due to their differing physical properties.

A general equation for addition polymerisation is shown in Figure 2.

Figure 2

Polymers have been formed from thousands of monomer units, so it is impossible to draw the full structural formula of any particular polymer. In order to represent polymers we use the term **repeat unit**.

Common polymers

The equations in Figures 3–6 show the formation of several common polymers. Notice the link between the monomer unit and the polymer.

- The monomer is *unsaturated* with a C=C double bond.
- The polymer is *saturated* with single bonds only.
- *n* monomer molecules ⟶ one polymer molecule with *n* repeat units.

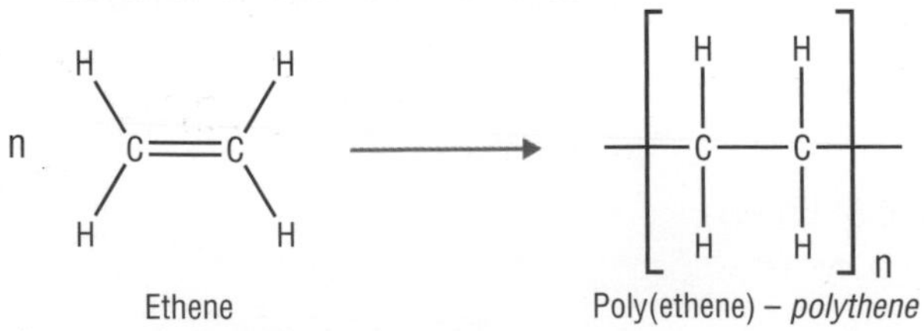

Ethene — Poly(ethene) – *polythene*

Figure 3 Formation of poly(ethene)

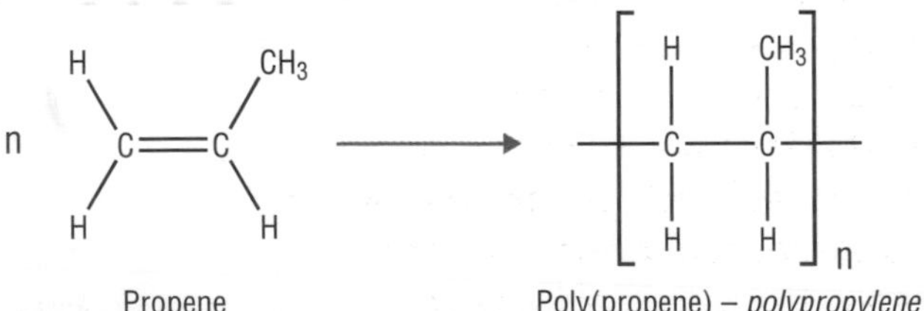

Figure 4 Formation of poly(propene)

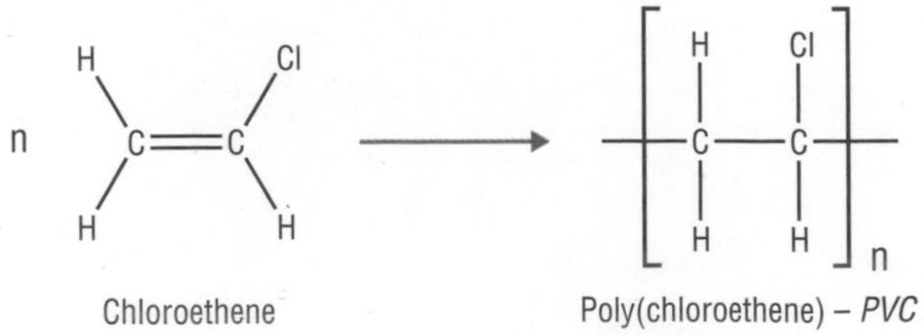

Chloroethene — Poly(chloroethene) – *PVC*

Figure 5 Formation of poly(chloroethene) PVC

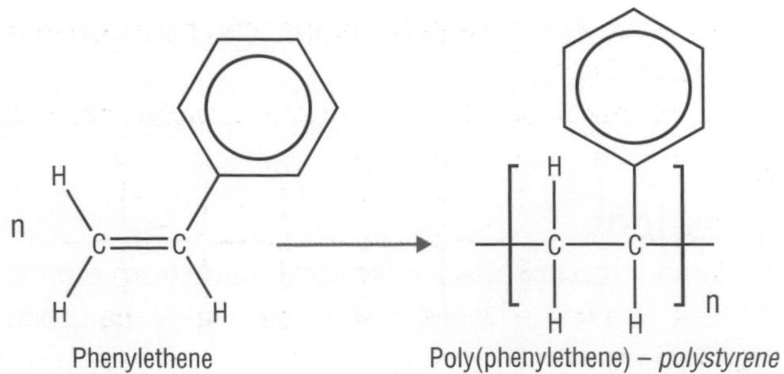

Figure 6 Formation of poly(phenylethene) (polystyrene)

Teflon – found by mistake

Roy Plunkett, an industrial chemist working for Du Pont, was cleaning out a cylinder of tetrafluoroethene gas when, to his surprise, he found that some of the gas had turned into a white sticky solid.

Further investigations into the curious substance found that it was resistant to heat and chemically unreactive. He could not get other substances to stick to it, as it had low surface friction. He realised that this substance was a polymer.

Du Pont later named it *Teflon* (see Figure 8).

In the 1950s, scientists discovered a method for bonding Teflon to aluminium. This increased sales of Teflon significantly and gave birth to the non-stick frying pan!

Figure 7 Roy Plunkett – discoverer of Teflon (PTFE)

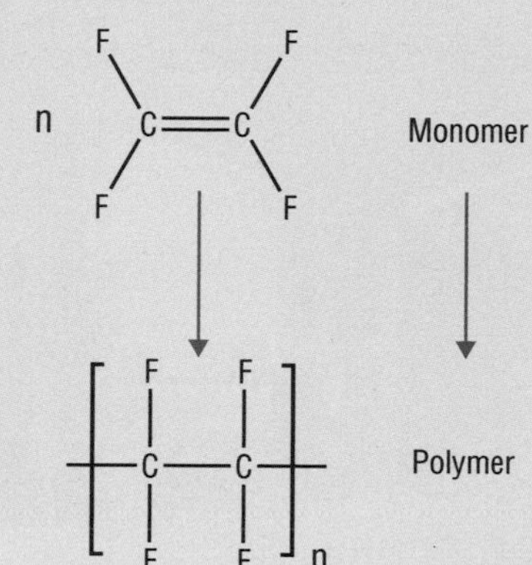

Poly(tetrafluoroethene) (Teflon or PTFE)

Figure 8 Poly(tetrafluoroethene) (Teflon or PTFE)

Identifying the monomer

If it is possible to identify the repeat unit in a polymer, then it should also be possible to identify the monomer from which the polymer has been formed.

Orlon is a polymer used as a synthetic fibre. It is resistant to sunlight and atmospheric gases, making it suitable for outdoor use as a tent material and sunshades. The structure of Orlon is shown in Figure 9 with its repeat unit shaded. The monomer is based on this shaded region and must include a carbon-to-carbon double bond.

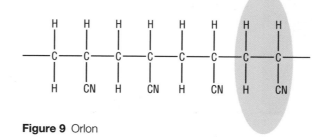

Figure 9 Orlon

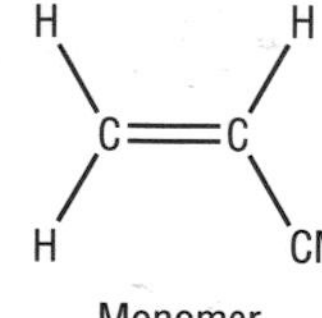

Questions

1 Draw a section of the polymer that can be made from 1-chloropropene, showing two repeat units.

2 The polymer shown below is called poly(vinyl alcohol), PVA.

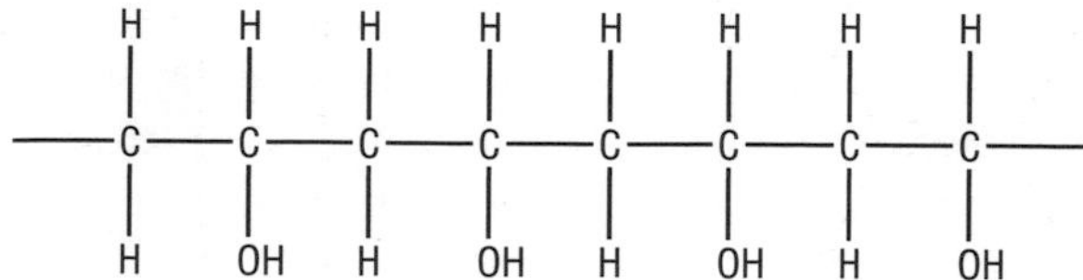

(a) Identify the repeat unit.

(b) Draw the structure of the monomer that forms poly(vinyl alcohol).

2.1 (19) Polymers – dealing with our waste

How Science Works

By the end of this spread, you should be able to . . .

* **Describe some of the uses for certain polymers.**
* **Outline how waste polymers are recycled, separated and processed.**

Living in the plastic age

Polymers have become an integral part of everyday life. In many instances, plastics have almost replaced traditional, natural materials such as wood, metal and wool. In particular, many household items are now made from plastic (see Figure 1).

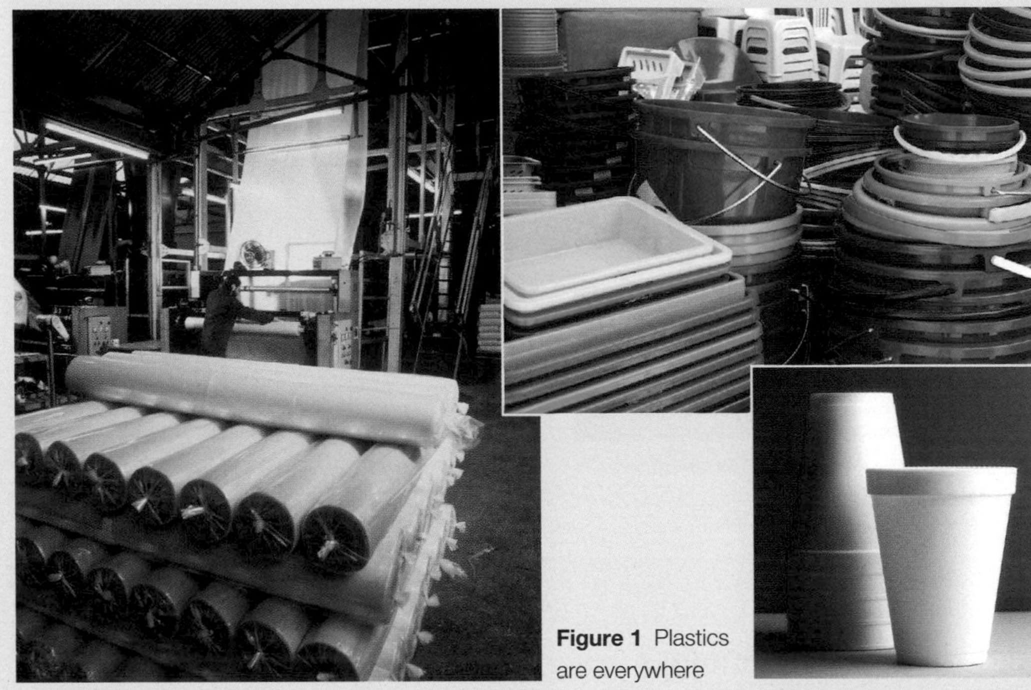

Figure 1 Plastics are everywhere

Figure 2 Plastic recycling. Discarded PET plastic drinks bottles being loaded into a compactor at a recycling facility. At the start of the process, the bottles are compacted into bales, then shredded. The resulting plastic chips are then cleaned and sent to be re-used in new bottles

Key definition

Biodegradable material is a material that is broken down naturally in the environment by living organisms.

Uses for different polymers

- Poly(phenylethene) or *polystyrene*: a relatively cheap plastic used in foam packaging, insulation, model-making and in the food retail trade. **Biodegradable materials** are now replacing polystyrene packing as environmental pressure grows to use less of this type of packaging.
- Poly(propene) or *polypropylene*: used in food packaging; for containers that are dishwasher-safe; as a fibre in carpets; and for making synthetic ropes. It is also used as a material for certain laboratory equipment, due to its resistance to chemical attack.
- Poly(ethene) or *polythene*: the most widely-used plastic; used to make grocery bags, shampoo bottles, toys, etc.

Addition polymers do not usually react with the substances placed inside them. They are fairly durable and do not break down naturally. These physical properties make polymers ideal for a broad range of uses, but their disposal poses problems.

Polymer waste

- Polymers represent just over 10% of municipal waste. However, overflowing landfill sites and unsightly dumped litter highlight the problem of non-biodegradable plastics. Recently, we have been warned that unless drastic action is taken, landfill space in the UK will run out within a decade.
- In 2001, the environment agency reported that 85% of plastic disposal in the UK was landfill, with approximately 8% incinerated and only 7% being recycled.

- The non-biodegradability of plastics is a major reason why plastics need to be recycled. However, much of the plastic waste we create includes additives such as colorants, stabilisers and plasticisers, which often contain poisonous heavy metal ions.
- New legislation may well restrict the use of polymers in packaging until methods of waste management are improved.

Recycling, incineration and the development of biodegradable plastic all provide some hope for the future.

Recycling polymers

The best way to protect the environment from non-biodegradable plastics is to re-use the polymer without processing. In recent years, supermarket delivery crates and display cases would have been used only once. They are now being reused, thus saving the Earth's precious oil resources.

For every tonne of plastic recycled, it is estimated that 1.8 tonnes of oil are saved. If this was as petrol, a car could travel 25 000 miles from the saving – that is once around the Earth!

Recycling started in the early 1980s, when materials such as poly(ethene) terephthalate (PET or PETE), used for plastic bottles, were first recycled. Today PET and high density poly(ethene) (HDPE) plastic bottles are the most common polymers to be recycled.

- Approximately 58% of recycled PET finds a market in the carpeting industry. Some is also used to make new food containers.
- The main use for HDPE is in the manufacture of new bottles.

Recycling involves two stages, sorting and reclamation.

Sorting

Before plastic waste can be processed and converted into new products, it has to be sorted. The polymer identification codes shown in Table 1 are used in this separation process.

Originally, recycling companies relied on labour-intensive, tedious hand-sorting as a method of separating plastic types. Newer systems use optical scanning techniques to separate PET bottles from HDPE. Other technological advances have enabled companies to distinguish polyvinyl chloride, PVC, from other plastics. If PVC is not separated effectively from PET, it undermines the recycling process. Even a small quantity of PVC in PET can render the product unsuitable for re-use. In addition, PVC is known to release poisonous dioxins when heated.

Reclamation

Once sorted, the polymers are now processed. This usually involves mechanically chopping the polymers into small flakes, before washing them to remove any impurities. These flakes are then sent to manufacturing companies, where the waste is turned into new materials by melting the pellets and remoulding them.

- PET bottles are converted into a wide range of new materials, such as carpets, clothing and of course new bottles.
- HDPE is often re-used to make hard plastic materials such as plastic boxes, water butts and bins.
- LDPE waste is frequently made into plastic refuse sacks.

Questions

1 Give two uses of polystyrene.
2 What do you understand by the term non-biodegradable?
3 Why is it important to replace polystyrene as a packaging material?
4 When PVC is burned poisonous gases may be released. Suggest two such gases.

 PETE	Poly(ethene) terephthalate is used for plastic soft drink bottles and pre-packed meals.
 HDPE	High density poly(ethene) is used for milk bottles, detergent bottles and butter tubs.
 V	Polyvinyl chloride is used for medical tubing, cable insulation and drainpipes.
 LDPE	Low density poly(ethene) is used in squeezy bottles and frozen food bags.
 PP	Poly(propene) is used in yoghurt tubs and medicine bottles.
 PS	Polystyrene is used for cups, plates and cutlery, for CD cases and to serve hot food for its insulating properties.
 OTHER	Made from none of the above or made from a mixture of resin.

Table 1 Polymer identification codes

Figure 3 Coloured poly(ethene) pellets

2.1 (20) Other uses of polymer waste

How Science Works

By the end of this spread, you should be able to . . .

* **Explain how polymers undergo combustion for energy production.**
* **Explain how polymers are used as a feedstock in cracking for producing plastics and other chemicals.**
* **Outline the role of chemists in minimising environmental damage by removal of toxic waste and development of biodegradable and compostable polymers.**

Figure 1 Second-hand tyres awaiting processing to generate electricity

Polymers as a fuel source

Burning polymers under controlled conditions produces heat energy, which can be harnessed to make electricity. Agricultural plastics, such as poly(ethene), have long been used to cover crops in the growing season. Now, these plastics are being turned into briquettes and burned to produce heat. A South Korean company is now working on similar technology to generate electricity.

A process that burns old rubber-based tyres is also being used to generate electricity (see Figure 1). This has the added bonus of removing unsightly car tyres from the environment. The first plant in the UK was in Wolverhampton and there are several more worldwide.

Figure 2 Research is carried out into ways of breaking down polymers into more useful materials

Feedstock recycling

In addition to producing monomers, processes are being developed to convert polymers into synthesis gas, a mixture of hydrogen and carbon monoxide. Hydrocarbons and synthesis gas can be used as a chemical feedstock for conversion into useful products or as a fuel at oil refineries.

In the US and Europe, many scientists are working to increase the efficiency and cost-effectiveness of feedstock recycling. if successful, it may one day be possible to convert common household plastics into other useful organic materials rather than being disposed of in landfill sites.

Recycling PVC

The recycling of poly(vinyl chloride), PVC, is problematic because of its high chlorine content. PVC recycling is uneconomical as the costs associated with the recycling process make the end product more expensive than that made from crude oil.

Recently, incineration has been used to prevent PVC going into landfill sites. Unfortunately, combustion of PVC releases toxic fumes into the environment and may also cause corrosion in the plant itself. Acidic hydrogen chloride fumes are frequently detected when PVC is heated to high temperatures as part of the recycling process. The incinerators have been equipped with pollution control apparatus to minimise the release of such harmful emissions.

However, new technological advances have led to the first commercial PVC recycling plant being built in Italy (see Figure 3). PVC is separated from other scrap by dissolving in solvents. High-quality PVC is then recovered by precipitation from the solvent. The solvent is recovered and used again. This process recycles the PVC coatings of electrical wiring and other PVC waste.

Figure 3 Solvay's Vinyloop PVC Plant

Biodegradable and compostable polymers

A new generation of plastics known as bioplastics are *biodegradable* or compostable. These plastics are derived from renewable raw materials such as starch, maize, cellulose and lactic acid (see Figure 4).

Bioplastics are manufactured by non-hazardous processes and are kinder to the environment. When disposed of, bioplastics degrade naturally in the environment to carbon dioxide and water.

- A biodegradable plastic will break down as a result of bacterial activity.
- Compostable plastics have to meet strict criteria. They must break down by a biological process during composting to carbon dioxide, water, inorganic compounds, and biomass. Compostable plastics break down in a similar time frame as natural compostable materials, such as cellulose.
- Biobags and cutlery made from cornstarch are compostable and could replace oil-derived poly(ethene) in the future. The latest generation of supermarket bags are now made from plant starch. Since these bags are compostable, they can also be used as bin liners for the collection of food waste. The bag and the waste can then be composted together (see Figures 5 and 6).
- Compostable disposal tableware such as plates, cups and food trays made from sugar cane fibre are also replacing materials made out of polystyrene (see Figure 8). Although compostable plastics are less resistant to breakdown than polystyrene, this new generation of polymers can withstand temperatures up to 100 °C.
- Poly(lactic acid) is used for cold-drink cups that biodegrade in 180 days.

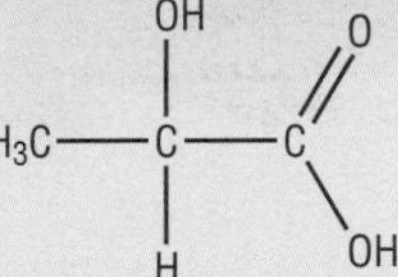

Figure 4 Lactic acid is now used to make compostable polymers

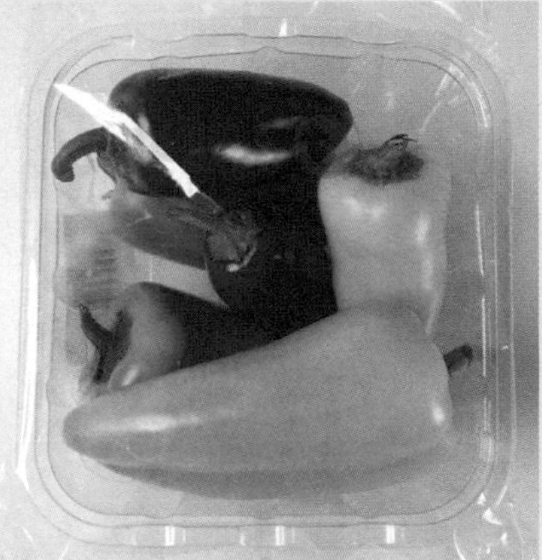

Figure 6 Peppers wrapped in poly(lactic acid) film

Figure 7 A new mobile phone has gone on sale in Japan that uses bioplastic based on polylactic acid for its casing. Polylactic acid is biodegradable and, with so many mobile phones becoming 'waste' each year, the casing can simply be thrown away to biodegrade in the soil. One biophone has even been produced with a sunflower seed in its case!

Figure 5 Carrier bag made from plant starch. Bag and waste can be composted together

Questions

1 A car tyre has a mass of 10 kg. Tyres can be disposed of using a process known as pyrolysis. 41% by mass of a car tyre can be converted into carbon and 22% by mass can be converted into gas, which is predominantly methane.

(a) Calculate the mass of carbon that can be collected from the pyrolysis of four car tyres.

(b) In the UK 90 000 tyres are treated in this way each year. What volume of methane, at room temperature and pressure, is produced in this process?

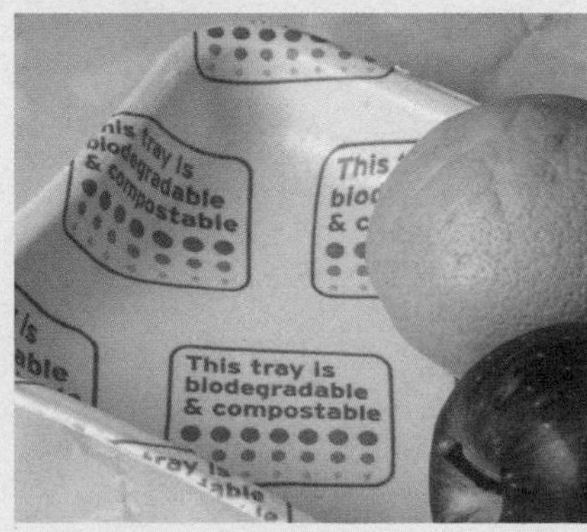

Figure 8 Compostable tray

2.1 Basic concepts and hydrocarbons summary

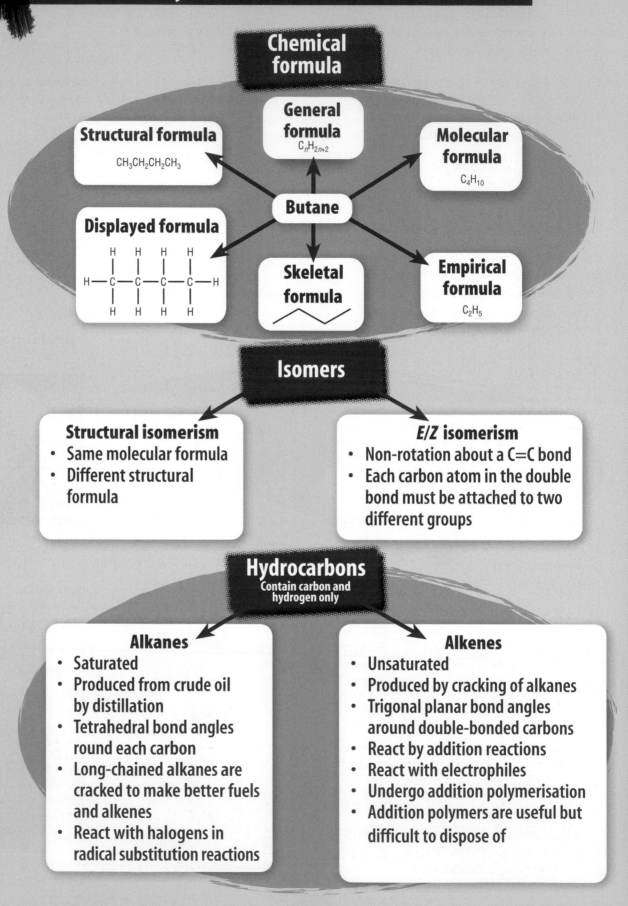

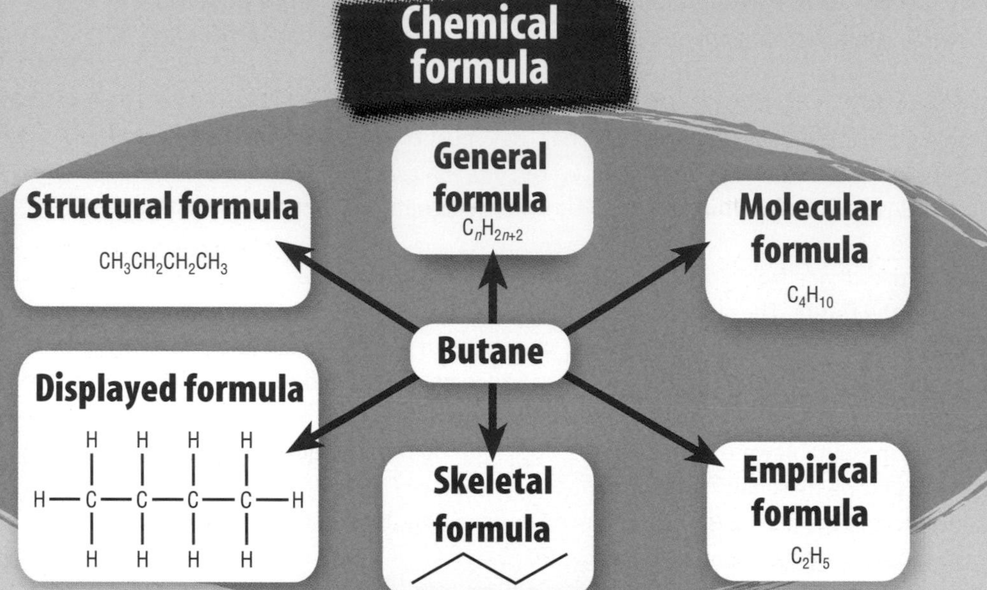

Isomers

Structural isomerism
- Same molecular formula
- Different structural formula

***E/Z* isomerism**
- Non-rotation about a C=C bond
- Each carbon atom in the double bond must be attached to two different groups

Hydrocarbons

Contain carbon and hydrogen only

Alkanes
- Saturated
- Produced from crude oil by distillation
- Tetrahedral bond angles round each carbon
- Long-chained alkanes are cracked to make better fuels and alkenes
- React with halogens in radical substitution reactions

Alkenes
- Unsaturated
- Produced by cracking of alkanes
- Trigonal planar bond angles around double-bonded carbons
- React by addition reactions
- React with electrophiles
- Undergo addition polymerisation
- Addition polymers are useful but difficult to dispose of

Practice questions

1 Define each of the following terms:
(a) saturated hydrocarbon; (b) homologous series;
(c) empirical formula; (d) structural formula.

2 Draw the displayed formula for the following hydrocarbons:
(a) 2-methylpentane;
(b) 2-chloro-3-methyloctane;
(c) 2,2-dichloro-3,4-dimethylnonane.

3 An organic compound of bromine, **X**, has a molecular mass of 136.9 and the following percentage composition by mass: C, 35.0%; H, 6.6%: Br, 58.4%.
(a) Calculate the empirical formula of **X**.
(b) Show that the molecular formula of **X** is the same as its empirical formula.
(c) Draw, and name, all possible isomers of **X**.

4 (a) What do you understand by the term functional group?
(b) Draw two molecules of your choice to show two different functional groups. Name the functional group present in each molecule.

5 An alkene, **Y**, has the empirical formula CH_2 and a molecular mass of 56.0.
(a) Identify the molecular formula of the alkene **Y**.
(b) There are four isomers of **Y** that are alkenes. Draw and name each of the isomers.

6 In organic reactions, single bonds can be broken homolytically or heterolytically.
Explain the terms:
(a) homolytic fission; (b) heterolytic fission.
Use equations in your answer.

7 Define the terms:
(a) cracking; (b) fractional distillation.

8 (a) Write an equation for the combustion of hexane in a plentiful supply of oxygen.
(b) Which different products would be formed if the supply of oxygen were limited?

9 Explain why the petrochemical industry is keen to produce cyclic and branched alkanes from straight-chained isomers.

10 Some reactions of the alkene pent-2-ene are shown below.

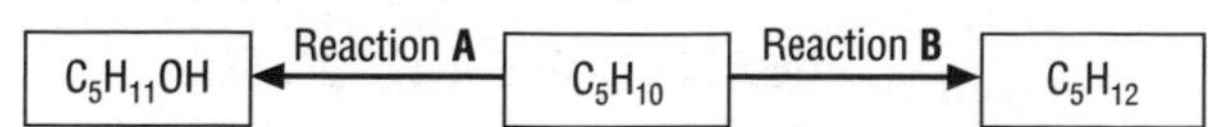

(a) For reaction **A**, state the reagents and conditions, and write a fully balanced equation.
(b) Draw the two structural isomers produced in reaction **A**.
(c) For reaction **B**, state the reagents and conditions, and write a fully balanced equation.

11 Butane and but-2-ene both react with bromine under different conditions.
(a) Discuss the mechanisms for the two reactions, stating any reaction conditions.
(b) State how the bond in bromine is broken in each of the two mechanisms.

12 Compound A is a chloroalkene with a percentage composition by mass: C, 24.7%; H, 2.1%; Cl, 73.2%.
(a) Calculate the empirical formula of compound A, showing all of your working.
(b) The relative molecular mass of A is 145.5.
(i) What does this tell you about the number of chlorine atoms in the compound?
(ii) Determine the molecular formula of compound A.

13 $C_3H_4Cl_2$ has five structural isomers that are chloroalkenes. Three of these isomers are shown below.

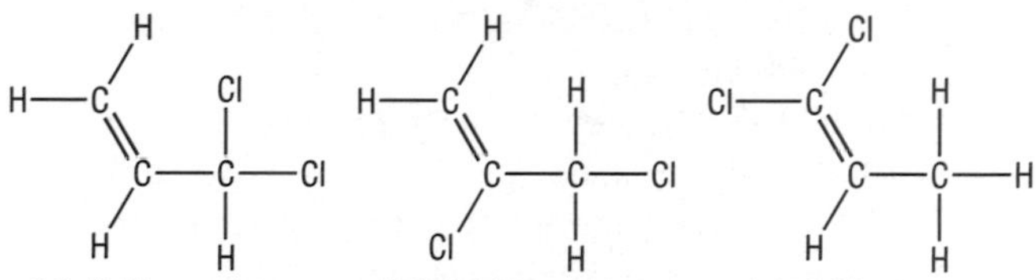

(a) Each of the other two structural isomers of $C_3H_4Cl_2$ have *E* and *Z* stereoisomers.
Draw and name these stereoisomers.
(b) Chloroalkenes can undergo addition polymerisation.
Draw a section of the polymer made from 2,3-dichloropropene, showing two repeat units.
(c) Addition polymers are difficult to dispose of.
State two general problems associated with the disposal of such polymers.

14 Alkenes undergo electrophilic reactions to form saturated compounds.
(a) Define the term electrophile.
(b) The reaction between hydrogen chloride and methylpropene is an electrophilic addition reaction.
Describe, with the aid of curly arrows, the mechanism for this reaction.
Show the intermediate, product, relevant dipoles and lone pairs of electrons.

15 Polymer A, shown here, can be formed from an alkene monomer.

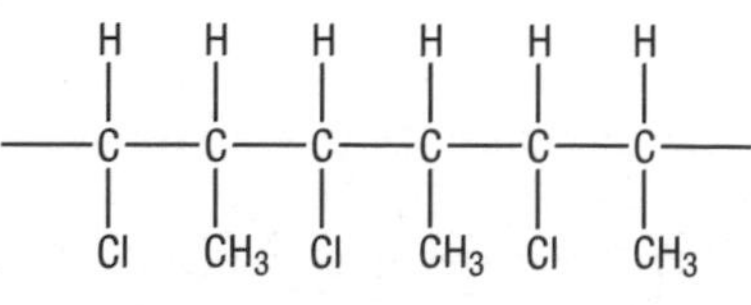

(a) State the type of polymerisation involved in the formation of A.
(b) Draw a circle around the repeat unit of polymer A.
(c) Identify the monomer responsible for the formation of polymer A.
(d) Name the monomer.

2.1 Examination questions

1 Bromine can react with both alkanes and alkenes. The type of reaction depends on whether the Br–Br bond breaks by homolytic or heterolytic fission.

(a) (i) Write a balanced equation to show the **homolytic** fission of the Br–Br bond. [1]

(ii) Write a balanced equation to show the **heterolytic** fission of the Br–Br bond. [1]

(iii) Each product of the fission of the Br–Br bond may be an electrophile, a free radical or a nucleophile. Choose from the products in parts **(i)** and **(ii)**
a free radical,
a nucleophile. [2]

(b) Hexane reacts with Br_2 in the presence of ultraviolet light.

$C_6H_{14} + Br_2 \rightarrow C_6H_{13}Br + HBr$

(i) State the type of reaction. [1]

(ii) Identify the three structural isomers of the product, $C_6H_{13}Br$, that could be formed from this reaction with hexane. [3]

(c) Hex-3-ene reacts with Br_2 to produce 3,4-dibromohexane.
Describe, with the aid of curly arrows, the movement of the electrons in the mechanism.
Show the intermediate, any relevant dipoles and lone pairs of electrons.

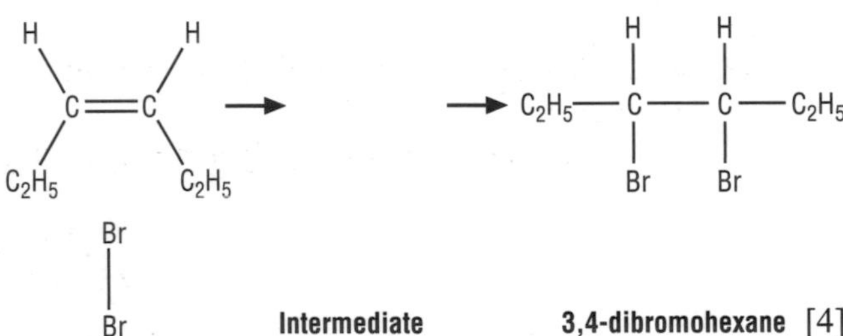

[4]

(d) The mechanism in **(c)** shows *cis*-hex-3-ene reacting with Br_2. *Trans*-hex-3-ene also reacts with Br_2 to produce 3,4-dibromohexane.

(i) How does the structure of *trans*-hex-3-ene differ from that of *cis*-hex-3-ene? [1]

(ii) Explain why both *cis* and *trans*-hex-3-ene react with Br_2 to produce the same structural isomer. [1]

[Total: 14]

(Jun 07 2812)

2 Crude oil is first separated by fractional distillation. The fractions can then be refined further by cracking, reforming and isomerisation.
The reaction sequence below shows the production of heptane, C_7H_{16}, from fractional distillation of crude oil, followed by cracking, reforming and isomerisation.

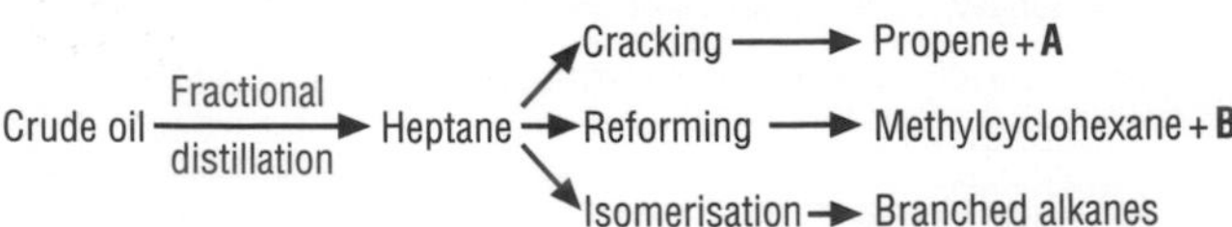

(a) What is meant by the term *fractional distillation*? [1]

(b) The cracking of heptane produces propene and **A**. Write a balanced equation for this cracking of heptane. [1]

(c) The reforming of heptane produces methylcyclohexane and **B**.

(i) Show the structural formula of methylcyclohexane. [1]

(ii) Write a balanced equation for this reforming. [1]

(d) The isomerisation of heptane produces **seven** branched alkanes, five of which are shown below.

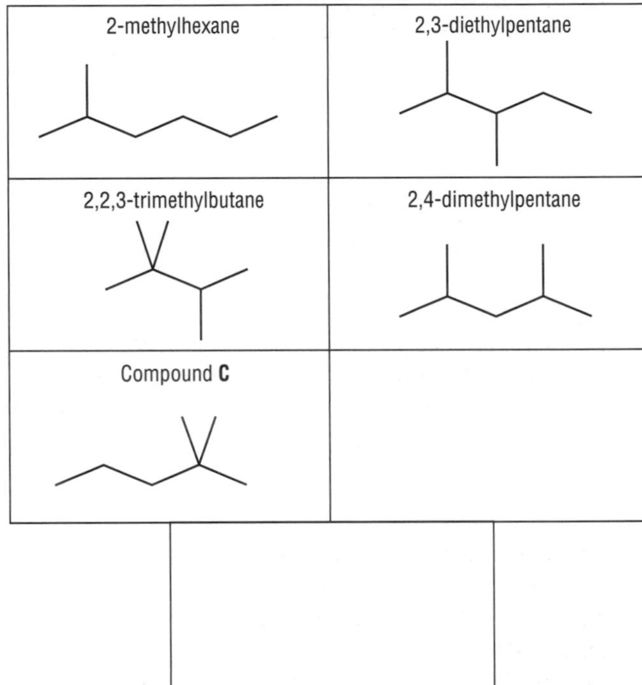

(i) Name compound **C**. [1]

(ii) In the boxes above, draw skeletal formulae for the other **two** branched alkanes formed by isomerisation of heptane. [2]

(iii) Predict which of 2-methylhexane, 2,3-dimethylpentane and 2,2,3-trimethylbutane has the lowest boiling point. [1]

(iv) Explain why 2-methylhexane, 2,3-dimethylpentane and 2,2,3-trimethylbutane have different boiling points. [2]

(e) Crude oil and its fractions are described as non-renewable fossil fuels. To reduce the demand for fossil fuels, ethanol can be mixed with petrol. Ethanol is an example of a renewable biofuel.

(i) What is meant by a *biofuel*? [1]

(ii) Why are fossil fuels *non-renewable* but ethanol *renewable*? [1]

[Total: 12]

(Jan 07 2812)

Answers to examination questions will be found on the Exam Café CD.

3 Butane, C_4H_{10}, under certain conditions, reacts with Cl_2 to form a mixture of chlorinated products.
One possible product is C_4H_9Cl.
$C_4H_{10} + Cl_2 \rightarrow C_4H_9Cl + HCl$

(a) **(i)** State the conditions. [1]
(ii) Write equations to show the mechanism of this reaction. [3]
(iii) Write **one** equation for a reaction that would terminate this mechanism. [1]
(iv) State the type of bond fission involved in the initiation step. [1]

(b) One other possible product of the reaction between butane and chlorine is compound **J**, $C_4H_8Cl_2$, shown below.

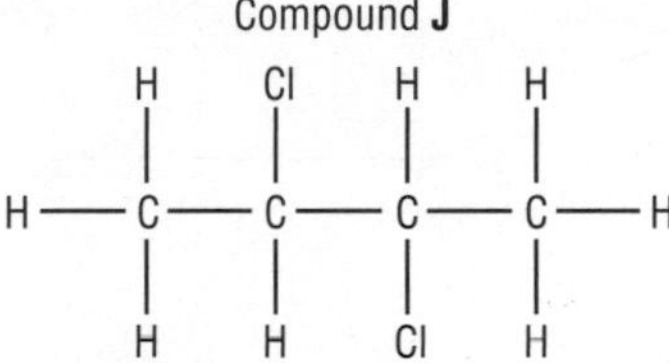
Compound **J**

(i) Name compound **J**. [1]
(ii) Draw the skeletal formula of compound **J**. [1]
(iii) In addition to compound **J**, suggest **one** other possible structural isomer of $C_4H_8Cl_2$ that could have been formed in this reaction. [1]

[Total: 9]
(Jun 07 2812)

4 Crude oil is a complex mixture of hydrocarbons. Initial separation is achieved by fractional distillation. The separate fractions are then further refined to produce hydrocarbons such as decane.

(a) **(i)** State what is meant by the term *hydrocarbon*. [1]
(ii) A molecule of decane contains ten carbon atoms. State the molecular formula of decane. [1]
(iii) Deduce the empirical formula of decane. [1]

(b) Dodecane, $C_{12}H_{26}$, is a straight chain alkane that reacts with chlorine to produce a compound with molecular formula $C_{12}H_{25}Cl$.
$C_{12}H_{26} + Cl_2 \rightarrow C_{12}H_{25}Cl + HCl$
The reaction is initiated by the formation of chlorine free radicals from chlorine.
(i) What is meant by the term *free radical*? [1]
(ii) State the conditions necessary to bring about the formation of the chlorine free radicals from Cl_2. [1]
(iii) State the type of bond fission involved in the formation of the chlorine free radicals. [1]
(iv) The chlorine free radicals react with dodecane to produce $C_{12}H_{25}Cl$. Write equations for the **two** propagation steps involved. [2]
(v) How many different structural isomers can be formed when chlorine reacts with dodecane to form $C_{12}H_{25}Cl$? [1]

(c) Dodecane, $C_{12}H_{26}$, can be cracked into ethene and a straight chain alkane such that the molar ratio ethene : straight chain alkane is 2 : 1.
(i) Write a balanced equation for this reaction. [2]
(ii) Name the straight-chain alkane formed. [1]

(d) Straight-chain alkanes such as heptane, C_7H_{16}, can be isomerised into branched-chain alkanes and reformed into cyclic compounds.
(i) Using **skeletal** formulae, write an equation to show the isomerisation of heptane into 2,2,3-trimethylbutane. [2]
(ii) Write a balanced equation to show the reforming of heptane into methylcyclohexane. [2]

[Total: 16]
(Jan 06 2812)

5 In this question, one mark is available for the quality of use and organisation of scientific terms.
Alkenes are used in the industrial production of many organic compounds.
Outline how alkenes are used in the manufacture of
- margarine
- polymers such as poly(propene).

State any essential conditions.
Write a balanced equation for the manufacture of poly(propene) and draw a section of the polymer to show two repeat units.
State **two** difficulties in the disposal of polymers like poly(propene).
Suggest **two** ways in which waste polymers may be treated in the future.

[Total: 10]
(Jan 06 2812)

6 **(a)** Alkenes are unsaturated hydrocarbons. The structures of but-1-ene and methylpropene are shown below.

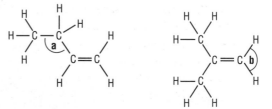

(i) What is meant by the terms *unsaturated* and *hydrocarbon*? [2]
(ii) Suggest values for the bond angle **a** in but-1-ene and the bond angle **b** in methylpropene. [2]
(iii) Explain, with the aid of a sketch, how p-orbitals are involved in the formation of the C=C double bond. [2]

[Total: 6]
(Jan 04 2812)

UNIT 2

Module 2 Alcohols, halogenoalkanes and analysis

Introduction

Throughout the centuries, chemists have synthesised new substances and investigated their properties in the search for more useful materials. In the recent past, organic chemists have developed a broad range of original and exciting materials, such as pharmaceuticals, refrigerants, solvents and plastics.

Halogenoalkanes are important starting materials for many synthetic routes. This is because they are readily converted into alcohols and other functional groups – you will become familiar with many of these compounds during your Advanced level work. For centuries, alcohols have been widely known and used. Even chimpanzees recognise how to get a buzz from the juices of rotting fruit! Today, common alcohol (ethanol) is seen as a potential replacement fuel for dwindling supplies of crude oil.

Advances in modern chemistry have seen the development of sophisticated techniques for the analysis of chemical compounds. Techniques such as infrared spectroscopy and mass spectrometry are now used to determine the structure of organic materials. They are also employed in forensic science as an aid to solve crimes, and in law enforcement to check the alcohol consumption of drivers.

In this module, you will study the physical properties and chemical reactions of two functional groups: alcohols and halogenoalkanes. The chemicals known as chlorofluorocarbons (CFCs) are halogenoalkanes. The image shows a satellite picture of the ozone hole (dark blue) over Antarctica in 2005; the ozone layer protects the planet from ultraviolet radiation, but has been disrupted by CFCs released by humans. You will also discover how to analyse a spectrum in order to determine the molecular mass and functional groups in an unknown organic compound.

Test yourself

1 How is ethanol for alcoholic drinks made?
2 What is the difference between a fossil fuel and a renewable fuel?
3 What does 'distillation' mean?
4 What does 'CFC' stand for?
5 Name the two types of radiation on either side of the visible spectrum.

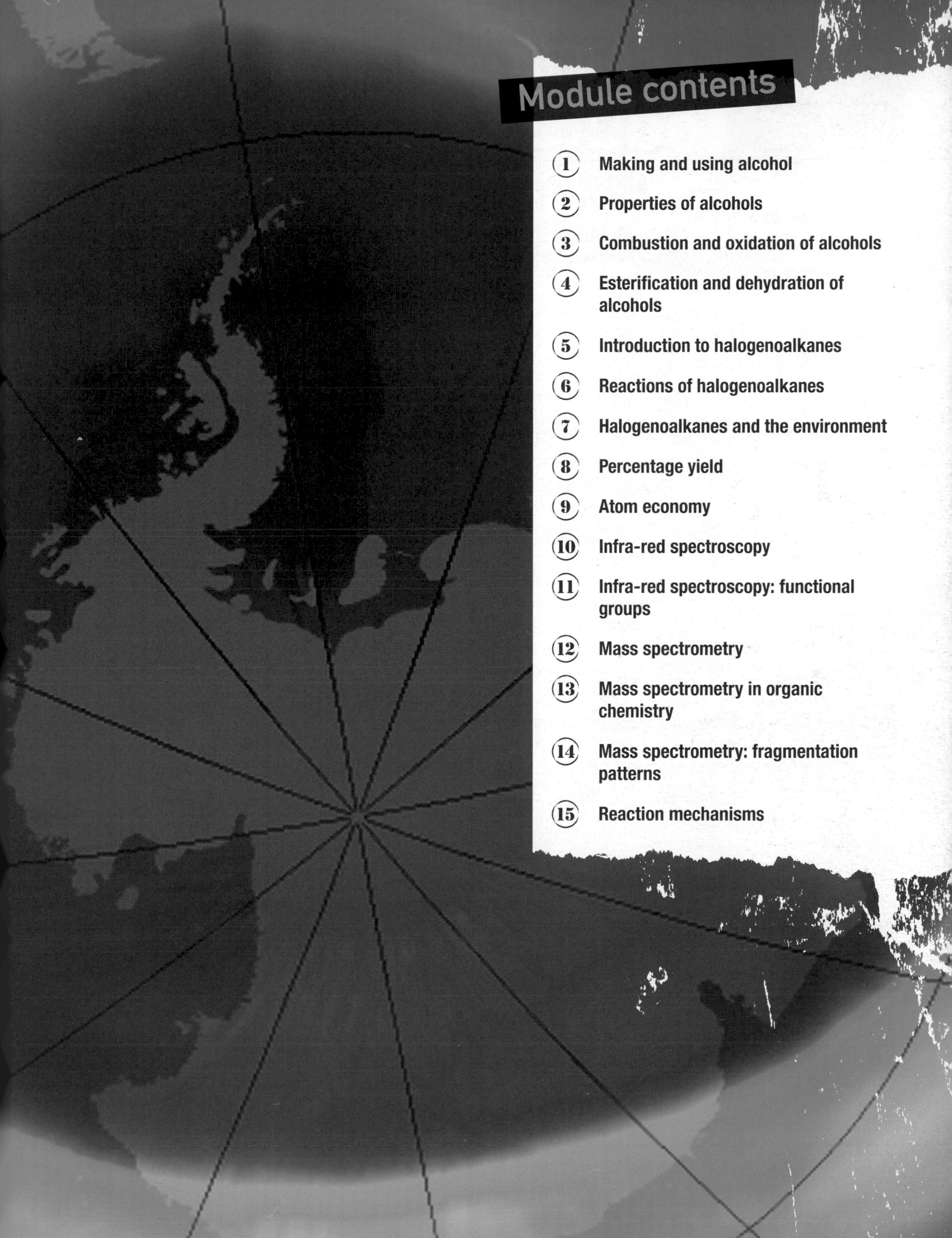

Module contents

1. Making and using alcohol
2. Properties of alcohols
3. Combustion and oxidation of alcohols
4. Esterification and dehydration of alcohols
5. Introduction to halogenoalkanes
6. Reactions of halogenoalkanes
7. Halogenoalkanes and the environment
8. Percentage yield
9. Atom economy
10. Infra-red spectroscopy
11. Infra-red spectroscopy: functional groups
12. Mass spectrometry
13. Mass spectrometry in organic chemistry
14. Mass spectrometry: fragmentation patterns
15. Reaction mechanisms

2.2 (1) Making and using alcohol

By the end of this spread, you should be able to . . .

* **Describe the industrial production of ethanol from both sugars and ethene.**
* **Outline the uses of ethanol and methanol.**

Making ethanol

Ethanol can be made by two different chemical processes:

- the hydration of ethene, C_2H_4;
- the fermentation of sugars.

Hydration of ethene

Ethanol is manufactured industrially by the catalytic hydration of ethene, C_2H_4, using steam in the presence of a phosphoric acid catalyst, H_3PO_4. The reaction is carried out at high temperature and moderate pressure and can be operated continuously.

$$H_2C{=}CH_2(g) + H_2O(g) \xrightarrow[300°C\ 60\ atm]{H_3PO_4} CH_3CH_2OH(l)$$

Figure 1 Hydration of ethene

The reaction is reversible and so conversion of ethene is incomplete. Each time the reagents pass through the reactor, only 5% of the ethene is converted into ethanol. Any unreacted gases are recycled and passed through the reactor again. Overall, a 95% conversion is achieved.

Examiner tip

Sometimes a longer question will ask you to compare the two different methods for making ethanol.

Make sure you know the reagents, conditions and equations for both processes.

You may also be asked about the advantages and disadvantages of the processes. Important points here would be finite and renewable resources, energy requirements and atom economy (see spread 2.2.9).

Fermentation

Fermentation is another method for manufacturing alcohol. Carbohydrates are converted into ethanol and carbon dioxide. Sugar or starch is usually used as the carbohydrate source. Ethanol solution produced in this way has a concentration of up to 14% alcohol by volume.

Fermentation is carried out in solution at relatively low temperatures, in the presence of yeast. The reaction is catalysed by an enzyme in the yeast called zymase. The reaction is slow at temperatures below 25 °C, while at temperatures above 37 °C, the enzyme starts to denature and lose its efficiency.

The toxicity of the alcohol also limits the concentration of the ethanol that can be made by fermentation. This is because the enzyme ceases to function above an alcohol concentration of about 14%.

Figure 2 Fermentation in action

Fermentation of carbohydrates to ethanol is anaerobic – it does not require oxygen. It is therefore important that fermentation is carried out in the absence of air. Such conditions prevent the oxidation of ethanol to undesirable compounds such as ethanal or ethanoic acid, which would affect the flavour of the product.

$$C_6H_{12}O_6(aq) \xrightarrow[37°C]{yeast} 2CH_3CH_2OH(aq) + 2CO_2(g)$$

Figure 3 Fermentation of the sugar glucose

Uses of alcohols

Ethanol in alcoholic drinks

When we hear the word alcohol, we automatically think of alcoholic drinks. However, ethanol has many uses apart from alcoholic beverages. Ethanol can be used in perfumes, aftershaves and cleaning fluids. It is also an important solvent in *methylated spirits* and is being developed as an alternative for petrol to fuel cars.

Ethanol used in alcoholic beverages is made via the fermentation process. In the brewing industry, barley, water and hops are combined together with yeast to produce beers and lagers of varying taste. The fermented sugars come from the barley grains and the hops provide the flavour. Hops also help to prevent the action of bacteria, which may spoil the beer-making process.

Spirits such as whisky, vodka and gin have a higher alcohol content than beer and are produced in a distillery. Here, the fermented drink is distilled by slowly heating the alcohol–water mixture. The alcohol boils off faster than water and is then allowed to condense. The distillate will now have a higher alcohol content than the original liquid. This is the method used by Scottish distilleries for the production of whisky.

Figure 4 Hops provide beer with both flavour and aroma

Ethanol as a fuel

The vast majority of ethanol produced in the USA is used for fuel. Up to 10% ethanol is blended with petroleum to increase the octane rating of the fuel. The resulting fuel burns more cleanly. Fuels based on ethanol are made from renewable resources, so these fuels have benefits both for the economy and the environment.

Figure 5 Alcohol fuel on sale in Brazil

Note

By law, the alcohol content of a drink has to be specified on the container. The % ABV (alcohol by volume) shows the percentage of the drink that is pure alcohol. Here are some common ABV values:

Tequila = 40%
Vodka = 38%
Wine = 11–14%
Cider = 4–8%
Alcopops = 5%
Beer = 3.5%

Ethanol in methylated spirits

Methylated spirits consists of ethanol mixed with small amounts of methanol and a coloured dye. These extra ingredients make the ethanol mixture toxic and undrinkable. Methylated spirits is exempt from tax and so is far cheaper than the ethanol used in alcoholic drinks. Methylated spirits is used as a solvent for removing paint or ink stains from clothing, and as a fuel in spirit burners and camping stoves.

Uses of methanol

Methanol can be used as a clean-burning fuel. It is often used as an additive in high-performance racing cars. Methanol tastes and smells like ethanol. However, it is extremely toxic and ingestion of small quantities of methanol can lead to severe organ damage and death. There have been a number of media stories of people dying as a result of drinking methanol.

Methanol is also an important feedstock for the chemical industry and can be easily converted into methanal and ethanoic acid.

Figure 6 High-performance racing car being refuelled

Questions

1 In the production of alcohol, suggest possible disadvantages of the hydration of ethene compared with fermentation.

2 Ethene is obtained from crude oil. Name the processes that are used to obtain ethene.

3 Why is the production of ethanol using either cane sugar or sugar beet as a raw material of long-term interest?

4 Suggest why a country such as Brazil would make ethanol by fermentation rather than hydration.

5 Write an equation for the combustion of methanol.

2.2 (2) Properties of alcohols

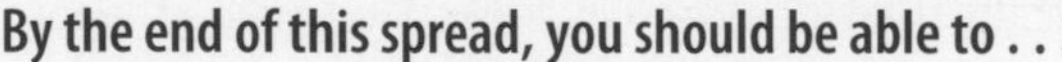

By the end of this spread, you should be able to . . .

* **Explain, in terms of hydrogen bonding, the water solubility and the relatively low volatility of alcohols.**
* **Classify alcohols into primary, secondary and tertiary alcohols.**

Physical properties

The physical properties of alcohols are influenced by their ability to form *hydrogen bonds* between the –O–H groups of neighbouring molecules. Hydrogen bonding is discussed in more detail in spread 1.2.13.

Volatility and boiling point

Hydrogen bonds are the strongest type of intermolecular forces. The relatively high melting and boiling points of alcohols result from the presence of hydrogen bonds between the alcohol molecules.

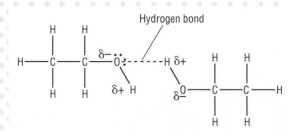

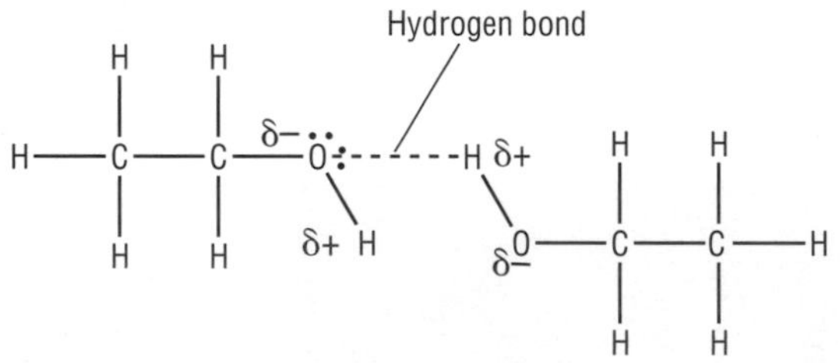

Figure 1 Hydrogen bonding between ethanol molecules

Key definition

Volatility is the ease that a liquid turns into a gas. Volatility increases as boiling point decreases.

Examiner tip

A *volatile* chemical evaporates readily into the atmosphere at room temperature and pressure. Make sure you understand the meaning of the word volatility!

The presence of hydrogen bonds in alcohol molecules results in alcohols having a lower **volatility** than alkanes of similar molecular mass. This is illustrated by the different boiling points of alkanes and alcohols shown in Table 1.

Alkane	Molecular mass	Boiling point /°C
ethane	30	–89
propane	44	–42
butane	58	–0.5
pentane	72	36
hexane	86	69

Alcohol	Molecular mass	Boiling point /°C
methanol	32	65
ethanol	46	78
propanol	60	97
butanol	74	118
pentanol	88	138

Table 1 Comparing boiling points of alkanes with alcohols

Solubility

Alcohols dissolve in water because hydrogen bonds form between the polar –O–H groups of the alcohol and water molecules, as shown in Figure 2.

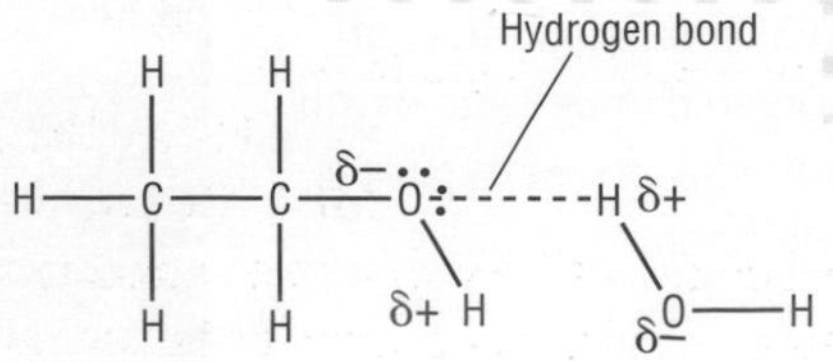

Figure 2 Hydrogen bonding between ethanol and water

The first three members of the alcohols homologous series are all soluble in water.

Solubility *decreases* as the chain length *increases*:
- a larger part of the alcohol molecule is made up of a non-polar hydrocarbon chain;
- the hydrocarbon chain does not form hydrogen bonds with water molecules.

Classification of alcohols

Alcohols can be classified as primary, secondary or tertiary alcohols. The classification depends on the number of *alkyl* groups attached to the carbon carrying the alcohol group, C–OH.

Primary alcohol

The –OH group is attached to a carbon atom with *no* alkyl groups or bonded to *one* alkyl group.

Methanol and butan-1-ol are both primary alcohols.
- Methanol has an –OH group attached to a carbon with *no* alkyl groups.
- Butan-1-ol has an –OH group attached to a carbon with *one* $CH_3CH_2CH_2$– alkyl group.

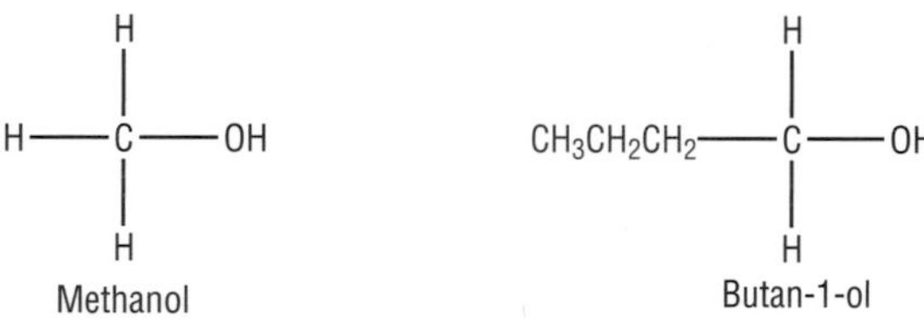

Figure 3 Structures of the primary alcohols methanol and butan-1-ol

Secondary alcohol

The –OH group is attached to a carbon atom bonded to *two* alkyl groups.

Propan-2-ol is a secondary alcohol (see Figure 4).
- Propan-2-ol has an –OH group attached to a carbon with *two* CH_3– alkyl groups.

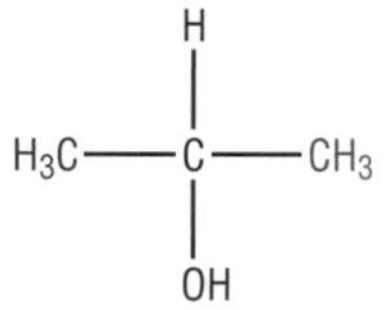

H
H3C—C—CH3
OH

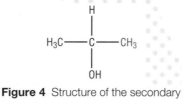

Figure 4 Structure of the secondary alcohol propan-2-ol

Tertiary alcohol

The –OH group is attached to a carbon atom bonded to *three* alkyl groups.

2-methylbutan-2-ol is a tertiary alcohol (see Figure 5).
- 2-methylbutan-2-ol has an –OH group attached to a carbon with a total of *three* alkyl groups: two CH_3– groups and one CH_3CH_2– group.

Disappearing alcohol

When 50 cm^3 of ethanol is mixed with 50 cm^3 of water in a 100 cm^3 measuring cylinder the total volume is only 98 cm^3.

Is this magic? Where has the missing 2 cm^3 gone?

When water and ethanol are mixed together, hydrogen bonds form between water and ethanol molecules. These draw the molecules closer together than in pure ethanol or pure water, reducing the volume.

Heat is also generated as the hydrogen bonds form between water and ethanol molecules.

Examiner tip

It is important that you can classify an alcohol as primary, secondary or tertiary. You can then decide how the alcohol will react with an oxidising agent (see spread 2.2.3).

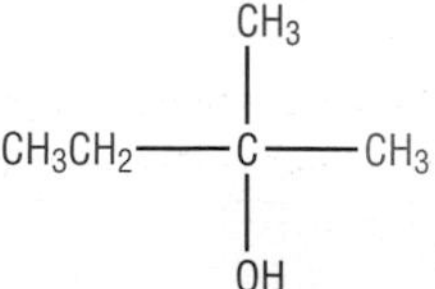

Figure 5 Structure of the tertiary alcohol 2-methylbutan-2-ol

Questions

1 Name and classify each of the following alcohols:
   (a) $CH_3CH_2CH_2OH$;
   (b) $CH_3CH_2CH(OH)CH_3$;
   (c) $(C_2H_5)_3COH$;
   (d) $CH_3CH_2CH_2CH(OH)CH_3$.
2 Draw the four structural isomers of $C_4H_{10}O$ that are alcohols. Name each alcohol and classify each structure as a primary, secondary or tertiary alcohol.
3 Show how hydrogen bonds act between methanol and water. Explain why methanol is more soluble than pentan-1-ol in water.

2.2  Combustion and oxidation of alcohols

By the end of this spread, you should be able to . . .

- ✱ **Describe the combustion of alcohols.**
- ✱ **Describe the oxidation of primary and secondary alcohols.**
- ✱ **Outline the resistance to oxidation of tertiary alcohols.**

The combustion of alcohols

Figure 1 Ethanol burning

In a plentiful supply of oxygen, alcohols burn completely to form carbon dioxide and water. This equation below shows the complete combustion of ethanol.

$$C_2H_5OH(l) + 3O_2(g) \longrightarrow 2CO_2(g) + 3H_2O(l)$$

Oxidation of alcohols

Primary and secondary alcohols can be oxidised using an oxidising agent.

- A suitable oxidising agent is a solution containing acidified dichromate ions, $H^+/Cr_2O_7^{2-}$.
- This oxidising mixture can be made from potassium dichromate, $K_2Cr_2O_7$, and sulfuric acid, H_2SO_4.

During the reaction, the acidified potassium dichromate changes colour from orange to green.

Primary alcohols

On gentle heating with acidified potassium dichromate, a primary alcohol can be oxidised to produce an aldehyde.

- The equation in Figure 2 shows the oxidation of the primary alcohol propan-1-ol to form the aldehyde propanal.
- We use the symbol [O] to represent the oxidising agent, as shown in the equation below.

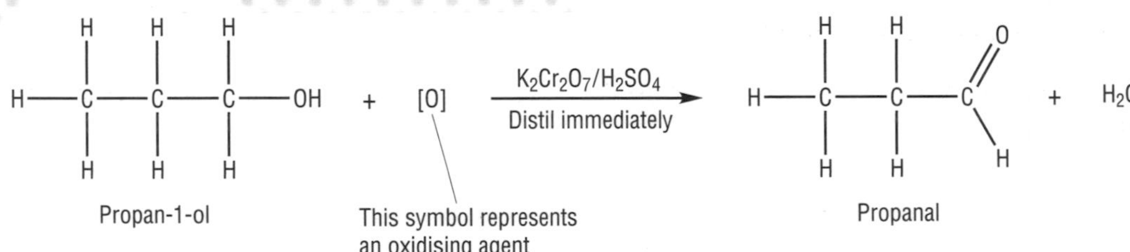

Figure 2 Oxidation of propan-1-ol to the aldehyde propanal

On stronger heating with excess acidified dichromate, the alcohol will be completely oxidised, passing through the aldehyde stage, to form a carboxylic acid. This is shown in the equation for Figure 3.

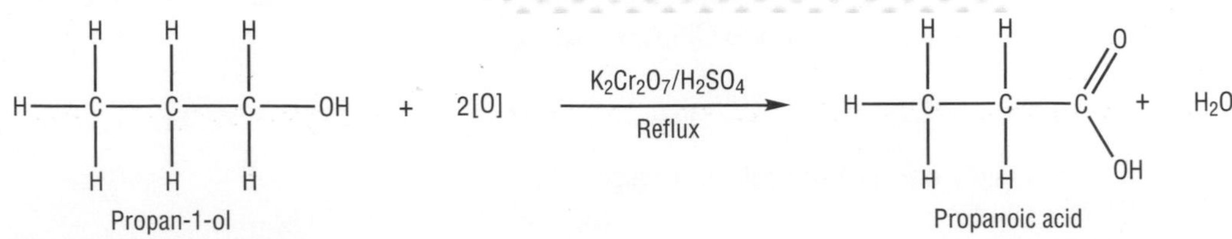

Figure 3 Oxidation of propan-1-ol to the carboxylic acid propanoic acid

- When preparing aldehydes in the laboratory, you will need to distil off the aldehyde from the reaction mixture as it is formed. This prevents the aldehyde being oxidised further to a carboxylic acid.

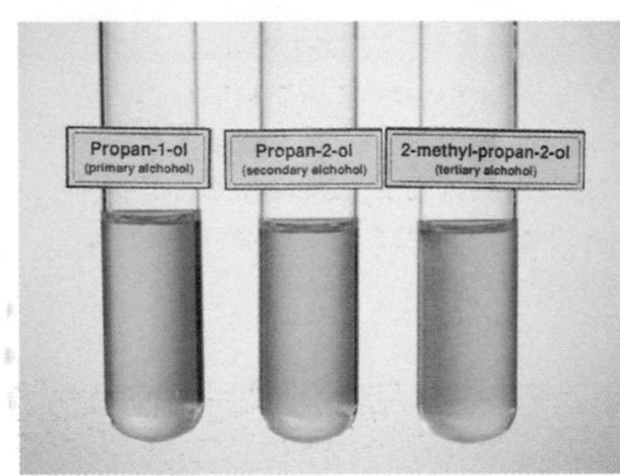

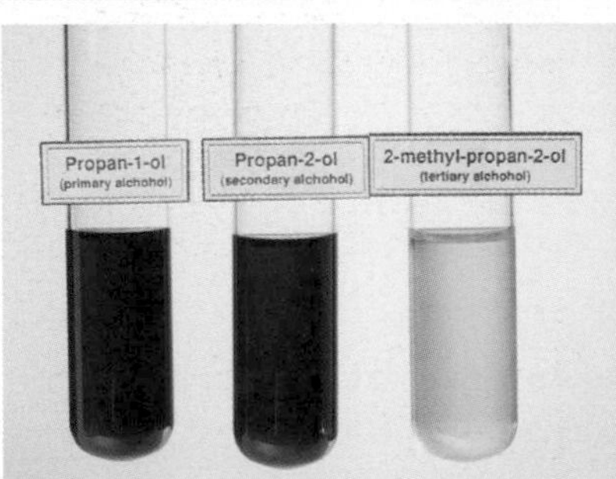

Figure 4 Oxidation reactions of three alcohols (labelled) by orange acidified potassium dichromate solution ($H_2SO_4/K_2Cr_2O_7$). The dichromate is reduced when it reacts, forming a green solution. Primary and secondary alcohols can be oxidised, forming carboxylic acids and ketones respectively. Tertiary alcohols are not oxidised and do not react

- When making the carboxylic acid, the reaction mixture is usually heated under **reflux** before distilling off the product.

The apparatus needed to carry out distillation and for heating under reflux are shown in Figures 5 and 6, respectively.

> **Key definition**
>
> **Reflux** is the continual boiling and condensing of a reaction mixture to ensure that the reaction takes place without the contents of the flask boiling dry.

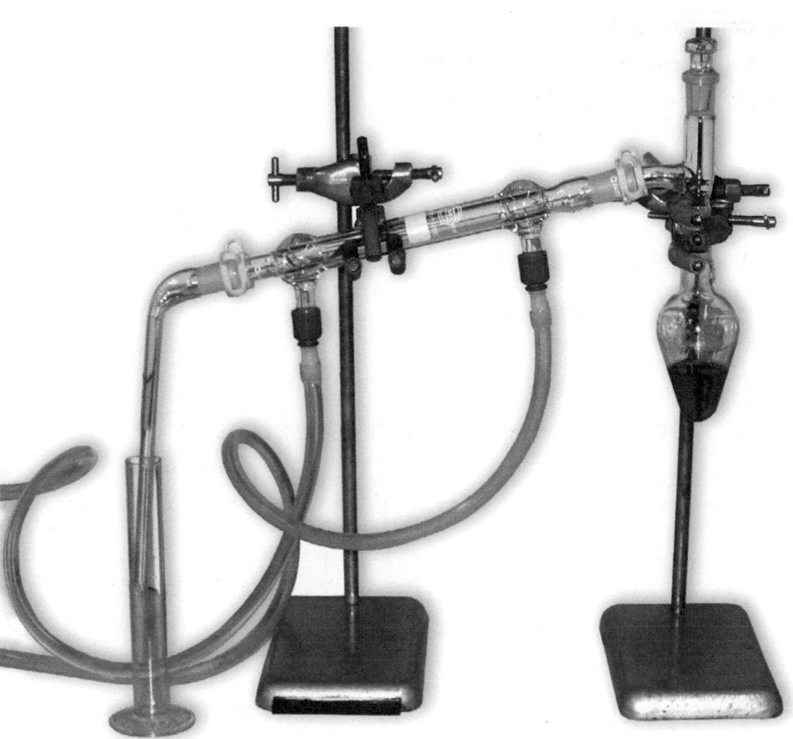

Figure 5 Distillation apparatus

> **Examiner tip**
>
> If you connect a condenser the wrong way around, the water cannot fill the outer water jacket of the condenser and it will not cool vapour efficiently. Water always enters lower than it leaves.

Figure 6 Reflux apparatus.
Reflux is the continual boiling and condensing of a solvent. A condenser is attached vertically above the reaction flask to prevent loss of the solvent by evaporation

Secondary alcohols

Secondary alcohols are oxidised by acidified dichromate ions to produce ketones. Ketones *cannot* be oxidised further.

The equation in Figure 7 shows the oxidation of the secondary alcohol butan-2-ol to form the ketone butanone.

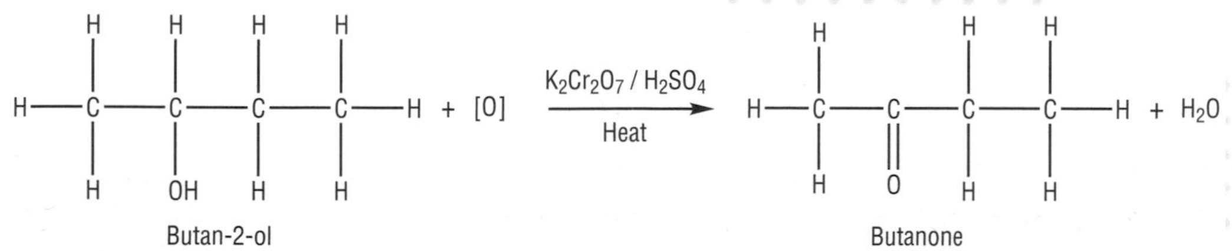

Figure 7 Oxidation of butan-2-ol to the ketone butanone

> **Examiner tip**
>
> A popular question in exams is to be asked for the structural isomers of C_4H_9OH that are alcohols. You are then asked for the products formed when each isomer is heated with excess acidified potassium dichromate.
>
> Remember:
> primary – aldehyde, then carboxylic acid;
> secondary – ketone;
> tertiary – *no* reaction.

Tertiary alcohols

Tertiary alcohols are *resistant* to oxidation. The oxidising agent remains orange in colour.

Questions

1 Write equations for the complete combustion of pentan-1-ol and 2-methylhexan-1-ol.

2 Draw the skeletal formula of the organic product formed when excess acidified potassium dichromate is heated under reflux with:
(a) 2-methylhexan-1-ol; **(b)** 3-methylpentan-2-ol.

3 What do you understand by the word *reflux*?

> **Examiner tip**
>
> The colour change of acidified dichromate from orange to green is often tested in examination papers.

Esterification and dehydration of alcohols

By the end of this spread, you should be able to . . .

- **Describe the esterification of alcohols with carboxylic acids.**
- **Describe elimination of H_2O from alcohols to form alkenes.**

Key definition

Esterification is the reaction of an alcohol with a carboxylic acid to produce an ester and water.

Examiner tip

When naming an ester, the *alcohol* provides the *alkyl* part of the name and the *carboxylic acid* provides the *alkanoate* part:

pentan-1-ol and ethanoic acid → pentyl ethanoate

Esterification

An ester is formed when an alcohol is warmed with a carboxylic acid in the presence of an acid catalyst.

- The reaction is known as **esterification**.
- Concentrated sulfuric acid is often used as the acid catalyst.

In the reaction shown in Figure 1, methanol reacts with propanoic acid to produce the ester methyl propanoate.

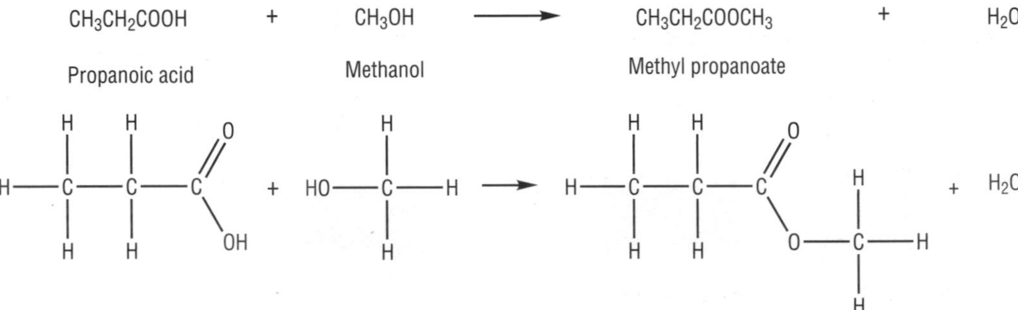

Figure 1 The esterification reaction between propanoic acid and methanol

In the reaction, the O–H bond in the alcohol is broken and water is formed. The water molecule comes from the OH of the carboxylic acid group and the H in the alcohol group, as highlighted in green in Figure 2.

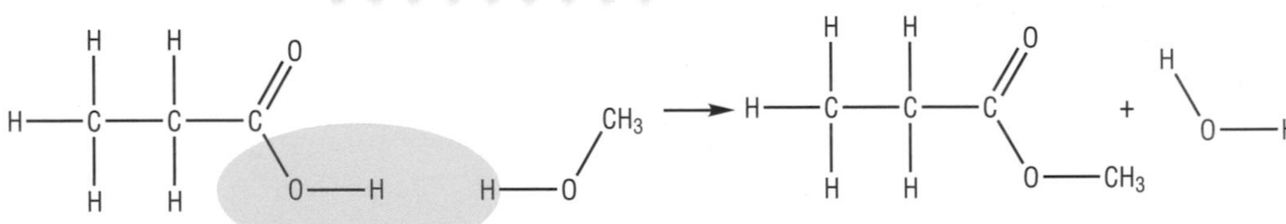

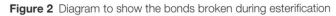

Figure 2 Diagram to show the bonds broken during esterification

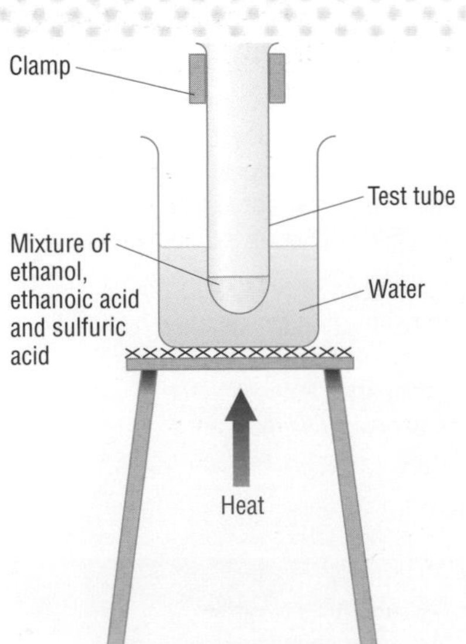

Figure 3 Simple laboratory preparation of an ester. The alcohol and carboxylic acid are heated in the presence of H_2SO_4

Esters

Esters are used as adhesives and solvents in the chemical industry. The flavours and fragrances of different esters are widely used to produce food flavourings and perfumes.

The ester responsible for the smell of orange is octyl ethanoate.

Pentyl ethanoate provides a banana smell.

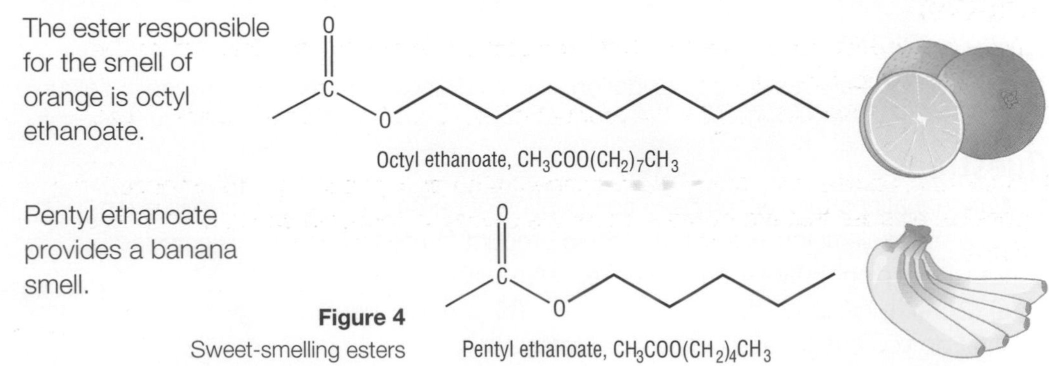

Figure 4 Sweet-smelling esters

Preparing an ester

Figure 3 shows how you can make a small sample of an ester in the laboratory quickly and easily.

- In a boiling tube, add 1 cm^3 of the carboxylic acid to 1 cm^3 of the alcohol. Carefully adding a few drops of concentrated sulfuric acid.
- Place the boiling tube in a hot water bath at about 80 °C for five minutes.
- Pour the product into a beaker of cold water.
- You will see an oil floating on the surface of the water. The oil is the sweet-smelling ester.

Dehydration of an alcohol

An alcohol can be **dehydrated** to form an alkene in the presence of an acid catalyst.

- This is an example of an *elimination* reaction.
- Concentrated phosphoric acid, H_3PO_4, or concentrated sulfuric acid, H_2SO_4, are suitable acid catalysts.

The alcohol is heated under reflux, in the presence of phosphoric acid, for about 40 minutes.

The equation for the dehydration of ethanol to ethane is shown in Figure 5.

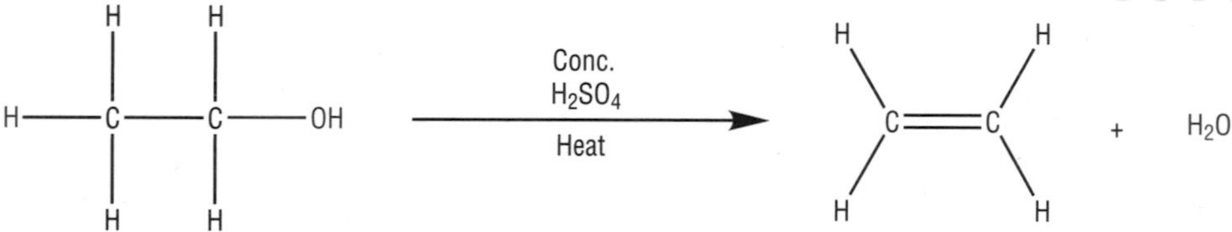

Figure 5 Dehydration of ethanol to form ethene

The dehydration of cyclohexanol to cyclohexene is shown in Figure 6. You may carry out this reaction as part of your AS practical course.

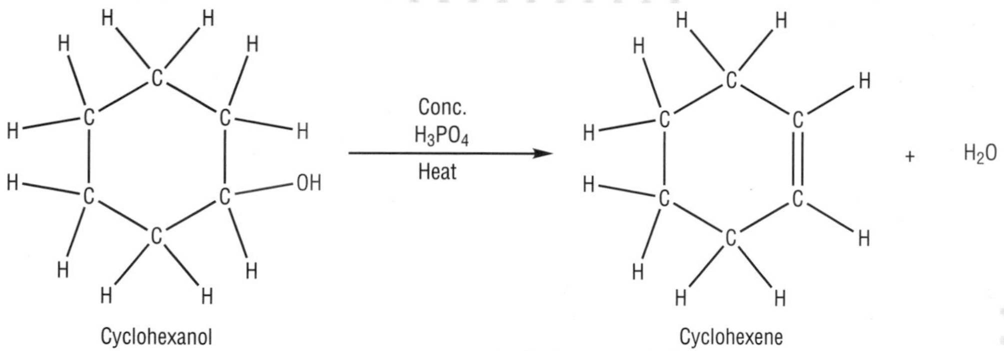

Figure 6 Dehydration of cyclohexanol to form cyclohexene

Key definition

Dehydration is an elimination reaction in which water is removed from a saturated molecule to make an unsaturated molecule.

Examiner tip

In an elimination reaction,

1 molecule ⟶ 2 molecules.

Examiner tip

When an alcohol is dehydrated the atoms lost are:

- an OH group from one carbon atom; and
- an H atom from the adjacent carbon atom.

The double bond forms between these two carbon atoms.

Questions

1 Write an equation for the reaction between ethanol and pentanoic acid. Name the organic product formed.

2 Draw the displayed formula for the product of the reaction between butan-2-ol and ethanoic acid.

3 Butan-2-ol can be dehydrated using concentrated phosphoric acid, to produce three possible alkenes that are isomers. Draw the displayed formulae and name the alkenes formed.

(5) Introduction to halogenoalkanes

By the end of this spread, you should be able to . . .

* **Describe the structure, general formula and uses of halogenoalkanes.**
* **Describe and explain the nature of the carbon–halogen bond and its susceptibility to nucleophilic attack.**
* **Define the term nucleophile as an electron pair donor.**

Halogenoalkanes

Figure 1 Halogenoalkanes were once used extensively in aerosol propellants

Halogenoalkanes are compounds in which a halogen atom has replaced at least one of the hydrogen atoms in an alkane chain.

In the past, halogenoalkanes have been used as refrigerants, aerosol propellants, degreasing agents and dry-cleaning solvents. However, they are no longer used in aerosols because of their damaging effect on the ozone layer. We will discuss in further detail the environmental effects of halogenoalkanes in spread 2.2.7. Halogenoalkanes are also important in organic synthesis and can be used to prepare many useful materials.

Halogenoalkanes have the general formula $C_nH_{2n+1}X$, where X represents the halogen atom.

Naming the halogenoalkanes

- Halogenoalkanes contain only single bonds, and their names are based on the alkane homologous series.
- The appropriate prefix is added to the name of the longest alkane chain.
- The position of the halogen on the chain is indicated.
- If there is more than one halogen atom in the compound, they are listed alphabetically.

Halogen	Prefix
F	fluoro-
Cl	chloro-
Br	bromo-
I	iodo-

Table 1 Prefix used for halogenoalkane

Example

For the halogenoalkane 3-bromo-2-chloropentane shown in Figure 2:

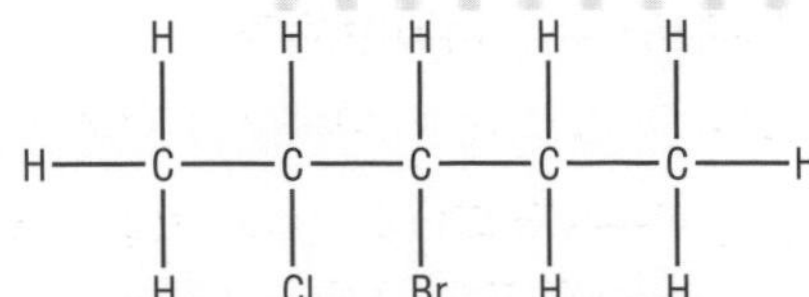

Figure 2 3-bromo-2-chloropentane

- there are five carbon atoms in the longest chain.
- the name is based on *pentane*.
- there is a *bromo* group on the *third* carbon and a *chloro* group on the *second* carbon.
- the name is 3-bromo-2-chloropentane.

Reactivity of the halogenoalkanes

Halogenoalkanes contain a polar carbon–halogen bond. The polarity arises from the different *electronegativities* of the carbon and halogen atoms.

- Halogen atoms are *more* electronegative than carbon atoms.
- The bonded electron pair is attracted more towards the halogen atom than towards the carbon atom.
- The result is a *polar* bond.

The diagram in Figure 3 illustrates the polarity in a molecule of bromomethane.

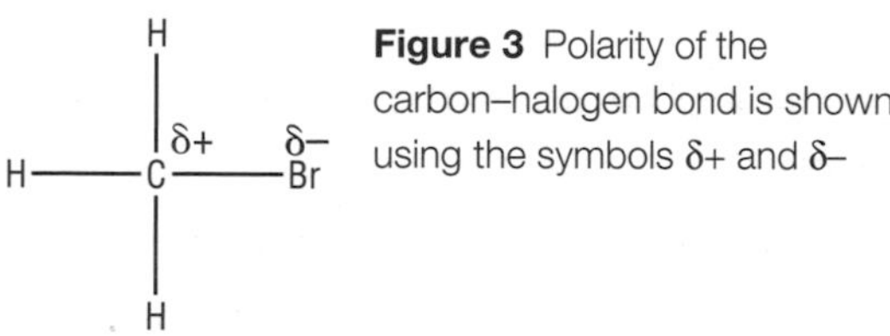

Figure 3 Polarity of the carbon–halogen bond is shown using the symbols δ+ and δ−

Examiner tip

Remember that a *nucleophile* is a reactant that attacks an electron deficient carbon atom, donating its electron pair.

Many nucleophiles contain an electronegative atom with a lone pair of electrons.

The electronegativity of the halogens *decreases* down the group, resulting in a decrease in polarity of the carbon–halogen bond from fluorine to iodine (see spread 1.2.11).

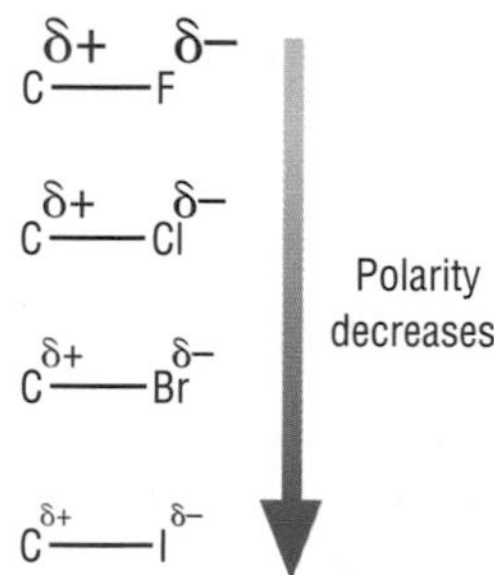

Figure 4 Polarity of the carbon–halogen bond decreases down the group

The electron-deficient carbon atom in halogenoalkanes attracts nucleophiles such as H_2O, OH^- and NH_3.

- Halogenoalkanes react with nucleophiles in substitution reactions.
- The nucleophile replaces the halogen atom in the halogenoalkane, forming a compound containing a different functional group.

Hydrolysis of halogenoalkanes

- When halogenoalkanes react with an aqueous solution of hot hydroxide ions, a nucleophilic substitution reaction occurs. The product of this reaction is an alcohol.
- Aqueous sodium hydroxide is commonly used, but any aqueous hydroxide is suitable for this reaction. This reaction is called **hydrolysis**.

Key definition

Hydrolysis is a reaction with water or aqueous hydroxide ions that breaks a chemical compound into two compounds.

The equation for the hydrolysis of 1-chloropropane is shown below:

$$CH_3CH_2CH_2Cl\,(aq) + OH^-(aq) \longrightarrow CH_3CH_2CH_2OH\,(aq) + Cl^-(aq)$$

1-chloropropane ⟶ propan-1-ol

The reaction is carried out by heating under reflux.

Examiner tip

Reflux is the continuous boiling and condensing of a reaction mixture to ensure that the reaction takes place without the contents of the flask boiling dry.

Questions

1 Name the following halogenoalkanes and draw their full displayed formulae:
   (a) $CH_3CH_2CH_2CHClCH_3$; (b) $CH_3CH_2CH_2CH_2CH(CH_3)CH_2CH_2Cl$;
   (c) $CH_3CH(Br)CH(I)CH_2CH_3$; (d) $CH_3CH_2CH_2CH_2CH_2I$.
2 Explain why a carbon–chlorine bond is polar.
3 Why are halogenoalkanes attacked by nucleophiles?

Reactions of halogenoalkanes

By the end of this spread, you should be able to . . .

* **Describe the hydrolysis of halogenoalkanes as a nucleophilic substitution reaction.**
* **Describe the mechanism of nucleophilic substitution in the hydrolysis of primary halogenoalkanes with hot aqueous alkali.**
* **Explain the rates of primary halogenoalkane hydrolysis in terms of the relative bond enthalpies of carbon–halogen bonds.**

Key definition

Nucleophilic substitution is a type of substitution reaction in which a nucleophile is attracted to an electron-deficient centre or atom, where it donates a pair of electrons to form a new covalent bond.

Nucleophilic substitution reactions

In **nucleophilic substitution**, an atom or group of atoms is replaced by a nucleophile (an electron pair donor).

During hydrolysis, the halogen atom is replaced by the hydroxide ion. The substitution takes place as follows:

- The hydroxide ion, OH^-, has a lone pair of electrons. These are attracted and donated to the electron-deficient carbon atom in the halogenoalkane. This is known as *nucleophilic* attack.
- The donation of the electron pair leads to the formation of a new covalent bond between the oxygen atom of the hydroxide ion and the carbon atom.
- The carbon–halogen bond breaks by heterolytic fission. Both electrons from the bond move to the halogen, forming a halide ion.

The mechanism for the nucleophilic substitution of 1-iodopropane by hydroxide ions is shown in Figure 1.

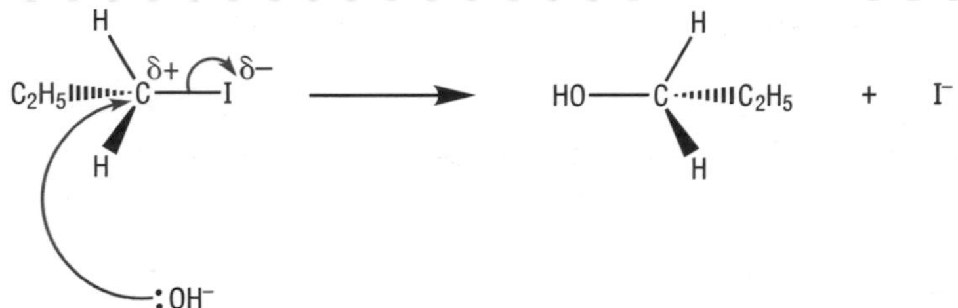
H
δ+
δ–
C_2H_5
C
I
H
:OH⁻
H
HO
C
C_2H_5
H
+
I⁻

Figure 1 Mechanism for nucleophilic substitution – the curly arrows represent the movement of a pair of electrons to form or break a covalent bond

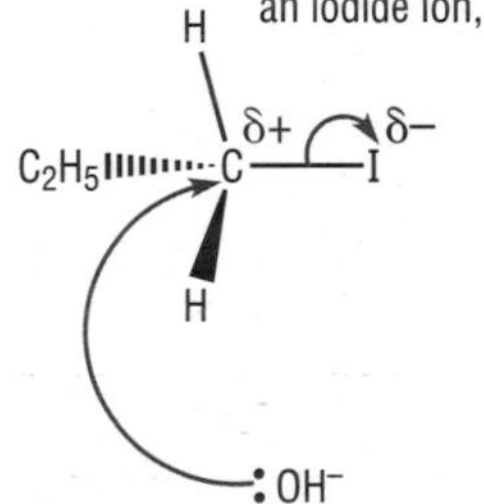
The carbon–iodine bond breaks as the two electrons in the covalent bond move to the iodine, forming an iodide ion, I⁻.
H
δ+
δ–
C_2H_5
C
I
H
:OH⁻
Two electrons from the hydroxide ion, OH⁻, form a covalent bond with the carbon atom.

Figure 2 Curly arrows are used to represent the movement of electron pairs

Rate of hydrolysis of primary halogenoalkanes

The rates of hydrolysis for different halogenoalkanes can be determined using the following experiment.

The halogenoalkane is heated with aqueous silver nitrate, with ethanol added.

- Water in the mixture behaves as the **nucleophile**.
- The ethanol acts as a common solvent, ensuring that the halogenoalkane and aqueous silver nitrate mix together and react.

As the hydrolysis reaction takes place, halide ions form. The following equation shows the hydrolysis of 1-chlorobutane:

$$CH_3CH_2CH_2CH_2Cl(aq) + H_2O(l) \longrightarrow CH_3CH_2CH_2CH_2OH(aq) + H^+(aq) + Cl^-(aq)$$

The aqueous silver nitrate, $AgNO_3(aq)$ reacts with any halide ions present, forming a precipitate of the silver halide. The equation below shows the *precipitation* of silver chloride:

$$Ag^+(aq) + Cl^-(aq) \longrightarrow AgCl(s)$$

You can find more details of this precipitation, used in halide tests, in spread 1.3.9.

The rate of hydrolysis can be found by calculating 1/time taken for the precipitate to occur.

In order to perform a controlled test, it is important:
- to use equal amounts, in mol, of each halogenoalkane;
- to use halogenoalkanes with the same chain length;
- to use a water bath to ensure a constant temperature.

This is the procedure:
- You measure each halogenoalkane into a separate test tube, which is then placed in a water bath at 50 °C.
- You put a solution of ethanol, water and aqueous silver nitrate into another test tube. This test tube is placed in the same water bath.
- Once all the tubes have reached the same temperature, you add equal volumes of each mixture to each halogenoalkane.
- You time how long it takes for each precipitate to form.

Figure 3 Hydrolysis of halogenoalkanes in a water bath in the presence of aqueous silver nitrate

Halogenoalkane	Precipitate	Colour of precipitate
chloroalkane	AgCl(s)	white precipitate
bromoalkane	AgBr(s)	cream precipitate
iodoalkane	AgI(s)	yellow precipitate

Table 1 Formulae and colours of silver halides

Polarity vs bond energy

Two factors can affect the rate of hydrolysis:
- *Polarity* – the carbon–fluorine bond is the most polar amongst the halogenoalkanes, so the $C^{\delta+}$ atom should attract the nucleophile most readily and give the fastest reaction.
- *Bond enthalpy* – the carbon–iodine bond is the weakest amongst the halogenoalkanes, so the C–I bond should be broken the most easily and give the fastest reaction.

Bond enthalpies

The strengths of the carbon–halogen bonds are shown in Table 2.

In the hydrolysis of halogenoalkanes:
- bond enthalpy is *more* important than bond polarity;
- the rate of reaction *increases* from the fluoroalkanes (slowest) to the iodoalkanes (fastest) as the carbon–halogen bond enthalpy weakens.

Bond	Bond enthalpy /kJ mol^{-1}
C–F	+467
C–Cl	+340
C–Br	+280
C–I	+240

Table 2 Bond enthalpies for C–halogen

1-fluorobutane | 1-chlorobutane | 1-bromobutane | 1-iodobutane

Slowest → Fastest

Rate of hydrolysis increases

Figure 4 Rates of hydrolysis of halogenoalkanes

Examiner tip

Many exam questions ask you to predict what happens to the rate of the reaction if the halogenoalkane is changed.

Remember that the rate of reaction depends on the strength of the carbon–halogen bond.

Questions

1 Using suitable examples, define the terms:
   (a) nucleophile;
   (b) substitution.
2 1-bromobutane is refluxed with aqueous potassium hydroxide to form an alcohol. Write a balanced equation for this reaction and name the alcohol formed.
3 Outline the mechanism for the hydrolysis of 1-chloropentane by hydroxide ions.

2.2 (7) Halogenoalkanes and the environment

By the end of this spread, you should be able to . . .

* **Outline the uses of chloroethene and tetrafluoroethene in the production of the plastics PVC and PTFE.**
* **Explain why CFCs were first developed and their subsequent effect on the ozone layer.**
* **Outline the role of 'green' chemistry in minimising damage to the environment by promoting alternatives to CFCs.**

Halogen-containing polymers

A number of polymers containing halogens can be manufactured, including poly(vinyl chloride) (PVC) and poly(tetrafluoroethene) (PTFE).

PTFE is made from the polymerisation of tetrafluoroethene as shown in Figure 1 and has the trade name *Teflon*.

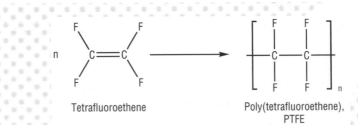

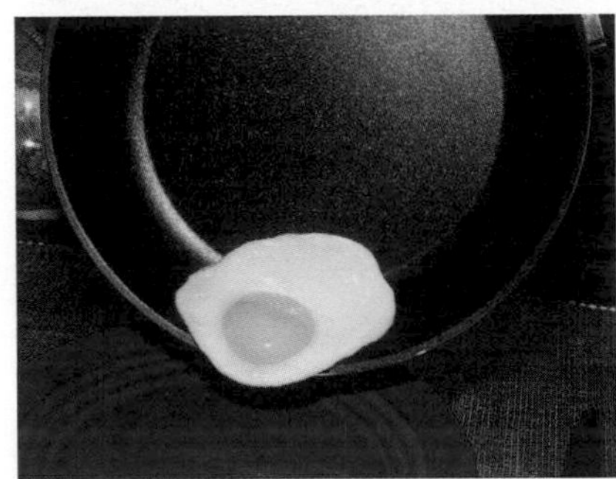

$$n\,F_2C{=}CF_2 \longrightarrow \left[-CF_2-CF_2-\right]_n$$

Tetrafluoroethene — Poly(tetrafluoroethene), PTFE

Figure 1 The polymerisation of tetrafluoroethene produces PTFE or Teflon, the non-stick coating used on frying pans

The properties of these polymers are influenced by the strength of the carbon–halogen bonds in the long-chained polymer structure.

Carbon–fluorine bonds are very strong, which makes PTFE inert and resistant to chemical attack. This chemical inertness, along with its heat resistance, electrical insulating properties and non-stick qualities, make PTFE an excellent choice for coating pans or metal surfaces to prevent chemical attack. It is even used as nail polish!

PVC is used in many varied applications such as drainpipes, plastic window frames, sports equipment, children's toys and packaging.

PVC is made from the polymerisation of chloroethene as shown in Figure 2.

$$n\,H_2C{=}CHCl \longrightarrow \left[-CH_2-CHCl-\right]_n$$

Chloroethene (vinyl chloride) — Poly(chloroethene) poly(vinyl chloride), PVC

Figure 2 The polymerisation of chloroethene to produce PVC, which has many uses

Chlorofluorocarbons, CFCs – enter Thomas Midgley

Thomas Midgley was born in Pennsylvania, USA, in 1889. Although he trained as a mechanical engineer, he had a keen interest in chemistry. He developed *tetraethyl lead* as a petrol additive to improve the combustion of fuels. He also worked extensively with halogenoalkanes.

In 1929, Midgley was challenged by his employers to develop a non-toxic refrigerant for household appliances. Refrigeration gases were introduced in the late 1800s. The first gases used were ammonia, chloromethane and sulfur dioxide. Unfortunately, these had been used in domestic refrigerators with fatal consequences. The gases often leaked, and many people were reported to have died in their sleep as a result of breathing in the fumes. Midgley developed *Freon* as a refrigerant. Freon is dichlorodifluoromethane, CCl_2F_2, an unreactive and non-toxic gas. Midgley had developed the first chlorofluorocarbon, or CFC.

In 1930, Midgley famously demonstrated that the gas was non-toxic by inhaling a lungful! He then exhaled the gas onto a candle flame, which was extinguished, illustrating the CFC's non-flammable properties.

Expanded polystyrene foams used to contain CFC bubbles. The CFCs were used as blowing agents to expand the foam, but have now been replaced with other gases, such as carbon dioxide. See also spread 2.4.8 for more details.

If only Midgley had known the damage his discoveries would create many years later!

Figure 3 Thomas Midgley, 'Father of CFCs', is shown here with his first refrigerator

Trouble with CFCs

The stability of CFCs arises from the strength of the carbon–halogen bonds within their structures – a property that had led Midgley to suggest their use as refrigerants many years before. In 1973, two chemists, Rowland and Molina, began to look into the impact of CFCs on the Earth's atmosphere. They discovered that CFCs remain stable until they reach the stratosphere, where they break down in the presence of ultraviolet radiation to form chlorine radicals. These chlorine radicals are thought to catalyse the breakdown of the ozone layer. Ozone absorbs much of the ultraviolet radiation from the Sun, preventing this radiation from reaching the Earth's surface. This depletion of the ozone layer has allowed harmful UV radiation to reach the lower atmosphere, increasing the prevalence of skin cancers.

Depletion of the ozone layer is discussed in greater detail later in spread 2.4.5.

The way forward – alternatives to CFCs

CFCs were originally developed as refrigerants, propellants, blowing agents and solvents for the dry-cleaning industry. CFCs now present research scientists with a challenge – finding replacements.

Hydrofluoroalkanes, HFCs, and hydrofluorohydrocarbons, HCFCs, are now being used as alternatives to CFCs. These replacements are also non-flammable and non-toxic.

However, HCFCs can still deplete the ozone layer. Although their depleting effect is only about one-tenth or less than that of CFCs, serious damage is still being caused. HCFCs are a short-term fix until better replacements can be developed.

Ozone-friendly products

Refrigerator gases based on hydrocarbons are currently being developed that cause no damage to the ozone layer. You may have seen propellants labelled *ozone-friendly*. These often contain hydrocarbons such as butane. Although these do not harm the ozone layer, they are flammable, so take care!

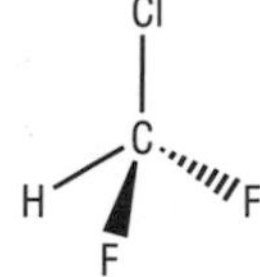

Figure 4 R-22 ($CHClF_2$) – the main HCFC used today as a refrigerant, but still an ozone depleter and set to be phased out by 2030

Questions

1 Why are fluoroalkanes used as non-stick coatings on frying pans?

2 Poly(chloroethene), poly(vinyl chloride), PVC is a polymer of chloroethene. Draw a molecule of chloroethene and draw two repeat units of poly(chloroethene).

3 State two uses of PVC.

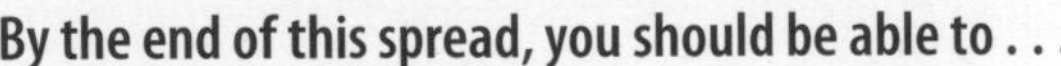

By the end of this spread, you should be able to . . .

* **Perform calculations to determine the percentage yield of a reaction.**

Percentage yield

When writing a fully balanced chemical equation, it is assumed that *all* of the reactants will be converted into products. If this were the case, then the yield would be 100%.

However, in practical work, yields of 100% are rarely obtained for various reasons:

- the reaction may be at equilibrium and may not go to completion
- other side reactions may occur, leading to by-products
- the reactants may not be pure
- some of the reactants or products may be left behind in the apparatus used in the experiment
- separation and purification may result in the loss of some of the product.

The **percentage yield** can be calculated to measure the success of a laboratory preparation.

Key definition

$$\%\ \text{yield} = \frac{\text{actual amount, in mol, of product}}{\text{theoretical amount, in mol, of product}} \times 100$$

Worked example 1

A student prepared some ethanoic acid, CH_3COOH, by heating ethanol, C_2H_5OH, with an oxidising agent.

The equation for the reaction is shown below:

$$CH_3CH_2OH + 2[O] \longrightarrow CH_3COOH + H_2O$$

The student used a mixture of sulfuric acid and potassium dichromate as the oxidising agent.

She reacted 9.20 g of ethanol with an excess of sulfuric acid and potassium dichromate.

The student obtained 4.35 g of ethanoic acid.

What is the percentage yield of ethanoic acid?

Answer

1. Calculate the amount, in mol, of ethanol that was used:

   $$\text{amount of ethanol, } n = \frac{\text{mass, } m}{\text{molar mass, } M} = \frac{9.20}{46.0} = 0.200 \text{ mol}$$

2. Using the equation, calculate the amount, in mol, of ethanoic acid product expected.
   Looking at the equation, you can see that *one* mole of ethanol reacts to produce *one* mole of ethanoic acid.
   So, 0.200 mol of ethanol should react to give 0.200 mol of ethanoic acid.
   This is the *theoretical* amount of ethanoic acid product, in mol.

3. Find the *actual* amount, in mol, of ethanoic acid product made in the experiment.

   $$\text{amount of ethanoic acid made, } n = \frac{m}{M} = \frac{4.35}{60.0} = 0.0725 \text{ mol}$$

4. Using your answers from steps 1 and 3, calculate the percentage yield.

   $$\%\ \text{yield} = \frac{\text{actual amount, in mol, of product}}{\text{theoretical amount, in mol, of product}} \times 100 = \frac{0.0725}{0.200} \times 100$$
   $$= 36.25\%$$

Examiner tip

If you are unsure about any terms here, then look back at spreads 1.1.6, 1.1.7 and 1.1.9. There are many worked examples in these spreads and further information about amount of substance and the mole.

Worked example 2

In some questions you have to calculate the amount, in mol, of each reactant to find the **limiting reagent**. The theoretical amount, in mol, of product is calculated based on the amount, in mol, of the limiting reagent. This worked example uses this principle.

A student prepared propyl methanoate, $HCOOCH_2CH_2CH_3$, from propan-1-ol, $CH_3CH_2CH_2OH$, and methanoic acid, HCOOH. The equation for the reaction is:

$$CH_3CH_2CH_2OH + HCOOH \longrightarrow HCOOCH_2CH_2CH_3 + H_2O$$

The student reacted 3.00 g of propan-1-ol with 2.50 g of methanoic acid in the presence of a sulfuric acid catalyst. He was disappointed to obtain only 1.75 g of propyl methanoate.

What is the percentage yield of propyl methanoate?

Answer

1 Calculate the amount, in mol, of each reagent.
There is a problem.
- Here, two chemicals are reacting together, but we do not know which is the limiting reagent and which is in excess.
- We need to find out the amount, in mol, of each reagent.
- We will then know which reagent is the limiting reagent and which to use in Step 2.

amount of propan-1-ol, $n = \frac{\text{mass, } m}{\text{molar mass, } M} = \frac{3.00}{60.0} = 0.0500 \text{ mol}$

amount of methanoic acid, $n = \frac{m}{M} = \frac{2.50}{46.0} = 0.0543 \text{ mol}$

- Propan-1-ol was the limiting reagent and was used up first in the reaction.

Once we know the limiting reagent, we use this for step 2.

2 Using the equation, calculate the amount, in mol, of propyl methanoate product expected.
The equation tells us that *one* mole of propan-1-ol reacts to form *one* mole of propyl methanoate.
So, 0.0500 mol of propan-1-ol should react to form 0.0500 mol of propyl methanoate.
The *theoretical* amount, in mol, of propyl methanoate product = 0.0500 mol.

3 Find the *actual* amount, in mol, of propyl methanoate product made in the experiment.
amount of propyl methanoate made, $n = \frac{m}{M} = \frac{1.75}{88.0} = 0.0199 \text{ mol}$

4 Using your answers from steps 1 and 3, calculate the percentage yield.

$$\% \text{ yield} = \frac{\text{actual amount, in mol, of product}}{\text{theoretical amount, in mol, of product}} \times 100 = \frac{0.0199}{0.0500} \times 100$$
$$= 39.8\%$$

Key definition

A **limiting reagent** is the substance in a chemical reaction that runs out first.

Examiner tip

Notice the importance of working in moles.
- the mass of propan-1-ol used was 3.00 g, greater than the mass of methanoic acid used, 2.50 g. So it appears the propan-1-ol is in excess.
- *But* there are *more* moles of methanoic acid. Although less mass of methanoic acid was used, its molar mass is also less.
- You will only know this by converting grams into moles.

Questions

1 3.925 g of 2-chloropropane reacted with an excess of aqueous sodium hydroxide. 2.955 g of propan-2-ol was formed, as shown in the equation below.
$$CH_3CH(Cl)CH_3 + NaOH \longrightarrow CH_3CH(OH)CH_3 + NaCl$$
Calculate the percentage yield of propan-2-ol.

2 Ethanol reacts with ethanoic acid to produce an ester and water, as shown in the equation below.
$$C_2H_5OH + CH_3COOH \longrightarrow CH_3COOC_2H_5 + H_2O$$
5.5 g of ethyl ethanoate is produced from 4.0 g of ethanol and 4.5 g of ethanoic acid. Calculate the percentage yield of ethyl ethanoate.

Atom economy

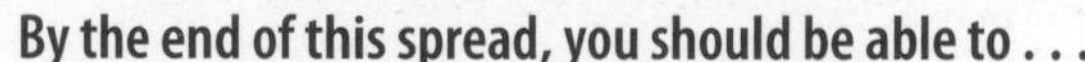

By the end of this spread, you should be able to . . .

* **Define the *atom economy* of a reaction and describe the benefits of developing chemical processes with a high atom economy.**
* **Explain that addition reactions have an atom economy of 100%, whereas substitution reactions are less efficient.**
* **Carry out calculations to determine the atom economy of a reaction.**
* **Explain that a reaction may have a high percentage yield but a low atom economy.**

Figure 1 Waste from chemical synthesis is often stored in containers for long periods of time before being sent to landfill sites

Atom economy

In the previous spread (2.2.8), we discussed how percentage yield is used to assess the efficiency of a chemical reaction. Calculating the percentage yield only tells *part* of the story. A reaction may not only produce the desired product, but by-products as well.

What do we do with the by-products?

- By-products are often considered as waste and have to be disposed of. This is costly and poses potential environmental problems.
- By-products may be sold on or used elsewhere in the chemical plant. This practice is likely to increase in the future, as we become increasingly concerned about preserving the Earth's resources and of minimising waste.

Atom economy considers not only the desired product, but also all the by-products of a chemical reaction. It describes the efficiency of a reaction in terms of *all* the atoms involved. A reaction with a high atom economy uses atoms with minimal waste.

Key definition

$$\text{atom economy} = \frac{\text{molecular mass of the desired product}}{\text{sum of molecular masses of all products}} \times 100$$

How atom economy can benefit society

We are now much more aware of our environment. By using processes with a *higher* atom economy, chemical companies can *reduce* the amount of waste produced. This is good news, especially as we are running out of landfill sites. It has also been suggested that about 5–10% of the total expenditure of a chemical company goes on waste treatment.

Calculating atom economy

Worked example 1

The reaction between propene and bromine is shown in Figure 2. The desired product is 1,2-dibromopropane. Calculate the atom economy.

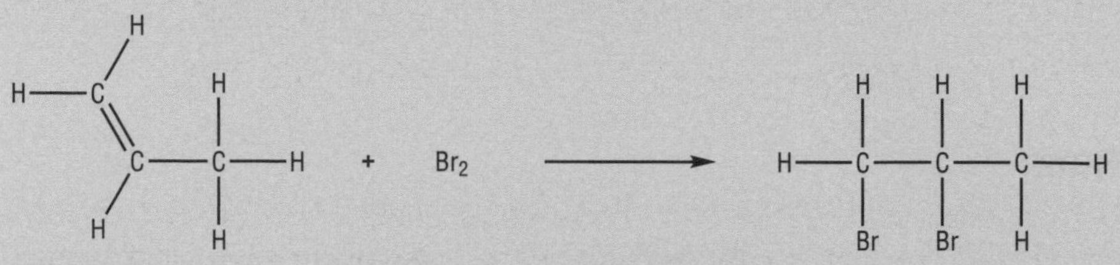

Figure 2 The reaction between propene and bromine

To calculate the atom economy you will need to:

- calculate the molecular mass of the desired product;
- divide this by the sum of the molecular masses of all products.

In this reaction, the desired product is the only product! So the atom economy is 100%.

Examiner tip

If an addition reaction produces a single addition product, the atom economy is 100%. *All* of the reactant molecules are converted into the desired product.

Worked example 2

A student decides to prepare a sample of butan-1-ol by reacting 1-bromobutane with aqueous sodium hydroxide. The equation for the reaction is given in Figure 3. Calculate the atom economy of the reaction.

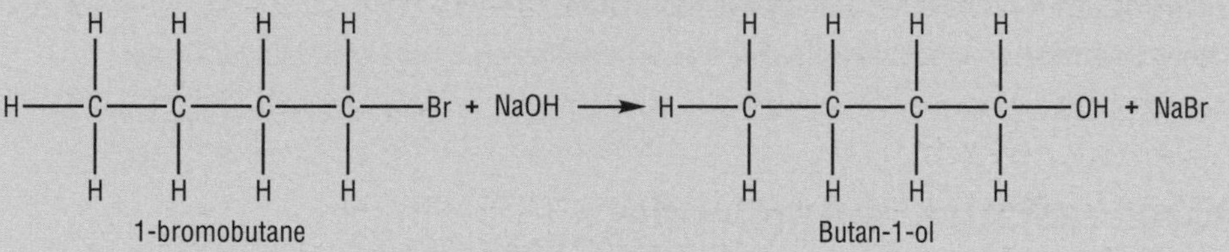

Figure 3 Preparation of butan-1-ol from 1-bromobutane

To calculate the atom economy you will need to calculate the molecular mass of the desired product and divide this by the sum of the molecular masses of *all* products.

Molecular mass of desired product: $C_4H_9OH = (4 \times 12.0) + (10 \times 1.0) + (1 \times 16.0) = 74.0$

All products: $C_4H_9OH + NaBr$

Molecular masses of all products $= 74.0 + (23.0 + 79.9) = 176.9$

$$\text{atom economy} = \frac{\text{molecular mass of the desired products}}{\text{sum of molecular masses of all products}} \times 100 = \frac{74.0}{176.9} \times 100$$

The reaction has an atom economy of 41.8%.
- This means the majority of the starting materials is turned into waste.
- Even if the reaction proceeded with a 100% percentage yield, more than half the mass of atoms used would be wasted.

Examiner tip

Make sure you show all of your working. Even if you get the final answer wrong you will still be awarded some marks provided that your method is correct.

Examiner tip

Remember that percentage yield and atom economy are different.

Percentage yield tells you the efficiency of converting reactants into products.

Atom economy tells you the proportion of desired products compared with all the products formed.

Atom economy and type of reaction

From the worked examples above, it is clear that the type of reaction used for a chemical process is a major factor in achieving a higher atom economy.
- Addition reactions have an atom economy of 100%.
- Reactions involving substitution or elimination have an atom economy less than 100%.

This is not to say that we should never carry out substitution or elimination reactions. However, to improve their atom economy, we need to find a use for *all* the products of a reaction.

In Example 2, we need to find a use for the NaBr product, rather just than throw it away.

If the undesired products are toxic, then we have an even bigger problem!

Questions

1 2-iodopropane can be hydrolysed using aqueous sodium hydroxide to form the product propan-2-ol. Write a balanced equation for this reaction and calculate the atom economy.

2 The following reaction was carried out to produce a sample of but-2-ene.

$$CH_3CH_2CH(OH)CH_3 \longrightarrow CH_3CH{=}CHCH_3 + H_2O$$

Calculate the atom economy of the reaction.

2.2 (10) Infrared spectroscopy

By the end of this spread, you should be able to . . .

- **State that absorption of infrared radiation causes covalent bonds to vibrate.**
- **Identify absorption peaks in an infrared spectrum.**
- **State that modern breathalysers measure ethanol levels by analysis using infrared spectroscopy.**

Infrared radiation and molecules

All molecules absorb infrared radiation. This absorbed energy makes bonds vibrate with either a stretching or bending motion, as shown in Figure 1.

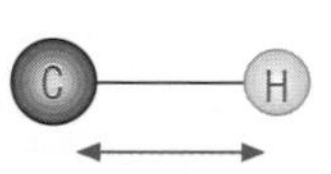

The C–H bond stretches when it absorbs infrared radiation.

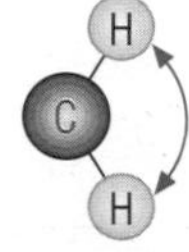

The C–H bond bends when it absorbs infrared radiation.

Figure 1 Stretching and bending vibrations resulting from the absorption of infrared radiation

Every bond vibrates at its own unique frequency.

The amount of vibration depends on:

- the bond strength
- the bond length
- the mass of each atom involved in the bond.

Most bonds vibrate at a frequency between 300 and 4000 cm^{-1}, in the infrared part of the electromagnetic spectrum.

The absorbed energies can be displayed as an infrared spectrum. By analysing this spectrum, we can determine details about a compound's chemical structure. In particular, the spectrum indicates the presence of functional groups in the compound under investigation.

In a modern infrared spectrometer, a beam of infrared radiation is passed through a sample of the material under investigation. This beam contains the full range of frequencies present in the infrared region. The molecule absorbs some of these frequencies and the emerging beam is analysed to plot a graph of transmittance against frequency. This is the infrared spectrum of the molecule. The frequency is measured using *wavenumbers*, with units of cm^{-1}.

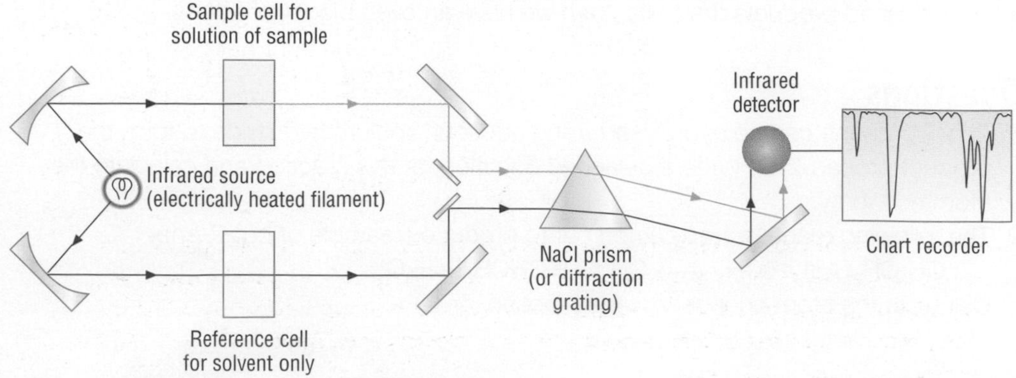

Figure 2 Schematic diagram of an infrared spectrometer

Examiner tip

You will not be expected to draw this diagram in exams.

What does the spectrum look like?

The molecule's spectrum has a number of troughs, surprisingly still called *peaks*. Each peak represents the absorbance of energy from infrared radiation that causes the vibration of a particular bond in the molecule under investigation.

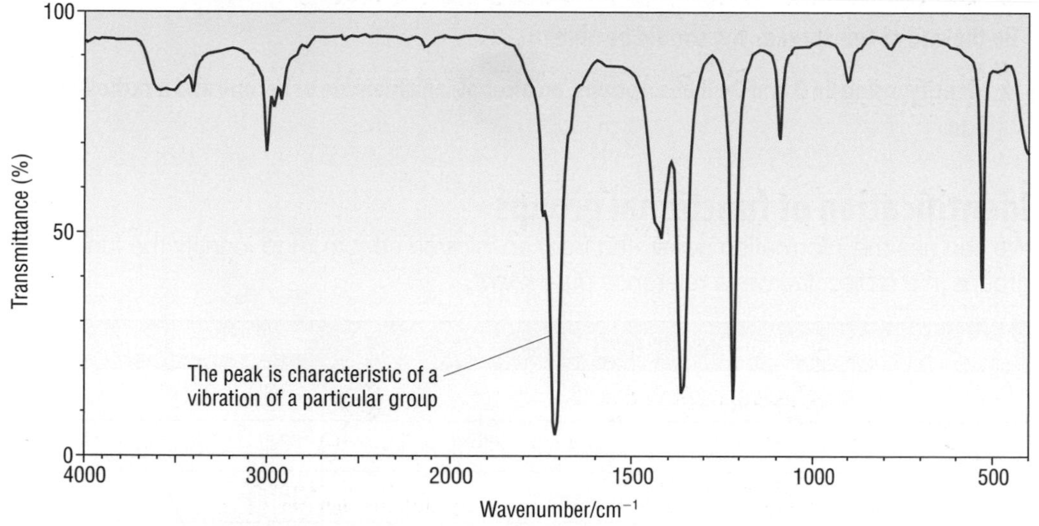

Figure 3 A typical infrared spectrum showing absorption peaks

Examiner tip

In an exam, you might find it easier to answer on the spectrum itself. If you do this make sure that you tell the examiner to look on the spectrum with a note in the answer space. Otherwise the examiner might just think that you have missed out the question.

Applications of infrared spectroscopy

Infrared spectroscopy has many everyday uses. It is used extensively in forensic science, for example, to analyse paint fragments from vehicles in hit-and-run offences. Other uses that rely on infrared spectrometry include:

- monitoring the degree of unsaturation in polymers
- quality control in perfume manufacture
- drug analysis (see Figure 4).

Infrared spectrometers are also used as one of the main methods for testing the breath of suspected drunken drivers for ethanol.

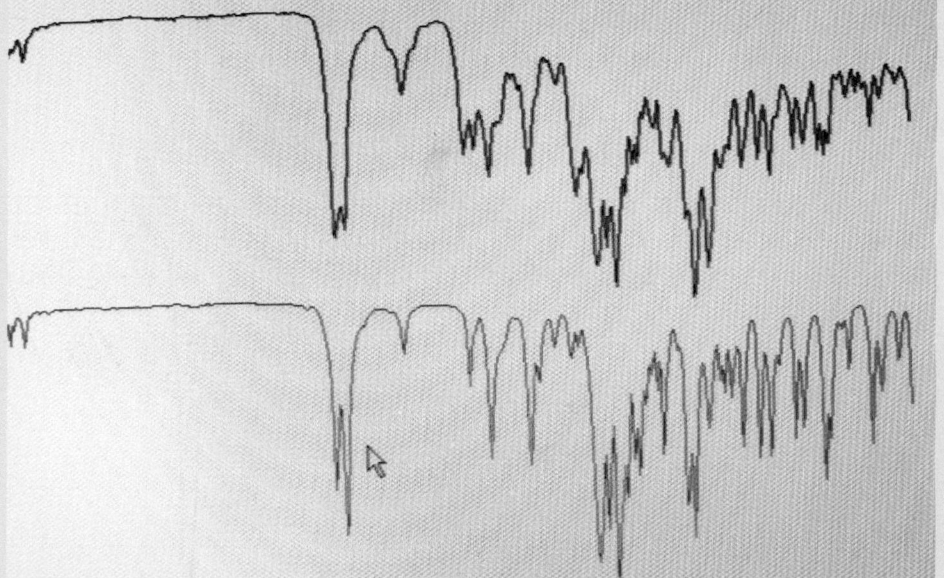

Figure 4 Infrared spectra of the pure illegal drug heroin (blue) compared with that of an unknown sample (black). The samples match almost perfectly, indicating that the sample contains a high percentage of heroin. Spectral analysis such as this can identify unknown compounds in mixtures or from samples taken from clothing or equipment. This technique is widely used in forensic science

Questions

1. What causes the bonds in a molecule to vibrate?
2. Name the two main types of vibrations of bonds caused by infrared radiation.
3. State two applications for infrared spectroscopy.

Infrared spectroscopy: functional groups

By the end of this spread, you should be able to . . .

* **Identify, using C=O and O–H absorptions, an alcohol, an aldehyde or ketone and a carboxylic acid.**

Examiner tip

Remember to use the information given in the *data sheet* in your answers. You must quote values to help you analyse the spectrum!

Identification of functional groups

We can use the information generated from an infrared spectrum to identify the functional groups in a molecule from a reference *data sheet*.

Bond	Functional group	Wavenumber /cm^{-1}
C–O	alcohols, esters, carboxylic acids	1000–1300
C=O	aldehydes, ketones, carboxylic acids, esters, amides	1640–1750
C–H	organic compound with a C–H bond	2850–3100
O–H	carboxylic acids	2500–3300 (very broad)
N–H	amines, amides	3200–3500
O–H	hydrogen bonded in alcohols, phenols	3200–3550 (broad)

Table 1 Characteristic absorptions for the functional groups required in AS/A2 chemistry

At AS level you only need to identify the presence of the following functional groups (blue in Table 1):

- The O–H (hydroxyl) group in alcohols.
- The C=O (carbonyl) group in aldehydes and ketones.
- The COOH (carboxyl) group in carboxylic acids.

You also need to be aware that most organic compounds produce a peak at approximately 3000 cm^{-1} due to absorption by C–H bonds. You need to be careful that you are not fooled into thinking that this is an O–H absorption peak.

Alcohols

The infrared spectrum of methanol, CH_3OH, is shown in Figure 1 below.

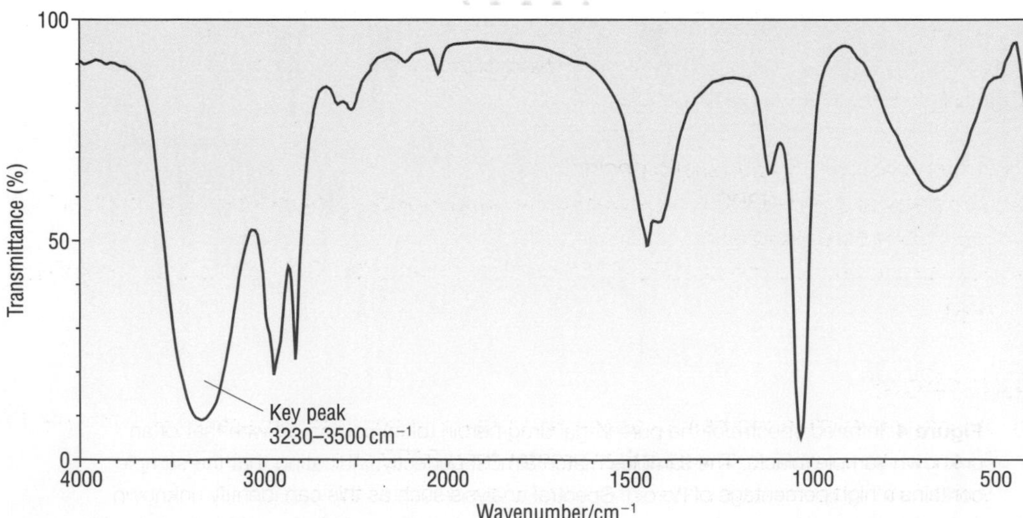

Figure 1 Infrared spectrum of methanol

The peak at 3230–3500 cm^{-1} represents an O–H group in alcohols.

Aldehydes and ketones

The spectrum of propanal, CH_3CHO, is shown in Figure 2.

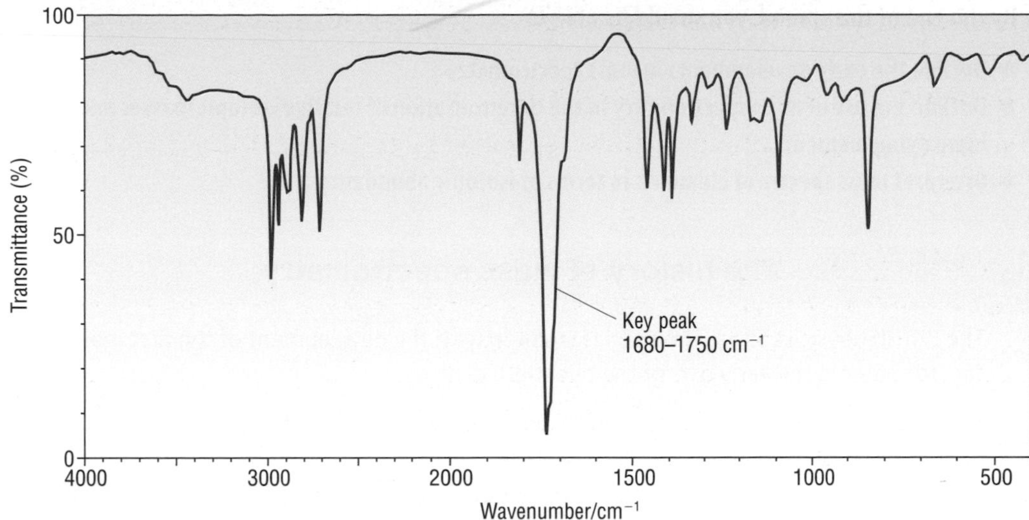

Figure 2 Infrared spectrum of propanal

The peak at 1680–1750 cm^{-1} represents a C=O bond in aldehydes and ketones.

Carboxylic acids

The spectrum of propanoic acid, CH_3CH_2COOH, is shown in Figure 3.

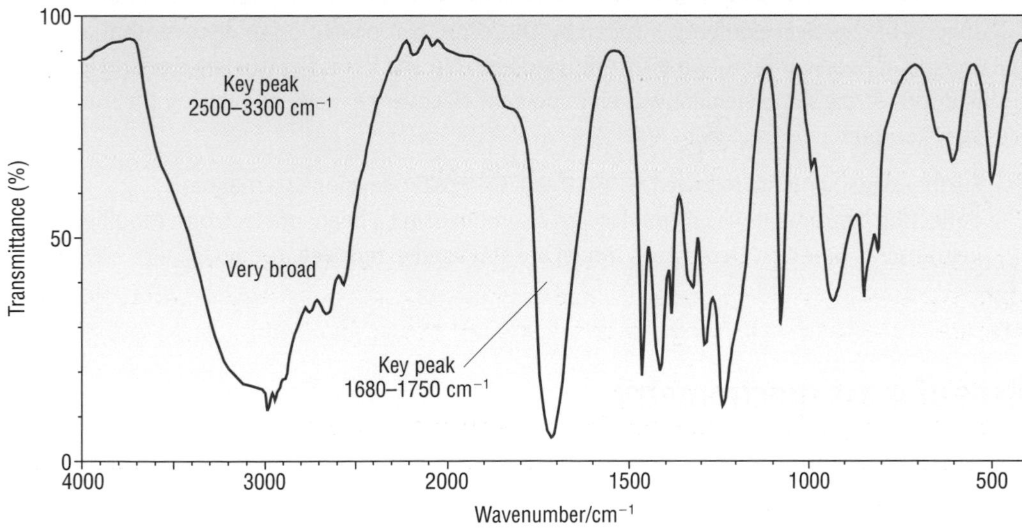

Figure 3 Infrared spectrum of propanoic acid

Here we have two characteristic peaks:

- The peak at 2500–3300 cm^{-1} is very broad and indicates the presence of the O–H group in a carboxylic acid.
- The strong sharp peak at 1680–1750 cm^{-1} represents the C=O group in a carboxylic acid.

Questions

1 An organic compound has an absorption peak in the infrared spectrum at 1700 cm^{-1}. There is no peak at a wavenumber greater than 3000 cm^{-1}.
What functional group must be present in the compound?

2 State the approximate wavenumbers for the key infrared absorptions for the following molecules:
(a) pentan-2-ol; **(b)** 2-hydroxypentanal; **(c)** butanoic acid.

2.2 (12) Mass spectrometry

By the end of this spread, you should be able to . . .

* **Outline the early developments in mass spectrometry.**
* **Outline the use of mass spectrometry in the determination of relative isotopic masses and for identifying elements.**
* **Interpret mass spectra of elements in terms of isotopic abundances.**

Figure 1 Francis Aston. Aston improved Thomson's early mass spectrometer to obtain spectra with higher resolution. He designed a *mass spectrograph*, which separated the ions present in a sample by their mass. Although Aston's work was disrupted by the outset of the First World War, he was awarded the Nobel Prize in Chemistry in 1922 for his study of naturally occurring isotopes

Figure 2 Life on Mars? The Beagle 2 space probe carried a mass spectrometer as one of its instruments to identify elements in rock samples. Unfortunately, contact with Beagle 2 was lost while descending to Mars on 25th December 2003 and the spectrometer was never able to analyse the planet! No-one knows whether Beagle 2 successfully landed on Mars or just crashed

The history of mass spectrometry

The British physicist Sir J.J. Thomson is credited with the development of the first mass spectrometer in the early part of the twentieth century.

The first mass spectrometer was used to determine the mass-to-charge ratios of ions.
- Ions were first generated in a discharge tube.
- The ions were then passed through a magnetic field and deflected from their original path.
- The ions were finally detected on a photographic plate.

In 1906, Thomson was awarded the Nobel Prize in Physics for his work. Thomson also carried out experiments in which he showed that an atom contains electrons. You can find out more about these in spread 1.1.1.

The British chemist and physicist Francis Aston carried out detailed work on the mass spectra of chemical elements. Inspired by Thomson, Aston went on to discover that a sample of neon contains two different isotopes, ^{20}Ne and ^{22}Ne. The existence of stable isotopes of the same element was an important discovery – made possible by the mass spectrometer.

Further developments followed. In 1920, A.J. Dempster developed a magnetic deflecting instrument which could ionise a sample using a beam of electrons. Modified instruments based on Dempster's model are still used extensively today.

Uses of mass spectrometry

Mass spectrometry is a powerful tool.

It can be used:
- to identify unknown compounds;
- to determine the abundance of each isotope in an element;
- to gain further information about the structure and chemical properties of molecules.

Mass spectrometry has a wide range of industrial and medical applications.

Examples of its use today include:
- monitoring the breath of patients during surgery whilst under anaesthetic;
- detecting banned substances such as steroids in athletics;
- analysing molecules in space;
- detecting traces of toxic chemicals in contaminated marine life.

How a mass spectrometer works

A mass spectrometer determines the mass of a molecule or isotope by measuring the mass-to-charge ratio of ions. Although mass spectrometers differ considerably in their operation, the same basic processes occur inside the instrument.
- The sample is introduced via a sample inlet.

- Sample molecules are converted into ions by an ionisation source. Methods of ionisation include electron impact, chemical ionisation, use of an electrospray and even lasers. These ionisation sources vary depending on the type of instrument being used.
- The ions are propelled into a mass analyser.
- The ions are separated according to their mass-to-charge ratio. Methods of separation include deflection with a magnetic field or *time of flight* (measurement of how long it takes for ions of different mass to reach the detector).
- The ions are detected. A mass spectrum is generated by computer software and displayed on a screen or printed.

The mass spectrum produced can be used to find the molecular mass, determine the structural features of a molecule or to find the abundance of isotopes.

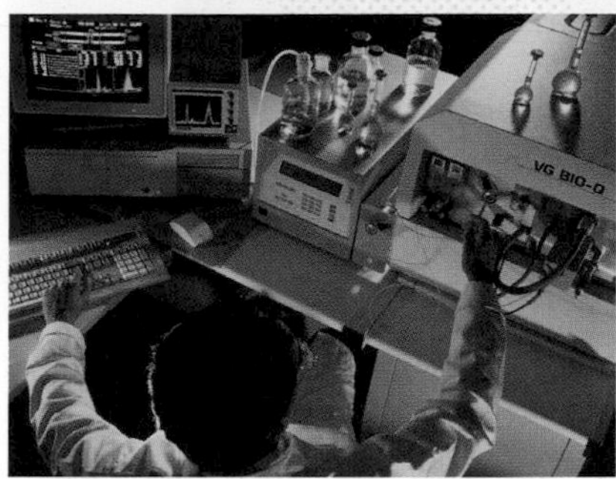

Figure 3 Mass spectrometry – sample is injected and its spectrum displayed on screen

Mass spectra of elements

One of the most important uses of mass spectrometry is to determine the isotopes present in an element. You will need to be able to analyse a mass spectrum to find the proportions of each isotope in the element being investigated. You can then calculate the *relative atomic mass* of the element.

A mass spectrum is a graph showing relative or percentage abundance on the *y*-axis and mass : charge ratio on the *x*-axis.

The mass to charge ratio in all mass spectra is shown as *m/z*.

- *m* is the mass
- *z* is the charge on the ion.

The mass spectrum of a sample of magnesium is shown in Figure 4.

Key features of the spectrum

- There are three peaks in the spectrum, so there are *three* isotopes of magnesium.
- The *heights* of the peaks gives the *relative abundances* of the isotopes in the sample. These are given as percentages in Figure 4.
- The results can be used to find the relative atomic mass.

The spectrum tells us that:

- 79% is magnesium-24
  10% is magnesium-25
  11% is magnesium-26.

$$\text{Relative atomic mass} = \frac{(24 \times 79) + (25 \times 10) + (26 \times 11)}{100} = 24.32$$

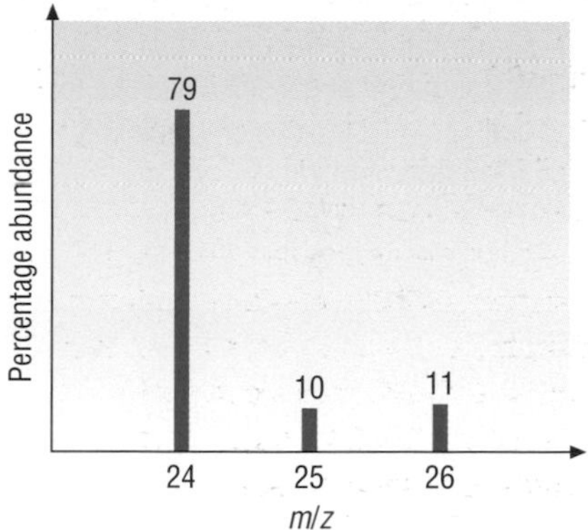

Figure 4 The mass spectrum of magnesium

Examiner tip

If you are unsure what to do here, look back to spread 1.1.3. You will then find worked examples to show how relative atomic mass can be calculated from isotopic abundances.

Take care with units – relative atomic mass doesn't have any!

Questions

1 The mass spectrum of lithium showed two peaks. The first peak was at *m/z* = 6 and had an abundance of 7.4%; the second peak was at *m/z* = 7 and had an abundance of 92.6%.
Calculate the relative atomic mass of the lithium sample to three significant figures.

2 A sample of strontium has four isotopes as shown in Table 1. Sketch a mass spectrum for the strontium sample and calculate its relative atomic mass.

Table 1 Strontium's four isotopes

Ion	Percentage abundance
$^{84}Sr^+$	1
$^{86}Sr^+$	9
$^{87}Sr^+$	7
$^{88}Sr^+$	83

2.2 (13) Mass spectrometry in organic chemistry

By the end of this spread, you should be able to . . .

* **Use the molecular ion peak in an organic molecule's mass spectrum to determine its molecular mass.**
* **Explain that a mass spectrum is essentially a molecule's fingerprint that can be identified using a spectral database.**

Mass spectrometry and molecules

When an organic compound is placed in a mass spectrometer, some molecules lose an electron and are *ionised*. The resulting positive ion is called the **molecular ion** and is given the symbol M^+.

Several different methods can be used for ionisation but electron impact is the oldest and best established method. In electron impact, the molecules being tested are bombarded with electrons. If an electron is dislodged from a molecule, a positive ion is formed. With ethanol, a C_2H_5OH molecule is ionised to form the molecular ion, $C_2H_5OH^+$.

$$C_2H_5OH + e^- \longrightarrow \underset{\textit{molecular ion}}{C_2H_5OH^+} + 2e^-$$

- The mass of the lost electron is negligible.
- The molecular ion has a molecular mass equal to the molecular mass of the compound.
- This molecular ion can be detected and analysed.

Excess energy from the ionisation process can be transferred to the molecular ion, making it vibrate. This causes bonds to weaken and the molecular ion can split into pieces by **fragmentation**.

Fragmentation results in a positive fragment ion and a neutral species.

A possible fragmentation of the ethanol molecular ion would be:

$$C_2H_5OH^+ \longrightarrow CH_3 + \underset{\textit{fragment ion}}{CH_2OH^+}$$

Fragment ions are often broken up further into smaller fragments.

The *molecular ion* and *fragment ions* are detected in the mass spectrometer.

The molecular ion, M^+, produces the peak with the highest *m/z value* in the mass spectrum. High-resolution mass spectrometry can produce a spectrum with extremely precise values for M^+, making accurate molecular mass determination relatively easy.

Key definition

Molecular ion, M^+ is the positive ion formed in mass spectrometry when a molecule loses an electron.

Key definition

Fragmentation is the process in mass spectrometry that causes a positive ion to split into pieces, one of which is a positive fragment ion.

Examiner tip

In the mass spectrum, you may see a small peak, one unit beyond the molecular ion, M^+, peak. This is often referred to as the M+1 peak. The M+1 peak arises from the presence of the isotope carbon-13, which makes up 1.11% of all carbon atoms. Don't be fooled into thinking that this is the M^+ ion.

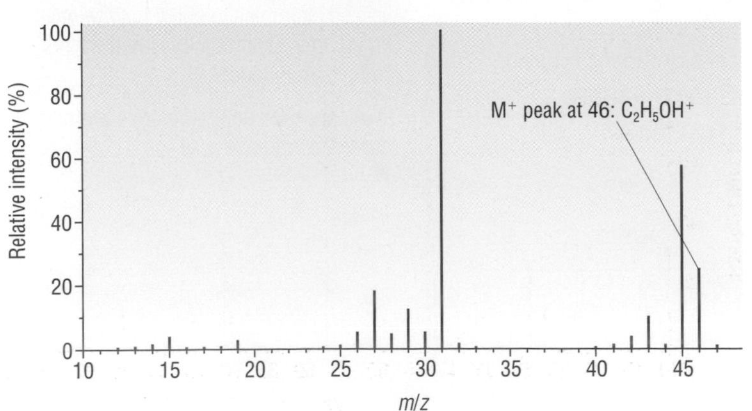

Figure 1 Mass spectrum of ethanol

Molecular mass determination

The molecular mass of a molecule can be determined using mass spectrometry by locating the M^+ peak. The mass spectrum of ethanol is shown in Figure 1.

The molecular ion peak is located at a *m/z* ratio of 46. This indicates that the molecular mass of the molecule is 46. The other peaks in the spectrum are a result of fragmentation.

Fragmentation patterns

Mass spectrometry can be used to determine the structure of an unknown compound and, in many cases, give its precise identity.

- Although the molecular ion peak of two isomers will have the same *m/z* value, the fragmentation patterns will be different.
- Each organic compound produces a unique mass spectrum, which can be used as a fingerprint for identification.

The mass spectra of two structural isomers of C_5H_{12}, pentane and 2-methylbutane, are shown in Figure 2. Each spectrum has a molecular ion peak at $m/z = 72$ for $C_5H_{12}^+$. However, the fragmentation patterns are very different.

Identification of an unknown compound

Although mass spectra can be analysed by simply viewing the spectra, modern mass spectrometers are often linked to a spectral database. When a mass spectrum for an unknown sample is produced, the spectral database is scanned automatically until a perfect match is found. This enables straightforward and immediate identification of an unknown compound.

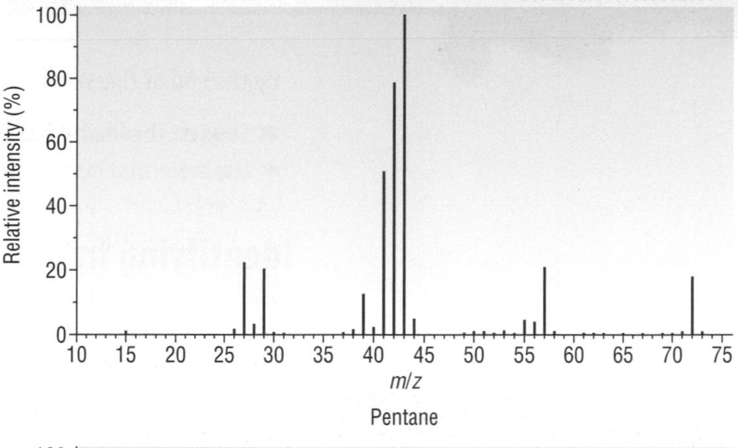

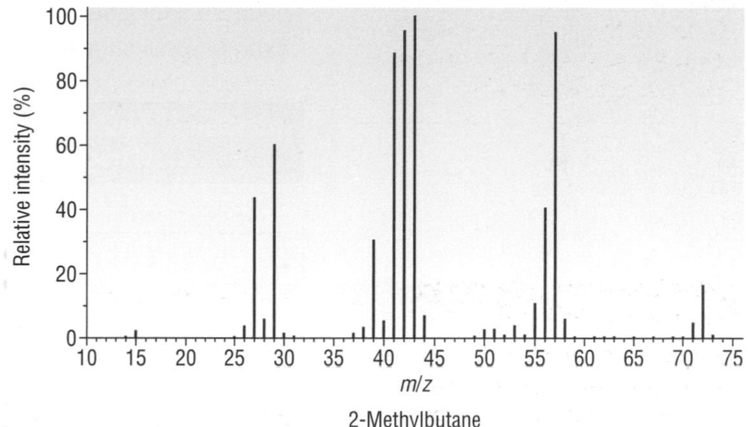

Figure 2 Mass spectra of the isomers of C_5H_{12}, pentane and 2-methylbutane

Questions

1 Write a balanced equation showing the ionisation of the following molecules to form the molecular ion. In each case, state the likely *m/z* value for the molecular ion peak:
   (a) propan-1-ol;
   (b) butane;
   (c) octane.

2 In the spectra of compounds **A** and **B** in Figure 3, identify the *m/z* value for each molecular ion peak.

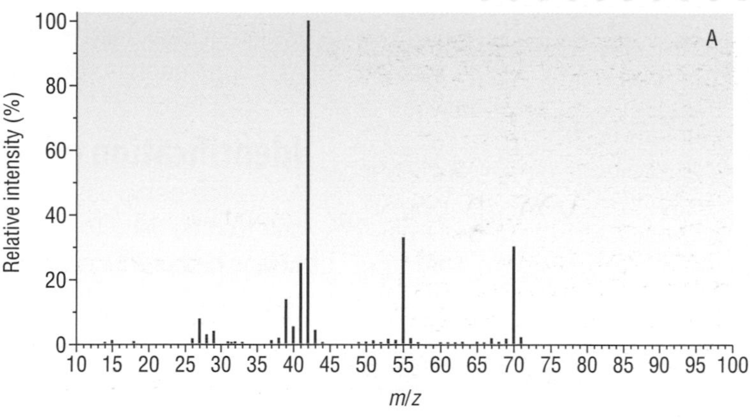

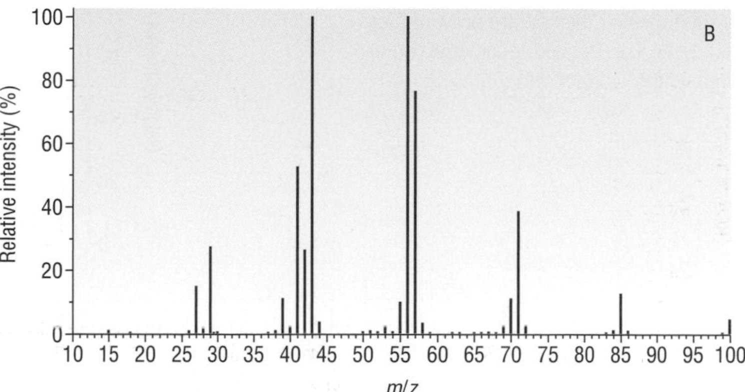

Figure 3 Spectra A and B

2.2 

Mass spectrometry: fragmentation patterns

By the end of this spread, you should be able to . . .

- ✱ **Suggest the identity of the major fragment ions in a given mass spectrum.**
- ✱ **Use molecular ion peaks and fragmentation peaks to identify structures.**

Identifying fragment ions

When looking at a mass spectrum, fragment peaks appear alongside the more important molecular ion peak. However, these fragment peaks give information about the structure of the compound.

Even with simple compounds, it is often impossible to identify every peak in the mass spectrum. However, there are a number of common peaks that can be identified. Common peaks for alkyl ions are shown in Table 1.

m/z value	Possible identity of fragment ion
15	CH_3^+
29	$C_2H_5^+$
43	$C_3H_7^+$
57	$C_4H_9^+$

Table 1 *m/z* values for common alkyl ions

In the mass spectrum for an alcohol, you will often see a peak with an *m/z* value of 17. This is due to the OH^+ ion and corresponds to the –OH group in the alcohol. Some of the fragment peaks are more difficult to recognise and may arise from molecular rearrangements – you will not be expected to know these!

Identification of organic structures

As well as being used to identify the molecular mass of a molecule, a mass spectrum can also be used to work out some of the molecule's structural detail. An unknown alkane has produced the mass spectrum shown in Figure 1. Some of the peaks have been labelled with the *m/z* ratio.

Examiner tip

In examinations feel free to draw or write notes on the spectra provided for analysis. This is good technique and will help you solve problems.

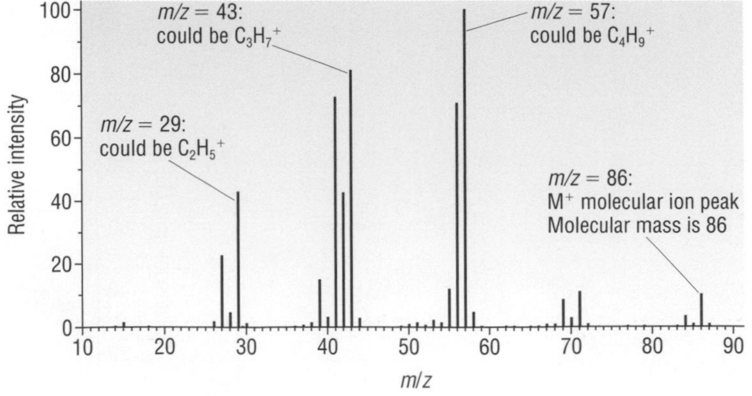

Figure 1 Mass spectrum for an unknown alkane

The mass spectrum shown in Figure 2 has been produced from hexane. The equations in Figure 2 illustrate how the molecular ion could be fragmented to form fragment ions with *m/z* values of 57 and 43.

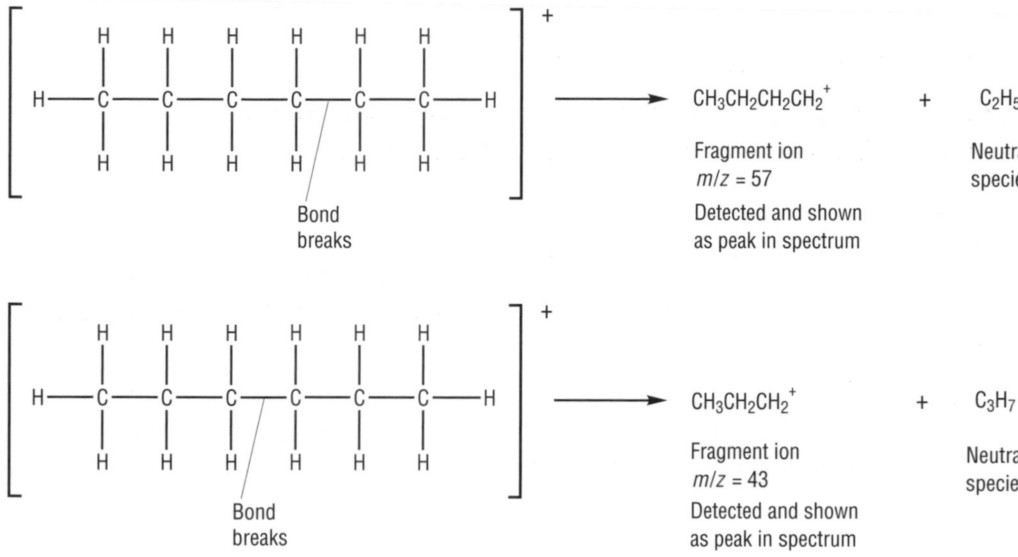

Figure 2 Equations explaining how some of the peaks arise in the mass spectrum for hexane

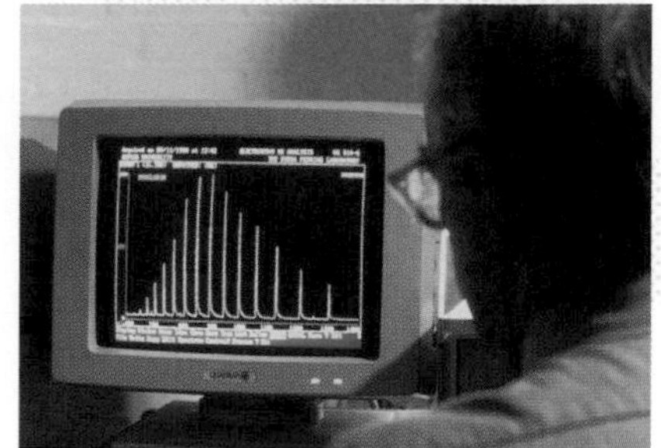

Figure 3 Mass spectrum of myoglobin. Scientist examining the mass spectrum of myoglobin, the oxygen-storage protein of red muscle. The spectrum obtained from the mass spectrometer is displayed on the computer screen used for protein analysis at Oxford University

Questions

1 The spectrum in Figure 4 on the right has been produced from the alkene pent-1-ene.
Suggest the ions responsible for the peaks labelled **A**, **B** and **C**.

2 Figure 5 below shows the mass spectrum of an alkane.
(a) Use the spectrum to identify the molecular ion peak and hence the molecular mass and molecular formula of the compound.
(b) The molecule is branched with one methyl group attached to the main chain at carbon two. Draw the skeletal formula for this alkane.
(c) Draw structures for the fragment ions represented by the peaks with *m/z* values of 43, 57 and 71.

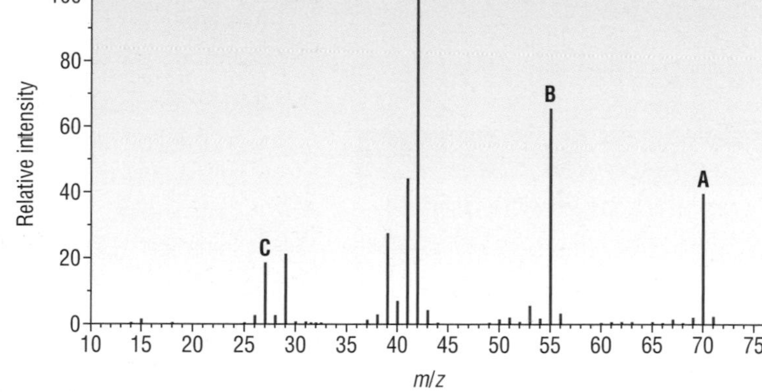

Figure 4 Mass spectrum of pent-1-ene

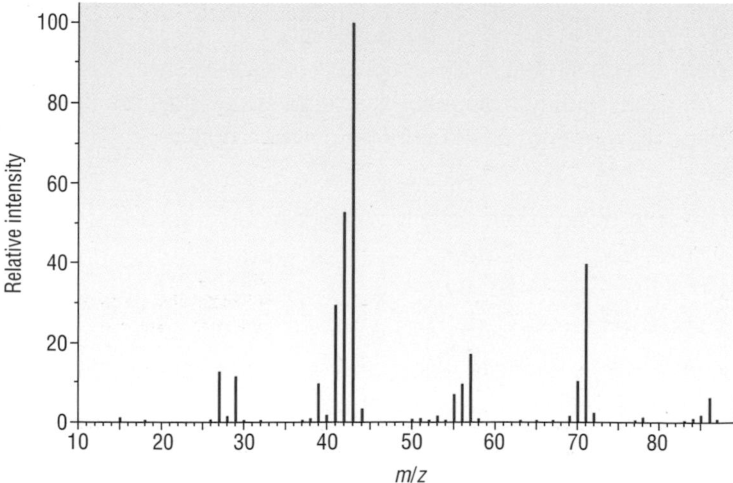

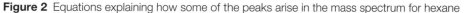

Figure 5 Mass spectrum of an alkane

2.2 (15) Reaction mechanisms

By the end of this spread, you should be able to . . .

*** Describe the three common mechanisms required for AS chemistry.**

Examiner tip

You must be able to draw each of these mechanisms. You must also be able to suggest mechanisms for similar or unfamiliar examples.

Mechanisms are always being asked on exam papers. You need to be able to draw out the mechanisms shown on this double-page spread. You must be absolutely precise with any curly arrows.

A mechanism is typically worth four marks, and if you are lucky, there may be two on the paper – eight easy marks for those who have been bothered to learn them.

Key definition

A **radical** is a species with an unpaired electron.

AS mechanisms

There are *three* mechanisms required in Unit F322 Chains, Energy and Resources. You should be able to draw and explain these mechanisms and suggest a mechanism for similar or unfamiliar examples. You will be required to produce mechanisms for:

- the reaction of an alkane with a halogen
- the reaction of an alkene with an electrophile
- the reaction of a halogenoalkane with a nucleophile.

Alkanes – radical substitution

Alkanes take part in *radical substitution* with halogens. In exams, the reaction of *bromine* or *chlorine* with *methane* is the one tested most often. However, you could be tested on the reaction of a halogen with a longer-chain alkane, or even a cycloalkane.

Worked example

Describe, with the aid of suitable equations, the mechanism for the reaction between methane and bromine in the presence of ultraviolet radiation.

Answer

Balanced equation: $CH_4 + Br_2 \longrightarrow CH_3Br + HBr$

Initiation

- In the first stage of the mechanism, **radicals** are formed.
- Ultraviolet radiation provides the energy to break the covalent bond in a bromine molecule, Br–Br. The bond breaks by *homolytic fission*.
  Initiation: $Br_2 \longrightarrow 2Br\cdot$

Propagation

- In the second stage, the products of the reaction are made.
  Propagation: $Br\cdot + CH_4 \longrightarrow CH_3\cdot + HBr$
  $CH_3\cdot + Br_2 \longrightarrow CH_3Br + Br\cdot$

Notice the numbers of radicals are the same on each side of the equations.

- A bromine radical is used in the first propagation step.
- A bromine radical is regenerated in the second propagation step.
- Bromine radicals catalyse the propagation stage. Although they are involved in the reaction steps, they are *not* used up in the overall reaction.

Termination

- In the final stage of the mechanism, any two radicals can combine.
  Termination: $CH_3\cdot + Br\cdot \longrightarrow CH_3Br$
  $Br\cdot + Br\cdot \longrightarrow Br_2$
  $CH_3\cdot + CH_3\cdot \longrightarrow C_2H_6$

Further substitution reactions can take place, as the bromoalkane product can react with further bromine radicals. A mixture of products is produced containing CH_3Br, CH_2Br_2, $CHBr_3$ and CBr_4.

Alkenes – electrophilic addition

Alkenes take part in *electrophilic addition* reactions with **electrophiles** such as chlorine, bromine or hydrogen halides. In exams, the reaction between *bromine* or *hydrogen bromide* and *ethene* is tested most often. However, alkenes with longer carbon chains or cycloalkenes could be used with any halogen, hydrogen halide or related compound.

Key definition

An **electrophile** is an electron pair acceptor.

Worked example

Describe, with the aid of suitable diagrams, the mechanism for the reaction between ethene and chlorine.

Answer

In this type of reaction the halogen adds *across* the double bond as shown in Figure 1.

Balanced equation: $C_2H_4 + Cl_2 \longrightarrow C_2H_4Cl_2$

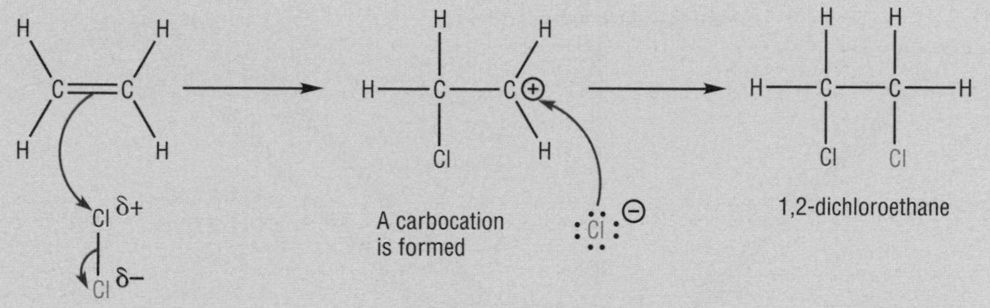

A carbocation is formed

1,2-dichloroethane

When Cl_2 approaches, the Cl–Cl bond becomes polar. A pair of electrons flows from the double bond to the slightly positive $Cl^{\delta+}$ and a bond formed.

Figure 1 The electrophilic addition mechanism

Halogenoalkanes – nucleophilic substitution

Halogenoalkanes take part in *nucleophilic substitution* reactions with **nucleophiles** such as OH^- or H_2O. In exams, the reaction between *hydroxide ions* and any halogenoalkane is likely to be tested.

Key definition

A **nucleophile** is an electron pair donor.

Worked example

Aqueous hydroxide ions, OH^-, react with chloromethane in a nucleophilic substitution reaction. Describe, with the aid of suitable diagrams, the mechanism for this reaction.

Answer

In this type of reaction the nucleophile (OH^-) attacks the back of the carbon atom, donating an electron pair, as shown in Figure 2. A new bond is formed and the carbon–halogen bond breaks.

Balanced equation: $CH_3Cl + OH^- \longrightarrow CH_3OH + Cl^-$

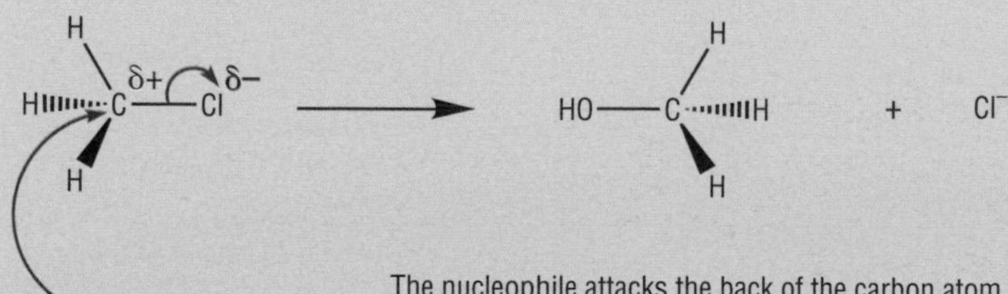

The nucleophile attacks the back of the carbon atom donating an electron pair. This causes a new bond to form and the carbon halogen bond to break.

Figure 2 The nucleophilic substitution mechanism

Questions

1 Describe, with the aid of diagrams, the mechanisms for the following reactions:
   (a) ethane and chlorine in ultraviolet radiation.
   (b) 1-iodobutane and hydroxide ions.
   (c) but-2-ene and bromine.

2.2 Alcohols and halogenoalkanes summary

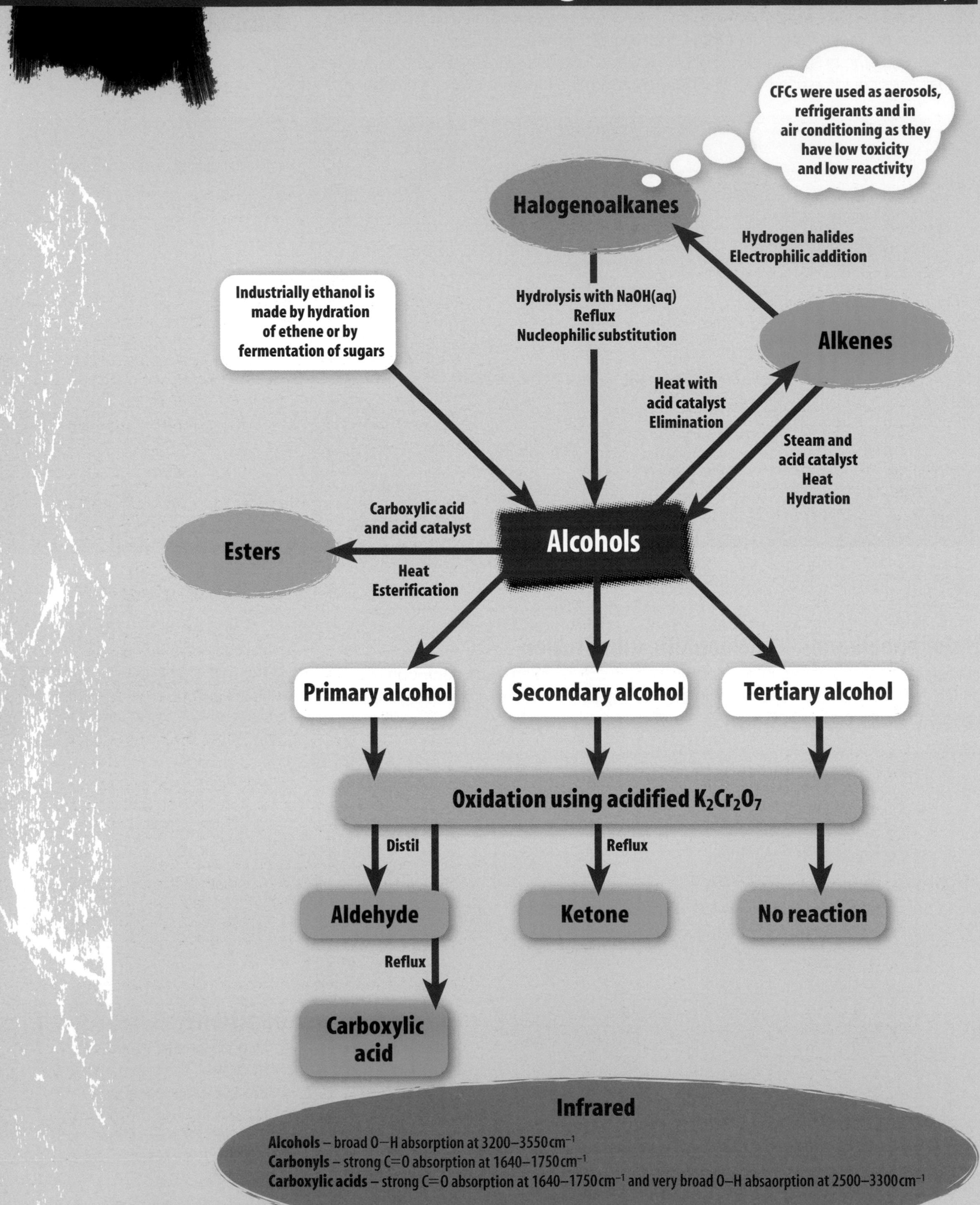

CFCs were used as aerosols, refrigerants and in air conditioning as they have low toxicity and low reactivity
Halogenoalkanes
Hydrogen halides
Electrophilic addition
Industrially ethanol is made by hydration of ethene or by fermentation of sugars
Hydrolysis with NaOH(aq)
Reflux
Nucleophilic substitution
Alkenes
Heat with
acid catalyst
Elimination
Steam and
acid catalyst
Heat
Hydration
Carboxylic acid
and acid catalyst
Esters
Alcohols
Heat
Esterification
Primary alcohol
Secondary alcohol
Tertiary alcohol
Oxidation using acidified $K_2Cr_2O_7$
Distil
Reflux
Aldehyde
Ketone
No reaction
Reflux
Carboxylic acid
Infrared
Alcohols – broad O–H absorption at 3200–3550 cm^{-1}
Carbonyls – strong C=O absorption at 1640–1750 cm^{-1}
Carboxylic acids – strong C=O absorption at 1640–1750 cm^{-1} and very broad O–H absaorption at 2500–3300 cm^{-1}

Practice questions

1. Draw the full structural formula for:
   (a) 2-bromopentane; **(b)** propan-2-ol;
   (c) hexane-1,2-diol; **(d)** 2-chlorohexan-3-ol.

2. There are two industrial methods for the preparation of ethanol – the fermentation of glucose and the direct hydration of ethene.
   (a) Write an equation for the preparation of ethanol from glucose.
   (b) Why does fermentation produce a maximum ethanol concentration of 15%?
   (c) State two advantages and two disadvantages of direct hydration compared with fermentation.

3. **(a)** What do you understand by the term *volatility*?
   (b) Why are the alcohols less volatile than alkanes of similar molecular mass?

4. Alcohols are soluble in water due to their ability to form hydrogen bonds with water molecules.
   Draw a diagram to show the hydrogen bonding between water and a molecule of propan-1-ol.

5. $C_4H_{10}O$ has four structural isomers that are alcohols.
   (a) Draw the structures of these four structural isomers and classify each as primary, secondary or tertiary.
   (b) For (i) one primary, (ii) one secondary and (iii) one tertiary alcohol from 5(a), describe the reaction, if any, with excess acidified potassium dichromate (VI). Name the organic products of each reaction.

6. **(a)** Write a balanced equation for the reaction between ethanol and butanoic acid.
   (b) Name the organic product formed.
   (c) Give a name for the type of reaction taking place and identify the catalyst for the reaction.

7. Give *two* uses of halogenoalkanes.

8. The hydrolysis of a bromoalkane with hot aqueous alkali is an example of nucleophilic substitution. Explain the term *nucleophilic substitution*.

9. **(a)** Write a balanced equation for the reaction of 2-bromopropane with aqueous hydroxide ions.
   (b) What would happen to the rate of the hydrolysis reaction if 2-bromopropane were replaced with 2-iodopropane? Explain your answer.

10. Describe the mechanism for the hydrolysis reaction that takes place between 3-chloropentane and aqueous hydroxide ions. Use curly arrows to show the mechanism.

11. Use of CFCs has been declining due to concern over the possible effects on the environment. It is thought that this damage is a result of a light-catalysed radical process involving chlorine radicals.
    (a) State *one* possible type of damage caused by CFCs to the environment.
    (b) Explain what is meant by a *radical*.
    (c) Give an equation for the formation of a radical from the chlorofluorocarbon CCl_2F_2.

12. Compound **A** is a hydrocarbon with a relative molecular mass of 56. Compound **A** contains 85.63% carbon by mass. Compound **A** reacts with hydrogen bromide to form a single monosubstituted compound **B**. Compound **B** reacts with hot aqueous sodium hydroxide to produce an alcohol, **C**. Alcohol **C** reacts with ethanoic acid in the presence of an acid catalyst to form a sweet-smelling ester **D**.
    Identify compounds **A**–**D**, showing displayed formulae for each structure.

13. The halogenoalkane **A**, C_4H_9Br, reacts with aqueous sodium hydroxide to produce alcohol **B**, C_4H_9OH. Alcohol **B** reacts with concentrated phosphoric acid to produce water and a mixture of three isomeric alkenes, **C**, **D** and **E**.
    (a) Draw the **four** structural isomers of C_4H_9Br.
    (b) Identify **A**, **B**, **C**, **D** and **E**, giving their full displayed formulae.
    (c) Write an equation for the reaction between the halogenoalkane **A** and aqueous sodium hydroxide. State the type of reaction taking place.
    (d) In the reaction sequence above, what is the function of the phosphoric acid?
    (e) Describe a chemical test to show the presence of an alkene.
    (f) If the halogenoalkane **A** were a chloroalkane, what difference would there be in the rate of reaction with aqueous sodium hydroxide? Explain your answer.

14. A student analysed a compound using infrared spectroscopy and mass spectrometry. He found that the compound gave a strong sharp IR absorbance at $1740\,cm^{-1}$ and peaks at m/z values of 15, 29 and 72. The compound had the following composition by mass: C, 66.63%; H, 11.18%; O, 22.19%.
    (a) Calculate the empirical formula of the compound.
    (b) What information can the student gain from the infrared absorbance?
    (c) What information is given by the peaks in the mass spectrum at 15 and 29?
    (d) The molecular ion is responsible for the peak at $m/z = 72$. Use this information and the empirical formula to deduce the molecular formula of the compound.
    (e) Draw the skeletal formula for two possible structures of the compound.

2.2 Examination questions

1 This question is about two alcohols, ethanol and propan-2-ol, $CH_3CH(OH)CH_3$.

(a) Ethanol can be formed by fermentation of glucose, $C_6H_{12}O_6$.

(i) Write a balanced equation, including state symbols, for the formation of ethanol by fermentation. [2]

(ii) Fermentation only occurs in the presence of yeast. State **two** other essential conditions. [2]

(iii) How would you know when fermentation of glucose is complete? [1]

(b) Propan-2-ol can be formed by the hydration of an alkene in the presence of a catalyst.

(i) Suggest a suitable catalyst for this reaction. [1]

(ii) This is an electrophilic addition reaction. What is meant by the term *electrophile*? [1]

(c) A mechanism for the reaction in **(b)** is shown below.

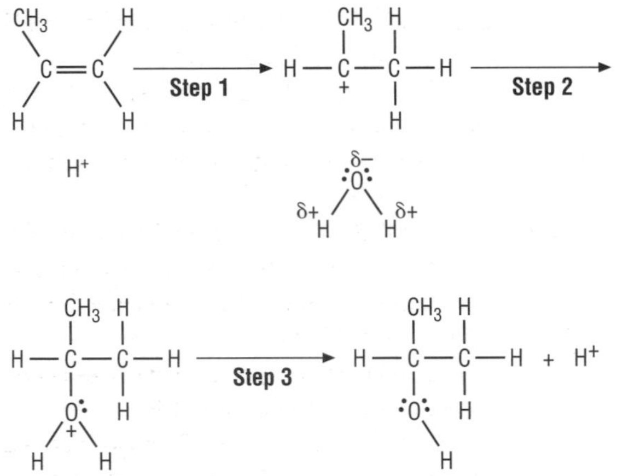

CH3, H, C=C, H, H → Step 1 → H—C(CH3)(+)—C(H)(H)—H → Step 2 →
H+
δ– :O: δ+ H H δ+

H—C(CH3)(O+ H H)—C(H)(H)—H → Step 3 → H—C(CH3)(:O: H)—C(H)(H)—H + H+

(i) Add 'curly arrows' to the mechanism to show the movement of electron pairs in steps **1**, **2** and **3**. [3]

(ii) Suggest, with a reason, the role of the H^+. [1]

(d) Propan-2-ol is flammable and readily burns. Write a balanced equation for the complete combustion of propan-2-ol. [2]

(e) Compound **D**, shown below, can be used as a solvent for plastics and fats and is also used in perfumery.

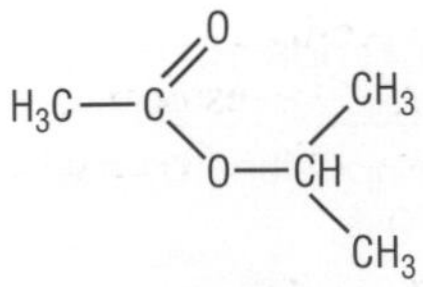

$H_3C—C(=O)—O—CH(CH_3)_2$

Compound **D**

Compound **D** can be prepared from propan-2-ol and another organic compound. Identify this other compound. [1]

[Total: 14]

(Jan 07 2812)

2 Halogenoalkanes are used in the production of pharmaceuticals, polymers and flame retardants.

(a) 1-Bromo-2-methylpropane is used in the production of ibuprofen and can be prepared from the reaction between 2-methylpropan-1-ol and HBr.

$(CH_3)_2CHCH_2OH + HBr \rightarrow (CH_3)_2CHCH_2Br + H_2O$

A student reacted 4.44 g of 2-methylpropan-1-ol with an excess of HBr. The student produced 5.48 g of 1-bromo-2-methylpropane.

(i) Calculate the number of moles of $(CH_3)_2CHCH_2OH$ used. [2]

(ii) Calculate the number of moles of $(CH_3)_2CHCH_2Br$ collected. $(CH_3)_2CHCH_2Br$, $M_r = 137$ [1]

(iii) Calculate the percentage yield. Quote your answer to three significant figures. [1]

(b) Chloroethene, CH_2CHCl, is polymerised to form poly(chloroethene) commonly known as PVC.

(i) Draw a section of PVC showing **three** repeat units. Put a bracket round one repeat unit. [2]

(ii) Polymers such as PVC are difficult to dispose of because they are non-biodegradable. Increasingly, they are disposed of by combustion.
State the problem associated with the combustion of polymers such as PVC. [1]

(iii) State **two** ways in which chemists are trying to minimise the damage to the environment caused by the disposal of halogenated plastics such as PVC. [2]

(c) F, C, Br, Cl, F

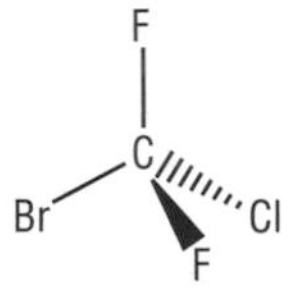

Bromochlorodifluoromethane has been used as a flame retardant.
When exposed to high temperatures, one of the C–halogen bonds undergoes homolytic fission to produce free radicals.
Suggest, with a reason, which C–halogen bond is most likely to be broken. [1]

[Total: 10]

(Jun 06 2812)

3 Acrolein, $CH_2{=}CHCHO$, and acrylic acid, $CH_2{=}CHCOOH$, are both used in industry for the manufacure of plastic resins and polymers. Both acrolein and acrylic acid can be made from prop-2-en-1-ol, $CH_2{=}CHCH_2OH$.

(a) (i) Draw the structures of prop-2-en-1-ol and acrolein. Clearly display the functional groups in each compound. [2]

(ii) Name the functional group common to **both** prop-2-en-1-ol and acrolein. [1]

(b) Prop-2-en-1-ol can be oxidised to form either acrolein or acrylic acid.

(i) Identify a suitable oxidising mixture. [2]

(ii) Write a balanced equation for the oxidation of prop-2-en-1-ol into acrolein. Use [O] to represent the oxidising agent. [1]

(c) A sample of prop-2-en-1-ol was oxidised and an infrared spectrum of the organic product was obtained.

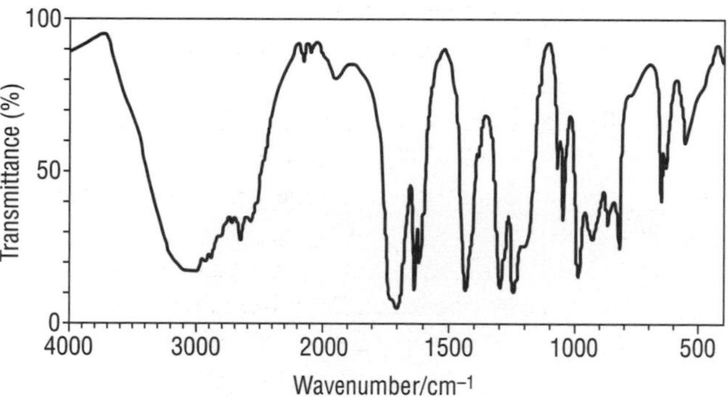

By referring to your *Data Sheet*, decide whether acrolein, CH_2=CHCHO, or acrylic acid, CH_2=CHCOOH, was formed. [3]

(d) Acrylic acid reacts with prop-2-en-1-ol to produce an ester.

(i) Complete the balanced equation for this reaction.
CH_2–CHCOOH + CH_2=$CHCH_2OH$ → . . . + . . . [2]

(ii) Draw the structure of the ester. Clearly display **all** of the functional groups.

[2]

[Total: 13]

(Jun 05 2812)

4 In this question, one mark is available for the quality of spelling, punctuation and grammar.

(a) The rates of hydrolysis of chloroethane, bromoethane and iodoethane are different.
Describe how you would monitor the reaction rates.
Explain why chloroethane, bromoethane and iodoethane react at different rates.
Use suitable equations in your answer. [6]

[Total: 6]

(Jun 05 2812)

5 Butan-1-ol, $CH_3CH_2CH_2CH_2OH$, reacts with sodium.

(a) There are four structural isomers of $C_4H_{10}O$ that are alcohols. One of the isomers has been drawn for you. Complete the table below to show the other structural isomers.

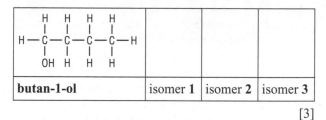

H H H H H—C—C—C—C—H OH H H H			
butan-1-ol	isomer **1**	isomer **2**	isomer **3**

[3]

(b) Butan-1-ol is oxidised by an acidified solution of potassium dichromate(VI) to form a carboxylic acid.

(i) State the colour change that you would see. [1]

(ii) Write a balanced equation for this oxidation of butan-1-ol to form a carboxylic acid. Use [O] to represent the oxidising agent. [2]

(iii) Identify which of the isomers, **1**, **2** or **3**, in **(a)** could also be oxidised to form a carboxylic acid. [1]

(c) Butan-1-ol reacts with hot concentrated sulfuric acid to form compound **B**.

(i) Compound **B** has an empirical formula of CH_2 and a relative molecular mass of 56. Use this information to deduce the molecular formula of compound **B**. Show your working. [2]

(ii) Write a balanced equation to show the conversion of butan-1-ol into compound **B**. [1]

(iii) One of the isomers, **1**, **2** or **3**, in **(a)** also reacts with hot concentrated sulfuric acid to form compound **B**. Identify which isomer. [1]

(d) The ester $CH_3COOCH_2CH_2CH_2CH_3$ was formed by reacting ethanoic acid with butan-1-ol.

$$H_3C—C(=O)—OH + CH_3CH_2CH_2CH_2—OH \xrightleftharpoons{\text{Catalyst}} H_3C—C(=O)—O—CH_2CH_2CH_2CH_3 + H_2O$$

Butan-1-ol **Ester** M_r = 116

(i) State a catalyst for this reaction. [1]

(ii) In an experiment, 6.96 g of the ester was produced from 0.100 mol of butan-1-ol. Calculate the number of moles of ester produced. [1]

(iii) Calculate the percentage yield. [1]

[Total: 14]

(Jan 05 2812)

UNIT 2

Module 3 Energy

Introduction

The chemical industry is one of the most successful and profitable sectors of the United Kingdom's economy. It employs directly a workforce of 215 000 and supports many more jobs. This sector has an annual turnover in excess of £50 billion, of which around £2 billion is used on capital investment.

Much of this success is the result of long-term research by dedicated professionals, working to improve the efficiency of industrial processes. The line between commercial success and failure is a fine one, and the ability to improve productivity through process developments is a highly valued skill.

In this module you will learn about energy, equilibrium and rates. You will study exothermic and endothermic reactions. You will discover how to calculate energy changes from both your own experiments and from theoretical data. You will explore the conditions required to increase the rate of a chemical reaction and the effect of these conditions on a chemical equilibrium.

The image shows a catalytic converter changing pollutants in a car exhaust into harmless gases.

Test yourself

1 Names two desirable properties of a fuel.
2 What do 'exothermic' and 'endothermic' mean?
3 State three ways of increasing the rate of a reaction.
4 What is meant by 'catalyst'?
5 What is meant by a 'reversible reaction'?

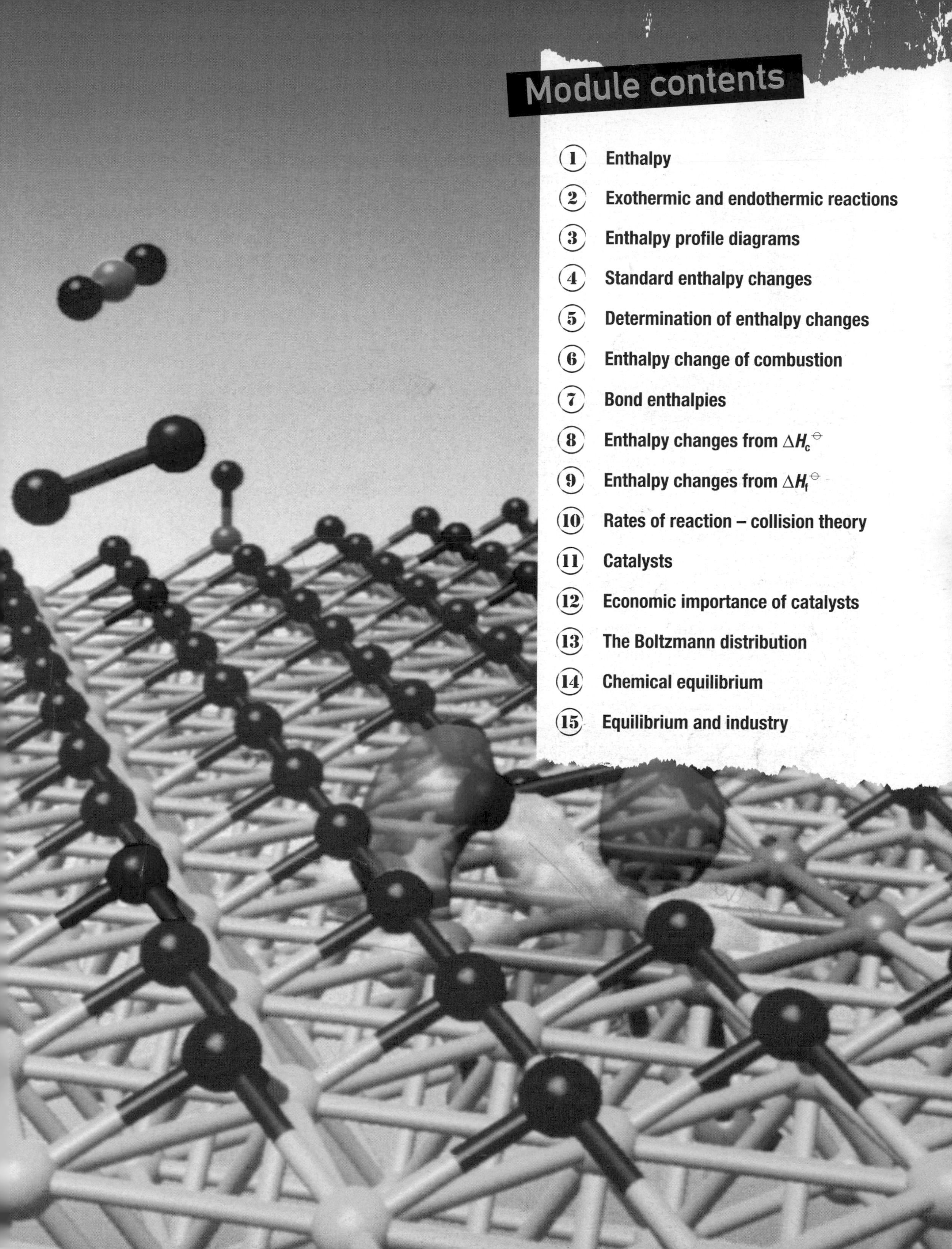

Module contents

2.3 (1) Enthalpy

By the end of this spread, you should be able to . . .

* **Understand that energy is conserved.**
* **Explain that all chemical reactions are accompanied by enthalpy changes.**
* **Explain that enthalpy change can be exothermic or endothermic.**

What is chemical energy?

Chemical energy is a special form of potential energy that lies within chemical bonds. Chemical bonds are the very forces of attraction that bind together atoms in compounds. When chemicals react to form new substances, bonds break in the reactants and new bonds are formed as products are made. This process changes the *chemical energy* of the atoms.

Enthalpy

Enthalpy, *H*, is the heat content that is stored in a chemical system.

It is actually impossible to measure directly the enthalpy of the reactants or products. But we *can* measure the energy absorbed or released to the surroundings during a chemical change. The form of this energy can vary. It sometimes appears as light, or electrical work, but most often occurs only as heat.

Key definition

Enthalpy, *H*, is the heat content that is stored in a chemical system.

Examiner tip

The *chemical system* is the reactants and products.

The *surroundings* are what is outside the chemical system.

We determine the heat exchange between the chemical system and the surroundings.

Conservation of energy

If heat is released, the amount of energy that leaves a chemical system is exactly the *same* as the amount that goes into the surroundings. No heat energy is lost. It just transfers from one place to another. Indeed some heat might change from one form to another. This is called the *law of conservation of energy*.

In this spread, energy changes will be measured as heat. If heat is released or absorbed, we can easily measure its effect by recording the temperature change with a thermometer.

This means that:

- heat loss in a chemical system = heat gain to surroundings (accompanied by a temperature increase)
- heat gain in a chemical system = heat loss from surroundings (accompanied by a temperature decrease).

Enthalpy change

In some reactions, the products of a reaction have *more* chemical energy than the reactants. In others, the products have *less* chemical energy than the reactants.

An enthalpy change, ΔH, is:

- the heat exchange with the surroundings during a chemical reaction, at constant pressure;
- the difference between the enthalpy of the products and the enthalpy of the reactants:
  $\Delta H = H_{products} - H_{reactants}$

In general, *all* chemical reactions either release heat (exothermic reactions) or absorb heat (endothermic reactions).

Key definition

Exothermic refers to a reaction in which the enthalpy of the products is smaller than the enthalpy of the reactants, resulting in heat loss to the surroundings (ΔH –ve).

Exothermic reactions

In an **exothermic** reaction:

- the enthalpy of the products is *smaller* than the enthalpy of the reactants
- there is a heat *loss* from the chemical system to the surroundings
- ΔH has a *negative* sign because heat has been lost by the chemical system.

An exothermic enthalpy change is shown in Figure 1.

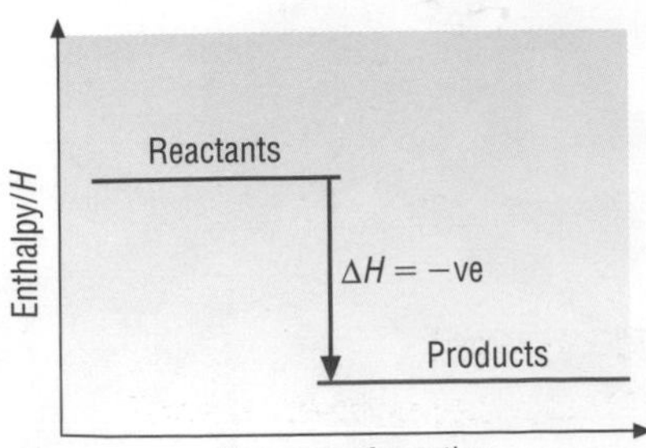

Figure 1 An exothermic enthalpy change

Self-heating cans

If two chemicals are mixed together in an exothermic reaction, the container will be warm to the touch. An interesting application of this has been in the development of self-heating cans.

A self-heating can is an extension of the common food can. It uses an inner chamber to hold food or drink and an outer chamber to hold certain chemicals, usually calcium oxide, CaO and water. The chemicals are separated by a barrier which can easily be broken. When you want to heat the food or drink, you break the barrier separating the chemicals in the outer chamber. This can be done by pulling a ring on the can, pressing a button, twisting the base, or other means. The chemicals react together, releasing heat into the food or drink (i.e. the *surroundings*).

$$CaO(s) + H_2O(l) \longrightarrow Ca(OH)_2(aq) \quad \Delta H = -ve$$

After the food or drink has absorbed the heat, you can enjoy a hot meal or drink.

Self-heating cans have been used for many years. In the 1940s, self-heating cans of soup were part of British troops' rations for the historic D-Day landings during the Second World War.

Endothermic

In an **endothermic** reaction:

- the enthalpy of the products is *greater* than the enthalpy of the reactants
- there is a heat *gain* to the chemical system from the surroundings
- ΔH has *positive* sign because heat has been gained by the chemical system.

An endothermic enthalpy change is shown in Figure 2.

Key definition

Endothermic refers to a reaction in which the enthalpy of the products is greater than the enthalpy of the reactants, resulting in heat being taken in from the surroundings (ΔH +ve).

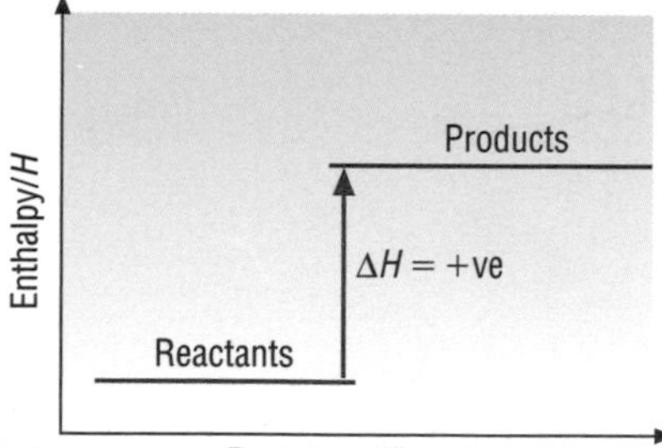

Figure 2 An endothermic enthalpy change

The self-cooling beer can

Self-cooling cans have also been developed to cool drinks such as beer!

In this instance, a simple endothermic process (the evaporation of water) does the job of absorbing heat from the beer.

$$H_2O(l) \longrightarrow H_2O(g) \quad \Delta H = +ve$$

As with self-warming cans, a ring, button or twist initiates the process.

The diagram in Figure 3 illustrates how this works.

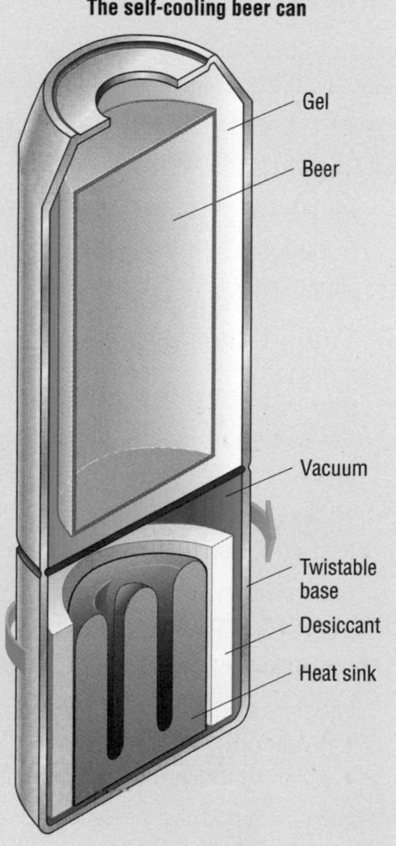

The beer is surrounded by an outer compartment containing a watery gel. The bottom section, sealed from the top, contains a desiccant in a vacuum.

The user twists the can to break the seal and the drop in pressure causes the water around the upper part to evaporate quickly, cooling the beer. The water is absorbed by the desiccant, and a sink absorbs the heat, stopping the can feeling warm to the touch.

It is claimed the beer can be cooled by 16.7°C in just three minutes.

Figure 3 Self-cooling beer can

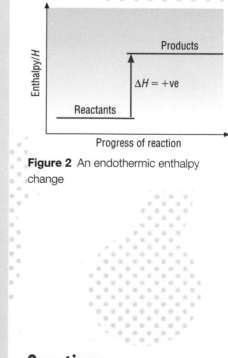

Questions

1 What does *enthalpy* mean?

2 What decides whether a reaction is exothermic or endothermic?

2.3  Exothermic and endothermic reactions

By the end of this spread, you should be able to . . .

* **Describe the importance of oxidation as an exothermic process.**
* **Explain that endothermic processes require an input of heat energy.**

Exothermic reactions

Oxidation of fuels

The oxidation of fuels by combustion is perhaps our most exploited exothermic reaction.

- A common example is the oxidation of methane in natural gas to form carbon dioxide and water (Figure 1).
- The products have *less* enthalpy than the reactants, so the excess energy is released as heat.

Figure 1 Combustion of methane. An exothermic reaction – exploiting energy from a fuel

The enthalpy change (ΔH) is included in the equation, as the *enthalpy change of reaction*.

$$CH_4(g) + 2O_2(g) \longrightarrow CO_2(g) + 2H_2O(l) \qquad \Delta H = -890 \text{ kJ mol}^{-1}$$

- Whenever you see the units kJ mol^{-1}, you know that the enthalpy change refers to molar quantities in the equation.
- So, 1 mol $CH_4(g)$ reacts with 2 mol $O_2(g)$ to form 1 mol $CO_2(g)$ and 2 mol $H_2O(l)$, with release of 890 kJ of heat.

Enthalpy change of reaction is discussed in more detail in spread 2.3.4.

Combustion reactions affect us all. We rely on the combustion of petrol and diesel in cars, buses and trains, and the combustion of kerosene in aircraft. There is much concern, however, not only about the sheer amount of fuel that we are burning, but also the levels of carbon dioxide produced. So just how much fuel do we burn and how much carbon dioxide and energy are we talking about?

Let's consider a car travelling on a two-mile journey. About 200–250 g of petrol will be needed. If the car is a 4 × 4, it will be nearer to 300–400 g of petrol.

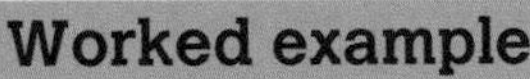

Worked example

Petrol is a mixture of hydrocarbons, such as octane, C_8H_{18}.

Equation: $C_8H_{18}(l) + 12\frac{1}{2}O_2(g) \longrightarrow 8CO_2(g) + 9H_2O(l) \quad \Delta H = -5470 \text{ kJ mol}^{-1}$
So, 1 mol $C_8H_{18}(l)$ will produce 8 x 24 = 192 dm^3 $CO_2(g)$ and will release 5470 kJ of energy.

200 g octane is 200/114 mol = 1.75 mol.
So, 200–250 g of petrol is about 2 mol C_8H_{18}.

So, this short trip will produce about 2 x 192 = 384 dm^3 of $CO_2(g)$ and 2 × 5470 = 10 940 kJ of energy.

That is roughly the volume of 150 footballs full of CO_2, and the amount of energy that your body uses in a whole day.

Now let's try a holiday to Singapore.

It has been calculated that every person on the flight will have produced the equivalent of a year's CO_2 from driving (about 10 000 miles).

Lifestyle changes to cut down on CO_2 production?

- Walk or cycle the two miles.
- Cut out long-haul holidays abroad.

Figure 2 Boeing 787 jet aircraft. Technology is moving to cut down on CO_2 production. The Boeing 787 is one of a new generation of aircraft, consuming 20% less fuel

Respiration

Respiration is probably the most important *exothermic* reaction for life.

- Sugars, such as glucose $C_6H_{12}O_6$, are oxidised to carbon dioxide and water.
- The overall equation is given below but, in a living system, respiration takes place over a series of steps.

$$C_6H_{12}O_6(aq) + 6O_2(g) \rightarrow 6CO_2(g) + 6H_2O(l) \qquad \Delta H = -2801 \text{ kJ mol}^{-1}$$

Without respiration, there would be no life.

Endothermic reactions

Thermal decomposition of limestone

Heat provides the energy that drives many reactions. These reactions are usually endothermic, the supplied energy being taken in by the chemical system.

Limestone contains calcium carbonate, $CaCO_3$. The decomposition of calcium carbonate by heat is an important endothermic reaction used to make calcium oxide, CaO.

$$CaCO_3(s) \longrightarrow CaO(s) + CO_2(g) \qquad \Delta H = +178 \text{ kJ mol}^{-1}$$

The calcium oxide product is commonly known as lime and was originally prepared by heating limestone in lime kilns. Nowadays, this process is carried out in large industrial plants. Lime has many uses, such as in the manufacture of cement and for treatment of acid soils by farmers. If water is added to lime, the heat is released again – see spread 1.3.7.

Photosynthesis

Photosynthesis is probably the most important *endothermic* reaction for life.

- During photosynthesis, sugars, such as glucose $C_6H_{12}O_6$ are made from carbon dioxide and water.
- Light from the Sun provides the energy for photosynthesis, which takes place in a series of steps.

$$6CO_2(g) + 6H_2O(l) \rightarrow C_6H_{12}O_6(aq) + 6O_2(g) \qquad \Delta H = +2801 \text{ kJ mol}^{-1}$$

As with respiration, without photosynthesis there would be no life.

Compare this equation with that for respiration. It is the reverse – even in terms of energy.

Figure 3 Test tube of pondweed in water exposed to light. The pondweed photosynthesises, producing the bubbles of oxygen seen on the leaves

Questions

1 How much energy would be released during combustion of 5 mol CH_4?

2 What do the reactions that take part in respiration and photosynthesis have in common?

3 Where has the energy for photosynthesis come from?

2.3 (3) Enthalpy profile diagrams

By the end of this spread, you should be able to . . .

* **Construct a simple enthalpy profile diagram for a reaction.**
* **Explain the term activation energy using enthalpy profile diagrams.**

Key definition

An **enthalpy profile diagram** is a diagram for a reaction to compare the enthalpy of the reactants with the enthalpy of the products.

Simple enthalpy profile diagrams

Reactions and their enthalpy changes can be illustrated in an **enthalpy profile diagram**. The diagram shows what happens to enthalpies during the course of a reaction. It also emphasises the exothermic or endothermic nature of the reaction. In spread 2.3.1 you were introduced to some simple enthalpy profile diagrams.

Exothermic reactions

The enthalpy profile diagram for the combustion of methane (see spread 2.3.2) is shown in Figure 1.

- H products < H reactants
- The enthalpy change, ΔH, is negative.

Examiner tip

In an *exothermic* reaction:

- ΔH is *negative*
- heat is *given* out to the surroundings
- the reacting chemicals *lose* energy
- heat lost by chemicals = heat gained by surroundings.

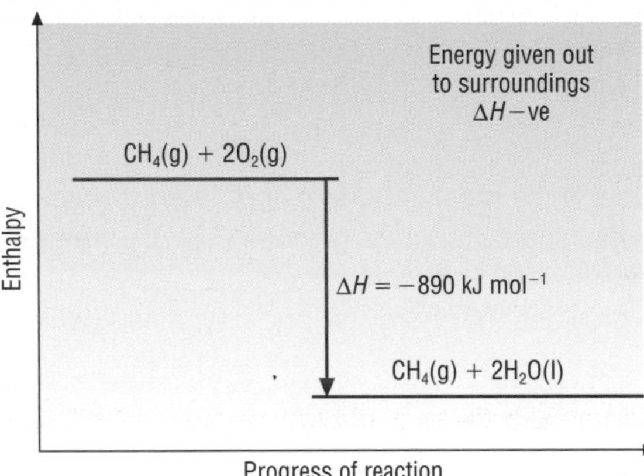

Figure 1 Enthalpy profile diagram for an exothermic reaction

Endothermic reactions

The enthalpy profile diagram for the decomposition of calcium carbonate (see spread 2.3.2) is shown in Figure 2.

- H products > H reactants
- The enthalpy change, ΔH, is positive.

Examiner tip

In an *endothermic* reaction:

- ΔH is *positive*
- heat is *taken* in from the surroundings
- the reacting chemicals *gain* energy
- heat gained by chemicals = heat lost by surroundings.

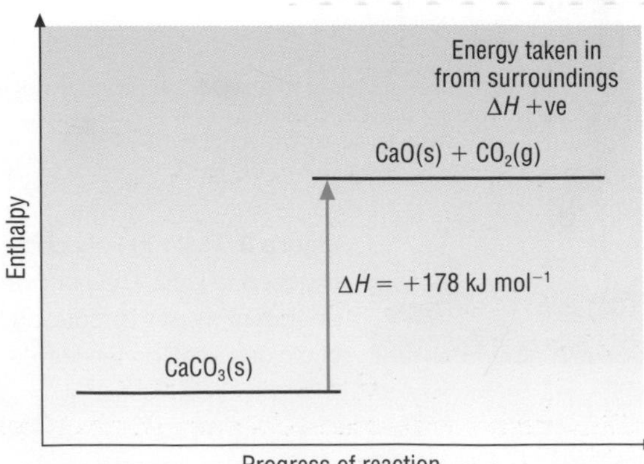

Figure 2 Enthalpy profile diagram for an endothermic reaction

Activation energy

Chemical reactions have an energy barrier that prevents many reactions from taking place spontaneously. This is called **activation energy** E_a and is required to break bonds in the reactants (see spread 2.3.7).

We can show activation energies in enthalpy profile diagrams.

Exothermic reactions

Even though the products have a *lower* energy than reactants, there still has to be an input of energy to break the first bond and start the reaction.

The activation energy is often supplied by a spark or by heating the chemicals.

Natural gas needs a spark to overcome the activation energy of the reaction. Once the energy barrier has been overcome, the net output of energy provides more energy that can be used to overcome the activation energy for the reaction to continue.

Once an *exothermic* reaction begins, the activation energy is regenerated and the reaction becomes self-sustaining.

If activation energy did not exist, exothermic reactions would take place spontaneously. Fuels would not exist as they would spontaneously combust!

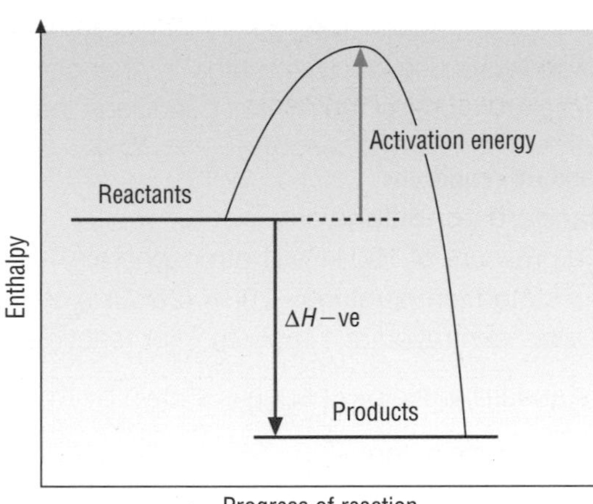

Figure 3 Activation energy for an exothermic reaction

Endothermic reactions

The diagram in Figure 4 shows the activation energy for an *endothermic* reaction.

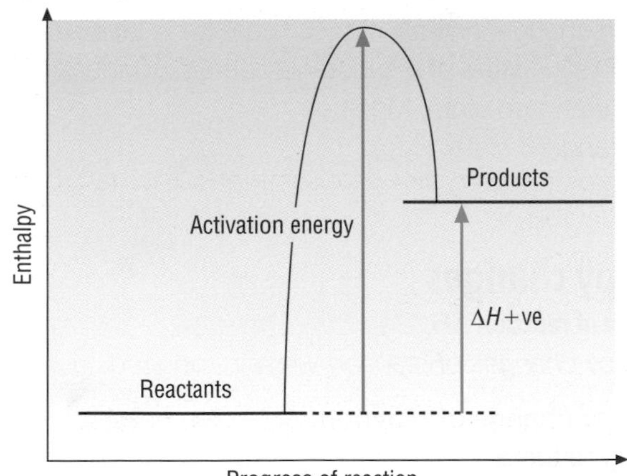

Figure 4 Activation energy for an endothermic reaction

Key definition

Activation energy is the minimum energy required to start a reaction by the breaking of bonds.

Examiner tip

It is important to show the activation energy and enthalpy change using arrows pointing in the correct direction.

- The activation energy is positive and the arrow points up.
- The reaction is exothermic with heat energy being lost to the surroundings so the arrow goes down.

Examiner tip

The activation energy is always the *input* of energy from reactants to the top of the energy barrier. This is easier to show for *exothermic* reactions than endothermic reactions so take care in exams!

Questions

1 You are given the data for the following reaction:
$CO(g) + NO_2(g) \longrightarrow CO_2(g) + NO(g)$
$\Delta H = -226 \text{ kJ mol}^{-1}$ $E_a = +134 \text{ kJ mol}^{-1}$
Draw an enthalpy profile diagram for this reaction showing both of these energy changes.

2 You are given the data for the following reaction:
$H_2(g) + I_2(g) \longrightarrow 2HI(g)$
$\Delta H = +53 \text{ kJ mol}^{-1}$ $E_a = +183 \text{ kJ mol}^{-1}$
Draw an enthalpy profile diagram for this reaction showing both of these energy changes.

2.3  Standard enthalpy changes

By the end of this spread, you should be able to . . .

* **Define and use the terms:**
  standard conditions;
  enthalpy change of reaction;
  enthalpy change of formation; and
  enthalpy change of combustion.

Standards

An enthalpy change varies depending on the conditions. Enthalpy changes for reactions are often listed in data books or databases and it is important that these are always the same value.

So chemists use *standard* enthalpy changes, measured under *standard* conditions. These are close to the *normal* conditions that are found in a science laboratory.

Standard conditions

Standard conditions are:

- a pressure of 100 kPa (1 atmosphere)
- a stated temperature – 298 K (25 °C) is usually used
- a concentration of 1 mol dm^{-3} (for reactions with aqueous solutions).

Key definition

Standard conditions are a pressure of 100 kPa (1 atmosphere), a stated temperature, usually 298 K (25 °C) and a concentration of 1.0 mol dm^{-3} (for reactions with aqueous solutions).

A standard enthalpy change is shown by the symbol: $\Delta H^{\ominus}$.

The symbols mean:

- H – enthalpy
- Δ – change
- $^{\ominus}$ – under standard conditions.

Standard states

For a standard enthalpy change, any substance must be in its **standard state**. A standard state is the physical state of a substance under standard conditions.

- Magnesium has the standard state: Mg(s).
- Hydrogen has the standard state: $H_2(g)$.
- Water has the standard state: $H_2O(l)$.

Key definition

Standard state is the physical state of a substance under the standard conditions of 100 kPa (1 atmosphere) and 298 K (25 °C).

Standard enthalpy changes

Standard enthalpy change of reaction $\Delta H_r^{\ominus}$

In spread 2.3.2, *enthalpy changes of reaction* were introduced.

For a **standard enthalpy change of reaction**, $\Delta H_r^{\ominus}$, we need:

- an *equation* – this gives the amounts, in mol, of reactants and products
- an *enthalpy change*, kJ mol^{-1} – this gives the enthalpy change for the molar quantities in the stated equation.

$$H_2(g) + \tfrac{1}{2}O_2(g) \rightarrow H_2O(l) \qquad \Delta H_r^{\ominus} = -286 \text{ kJ mol}^{-1}$$

1 mol ½ mol 1 mol

Key definition

The **standard enthalpy change of reaction**, $\Delta H_r^{\ominus}$, is the enthalpy change that accompanies a reaction in the molar quantities expressed in a chemical equation under standard conditions, all reactants and products being in their standard states.

Examiner tip

It is important that you can quote standard conditions as 298 K and 100 kPa. This may be worth two marks on the examination paper.

Standard enthalpy change of combustion

Combustion is such a common process that it is given its own special enthalpy change. Chemists have measured the **standard enthalpy change of combustion**, $\Delta H_c^{\ominus}$, for many substances, and these are listed in data books.

$\Delta H_c^{\ominus}$ for ethane is:

$$C_2H_6(g) + 3\tfrac{1}{2}O_2(g) \rightarrow 2CO_2(g) + 3H_2O(l) \qquad \Delta H_c^{\ominus} = -1560 \text{ kJ mol}^{-1}$$

- The $\Delta H_c^{\ominus}$ value of -1560 kJ mol^{-1} applies to the equation above.
- With no balancing number added to the ethane, the equation shows *1 mol* $C_2H_6(g)$.
- The equation then matches the definition for the enthalpy change of combustion – the complete combustion of 1 mol of $C_2H_6(g)$.

Standard enthalpy change of formation

The formation of a compound from its elements is another common process. Formation is also given its own special enthalpy change, the **standard enthalpy change of formation**, $\Delta H_f^{\ominus}$.

$\Delta H_f^{\ominus}$ for water is:

$$H_2(g) + \tfrac{1}{2}O_2(g) \rightarrow H_2O(l) \qquad \Delta H_f^{\ominus} = -286 \text{ kJ mol}^{-1}$$

- The $\Delta H_f^{\ominus}$ value of -286 kJ mol^{-1} applies to the equation above, showing formation of *1 mol* $H_2O(l)$.
- The equation then matches the definition for the enthalpy change of formation – 1 mol of $H_2O(l)$ being formed from its elements.

However, $\Delta H_f^{\ominus}$ of an element appears problematic:

- If we are forming an element, such as $H_2(g)$, from the element, $H_2(g)$, there is no chemical change.
- So, all elements have a standard enthalpy change of formation of 0 kJ mol^{-1}.

Key definition

The **standard enthalpy change of combustion**, $\Delta H_c^{\ominus}$, is the enthalpy change that takes place when one mole of a substance reacts completely with oxygen under standard conditions, all reactants and products being in their standard states.

Key definition

The **standard enthalpy change of formation**, $\Delta H_f^{\ominus}$, of a compound is the enthalpy change that takes place when one mole of a compound is formed from its constituent elements in their standard states under standard conditions.

Remember that

The enthalpy change of formation of an element is defined as 0 kJ mol^{-1}.

Questions

1 Two enthalpy changes of reaction are shown below.

$$N_2(g) + 3H_2(g) \longrightarrow 2NH_3(g) \qquad \Delta H = -92 \text{ kJ mol}^{-1}$$
$$N_2O_4(g) \longrightarrow 2NO_2(g) \qquad \Delta H = +58 \text{ kJ mol}^{-1}$$

What is the enthalpy change of reaction for the following:

(a) $\tfrac{1}{2}N_2(g) + 1\tfrac{1}{2}H_2(g) \longrightarrow NH_3(g)$

(b) $\tfrac{1}{2}N_2O_4(g) \longrightarrow NO_2(g)$

2 (a) Write equations for the change that occurs during the enthalpy change of combustion for each of the following compounds:

(i) $CH_4(g)$; **(ii)** $C_3H_8(g)$; **(iii)** $CS_2(l)$;
(iv) $CH_3OH(l)$; **(v)** $C_2H_5OH(l)$.

3 (a) Write equations for the change that occurs during the enthalpy change of formation for each of the following compounds:

(i) $C_2H_4(g)$; **(ii)** $C_2H_6(g)$; **(iii)** $C_2H_5OH(l)$;
(iv) $CaO(s)$; **(v)** $Al_2O_3(s)$.

2.3 Determination of enthalpy changes

By the end of this spread, you should be able to . . .

* Calculate enthalpy changes directly from appropriate experimental results, including the use of the relationship: energy change = $mc\Delta T$.

Determination of enthalpy changes

You cannot determine directly the actual enthalpy of the reactants or products. But you *can* determine the heat exchange with the surroundings.

Remember:

- heat loss in a chemical system = heat gain by surroundings
- heat gain in a chemical system = heat loss by surroundings.

So provided that we know what happens to the energy in the surroundings, we automatically know what has happened to the energy of the chemical system.

To determine the heat exchange during a reaction, you need to know the following:

- m – the mass of the surroundings involved in the heat exchange;
- c – the **specific heat capacity** of the surroundings;
- ΔT – the temperature change of the surroundings: $\Delta T = T_{final} - T_{initial}$.

The heat exchanged with the surroundings, Q, can be calculated using the following relationship:

$Q = mc\Delta T$ Joules

How the thermometer responds to heat being released or absorbed by the chemical system is illustrated in Figure 1.

Key definition

Specific heat capacity, c, is the energy required to raise the temperature of 1 g of a substance by 1 °C.

Examiner tip

If the temperature increases on the thermometer, then the reaction is exothermic and ΔH is negative. If the temperature decreases on the thermometer, then the reaction is endothermic and ΔH is positive.

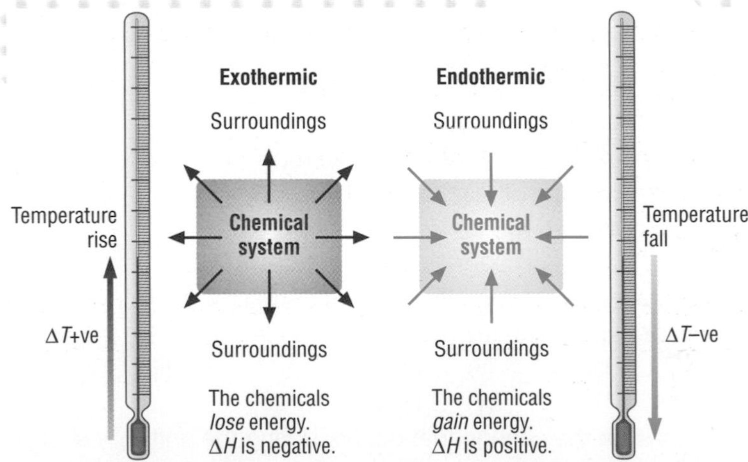

Figure 1 Types of heat exchange with the surroundings and their effects

Direct determination of enthalpy changes

For many reactions, you can use direct experimental results to calculate the enthalpy change.

Many reactions take place when two chemicals in solution are simply mixed together. You can carry out this type of reaction in a *calorimeter*. The simplest calorimeter is a plastic coffee cup, such as that shown in Figure 2. The expanded polystyrene insulates the solution inside the cup.

Examiner tip

Sometimes candidates get confused about what the surroundings are. It is not always obvious!

In experiments that you carry out, the surroundings are really 'where the thermometer is'.

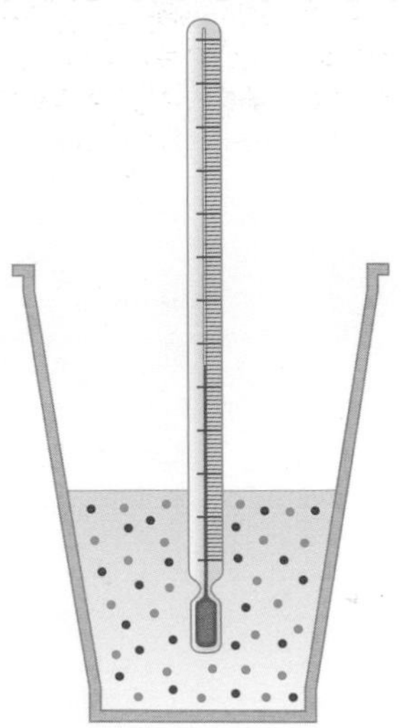

The particles are dissolved in the solution. The heat is exchanged from the particles into the solution. The heat exchange is measured using the thermometer.

Figure 2 A coffee cup calorimeter

In an *exothermic* reaction, the heat produced is trapped in the calorimeter, *increasing* the temperature of the solution.

For an *endothermic* reaction, the heat required for the reaction is removed from the solution, *decreasing* the temperature of the solution.

The experiment is carried out and the temperature change in the solution is measured.

Worked example

An excess of magnesium is added to 100 cm^3 of 2.00 mol dm^{-3} $CuSO_4$(aq). The temperature increases from 20.0 °C to 65.0 °C.
Find the enthalpy change of reaction for the following equation:

$Mg(s) + CuSO_4(aq) \rightarrow MgSO_4(aq) + Cu(s)$

- specific heat capacity of solution $c = 4.18\ J\ g^{-1}\ K^{-1}$
- density of solution = 1.00 g cm^{-3}

Answer

1 Find the energy change.
100 cm^3 of solution has a mass of 100 g.
Temperature change, $\Delta T = (65.0 - 20.0)\ °C = +45.0\ °C$.
Heat *gained* by surroundings, $Q = mc\Delta T = 100 \times 4.18 \times 45.0\ J = +18810\ J$.
This heat has been released by the chemical system.
So, heat *lost* by chemical system = –18810 J.

2 Find out the amount, in mol, that reacted.

$$\text{Amount, in mol, of } CuSO_4 \text{ that reacted} = \frac{c \times V\ (\text{in } cm^3)}{1000} = \frac{2.00 \times 100}{1000} = 0.200\ \text{mol.}$$

3 Scale the quantities to match the molar quantities in the equation.

$Mg(s)$ +	$CuSO_4(aq) \rightarrow$	$MgSO_4(aq)$ +	$Cu(s)$
1 mol	1 mol	1 mol	1 mol

For 0.200 mol $CuSO_4$, $\Delta H = -18810$ J.
1 mol of $CuSO_4$ = 5 × 0.200 mol.
For 1 mol of $CuSO_4$, $\Delta H = 5 \times -18810$
= –94 050 J = –94.05 kJ.

4 Finally, write down the equation together with the enthalpy change, in kJ mol^{-1}.
$Mg(s) + CuSO_4(aq) \rightarrow MgSO_4(aq) + Cu(s)$ $\quad \Delta H = -94.05\ kJ\ mol^{-1}$

Examiner tip

The temperature change ΔT is the same in °C as it is in in K.

Examiner tip

The mass is not always obvious – the *m* here is the mass of substance that is changing temperature.

Examiner tip

All working should be shown. Each step of the calculation may be worth a mark – even if the final answer is wrong, you may get most of the marks if most of your steps use the correct method.

Questions

1 0.327 g of zinc powder was added to 55.0 cm^3 of aqueous copper(II) sulfate at 22.8 °C. The temperature rose to 32.3 °C. The aqueous copper(II) sulfate is in excess.
Find the enthalpy change of reaction for the following equation:
$Zn(s) + CuSO_4(aq) \rightarrow ZnSO_4(aq) + Cu(s)$

2 25 cm^3 of 2.00 mol dm^{-3} HCl(aq) was mixed with 25 cm^3 of 2.00 mol dm^{-3} NaOH(aq).
The temperature increases from 22.5 °C to 34.5 °C.
Find the enthalpy change of reaction for the following equation:
$HCl(aq) + NaOH(aq) \rightarrow NaCl(aq) + H_2O(l)$

2.3  Enthalpy change of combustion

By the end of this spread, you should be able to . . .

* **Calculate an enthalpy change of combustion from experimental results.**

Combustion

Combustion is a reaction involving oxygen to form oxides.

The enthalpy change of combustion is the enthalpy change that accompanies the *complete* combustion of *one mole* of a substance. Equations representing the standard enthalpy change of combustion for CH_4, H_2 and Al are shown below. In each equation, *one mol* of the substance has been burnt.

$CH_4(g) + 2O_2(g) \longrightarrow CO_2(g) + 2H_2O(l)$ $\quad \Delta H_c^{\ominus} = -890\ kJ\ mol^{-1}$

$H_2(g) + \frac{1}{2}O_2(g) \longrightarrow H_2O(l)$ $\quad \Delta H_c^{\ominus} = -286\ kJ\ mol^{-1}$

$Al(s) + \frac{3}{4}O_2(g) \longrightarrow \frac{1}{2}Al_2O_3(s)$ $\quad \Delta H_c^{\ominus} = -1676\ kJ\ mol^{-1}$

Examiner tip

In an equation to represent $\Delta H_c^{\ominus}$, you must not put a balancing number in front of the substance being *burnt*. If you do, then more than 1 mol would have been combusted.

Experimental determination of ΔH_c

The experimental determination of ΔH_c for a liquid fuel is generally one of the easiest enthalpy experiments to carry out.

All you need to do is:

- Burn a known mass of a substance in air;
- to heat a known mass of water;
- and measure the temperature change in the water.

You only require a simple apparatus to determine the enthalpy change of combustion. An example is shown in Figure 1.

- You measure a volume of water into the beaker.
  1 cm^3 of water weighs 1 g, so the mass of water is easy to deduce.
- The burner containing the fuel is weighed.
- The initial temperature of the water is taken.
- The burner is lit and the water heated until the temperature has risen by a reasonable amount. The maximum temperature is taken and the temperature change, ΔT, can then be determined.
- The flame is extinguished and the burner re-weighed to find the mass of the fuel that has been burnt.

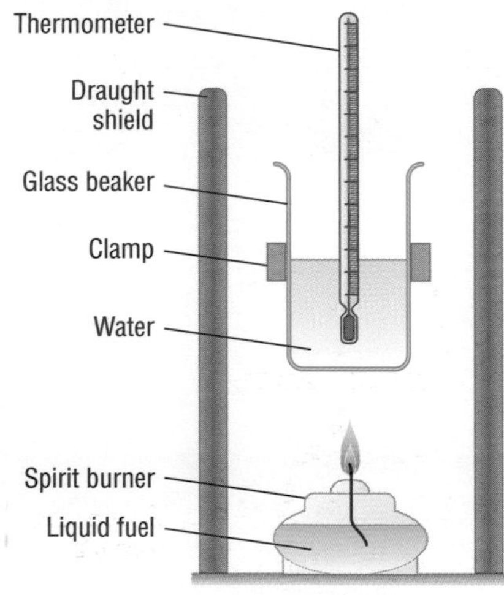

Figure 1 A calorimeter suitable for the determination of the enthalpy of combustion of a fuel

Worked example

During combustion, 1.50 g of propan-1-ol, $CH_3CH_2CH_2OH$, heated 250 cm^3 of water by 45 °C.
Find the enthalpy change of combustion of propan-1-ol.

- specific heat capacity of water, $c = 4.18\ J\ g^{-1}\ K^{-1}$
- density of water = $1\ g\ cm^{-3}$

Answer

1 Find the heat change during the experiment.
In spread 2.3.5, you found that a heat change can be calculated using the equation: $Q = mc\Delta T$.
250 cm^3 water weighs 250 g.
So, heat *gained* by the water, $Q = 250 \times 4.18 \times 45\ J = 47\,025\ J = +47.025\ kJ$.
The same quantity of heat will have been *lost* by the fuel, propan-1-ol, and oxygen during combustion.
So, heat *loss* from chemicals = -47.025 kJ.

Examiner tip

Remember that *m* is the mass of the substance which is changing temperature. Heat is being transferred to the water.

2 Find the amount, in mol, of propan-1-ol that reacted.
molar mass, M, of $CH_3CH_2CH_2OH$ = 60.0 g mol^{-1}
So, amount, n, of $CH_3CH_2CH_2OH$ that reacted = $\frac{m}{M} = \frac{1.50}{60.0}$ = 0.025 mol

3 Work out the heat loss in kJ mol^{-1}.
Simply divide the heat loss by the amount, in mol, of fuel that was burnt.
On combustion, 0.025 mol of $CH_3CH_2CH_2OH$ loses 47.025 kJ.
So, 1 mol of $CH_3CH_2CH_2OH$ loses $\frac{47.025}{0.025}$ = 1881 kJ.
So, $\Delta H_c(CH_3CH_2CH_2OH)$ = –1881 kJ mol^{-1}.

Examiner tip

Remember: amount, $n = \frac{\text{mass, } m}{\text{molar mass, } M}$

Examiner tip

You may prefer to work out the scaling factor.

Here, the scaling factor is the number of times that 0.025 divides into 1. This gives 40.

Dividing by 0.025 or multiplying by 40 gives the same answer. Use the method with which you are most comfortable.

Examiner tip

Remember to include the sign for the enthalpy change. Here the reaction is exothermic and the '–' sign is required.

In exams, many candidates forget to do this!

Comparison of experimental value with standard enthalpy change

The calculated value in the above worked example for the direct determination of ΔH_c for $CH_3CH_2CH_2OH$ is –1881 kJ mol^{-1}.

The data book value for the standard enthalpy change of combustion of $CH_3CH_2CH_2OH$ is –2021 kJ mol^{-1}.

If we compare the values, the standard value from a data book is more exothermic.

So why is there such a big difference?

- There may have been incomplete combustion.
- There may have been heat loss to the surroundings.

Both factors would reduce the amount of heat being released into the water.

To get a better agreement, a more sophisticated set of apparatus is required that:

- cuts down on heat loss
- ensures complete combustion.

Figure 2 shows a bomb calorimeter. This sophisticated apparatus can be used to reduce errors from heat transfer with the surroundings.

A bomb calorimeter allows accurate measurements of energy changes.

- The fuel burns using oxygen to ensure *complete* combustion.
- Heat is transferred to the water which is well insulated to reduce heat losses to the surroundings.

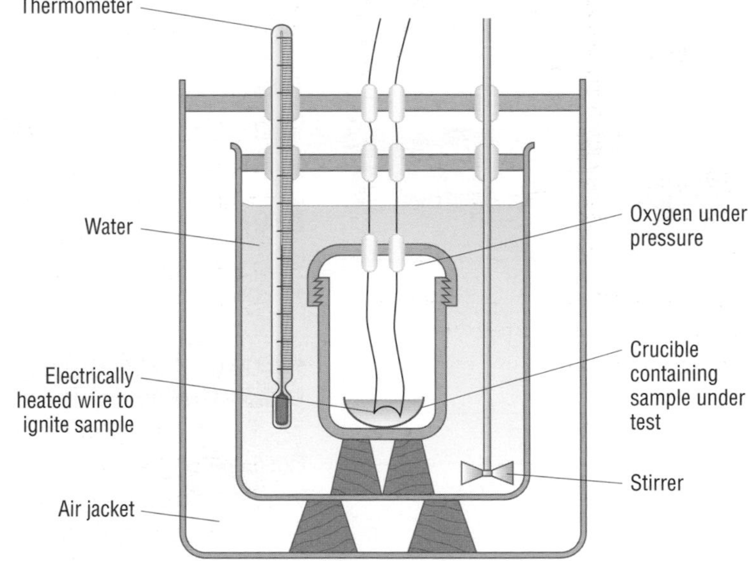

Figure 2 Bomb calorimeter

Questions

1 Calculate the enthalpy changes of combustion for butan-1-ol and for decane using the experimental results below. Give both answers to three significant figures.
 (a) 1.51 g of butan-1-ol, $CH_3CH_2CH_2CH_2OH$, heated 300 cm^3 of water by 42.0 °C.
 (b) 0.826 g of decane heated 180 cm^3 of water from 22.0 °C to 71.0 °C.

2 Simple experiments to determine the enthalpy change of combustion often produce a result that is less exothermic that the accepted value. Suggest two reasons why there is a difference.

2.3 (7) Bond enthalpies

By the end of this spread, you should be able to . . .

- ✱ **Explain exothermic and endothermic reactions in terms of enthalpy changes associated with the breaking and making of chemical bonds.**
- ✱ **Define and use the term average bond enthalpy.**
- ✱ **Calculate an enthalpy change of reaction from average bond enthalpies.**

Key definition

Bond enthalpy is the enthalpy change that takes place when breaking by homolytic fission 1 mol of a given bond in the molecules of a gaseous species.

Key definition

Average bond enthalpy is the average enthalpy change that takes place when breaking by homolytic fission 1 mol of a given type of bond in the molecules of a gaseous species.

Bond	Average bond enthalpy/kJ mol^{-1}
C–H	+413
O=O	+497
O–H	+463
C=C	+612
H–H	+436

Table 1 Values of some average bond enthalpies

Examiner tip

Bond enthalpy is an *endothermic* change (ΔH +ve) with bonds being *broken.*

When the same bonds are *made*, the enthalpy change will be the same magnitude but the opposite sign (ΔH –ve), that is, *exothermic.*

Bond enthalpy

Chemical bonds are the storehouses for chemical energy.

You get information about the strength of a chemical bond from its **bond enthalpy**. Bond enthalpies tell you how *much* energy is needed to break different bonds. You can then compare the strengths of different bonds.

Note that energy is needed to *break* bonds – the change is *endothermic*.

When bonds *form*, the same quantity of energy is released – the change is *exothermic*.

The equations below show the bond enthalpies for H–H and H–Cl bonds.

$$\text{H–H(g)} \longrightarrow \text{2H(g)} \qquad \Delta H = +436 \text{ kJ mol}^{-1}$$
$$\text{H–Cl(g)} \longrightarrow \text{H(g)} + \text{Cl(g)} \qquad \Delta H = +432 \text{ kJ mol}^{-1}$$

The H–H bond enthalpy value is always the same because a H–H bond can only ever exist in a H_2 molecule. Similarly, the H–Cl bond enthalpy applies only to a HCl molecule.

Average bond enthalpies

Unlike H–H and H–Cl bonds, some bonds can occur in different molecules. For example, almost every organic molecule contains C–H bonds. The C–H bond strength will vary across the different environments in which it is found.

Table 1 shows some **average bond enthalpies**. These are averaged over a number of typical chemical species containing that type of bond.

Breaking and making bonds

In a chemical reaction, bond *breaking* is followed by bond *making*.

- Energy is first needed to break bonds in the reactants.
  Bond *breaking* is an *endothermic* process and requires energy.
- Energy is then released as new bonds are formed in the products.
  Bond *making* is an *exothermic* process and releases energy.

Exo or endo?

What decides whether a reaction is exothermic or endothermic overall? The answer lies with bond enthalpies – the relative strengths of bonds being broken and bonds being made.

- In an *exothermic* reaction, the bonds that are formed are *stronger* than the bonds that are broken.
- In an *endothermic* reaction, the bonds that are formed are *weaker* than the bonds that are broken.

Using bond enthalpies to determine enthalpy changes

For a reaction involving simple gaseous molecules (Σ is short for 'sum of'):

- Energy required to break bonds = Σ(bond enthalpies of bonds broken).
- Energy released when bonds are made = $-\Sigma$(bond enthalpies of bonds made).
- $\Delta H = \Sigma$(bond enthalpies of bonds broken) $-\Sigma$(bond enthalpies of bonds made).

Example

We can use values of average bond enthalpies to work out the enthalpy change of reaction for reactions involving gases.

$$CH_4(g) + 2O_2(g) \longrightarrow CO_2(g) + 2H_2O(g)$$

The bond breaking and bond making processes are shown in Figure 1.

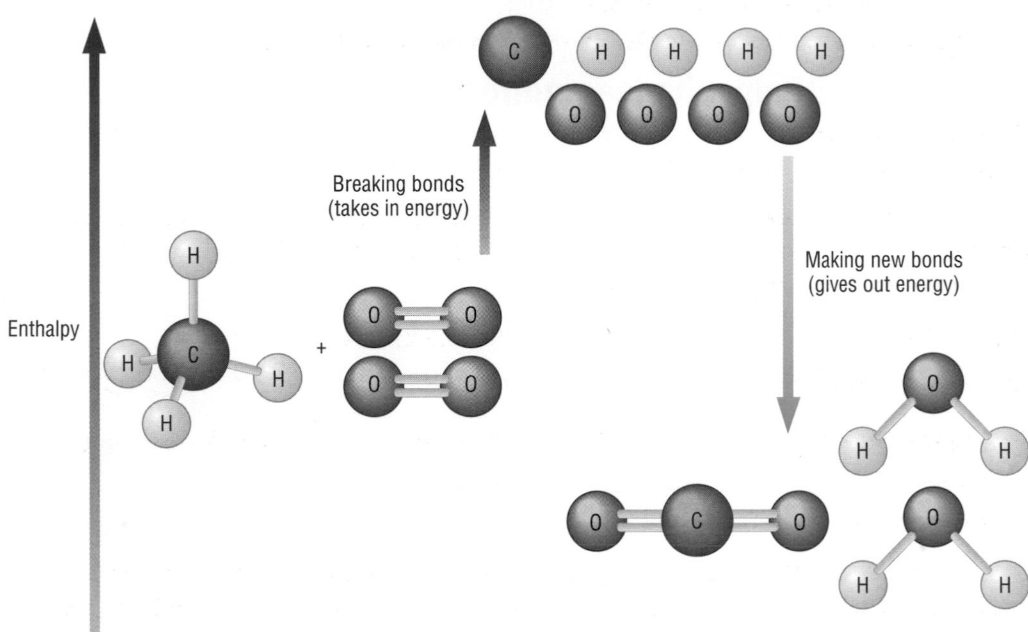

Figure 1 Bond breaking and bond making for reaction of methane with oxygen

Average bond enthalpies:

C–H: +413 kJ mol^{-1}; O=O: +497 kJ mol^{-1}; C=O: +805 kJ mol^{-1}; O–H: +463 kJ mol^{-1}.

Using Figure 1:

Bonds broken = 4 (C–H) + 2 (O=O)
Bonds made = 2 (C=O) + 4 (O–H)

ΔH = Σ(bond enthalpies of bonds broken) – Σ(bond enthalpies of bonds made)
So, ΔH = [(4 x 413) + (2 x 497)] – [(2 x 805) + (4 x 463)] kJ mol^{-1}
Bonds broken (endothermic) Bonds made (exothermic)
So, ΔH = –816 kJ mol^{-1}

Questions

1 How can exothermic and endothermic reactions be explained in terms of breaking and making of chemical bonds?

2 You are provided with the following average bond enthalpies:
C–H: +413 kJ mol^{-1}; C–C: +347 kJ mol^{-1}; O=O: +497 kJ mol^{-1}; C=O: +805 kJ mol^{-1};
O–H: +463 kJ mol^{-1}; C=C: +612 kJ mol^{-1}; H–H: +436 kJ mol^{-1};
N≡N: +945 kJ mol^{-1}; N–H: +391 kJ mol^{-1}.
Calculate the enthalpy changes of reaction for each of the following reactions:
(a) $C_2H_4(g) + H_2(g) \longrightarrow C_2H_6(g)$
(b) $C_3H_8 + 5O_2(g) \longrightarrow 3CO_2(g) + 4H_2O(g)$
(c) $N_2(g) + 3H_2(g) \longrightarrow 2NH_3(g)$

Examiner tip

Notice that every species must be a *gas*.

Here H_2O is gaseous.

If we wanted to work out an enthalpy change forming $H_2O(l)$, we would also have to consider the enthalpy change in converting $H_2O(g)$ into $H_2O(l)$.

Examiner tip

When tackling problems to calculate the enthalpy change from bond enthalpies, always draw out each molecule so that you can see the bonds broken and bonds made. You then just have to count up the bonds of each type.

In exams, many mistakes are made by candidates who don't bother to do this!

2.3 (8) Enthalpy changes from $\Delta H_c^{\ominus}$

By the end of this spread, you should be able to . . .

- **Use Hess' law to construct enthalpy cycles.**
- **Use enthalpy changes of combustion in calculations to determine an enthalpy change of reaction indirectly.**

Figure 1 Germain Hess (1802–1850). Hess was a Swiss-born Russian chemist, a pioneer in the field of thermochemistry. Hess' law, shown below, is an application of the law of conservation of energy – that energy cannot be created nor destroyed

Key definition

Hess' law states that, if a reaction can take place by more than one route and the initial and final conditions are the same, the total enthalpy change is the same for each route.

Examiner tip

If *two* of the enthalpy changes *A*, *B* and *C* are known, the *third* can always be calculated.

In exams, questions may be set in which you need to find *A*, *B* or *C*.

Substance	$\Delta H_c^{\ominus}$/kJ mol^{-1}
C(s)	–394
H_2(g)	–286
C_3H_8(g)	–2219

Table 1

Measuring enthalpy changes indirectly

Unfortunately, it is not always possible to measure the enthalpy change of a reaction *directly*.

This might be due to a number of factors such as:

- a high activation energy
- a slow reaction rate
- more than one reaction taking place.

Luckily chemists have **Hess' law** at their disposal. This law provides a method for finding an enthalpy change *indirectly*. Using Hess' law, we can use enthalpy values that we *can* measure in order to calculate the ones we *can't* measure. Hess' law is an extension of the law of conservation of energy.

Imagine a reaction route for converting reactants into products. The enthalpy change is labelled *A*.

Route 1: reactants $\xrightarrow{A}$ product

It may also be possible to convert reactants into products by another route. This will often involve an *intermediate*. The enthalpy changes are labelled *B* and *C*.

Route 2: reactants $\xrightarrow{B}$ intermediate $\xrightarrow{C}$ product

Now we have *two* routes for converting reactants into products:

- Route 1: *A*
- Route 2: *B* + *C*.

By Hess' law, the total enthalpy change is the same for each route.

So, $A = B + C$.

Determining an enthalpy change indirectly, using ΔH_c values

Many elements and compounds can be reacted with oxygen and the enthalpy changes determined comparatively easily. Chemists have measured accurately the enthalpy changes of combustion for hundreds of chemicals. These enthalpy changes can be looked up in tables and used to calculate the enthalpy changes of other reactions.

Worked example

The enthalpy change for the reaction below is virtually impossible to measure directly.

$3C(s) + 4H_2(g) \longrightarrow C_3H_8(g)$

- Just think of the number of compounds of carbon and hydrogen.
- It is impossible that a reaction could take place between carbon and hydrogen to form just one hydrocarbon!

We can calculate this enthalpy change *indirectly* using the enthalpy changes of combustion of the three chemicals in the equation. These are shown in Table 1.

By using enthalpy changes of combustion, we have a link between reactants and products. *Reactants* and *products* react to *form* combustion products (our intermediates in Route 2 above). The resulting **enthalpy cycle** is shown in Figure 2.

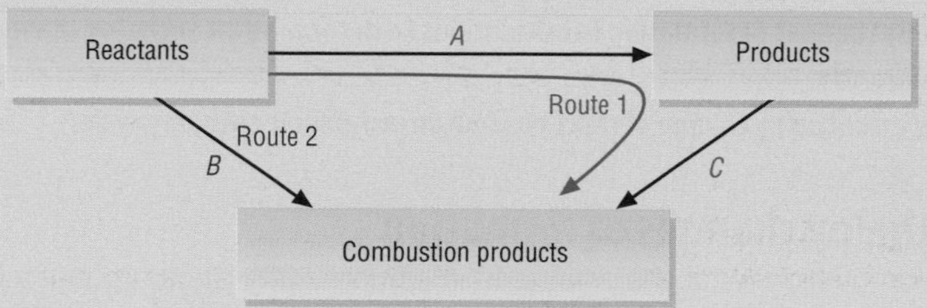

Figure 2 Enthalpy cycle using ΔH_c

1 We can identify two routes by following the arrows:
Route 1: $A + C$
Route 2: B
By Hess' law, the total enthalpy change is the same for each route.
$\therefore A + C = B$

2 Construct an energy cycle linking the reactants with the products. Add values of $\Delta H_c^{\ominus}$ from Table 1 to the energy cycle as shown in Figure 3.

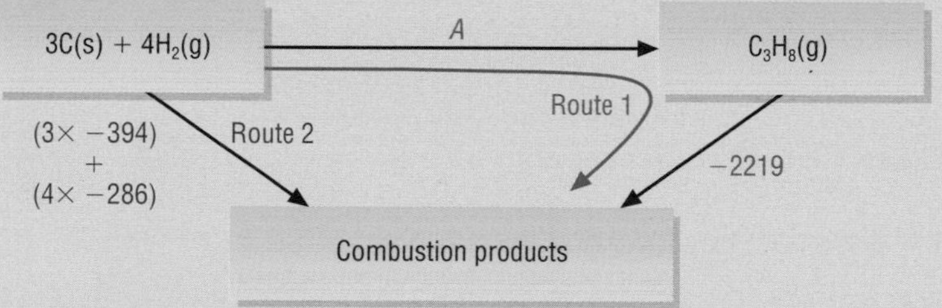

Figure 3 Enthalpy cycle with ΔH_c values added

3 Calculate the unknown enthalpy change from the two routes.
Route 1: $A + [(-2219)]$
Route 2: $[(3 \times -394) + (4 \times -286)]$

In step 1 above, we showed that $A + C = B$
By Hess' law, the enthalpy change for each route is the same:
$\therefore A + [(-2219)] = [(3 \times -394) + (4 \times -286)]$
$A = [(3 \times -394) + (4 \times -286)] - [(-2219)]$
$A = -107 \text{ kJ mol}^{-1}$
Thus, $3C(s) + 4H_2(g) \longrightarrow C_3H_8(g)$ $\Delta H = -107 \text{ kJ mol}^{-1}$

Key definition

An **enthalpy cycle** is a diagram showing alternative routes between reactants and products which allows the indirect determination of an enthalpy change from other known enthalpy changes using Hess' law.

Examiner tip

- The direction of the arrows goes *from* the reactants and products *to* the common combustion products. This is ΔH_c.
- Look at the diagram to see how we get the two routes: just follow the arrows!

Examiner tip

- The standard enthalpy changes of combustion are for the amount, in mol, of each chemical in the equation, so here the $\Delta H_c^{\ominus}$ value for $H_2(g)$ must be multiplied by *two* because there are 2 mol of $H_2(g)$.

Examiner tip

Using enthalpy changes of combustion:

$\Delta H = \sum \Delta H_c^{\ominus}(\text{reactants}) - \sum \Delta H_c^{\ominus}(\text{products})$

Questions

1 You are provided with the following enthalpy changes of combustion.

Substance	C(s)	$H_2(g)$	$C_4H_{10}(g)$	$C_2H_5OH(l)$
$\Delta H_c^{\ominus}$ kJ mol^{-1}	−394	−286	−2877	−1367

Determine the enthalpy change for the following reactions:

(a) $4C(s) + 5H_2(g) \longrightarrow C_4H_{10}(g)$.

(b) $2C(s) + 3H_2(g) + \frac{1}{2}O_2(g) \longrightarrow C_2H_5OH(l)$.

2.3 Enthalpy changes from $\Delta H_f^{\ominus}$

By the end of this spread, you should be able to . . .

* **Use enthalpy changes of formation in calculations to determine an enthalpy change of reaction indirectly.**
* **Determine an enthalpy change of reaction from an unfamiliar enthalpy cycle.**

Using enthalpy changes of formation

As with the combustion reactions discussed in spread 2.3.8, standard enthalpy changes of formation have been measured for hundreds of chemicals. Values for these enthalpy changes can also be looked up in tables.

By using enthalpy changes of formation, we have a link between reactants and products. *Reactants* and *products form* from the elements. The resulting energy cycle is shown in Figure 1.

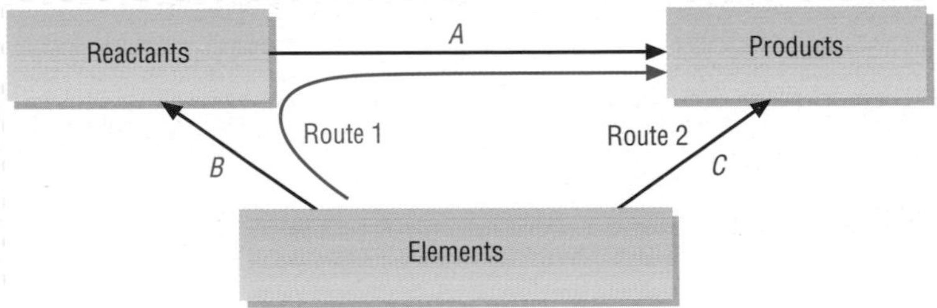

Figure 1 Enthalpy cycle using ΔH_f

We can identify two routes by following the arrows:

Route 1: $B + A$
Route 2: C

By Hess' Law, the total enthalpy change is the same for each route.

$\therefore B + A = C$

Worked example

Sulfur dioxide, SO_2, reacts with oxygen in the presence of a catalyst to form sulfur trioxide, SO_3.

$2SO_2(g) + O_2(g) \longrightarrow 2SO_3(l)$

We can calculate this enthalpy change *indirectly* using the enthalpy changes of formation of the three chemicals in the equation. Two of these values are shown in Table 1.

Substance	$\Delta H_f^{\ominus}$/kJ mol^{-1}
$SO_2(g)$	–297
$SO_3(l)$	–441

Table 1

The enthalpy change of formation of an element is defined as zero kJ mol^{-1}.

So for $O_2(g)$, there is no change and $\Delta H_f^{\ominus} = 0$ kJ mol^{-1}.

Examiner tip

Compare this energy cycle with the one you used with ΔH_c values in spread 2.3.8.

- The direction of the arrows goes *from* the common elements *to* the reactants and products. This is ΔH_f.
- The arrows point in opposite directions from the cycles in Figure 2 of spread 2.3.8.
- Try to understand why the arrows point the way they do. This is the *key* to constructing an energy cycle.

Examiner tip

- Look at the diagram to see how we get the two routes – just follow the arrows!

Examiner tip

For elements: $\Delta H_f^{\ominus} = 0$ kJ mol^{-1}.

For oxygen $\Delta H_f^{\ominus}$ would be for the *change*:

1 mol $O_2(g) \longrightarrow$ 1 mol $O_2(g)$.

There is no change!

1 Construct an energy cycle linking the reactants with the products. Add values of $\Delta H_f^{\ominus}$ from Table 1 to the energy cycle as shown in Figure 2.

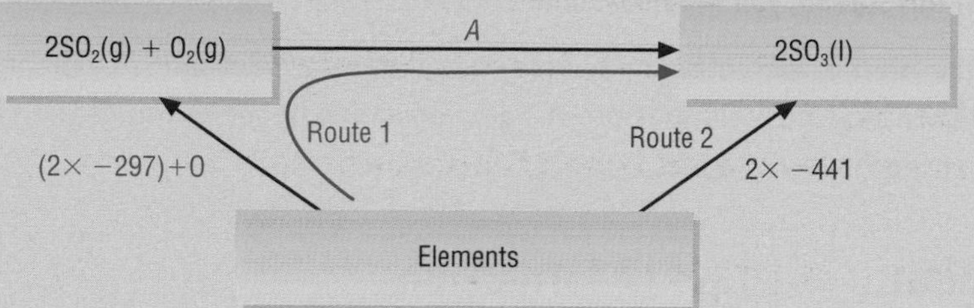

Figure 2 Enthalpy cycle with $\Delta H_f^{\ominus}$ values added

2 Calculate the unknown enthalpy change from the two routes.
By Hess' law:
Route 1: $[(2 \times -297) + 0] + A$
Route 2: $[(2 \times -441)]$

By Hess' law, the enthalpy change for each route is the same.
$[(2 \times -297) + 0] + A = [(2 \times -441)]$

$\therefore A = [(2 \times -441)] - [(2 \times -297)]$
$A = -288 \text{ kJ mol}^{-1}$

$\therefore 2SO_2(g) + O_2(g) \longrightarrow 2SO_3(l) \qquad \Delta H = -288 \text{ kJ mol}^{-1}$

Examiner tip

Using enthalpy changes of formation:

$\Delta H = \sum \Delta H_f^{\ominus}$ (products) − $\sum \Delta H_f^{\ominus}$ (reactants)

Other enthalpy cycles

Enthalpy cycles can be constructed for many other types of reaction provided there is a *link*.

In your AS course, the enthalpy cycles that use combustion and formation are very important. You will need to practise calculations using these energy cycles. But the same principles can be applied to *any* energy cycle.

Provided that an energy cycle is constructed correctly, you can apply Hess' law to find an unknown enthalpy change. Just follow the arrows and you can't go wrong!

Examiner tip

Get the arrows the right way around!

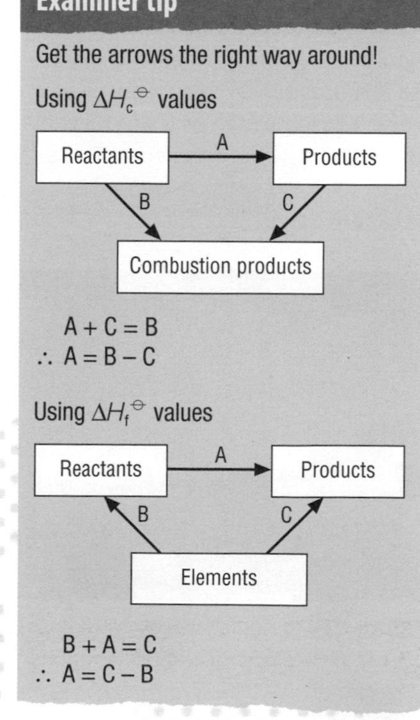

Questions

1 You are provided with the following enthalpy changes of formation.

Substance	$NO_2(g)$	$NH_3(g)$	$NO(g)$	$H_2O(l)$	$HNO_3(l)$
$\Delta H_f^{\ominus}$/kJ mol^{-1}	33	−46	90	−286	−174

Determine the enthalpy change for the following reactions:
(a) $2NO + O_2 \longrightarrow 2NO_2(g)$.
(b) $4NH_3(g) + 5O_2(g) \longrightarrow 4NO(g) + 6H_2O(l)$.
(c) $2H_2O(l) + 4NO_2(g) + O_2(g) \longrightarrow 4HNO_3(l)$.

2 You are provided with the following standard enthalpy changes.
$\Delta H_f^{\ominus}(CO_2) = -394 \text{ kJ mol}^{-1}$; $\Delta H_f^{\ominus}(H_2O) = -286 \text{ kJ mol}^{-1}$; $\Delta H_f^{\ominus}(C_5H_{12}) = -173 \text{ kJ mol}^{-1}$;
$\Delta H_c^{\ominus}(C_6H_{14}) = -4163 \text{ kJ mol}^{-1}$; $\Delta H_c^{\ominus}(C) = -394 \text{ kJ mol}^{-1}$; $\Delta H_c^{\ominus}(H_2) = -286 \text{ kJ mol}^{-1}$.
(a) Write equations, including state symbols, to represent:
(i) the standard enthalpy change of formation of pentane, C_5H_{12};
(ii) the standard enthalpy change of combustion of hexane, C_6H_{14}.
(b) Determine the following enthalpy changes.
(i) The standard enthalpy change of combustion for pentane, C_5H_{12};
(ii) The standard enthalpy change of formation for hexane, C_6H_{14}.

2.3 (10) Rates of reaction – collision theory

By the end of this spread, you should be able to . . .

* **Describe the effect of concentration changes on the rate of a reaction.**
* **Explain why an increase in the pressure of a gas, increasing its concentration, may increase the rate of a reaction involving gases.**

Figure 1 Ammonium nitrate burning – the true destructive power of ammonium nitrate was realised when a ship full of the compound blew up in harbour in Texas City in 1947. Explosions have very fast rates of reaction

Figure 2 An upturned car body lies rusting, an example of a reaction with a slow rate

Introduction

The rates of chemical reactions vary considerably.

- Combustion reactions and explosions have very high rates, taking place extremely quickly. An explosion takes place in a fraction of a second.
- The rusting of iron has a very slow rate. Rusting takes place over many days, even years.
- The rate of reaction for diamond to form graphite is very slow and is likely to take millions of years. This is just as well as you would not want your shiny diamond ring to turn black in front of your eyes!

Rate of reaction

The **rate of reaction** is defined as the change in concentration of a reactant or a product in a given time.

$$\text{rate} = \frac{\text{change in concentration}}{\text{time}} \qquad \text{Units: } \frac{\text{mol dm}^{-3}}{\text{s}} = \text{mol dm}^{-3}\,\text{s}^{-1}$$

For many reactions:

- at the start of a reaction, each reactant has its greatest concentration – the rate of the reaction is at its *fastest*
- as the reaction proceeds the concentrations of the reactants decrease – the rate of reaction *slows down*
- when one of the reactants has been used up the rate becomes zero – the reaction *stops*.

Key definition

The **rate of reaction** is the change in concentration of a reactant or a product in a given time.

Factors that alter the rate of a chemical reaction

Reaction kinetics is the study of the factors that alter the rate of a chemical reaction. You may remember some of these factors from your previous study of science.

The rate of a chemical reaction can often be altered by:

- temperature
- pressure, when the reactants are gases
- concentration
- surface area
- adding a catalyst.

We explain how these factors change the rate of a chemical reaction using *collision theory*. This states that a chemical reaction can take place only when the reacting molecules collide.

- When two molecules collide, a reaction *might* take place if certain conditions are met:
- The molecules must have sufficient energy to overcome the *activation energy* of the reaction as illustrated in Figure 3.

Head-on collision: more energetic – reaction takes place.
Glancing blow: less energetic – no reaction.

Figure 3 Reaction only occurs if the collision has energy equal to or greater than the activation energy of the reaction

- The molecules must also collide in the correct orientation as shown in Figure 4.

In a reaction: A + BC ⟶ AB + C

A must collide with B for a reaction to take place.

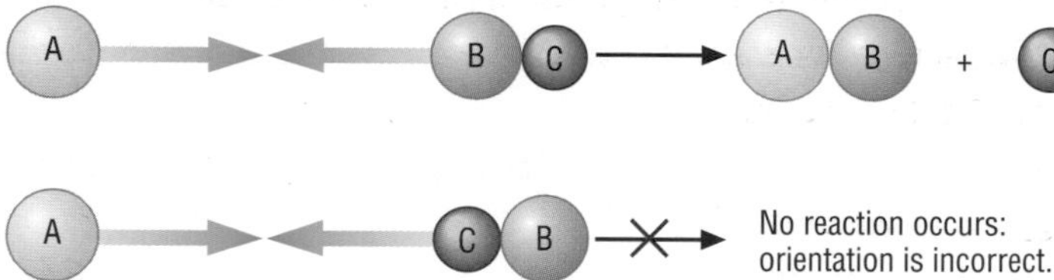

Figure 4 Even when the molecules collide, the orientation must be correct for a reaction to take place

The effect of concentration on reaction rate

If the concentrations of the reactants are increased, the rate of reaction also increases.

- Increased concentration gives more molecules in the same volume.
- The molecules will be closer together and there is a greater chance of the molecules colliding.
- Collisions will be more frequent – more collisions will occur in a certain length of time.

So, for a particular length of time *more* collisions will take place and there will be *more* collisions with energy *greater* than the activation energy. Collisions will be *more* frequent and the rate of reaction will *increase*.

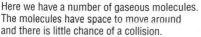

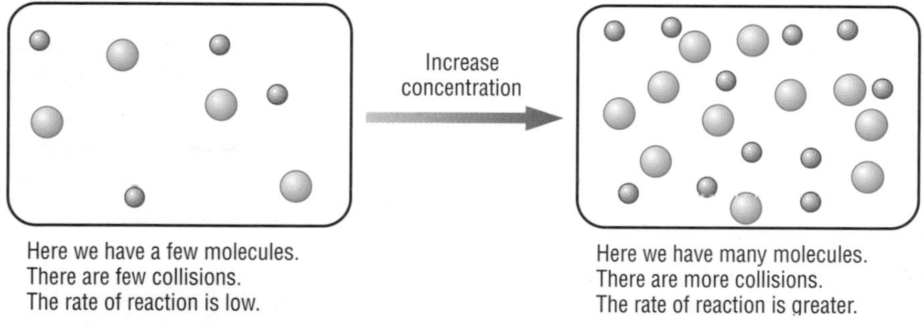

Increase
concentration
Here we have a few molecules.
There are few collisions.
The rate of reaction is low.
Here we have many molecules.
There are more collisions.
The rate of reaction is greater.

Figure 5 Increasing the concentration increases the number of molecules in the same volume, leading to more collisions

The effect of pressure on reaction rate

When the pressure of a gas is increased, the molecules are pushed closer together.

- The same number of molecules occupy a smaller volume.
- For a gaseous reaction, increasing the pressure is the same as increasing the concentration.

So, for a particular length of time *more* collisions will take place and there will be *more* collisions with energy *greater* than the activation energy. There will be *more* frequent collisions and the rate of reaction will *increase*.

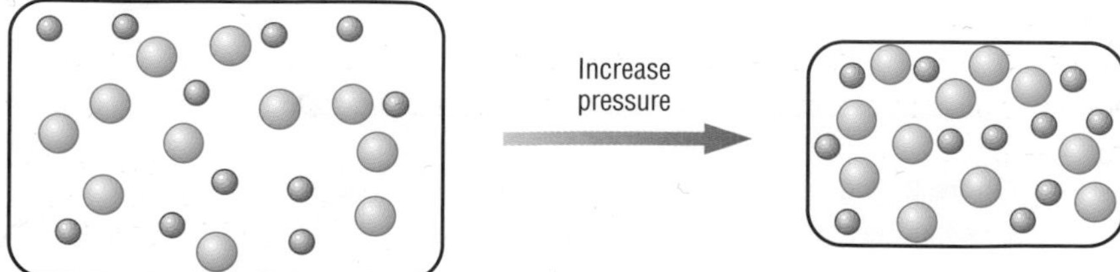

Increase
pressure
Here we have a number of gaseous molecules.
The molecules have space to move around
and there is little chance of a collision.
Increasing the pressure decreases the
volume and increases the concentration.
The molecules have less space to move in
and are more likely to collide.

Figure 6 Increasing the pressure means less volume so the molecules collide more often

Examiner tip

It is important to state and explain the effect of changing conditions on the number of molecules in the system and on their energy. These factors will have an effect on the reaction rate.

Use the collision theory in your answers – the bullet points are useful.

Questions

1 Using the collision theory, explain how the following changes affect the rate of a reaction:
   (a) An increase in concentration of a reactant.
   (b) An increase in pressure.
2 Define the term *activation energy* for a chemical reaction.

2.3 (11) Catalysts

By the end of this spread, you should be able to . . .

* **State that a catalyst speeds up a reaction, without being consumed by the overall reaction.**
* **Explain, using enthalpy profile diagrams, how the presence of a catalyst gives rise to an increased reaction rate.**
* **Understand that catalysts affect conditions, often requiring lower temperatures, reducing energy demand and emissions.**
* **Explain that catalysts enable different reactions to be used, with better atom economy and with reduced waste.**

Figure 1 Catalase, an enzyme found in raw liver, accelerates the decomposition of hydrogen peroxide, H_2O_2. An enzyme is a biological catalyst, increasing reaction rates without itself reacting or changing. The equation for this catalysed reaction is: $2H_2O_2 \rightarrow 2H_2O + O_2$

What is a catalyst?

The Swedish scientist Jöns Jakob Berzelius (1779–1848) was the first person to use the term *catalyst*. He was carrying out experiments on hydrogen peroxide, H_2O_2, which breaks down slowly into water and oxygen.

$$2H_2O_2(l) \longrightarrow 2H_2O(l) + O_2(g)$$

Berzelius discovered that the addition of certain compounds broke down hydrogen peroxide faster than if the hydrogen peroxide was left alone. He thought that these *catalysts* broke down hydrogen peroxide by loosening their bonds, hence the term catalyst from the Greek *kata*, down, and *lyein*, loosen.

We have known about catalysis for many centuries. The fermentation of wine and bread-making both involve catalysis, although the people who developed them probably knew little about the chemistry involved!

A *catalyst* increases the rate of a chemical reaction *without* being used up in the process.

- A catalyst may react to form an intermediate.
- The catalyst is later regenerated, so that the catalyst does not undergo any permanent change.

A catalyst *lowers* the activation energy of the reaction by providing an alternative route for the reaction to follow. The alternative route has a lower energy. This is shown in the reaction pathway diagram in Figure 2.

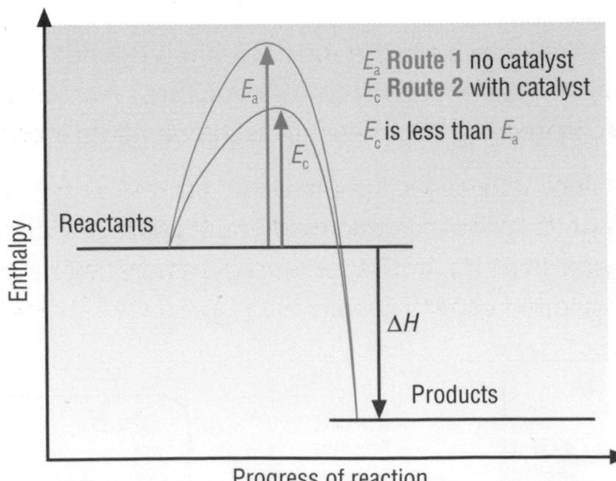

Figure 2 Reaction pathway diagram showing a catalyst lowering the activation energy

Reducing energy consumption and helping the environment

Many industrial processes rely on catalysis to reduce costs. It is estimated that about 80% of industrial processes involve catalysts. There are numerous catalysts currently in use; in fact about half of the Periodic Table's elements are either catalysts themselves or can be found in compounds that are used in catalysis. This is especially true for the transition elements.

In many industrial processes, the catalyst will significantly speed up the process by lowering the activation energy of the reaction. Less energy is then required for the molecules to react. Much of this energy is taken from electricity supplies or by burning crude oil. If a process can be run using *less* energy, then this *saves* energy costs.

A catalyst also has benefits for the environment. With less energy required, less fossil fuel is burnt and less carbon dioxide will be released into the atmosphere during energy production.

Some catalysts can be used to make a process more effective, improving the percentage yield of an industrial preparation.

Examiner tip

You should be able to state examples of transition metal catalysts in use:

- Iron in the Haber process for NH_3 production.
- V_2O_5 in the contact process within H_2SO_4 production.
- Ni in hydrogenation of margarine.

Producing ethanoic acid from catalysts

Most ethanoic acid, CH_3COOH, used to be made by oxidising butane or light naphtha or via the hydration of ethene.

Butane was heated in air, using a catalyst of manganese, cobalt or chromium ions, to produce ethanoic acid.

$$2C_4H_{10}(g) + 5O_2(g) \longrightarrow 4CH_3COOH(l) + 2H_2O(l)$$

This process was inefficient, with a low percentage yield of ethanoic acid.

Today, the Monsanto process is one of the commercial processes used for the production of ethanoic acid. In this process, methanol is reacted with carbon monoxide.

$$CH_3OH(l) + CO(g) \longrightarrow CH_3COOH(l)$$

Originally developed by BASF in 1960, the Monsanto process was carried out at a temperature of 300 °C and a pressure of 700 atmospheres with a catalyst of cobalt and iodide ions. At 700 atmospheres, safety would have to be a serious concern, and there are high energy costs to generate the high temperature and pressure.

Catalyst research led to the development of rhodium, rather than cobalt, as the catalyst. This new catalyst needs a lower temperature of 150 ° to 200 °C and a much lower pressure of about 30 atmospheres. The process has an atom economy of 100%..

The Cativa process has since superceded the Monsanto process. In this process, an iridium catalyst is used which is cheaper than rhodium, thereby cutting costs. The Cativa process also releases less CO_2 into the atmosphere with obvious environmental benefits. However, the operating conditions for this process are only known by the developers, BP Chemicals.

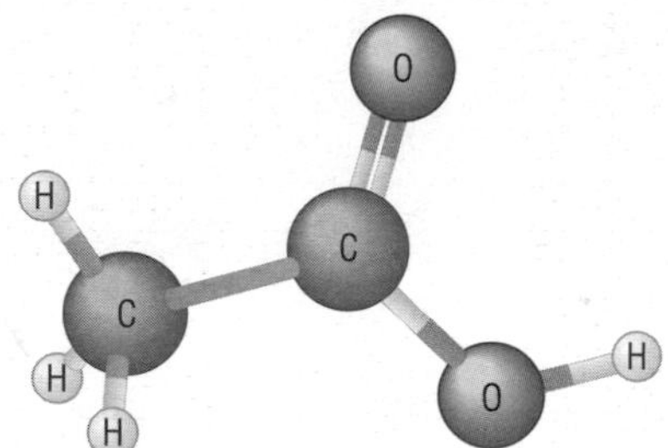

Figure 3 The structure of ethanoic acid, an important organic material used as a starting material for many other important products

Figure 4 Ethanoic acid is used in the production of all these goods. Low-cost synthesis of ethanoic acid is therefore important for the chemical industry

Questions

1 Define the term *catalyst*.
2 Explain how a catalyst increases the rate of a chemical reaction.
3 Write an equation for a reaction that relies on a catalyst and name the catalyst used.
4 Draw an enthalpy profile diagram for an exothermic reaction. Clearly label the axes and indicate both the enthalpy change and the activation energy. Show how the activation energy changes when a catalyst is added.

2.3 (12) Economic importance of catalysts

By the end of this spread, you should be able to . . .

* **State that catalysts have great economic importance.**
* **Understand that catalysts are often enzymes, operating close to room temperatures and pressures.**

Economic importance

Catalysts are used extensively in the chemical industry to transform chemical feedstock into materials. They are also required for the production of poly(ethene) and nylon, and for making many of the household items we are familiar with today.

Catalyst development and improvement often leads to greater profitability. Products can be made more quickly and easily, requiring less energy, cutting fuel costs and reducing waste.

The Haber process (spread 2.3.15) for the production of ammonia is of great economic importance. The ammonia produced is used as the basis of fertiliser manufacture, improving crop yields to feed the ever-increasing world population.

Ammonia is made by reacting together nitrogen and hydrogen.

$$N_2(g) + 3H_2(g) \rightleftharpoons 2NH_3(g)$$

The triple bond in nitrogen, N≡N, has to be broken. This requires a large input in energy, contributing to a high activation energy. Iron is used to catalyse this reaction, weakening the N≡N bond, and lowering the activation energy.

The Ziegler–Natta catalyst discussed in spread 2.1.17 was an important development in the production of non-branched poly(ethene).

Catalytic converters play an important role in improving our air quality by reducing toxic emissions from vehicles and preventing photochemical smog. Further information on catalytic converters can be found in spread 2.4.6.

Whole branches of chemistry are based on the development of successful catalytic processes. It is a profitable business!

Figure 1 Tanks containing anhydrous ammonia on a farm. Ammonia is made by reacting together nitrogen and hydrogen with an iron catalyst. Anhydrous ammonia is used as a fertiliser to add the plant nutrient nitrogen to soil

Key definitions

Heterogeneous catalysis is catalysis of a reaction in which the catalyst has a different physical state from the reactants; frequently, reactants are gases whilst the catalyst is a solid.

Homogeneous catalysis is catalysis of a reaction in which the catalyst and reactants are in the same physical state, which is most frequently the aqueous or gaseous state.

Enzymes as biological catalysts

Biocatalysis is any process in which the catalyst is an enzyme. Biocatalysis is a growing branch of the biochemical industry but has been known for many years. As long ago as 5000 BC, the Babylonians and Ancient Egyptians used enzymes in yeast to ferment fruit products to make alcohol (see also spread 2.2.1).

In 1858, Louis Pasteur carried out some experiments that led to a greater understanding of how enzyme-catalysed reactions can lead to the formation of a single pure product – a concept known today as *specificity*.

Later, Emil Fischer discovered that specificity could be explained by considering the structure of the enzyme and the molecule being catalysed – known as the *substrate*. He proposed the idea of the *lock-and-key* mechanism for enzyme action, which relies upon the enzyme and substrate being a perfect fit for one another.

Enzymes are large protein molecules, able to catalyse the reactions of large quantities of biological molecules in very short periods of time. Enzymes operate under mild conditions, such as low temperatures, atmospheric pressure and at an optimum pH value.

Enzymes in industry

Enzymes also find a use outside of living organisms as they are often used in the synthesis of organic chemicals by industry.

The industrial use of enzymes has many benefits:

- Lower temperatures and pressures can be used than with conventional inorganic catalysts, saving energy and costs.
- Enzymes often allow a reaction to take place which forms *pure* products, with no side reactions. They often remove the need for complex separation techniques, thereby *reducing* costs.
- Conventional catalysts are often poisonous and can pose disposal problems at the end of their industrial life. Enzymes are *biodegradable*.

Enzymes are used in a variety of important industrial processes. Manufacturing companies involved in enzyme use are often those involved in food and drink production. Dairy products, alcoholic drinks and fruit juices are all produced using enzymes. However, other materials, such as detergents, cleaning agents, and textile and leather processing agents, also rely on enzyme catalysis. Ibuprofen is widely used as a painkiller and anti-inflammatory drug. In the production of ibuprofen, enzyme action is used to separate the drug from a reaction mixture.

Figure 2 Bread, cheese and wine, all food produced using enzymes. Enzymes in yeast produce the alcohol in wine and to make the bread rise. Dairy enzymes convert milk into cheese

Detergents and washing powders

One of the most extensive uses of enzymes is in the home, with washing powders. Before biological washing powders, clothes were often washed at very high temperatures – this was very inefficient as water had to be heated to high temperatures, increasing energy costs. After washing, the clothes were often faded and mis-shaped.

The use of enzymes in washing powders and detergents significantly reduces the temperature required for washing. Milder conditions can be used, causing less damage to the clothes being washed.

The first recorded use of an enzyme being used for cleaning clothes was in 1913. Nowadays, many different enzymes are used in a washing powder. Stains are made of different types of molecules, and a range of enzymes are needed to break them down.

- *Proteases* break down proteins and remove stains caused by blood, egg, gravy and other proteins;
- *Amylases* break down starches.
- *Lipases* break down fats and grease.

More recently cellulase enzymes have been added to washing powders. These attack cellulose fibres to make clothes feel soft whilst maintaining their colour.

Chemists have also modified enzyme technology to make washing powders and detergents operate at higher temperatures for industrial and commercial washing operations which demand higher temperatures and faster washing. Dishwashers operate under alkaline conditions and the enzymes added to dishwasher powders have now been developed to work effectively over a wide pH range.

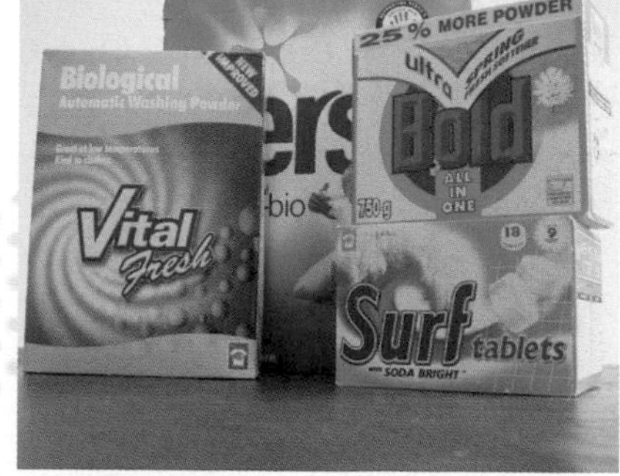

Figure 3 Many washing powders contain enzymes

Question

1 State three uses of the ammonia produced in the Haber Process.
2 What do you understand by the term 'activation energy'.
3 Enzymes are referred to as being specific. What do you understand by *specificity*?
4 State three important advantages of washing clothes using a biological washing powder.

2.3 (13) The Boltzmann distribution

By the end of this spread, you should be able to . . .

- **Explain the Boltzmann distribution and its relationship with activation energy.**
- **Describe qualitatively, using the Boltzmann distribution, the effect of temperature changes on reaction rate.**
- **Interpret catalytic behaviour in terms of the Boltzmann distribution.**

Figure 1 Portrait of Ludwig Boltzmann (1844–1906). Boltzmann studied the kinetic theory of gases. He was introduced to the work of James Clerk Maxwell, a Scottish scientist who was investigating how the distribution of velocities for the molecules in a gas depended on the temperature of the gas. Boltzmann's most important scientific contribution was in the field of reaction kinetics and, in particular, the development of the Boltzmann distribution

The Boltzmann distribution

For a gas or liquid, the molecules move around inside a container, colliding with each other and the walls of their container. The collisions are assumed to be elastic and so, when the molecules do collide, no energy is lost from the system. As the molecules are moving, they have kinetic energy.

In a sample of a gas or liquid:

- some molecules move *fast* and have *high* energy
- some molecules move *slowly* and have *low* energy
- the majority of the molecules have an *average* energy.

The **Boltzmann distribution** shows the distribution of molecular energies in a gas at constant temperature as illustrated in Figure 2.

Key definition

The **Boltzmann distribution** is the distribution of energies of molecules at a particular temperature, often shown as a graph.

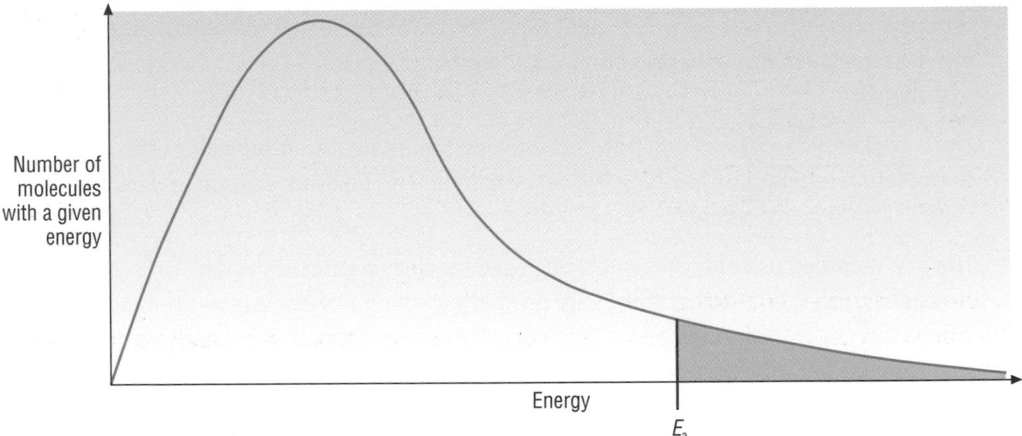

Figure 2 Distribution of molecular energies

Examiner tip

The graph shown in Figure 2 is very important as it features on many examination papers. You may have to draw this graph, so learn the labels carefully and practise drawing it.

Important features of the Boltzmann distribution

- The area under the curve is *equal* to the total number of molecules in the sample. The area does *not* change with conditions.
- There are *no* molecules in the system with zero energy – the curve starts at the origin.
- There is *no* maximum energy for a molecule – the curve gets close, but does not touch or cross the energy axis.
- Only the molecules with an energy *greater* than the activation energy, E_a, are able to react.

This last bullet point is extremely important and forms the basis of our understanding of how chemical reactions take place following collision. It also helps us to understand why not *all* collisions lead to a reaction. If you look back to spread 2.3.10, the importance of collisions to a chemical reaction is discussed in more detail.

The effect of temperature on reaction rate

At higher temperatures, the kinetic energy of all the molecules increases. The Boltzmann distribution flattens and shifts to the right. The number of molecules in the system does *not* change, so the area under the curve remains the same.

In the diagram shown in Figure 3, temperature T_2 > temperature T_1.

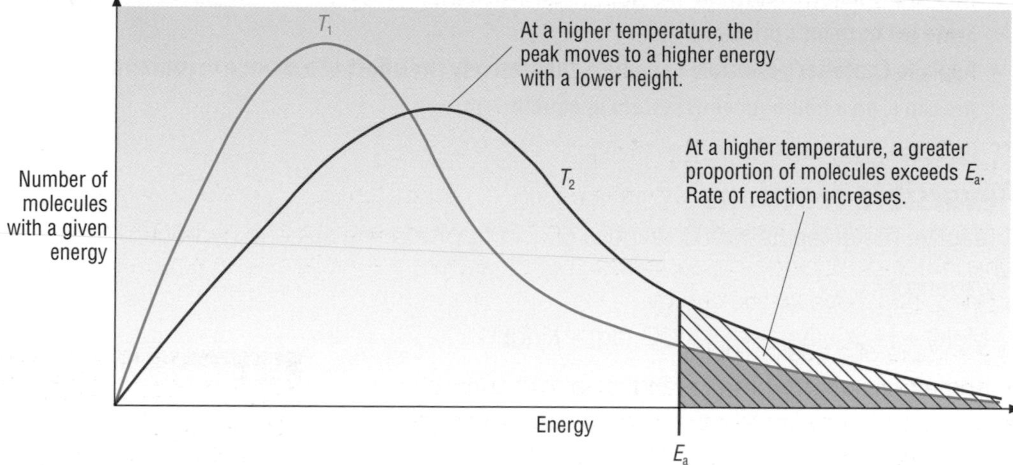
T_1
At a higher temperature, the peak moves to a higher energy with a lower height.
T_2
At a higher temperature, a greater proportion of molecules exceeds E_a. Rate of reaction increases.
Number of molecules with a given energy
Energy
E_a

Figure 3 The effect of temperature on reaction rate

As the temperature of the reaction is increased, the rate of reaction *increases*.

The rate increases because:

- More collisions take place in a certain length of time – the molecules are moving *faster* and have *more* kinetic energy;
- a higher proportion of molecules have an energy that is *greater* than the activation energy – *more* collisions will lead to a chemical reaction.

Thus, on collision, *more* molecules in the system will overcome the activation energy of the reaction. There will be *more* successful collisions in a certain length of time and the rate of reaction will *increase*.

The effect of a catalyst on reaction rate

The activation energy of the reaction is *reduced* in the presence of a catalyst. This can be seen in the diagram in Figure 4.

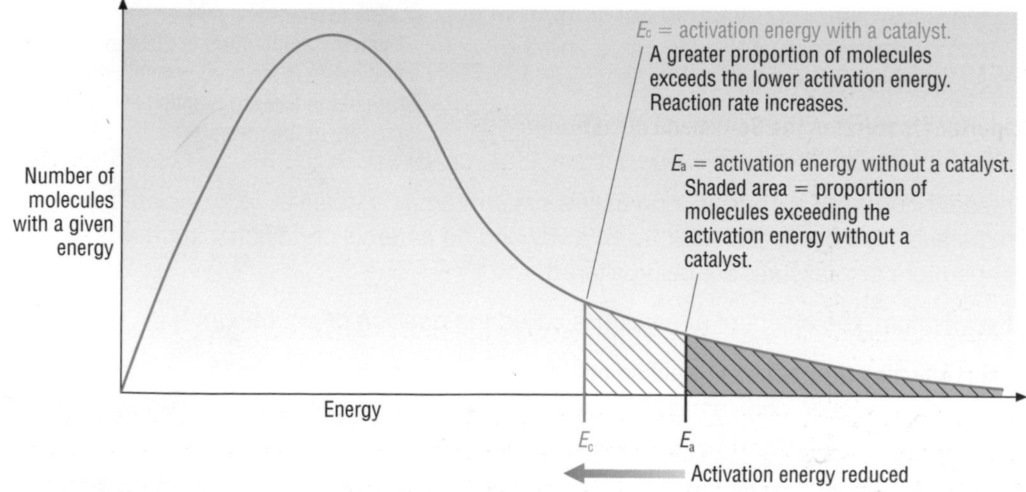
E_c = activation energy with a catalyst.
A greater proportion of molecules exceeds the lower activation energy. Reaction rate increases.
E_a = activation energy without a catalyst.
Shaded area = proportion of molecules exceeding the activation energy without a catalyst.
Number of molecules with a given energy
Energy
E_c
E_a
Activation energy reduced

Figure 4 Effect of a catalyst on reaction rate

On collision, *more* molecules in the system will overcome the new *lower* activation energy of the reaction. There will be *more* successful collisions in a certain length of time and the rate of reaction will *increase*.

Examiner tip

When you draw a Boltzmann distribution, it is important that the curve does not touch the x axis.

Examiner tip

Notice how the change of rate has been explained using the bullet points from the collision theory – the number of collisions and their energy have both been included in the explanation.

Examiner tip

Enthalpy profile diagrams can also be used to show the reduced activation energy in the presence of a catalyst. You can see this in spread 2.3.11. In examinations, questions are often set that link together the Boltzmann distribution and enthalpy profile diagrams.

Questions

1 What do you understand by the term *activation energy, E_a?*

2 How and why is the rate of a chemical reaction affected by an increase in temperature?

3 **(a)** Sketch a Boltzmann distribution curve for a sample of a gas at constant temperature. Label the axes.

(b) Add to your diagram a second distribution curve for a lower temperature.

2.3 (14) Chemical equilibrium

By the end of this spread, you should be able to . . .

- **Explain that a dynamic equilibrium exists when the rate of the forward reaction is equal to the rate of the reverse reaction.**
- **State le Chatelier's principle.**
- **Apply le Chatelier's principle to deduce qualitatively the effect of a change in concentration or pressure, on a homogeneous system in equilibrium.**

Reversible reactions

A reactive metal reacts with an excess of acid to produce a salt and hydrogen.

For example:

$$Mg(s) + H_2SO_4(aq) \longrightarrow MgSO_4(aq) + H_2(g)$$

- The reaction stops when *all* of the metal has reacted.
- The reaction is said to have gone to completion, indicated by the $\longrightarrow$ sign.

This is similar to many of the reactions that you have met during your AS studies.

However, some reactions are *reversible*. They can take place in either the forward or the reverse direction.

For example:

$$2SO_2(g) + O_2(g) \rightleftharpoons 2SO_3(g)$$

This reaction is reversible. At *equilibrium*, this is indicated by the $\rightleftharpoons$ sign. The following section explains what is meant by equilibrium.

Figure 1 Henri le Chatelier (1850–1936), a French chemist most famous for devising le Chatelier's principle. Chemists use this principle to predict the optimum conditions for carrying out reactions that exist in equilibrium

Dynamic equilibrium

When a system is in a state of equilibrium, there is no observable change – nothing appears to be happening. However, the system is *dynamic*, i.e. in constant motion. As fast as the reactants are converted into products, the products are being converted back into reactants. Although you cannot see anything happening, reactions are still taking place in both directions!

A chemical system is in **dynamic equilibrium** when:

- The *concentrations* of the reactants and the products remain the *same*.
- The rate of the *forward* reaction is the *same* as the rate of the *reverse* reaction.

Key definition

Dynamic equilibrium is the equilibrium that exists in a closed system when the rate of the forward reaction is equal to the rate of the reverse reaction.

An equilibrium only applies as long as the system remains *isolated*. In an isolated system, *no* materials are being added or taken away and *no* external conditions, such as temperature or pressure, are being altered.

At equilibrium, the extent of a reaction is called the *position of equilibrium*.

Factors affecting the position of equilibrium

A reversible reaction only remains in equilibrium when it is *isolated* in a closed system. This means that *nothing* must be allowed in or out.

The position of equilibrium can be altered by changing the following in the system:

- Concentrations of the reactants or products;
- pressure in reactions involving gases;
- temperature.

The effect of a change can be predicted using **le Chatelier's principle**.

Key definition

le Chatelier's principle states that when a system in dynamic equilibrium is subjected to a change, the position of equilibrium will shift to minimise the change.

The effect of concentration on equilibrium

For a system at equilibrium, a change in the concentration of reactants or products results in a shift in the position of equilibrium.

Consider this equilibrium:

$$CH_3CH_2OH(l) + CH_3CH_2COOH(l) \rightleftharpoons CH_3CH_2COOCH_2CH_3(l) + H_2O(l)$$

ethanol — propanoic acid — ethyl propanoate — water

Increasing the concentration of a *reactant* (either the ethanol or propanoic acid) causes the position of equilibrium to move in the direction that *decreases* this increased reactant concentration.

- The system opposes the change by *decreasing* the concentration of the reactant by removing it.
- The position of equilibrium moves to the *right-hand* side, forming *more* products.

Increasing the concentration of a *product* (either ethyl propanoate or water) causes the position of equilibrium to move in the direction that *decreases* this increased product concentration.

- The system opposes the change by *decreasing* the concentration of the product by removing it.
- The position of equilibrium will move to the *left-hand* side, forming *more* reactants.

The effect of pressure on equilibrium

Changing the total pressure of a system will only change the position of equilibrium if there are *gases* present.

Consider this equilibrium:

$$\underbrace{N_2(g) + 3H_2(g)}_{\text{4 mol}} \rightleftharpoons \underbrace{2NH_3(g)}_{\text{2 mol}}$$

higher pressure — lower pressure

- In total, there are *four* moles of gas on the left-hand side and *two* moles of gas on the right-hand side.
- The side with the *greater* moles of gas is the side at the higher pressure.

Increasing the *total* pressure of the system causes the position of equilibrium to move to the side with *fewer* gas molecules, as this will *decrease* the pressure. In our equation, the position of equilibrium will move to the *right*.

Decreasing the *total* pressure of the system causes the position of equilibrium to move to the side with the *greater* number of gas molecules, as this will *increase* the pressure. In our equation, the position of equilibrium will move to the *left*.

Increasing the concentration of one of the gases in the system is the same as increasing its concentration.

Examiner tip

Some examination questions ask what you would *observe* when an equilibrium is affected by a change in conditions. Check to see whether you have been given the colours of the reactants and products before answering. You might be expected to state any *colour changes* resulting from an equilibrium shift.

Questions

1 The following equilibrium system is set up. State and explain what you would *observe* if excess chloride ions were added to the equilibrium.

$$[Cu(H_2O)_6]^{2+}(aq) + 4Cl^-(aq) \rightleftharpoons [CuCl_4]^{2-}(aq) + 6H_2O(l)$$

pale blue solution — yellow solution

2 The production of nitric acid from ammonia involves the oxidation process shown below as one of the essential steps.

$$4NH_3(g) + 5O_2(g) \rightleftharpoons 4NO(g) + 6H_2O(g)$$

(a) State and explain how an increase in the total pressure of the system would affect the position of equilibrium.

(b) As it is made, nitrogen monoxide, NO, is removed from the system. How would this change affect the position of equilibrium?

2.3 (15) Equilibrium and industry

By the end of this spread, you should be able to . . .

- **Apply le Chatelier's principle to deduce qualitatively the effect of a change in temperature, on a homogeneous system in equilibrium.**
- **Explain, from given data, the importance in the chemical industry of a compromise between chemical equilibrium and reaction rate.**

The effect of temperature on equilibrium

The effect of changing the temperature on the position of equilibrium depends on the enthalpy sign.

For the equilibrium shown in Figure 1:

- the forward reaction is exothermic (gives out heat);
- the reverse reaction is endothermic (takes in heat).

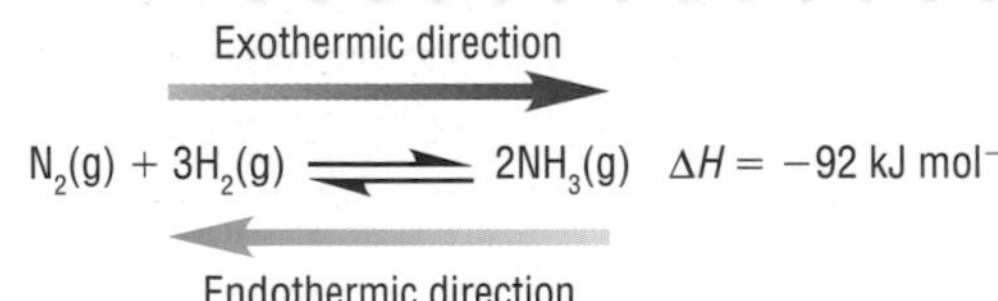

Figure 1 Enthalpy changes during equilibrium

Examiner tip

In this equilibrium, the *forward* reaction is exothermic with a ΔH value of $-92\ kJ\ mol^{-1}$.

The *reverse* reaction will be endothermic, with an enthalpy change of $+92\ kJ\ mol^{-1}$.

The forward and reverse reactions have the *same* magnitude of ΔH, but opposite signs.

Examiner tip

When discussing changes in temperature, it is important to state whether the reaction is exothermic or endothermic in the *forward* direction. Writing that the reaction is exothermic does not score any marks as the system is at *equilibrium* and can work both ways!

Increasing the *temperature* of the system causes the position of equilibrium to move in the direction that *decreases* the temperature.

- The system opposes the change by *taking* in heat and the position of equilibrium moves to the *left*.
- The position of equilibrium moves in the *endothermic* (ΔH +ve) direction.

Decreasing the *temperature* of the system causes the position of equilibrium to move in such a way as to *increase* the temperature.

- The system opposes the change by *releasing* heat and the position of equilibrium moves to the right.
- The position of equilibrium moves in the *exothermic* (ΔH –ve) direction.

Figure 2 Cobalt chloride equilibrium. Each flask contained the same solution of cobalt(II) chloride, dissolved in water containing chloride ions.
The cobalt(II) *complexes* with water forming pink-coloured $[Co(H_2O)_6]^{2+}$ ions; cobalt(II) *complexes* with chloride ions forming blue-coloured $[Co(Cl)_4]^{2-}$ ions.
The ions present in solution can be controlled using temperature:

- Formation of pink $[Co(H_2O)_6]^{2+}$ ions is exothermic, favoured by low temperatures.
- Formation of blue $[Co(Cl)_4]^{2-}$ ions is endothermic, favoured by high temperatures.

The effect of a catalyst on equilibrium

A catalyst does *not* alter the position of equilibrium or the composition of an equilibrium system.

- A catalyst speeds up the rate of the forward and reverse reactions *equally*.
- A catalyst increases the rate at which equilibrium is *established*.

Equilibrium and industry

Many important chemical processes exist as equilibrium systems.

Examples include:

- The preparation of ammonia from nitrogen and hydrogen in the Haber process.
- The conversion of sulfur dioxide into sulfur trioxide in the Contact process.

In industry, chemists strive to make the highest possible yield of a desired product as quickly and as cheaply as possible.

The Haber process

Chemical equation: $N_2(g) + 3H_2(g) \rightleftharpoons 2NH_3(g)$ $\Delta H = -92\ \text{kJ mol}^{-1}$

The raw materials, nitrogen and hydrogen, must be readily available.

- Nitrogen is obtained from the air by fractional distillation.
- Hydrogen is prepared by reacting together methane (from natural gas) and water.

What conditions favour production of ammonia?

Ammonia is produced by the *forward* reaction in this equilibrium.

- The *forward* reaction produces *fewer* gas molecules (4 molecules → 2 molecules), favoured by using *high* pressure.
- The *forward* reaction is *exothermic* (ΔH –ve), favoured by using *low* temperature.

However, there are drawbacks in using these theoretical conditions.

- Although a low temperature should produce a high equilibrium yield, the reaction would take place at a very low rate. At low temperatures, comparatively few N_2 and H_2 molecules have energy equal to, or greater than, the required activation energy.
- A high pressure increases the concentration of the gases, increasing the rate. So, a high pressure should produce both a high equilibrium yield and rate. However, large quantities of energy are required to compress gases, adding significantly to the process's running costs. There are also safety implications – any failure in the systems could potentially allow chemicals to leak into the environment, endangering those working on site.

Figure 3 Fritz Haber (1868–1934), German chemist. Haber is famous for developing a method for making ammonia

The modern ammonia plant

A modern ammonia plant needs to produce a sufficient yield of ammonia at a reasonable cost and in as short a time as possible.

In practice, a *compromise* is made between yield and rate.

- Temperature – this must be high enough to allow the reaction to proceed at a realistic rate, whilst still producing an acceptable equilibrium yield. A temperature of 400–500 °C is typically used.
- Pressure – a high pressure must still be used, but it must not be so high that the workforce is put in danger or the environment threatened. A pressure of 200 atmospheres is typically used.
- Catalyst – an iron catalyst is added to speed up the rate of reaction, allowing the equilibrium to be established faster and for lower temperatures to be used. Less energy is used to generate heat, reducing costs.

The actual compromise conditions used convert only 15% of nitrogen and hydrogen into ammonia. The ammonia produced is liquefied and removed. Unreacted nitrogen and hydrogen gases are passed through the reactor again. Eventually, virtually all the nitrogen and hydrogen will have been converted into ammonia.

Examiner tip

Using le Chatelier's principle, the theoretical conditions that favour ammonia production are:

- A *high* pressure; and
- a *low* temperature.

Questions

1 Synthesis gas, a mixture of CO and H_2 gases, is made by passing steam over hot coke. The equation for the process is:

$C(s) + H_2O(g) \rightleftharpoons CO(g) + H_2(g)$ $\Delta H = +132\ \text{kJ mol}^{-1}$

State and explain the effect of a temperature rise on the composition of the equilibrium mixture?

2 The Contact process is used in the manufacture of sulfuric acid, H_2SO_4. The equilibrium in the Contact process is shown below.

$2SO_2(g) + O_2(g) \rightleftharpoons 2SO_3(g)$ $\Delta H = -196\ \text{kJ mol}^{-1}$

(a) State and explain the effect of a decrease in pressure on the position of equilibrium?

(b) What conditions would be used to obtain the maximum theoretical yield of $SO_3(g)$?

(c) In the industrial process, a catalyst of vanadium pentoxide is used. What is the effect of the catalyst on the position of equilibrium and on the time taken to reach equilibrium?

2.3 Energy summary

Enthalpy cycles

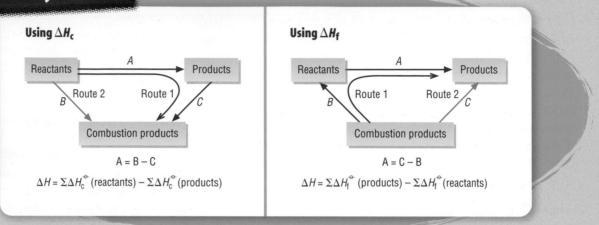

Using ΔH_c

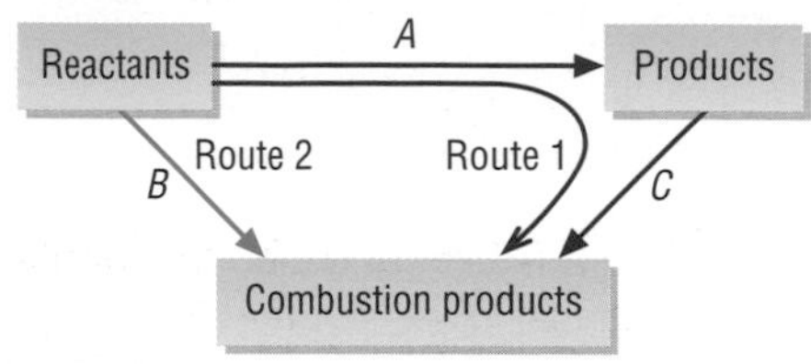

Reactants
A
Products
Route 2
Route 1
B
C
Combustion products

A = B − C

$\Delta H = \Sigma\Delta H_c^{\ominus}\text{ (reactants)} - \Sigma\Delta H_c^{\ominus}\text{ (products)}$

Using ΔH_f

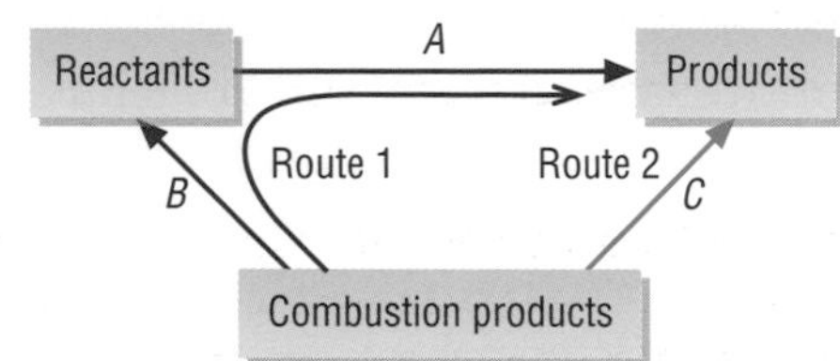

Reactants
A
Products
Route 1
Route 2
B
C
Combustion products

A = C − B

$\Delta H = \Sigma\Delta H_f^{\ominus}\text{ (products)} - \Sigma\Delta H_f^{\ominus}\text{ (reactants)}$

Rates

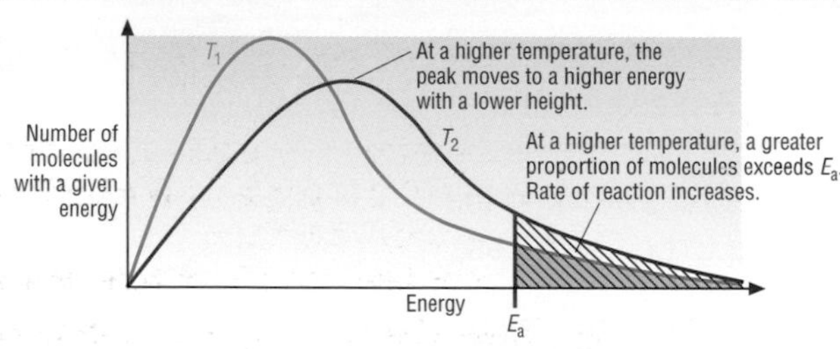

Number of molecules with a given energy
T1
T2
At a higher temperature, the peak moves to a higher energy with a lower height.
At a higher temperature, a greater proportion of molecules exceeds Ea. Rate of reaction increases.
Energy
Ea

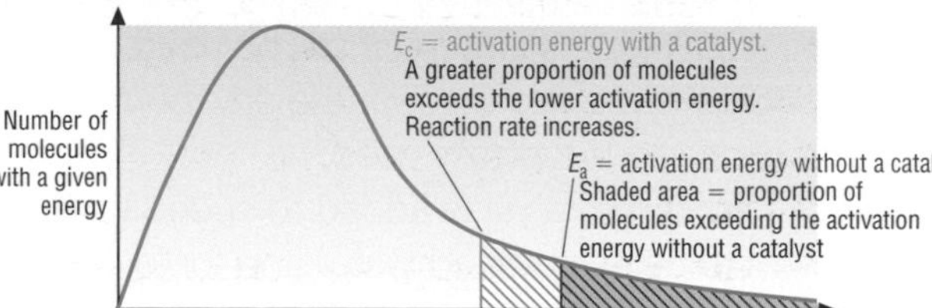

Number of molecules with a given energy
Ec = activation energy with a catalyst.
A greater proportion of molecules exceeds the lower activation energy. Reaction rate increases.
Ea = activation energy without a catalyst.
Shaded area = proportion of molecules exceeding the activation energy without a catalyst
Energy
Ec
Ea
Activation energy reduced

Equilibrium

An equilibrium system opposes changes

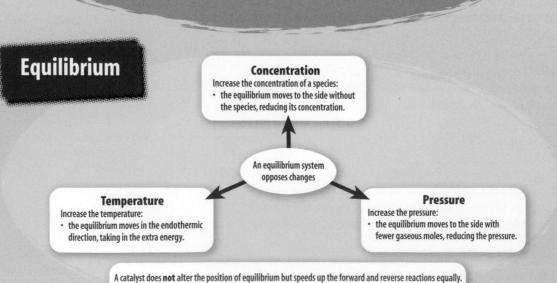

Concentration

Increase the concentration of a species:

- the equilibrium moves to the side without the species, reducing its concentration.

Temperature

Increase the temperature:

- the equilibrium moves in the endothermic direction, taking in the extra energy.

Pressure

Increase the pressure:

- the equilibrium moves to the side with fewer gaseous moles, reducing the pressure.

A catalyst does **not** alter the position of equilibrium but speeds up the forward and reverse reactions equally.

Practice questions

1. Define the term *exothermic* and give an equation to represent an exothermic reaction.

2. Define the term *enthalpy change of combustion*, stating the standard conditions for temperature and pressure.

3. Write equations to show the standard enthalpy change of formation of :
   (a) butane, $C_4H_{10}(g)$
   (b) sodium hydrogencarbonate, $NaHCO_3(s)$.

4. When burnt, 1.24 g of pentan-1-ol, $C_5H_{11}OH$, heated 500 g of water by 19 °C. Calculate the enthalpy change of combustion of pentan-1-ol. For water, $c = 4.18\ J\ g^{-1}\ K^{-1}$.

5. Draw an energy profile diagram for a reaction with an enthalpy change of –850 kJ mol^{-1} and an activation energy of +220 kJ mol^{-1}.

6. **(a)** Define the term average bond enthalpy.
   (b) The bond enthalpies for some common bonds are shown below.
   C–H: +413 kJ mol^{-1}; C–C: +347 kJ mol^{-1};
   C–O: +358 kJ mol^{-1}; O=O: +497 kJ mol^{-1};
   C=O: +805 kJ mol^{-1}; O–H: +463 kJ mol^{-1};
   C=C: +612 kJ mol^{-1}.
   Calculate the enthalpy change for the following reactions:
   (i) $CH_2{=}CH_2(g) + H_2O(g) \longrightarrow CH_3CH_2OH(g)$
   (ii) $(CH_3)_3COH(l) + 6O_2(g) \longrightarrow 4CO_2(g) + 5H_2O(g)$

7. From the following equations and enthalpy changes, calculate the standard enthalpy of formation of but-1-yne, $C_4H_4(g)$.
   $C(s) + O_2(g) \rightarrow CO_2(g)$ $\Delta H = -394$ kJ mol^{-1}
   $H_2(g) + \frac{1}{2}O_2(g) \rightarrow H_2O(l)$ $\Delta H = -286$ kJ mol^{-1}
   $C_4H_4(g) + 5O_2(g) \rightarrow 4CO_2(g) + 2H_2O(l)$ $\Delta H = -2597$ kJ mol^{-1}

8. Use the collision theory to explain how the following can affect the rate of a chemical reaction:
   (a) increasing the concentration of a reactant
   (b) increasing the temperature of the system.

9. Sketch a Boltzmann distribution curve to represent a range of energies of the molecules in a gas mixture. Indicate the activation energy on your diagram.

10. State le Chatelier's principle.

11. The Haber process is an important industrial process used to manufacture ammonia.
    The process is based on the following equilibrium system.
    $N_2(g) + 3H_2(g) \rightleftharpoons 2NH_3(g)\ \Delta H = -93$ kJ mol^{-1}
    Use le Chatelier's principle to predict the approximate conditions of temperature and pressure best suited to obtain the *maximum* equilibrium yield of ammonia. Explain fully how any suggested conditions will affect the position of equilibrium.

12. The following equilibrium exists between two of the oxides of nitrogen, NO_2 and N_2O_4.
    $2NO_2(g) \rightleftharpoons N_2O_4(g)$ $\Delta H = -58$ kJ mol^{-1}
    brown colourless
    (a) State le Chatelier's principle.
    (b) Describe and explain what you would see after the following changes have been made to the equilibrium system.
    (i) The temperature is increased.
    (ii) The pressure is increased.
    (iii) A catalyst is added.

13. **(a)** Define the term *standard enthalpy change of combustion*.
    (b) A student heated 50.0 cm^3 of water over a spirit burner. 0.100 g of propan-1-ol raised the temperature of the water by 13.5 °C.
    The specific heat capacity of water is $4.18\ J\ g^{-1}\ K^{-1}$.
    (i) Write a balanced equation, including state symbols, for the complete combustion of propan-1-ol.
    (ii) Calculate the energy produced, in kJ, when 0.100 g of propan-1-ol is burned.
    (iii) Calculate the amount, in moles, of propan-1-ol burned.
    Hence calculate the enthalpy change of combustion of propan-1-ol, in kJ mol^{-1}.
    (iv) The student checked the answer in a data book. He found that the actual enthalpy change of combustion was more exothermic than his answer. State three possible reasons for this difference.

2.3 Examination questions

1 Alkanes are important hydrocarbons since they are used as fuels in homes and in industry. It is important that the enthalpy changes involved in alkane reactions are known.

(a) **(i)** Define the term *enthalpy change of formation of a compound.* [2]

(ii) Write the equation, including state symbols, that accompanies the enthalpy change of formation of hexane, $C_6H_{14}(l)$. [2]

(iii) What conditions of temperature and pressure are used when measuring the **standard** enthalpy change of formation? [1]

(b) The standard enthalpy change of formation of hexane is -199 kJ mol^{-1}.

Using the axes below, show the enthalpy profile diagram for the formation of hexane.

On your diagram label the enthalpy change of reaction, ΔH, and the activation energy, E_a.

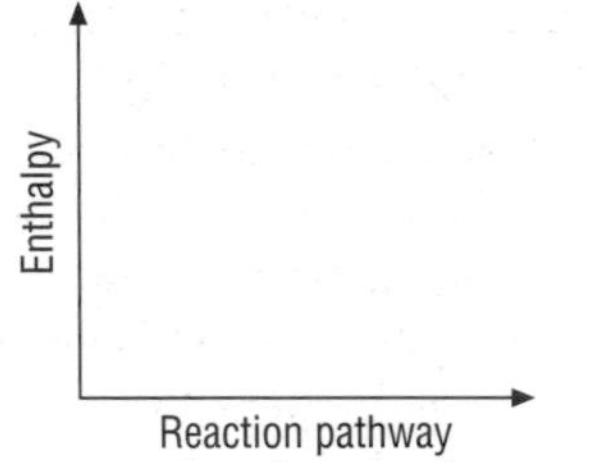

[3]

(c) Enthalpy changes can be calculated using enthalpy changes of combustion. The table below show some values for standard enthalpy changes of combustion.

substance	$\Delta H_c^{\ominus}$/**kJ mol**$^{-1}$
C(s)	−394
$H_2(g)$	−286
$CH_4(g)$	−890

Use these values to calculate the standard enthalpy change of the reaction below.

$C(s) + 2H_2(g) \rightarrow CH_4(g)$ [3]

(d) The equations for the combination of gaseous atoms of carbon and hydrogen to form methane, CH_4, and ethane, C_2H_6, are shown below.

$C(g) + 4H(g) \rightarrow CH_4(g)$ $\quad \Delta H = -1652$ kJ mol^{-1}

$2C(g) + 6H(g) \rightarrow C_2H_6(g)$ $\quad \Delta H = -2825$ kJ mol^{-1}

Use these data to calculate:

(i) the bond enthalpy of a C–H bond, [1]

(ii) the bond enthalpy of a C–C bond. [2]

[Total: 14]

(Jun 07 2831/1)

2 Many industrial processes, used to manufacture important chemicals, involve equilibrium reactions.

Chemists use their understanding of rates of reaction and of yields at equilibrium to find the most economic conditions for the reactions.

Chemists were investigating the production of a chemical, X_2Y, that could be formed from X_2 and Y_2 as shown in **equilibrium 2.1** below.

$2X_2(g) + Y_2(g) \rightleftharpoons 2X_2Y(g)$ $\quad$ **equilibrium 2.1**

(a) State le Chatelier's principle. [2]

(b) State and explain the effect on **equilibrium 2.1** of a **decrease** in pressure on:

(i) the equilibrium position of the reaction, [2]

(ii) the rate of the reaction. [2]

(c) The chemists measured the percentage conversion of X_2 at various temperatures. The results are shown in the graph below.

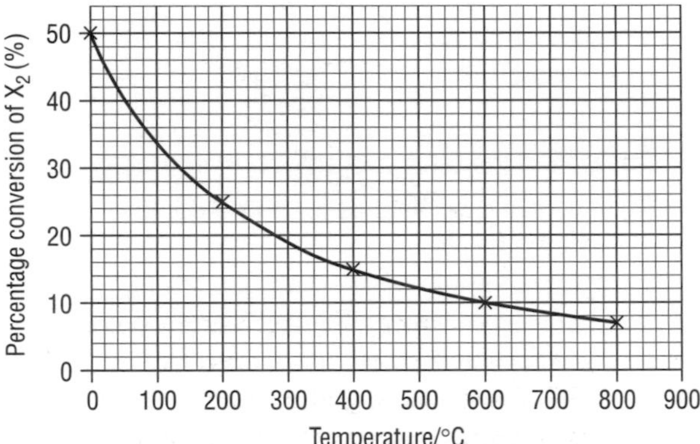

(i) Use the graph to predict the percentage conversion at 350 °C. [1]

(ii) The forward reaction in **equilibrium 2.1** is exothermic. Explain how the graph supports this statement. [2]

(d) The chemists decided to use a catalyst in the process. State, and explain, the effect of using a catalyst on:

(i) the rate of conversion of X_2 and Y_2 into X_2Y, [2]

(ii) the percentage conversion at equilibrium of X_2 and Y_2 into X_2Y. [2]

[Total: 13]

(Jun 07 2831/1)

Answers to examination questions will be found on the Exam Café CD.

3 Propane, C_3H_8, is a gas at room temperature and pressure. It is used in blow torches to melt the bitumen needed to apply the felt to flat roofs.

(a) Write the equation for the complete combustion of propane. [2]

(b) Define the term *standard enthalpy change of combustion*. [3]

(c) A blow torch was used to determine the enthalpy change of combustion of propane. The apparatus is shown below.

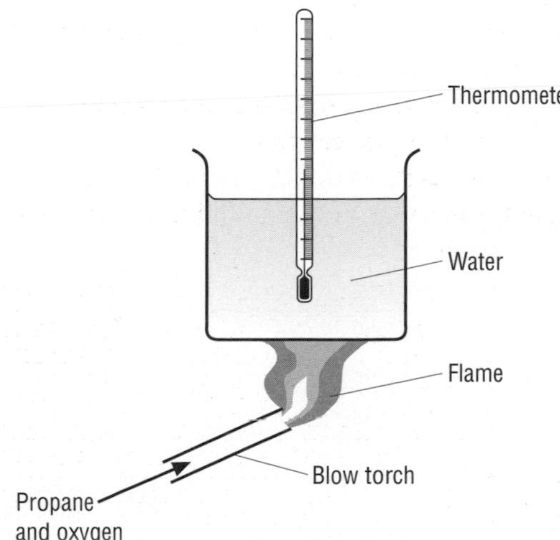

In the experiment, 200 g of water were used. The temperature of the water changed from 18.0 °C to 68.3 °C when 1.00 g of propane was burnt.

(i) Calculate the energy produced in kJ. The specific heat capacity of water is 4.18 J g^{-1} K^{-1}. [2]

(ii) Calculate the number of moles of C_3H_8 burnt during the experiment. [1]

(iii) Deduce the enthalpy change of combustion, in kJ mol^{-1}, of C_3H_8. [2]

[Total: 10]

(Jan 07 2831/1)

4 **(a)** Energy changes during reactions can be considered using several different enthalpy changes.
These include average bond enthalpies and enthalpy changes of combustion.
Table 4 shows the values of some average bond enthalpies.

bond	average bond enthalpy/kJ mol^{-1}
C–H	+410
O–H	+465
O=O	+500
C=O	+805
C–O	+336

Table 4

(i) Why do bond enthalpies have positive values? [1]

(ii) Define the term *bond enthalpy*. [2]

(iii) The equation below shows the combustion of methanol, CH_3OH, in the gaseous state.

$$H{-}\overset{\overset{\displaystyle H}{|}}{\underset{\underset{\displaystyle H}{|}}{C}}{-}O{-}H \;+\; 1\tfrac{1}{2}\,O{=}O \;\rightarrow\; O{=}C{=}O \;+\; \begin{matrix} H{-}O{-}H \\ H{-}O{-}H \end{matrix}$$

Use the average bond enthalpies from Table 1 to calculate the enthalpy change of combustion of gaseous methanol, ΔH_c. [3]

(iv) Suggest **two** reasons why the **standard** enthalpy change of combustion of methanol will be different from that calculated in part **(iii)**. [2]

(b) Methanol can be used as a fuel or as the feedstock for a variety of organic compounds. It is manufactured from carbon monoxide and hydrogen.

$CO(g) + 2H_2(g) \rightleftharpoons CH_3OH(g)$ $\quad \Delta H = -129$ kJ mol^{-1}

(i) Describe and explain how the composition of the equilibrium mixture is affected by:
increasing the temperature,
increasing the pressure in the reaction. [4]

(ii) Describe and explain the effect of increasing the pressure on the rate of reaction. [2]

(iii) Suggest and explain the effect of catalyst on the equilibrium position. [2]

[Total: 17]

(Jun 06 2831/1)

5 Methane can be reacted with steam to produce carbon monoxide and hydrogen. The equation for this process is given below.

$CH_4(g) + H_2O(g) \rightarrow CO(g) + 3H_2(g)$ **equation 5**

Table 5 below shows the enthalpy changes of formation for methane, steam and carbon monoxide.

compound	ΔH_f/kJ mol^{-1}
CH_4	–75
H_2O	–242
CO	–110

Table 5

(a) Define the term *enthalpy change of formation*. [2]

(b) Write the equation, including state symbols, representing the enthalpy change of formation for methane, CH_4. [2]

(c) Use the ΔH_f values in Table 5 to calculate the enthalpy change for the reaction shown in **equation 5**. [3]

(d) State one important manufacturing process in which hydrogen is used. [1]

[Total: 8]

(Jan 06 2813/1)

UNIT 2 Module 4

Resources

Introduction

For much of the twentieth century, humanity's effect on the environment went relatively unchecked while the development of new materials and industrial processes transformed the way we live, and inventions such as the motorcar and the aeroplane, coupled with the availability of cheap fuel, revolutionised travel.

The desire to travel and the growing use of domestic electricity has led to the depletion of fossil fuel. It is the increase in the amount of fossil fuel consumption, in particular, that has had such an enormous impact on our environment.

Industry is now striving to be more environmentally friendly, and this has led to the emergence of a new branch of chemistry – one that is growing rapidly, focussing on sustainability, atom economy and green issues.

In this module, you will examine the causes of the greenhouse effect as well as the reasons for ozone depletion. You will also investigate the role of the chemist in minimising further damage. The study of *green chemistry* will help you understand how the chemical industry is having a positive effect on our environment. In the future, the *green chemist* may well save the planet!

Test yourself

1 Name two uses of CFCs before they were banned.
2 What environmental damage is caused by CFCs?
3 What causes global warming?
4 What can be done to combat global warming?
5 What toxic gases are removed by a catalytic converter?
6 What does 'green' mean when applied to chemistry?

Module contents

1. The greenhouse effect – global warming
2. Climate change
3. Solutions to the greenhouse effect
4. The ozone layer
5. Ozone depletion
6. Controlling air pollution
7. Green chemistry
8. CO_2 – villain to saviour

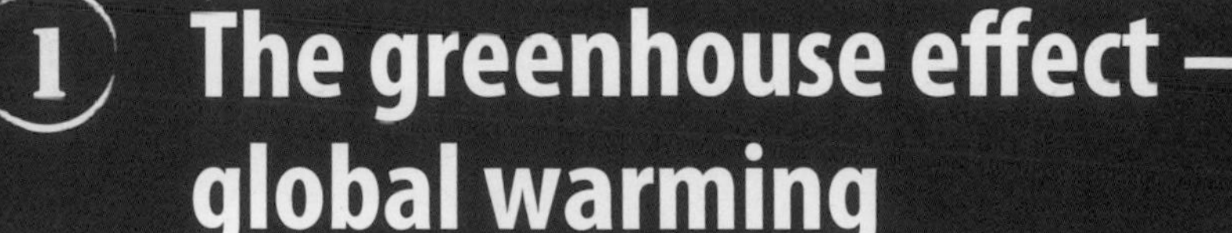

2.4 (1) The greenhouse effect – global warming

By the end of this spread, you should be able to . . .

* **Explain that infrared radiation is absorbed by C=O, O–H and C–H bonds in CO_2, H_2O and CH_4, and this contributes to global warming.**
* **Explain that the *greenhouse effect* of a given gas is dependent both on its atmospheric concentration and its ability to absorb infrared radiation.**

The greenhouse effect – it's not all bad!

When the term *greenhouse effect* is used in conversation or by the media, it usually has a negative connotation. However, without greenhouse gases, our planet would be covered in ice with an average temperature 35 °C lower than at present. In fact, the temperature of our planet has more to do with the greenhouse effect than with our proximity to the Sun! This is also the case for other planets in our solar system.

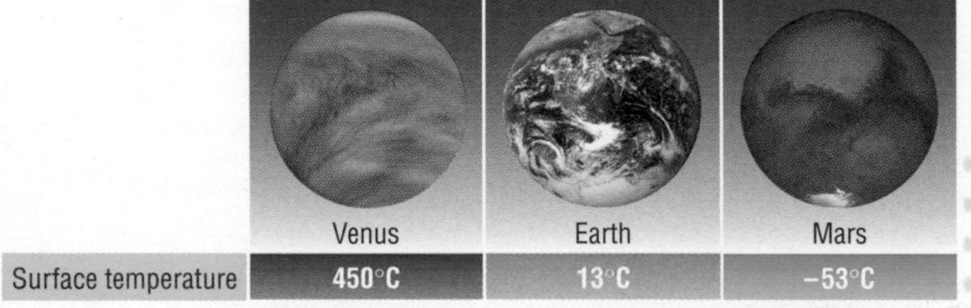

Figure 1 Average surface temperatures of the planets Venus, Earth and Mars

The Earth receives most of its energy in the form of electromagnetic radiation from the Sun. Most of this radiation is from the visible region of the spectrum, with small amounts of radiation from the infrared and ultraviolet regions. The incoming radiation is relatively unaffected by the gases in the Earth's atmosphere, and passes straight through to the Earth's surface. Here, the solar energy is absorbed and some is released back into the atmosphere as longer-wave infrared radiation.

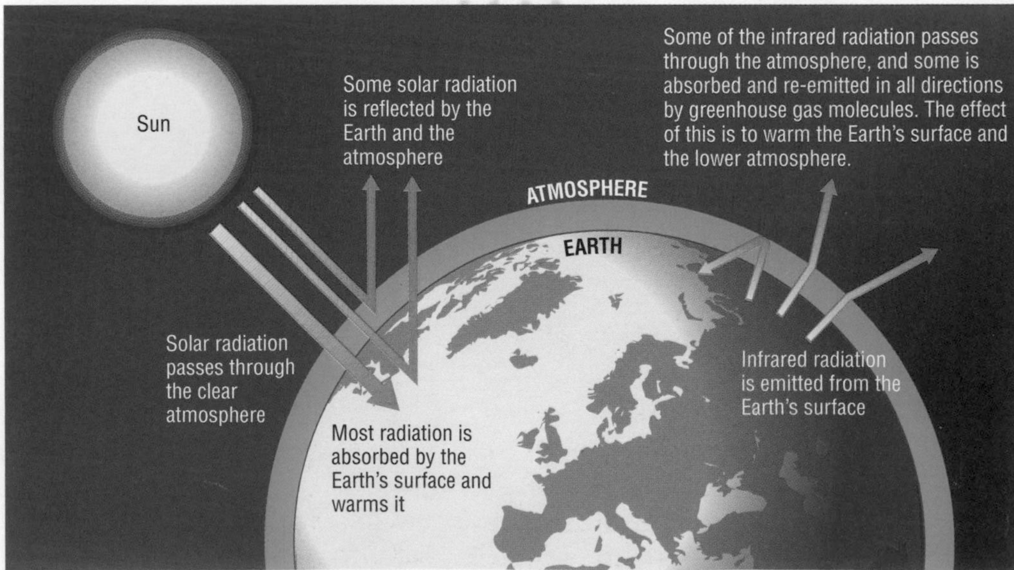

Figure 2 The greenhouse effect is a natural process, keeping our planet at a temperature capable of supporting life. Human activity is producing more greenhouse gases, which threaten to upset this fine natural balance, resulting in global warming

Most infrared radiation emitted by the Earth's surface goes back into space. However, certain gases in the atmosphere absorb some of this infrared radiation. It is then re-emitted as energy, with some passing back towards the Earth. This process effectively traps much of the heat in the lower atmosphere.

The **greenhouse effect** creates an equilibrium. The Earth's surface and atmospheric gases absorbs energy at the *same* rate as it radiates energy, thus maintaining a steady temperature.

Key definition

The **greenhouse effect** is the process in which the absorption and subsequent emission of infrared radiation by atmospheric gases warms the lower atmosphere and the planet's surface.

Greenhouse gases

Greenhouse gases occur naturally in the atmosphere. Water vapour – from evaporation of lakes and oceans – is by far the most abundant greenhouse gas, with carbon dioxide the next most abundant. Carbon dioxide is produced by many natural processes:

- volcanic eruptions
- respiration of animals
- burning or decay of organic matter, such as plants.

The third most abundant greenhouse gas is methane. Although present in smaller quantities than carbon dioxide, methane makes a *greater* contribution to the greenhouse effect than the same amount of carbon dioxide. Methane is:

- emitted during the production of coal, natural gas and oil
- a product of rotting organic waste in landfill sites
- released from certain animals, especially cows, as a by-product of digestion.

There are also large quantities of methane trapped in ice-like structures below the cold northern seas. These structures, called *clathrates*, contain 3000 times as much methane as is in the atmosphere. If released, this methane would have a huge impact on the greenhouse effect. There has been much speculation about whether '*methane burps*' from this source have contributed to global warming, but as yet there has been no concrete proof.

There are other greenhouse gases, but their concentrations in the atmosphere are much lower than those of water, carbon dioxide and methane.

Figure 3 Methane bubbles trapped in a frozen pond. These are produced by rotting organic matter at the bottom of the pond

How do gases absorb radiation?

Carbon dioxide is a linear molecule, O=C=O. The carbon dioxide molecule can absorb infrared radiation, causing the molecule to vibrate (see Figure 4 and spread 2.2.10). Eventually, the vibrating molecule emits some of this energy in the form of radiation. This can then be absorbed by another greenhouse gas molecule or at the Earth's surface.

Water vapour and methane absorb energy in a similar process:

- In H_2O, the O–H bonds absorb infrared radiation.
- In CH_4, the C–H bonds absorb infrared radiation.

The greenhouse effect of a gas depends not only on its concentration in the atmosphere but also on its ability to absorb infrared radiation.

The absorption–emission process keeps the heat close to the Earth's surface.

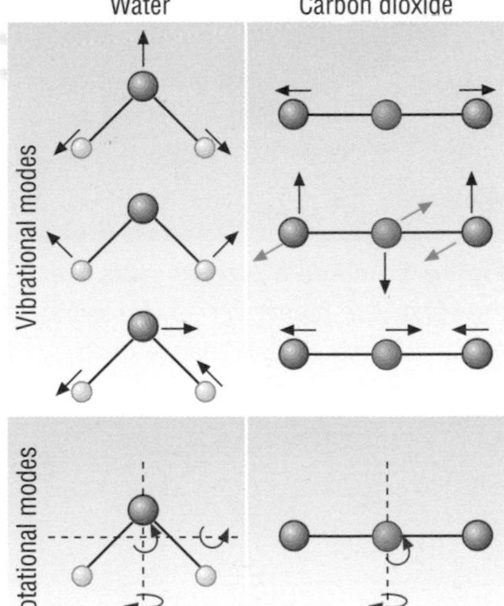

Figure 4 Carbon dioxide and water molecules will vibrate as energy is absorbed – a number of different vibrations are possible

Global warming potential

The ability of a trace gas to cause global warming is described by its *Global Warming Potential* (GWP). GWP is related to the lifetime of a gas in the atmosphere as well as the ability of the gas to absorb infrared radiation. Some chlorofluorocarbons (CFCs) are 25 000 times more efficient at absorbing infrared radiation than carbon dioxide but, since legislation banned the large-scale use of CFCs, their effect is diminishing. Luckily, the atmospheric concentrations of CFCs and other molecules with a high GWP have always been much lower than for carbon dioxide and water.

Questions

1 Name the three main greenhouse gases.

2 Outline the molecular processes that enable heat to be kept close to the Earth.

3 What factors contribute to a gas's contribution to global warming?

2.4 ② Climate change

How Science Works

By the end of this spread, you should be able to . . .

* **Explain the importance of controlling global warming caused by atmospheric increases in greenhouse gases.**
* **Outline the role played by chemists through the provision of scientific evidence that global warming is taking place.**
* **Understand the role of chemists in monitoring progress of initiatives such as the Kyoto protocol.**

Figure 1 An Adelie penguin looks over the edge of an iceberg on Paulet Island, Antarctica. Will climate change destroy this habitat?

Global warming

Climate change is a reality. The effect of global warming is there for all to see.

- In the Arctic, ice and permafrost are disappearing.
- In the Antarctic, the ice sheet is melting into the oceans.
- Tropical areas are experiencing more frequent and destructive storms and floods.
- In Europe, glaciers are disappearing and there have been extreme heat waves.

The Earth has not experienced this level of increase in temperature for over 1000 years. In 2005, atmospheric levels of carbon dioxide were 379 parts per million, higher than at any time in the past 650 000 years! Between 1995 and 2006, we have experienced 11 of the 12 warmest years on record.

It is now widely accepted that global warming is taking place. The cause? The Earth's supply of coal, natural gas and oil is being burnt in the thirst for energy. This combustion is releasing billions of tonnes of carbon dioxide into our atmosphere every year. The vast majority of scientists now agree that human activity is contributing to the pace of climate change, although a few deny this, maintaining that current trends are part of a natural cycle.

Warming brings with it unpredictable changes:

- Rivers overflow due to excessively heavy rainfall and melting glaciers. In other areas, drought-like conditions lead to water shortages.
- In some regions, longer growing seasons improve crop yields, whereas others experience drought and disease resulting in disastrous harvests.
- Storms and hurricanes are becoming more extreme. Hurricane Katrina devastated New Orleans, USA, in August 2005, and these violent storms are becoming even more frequent and destructive.
- Sea ice in the Arctic is melting faster each year. There are fears that the Gulf Stream current may be shut off, leading to severe winters in northern Europe, including the UK.
- Increasing temperatures expand the water in the oceans. Together with extra water from melting land ice, sea levels are rising. If the Greenland ice sheet and the Antarctic glaciers were to melt, sea levels would rise by six metres, flooding land populated by millions of people.

The list goes on . . .

Figure 2 Global warming? Upsala Glacier, Patagonia, Argentina – the top image was taken in 1928 and the bottom one in January 2004

Reducing greenhouse emissions

At the Earth Summit in 1992, there was international agreement that dangerous climate change must be prevented. But unfortunately there was little immediate action! In 1997, over 100 countries signed up to the Kyoto Protocol. This committed countries to reduce their emissions of six greenhouse gases by 5% by 2012.

The world's biggest emitter of greenhouse gases – the USA – has refused to sign. In 2006, global carbon dioxide emissions approached 32 billion tonnes, with about 25% coming from the USA. If appreciable reductions are to be made, then the USA, as the largest national producer of greenhouse gases, must sign up.

Other solutions stem from the development of non-carbon fuels, such as wind, solar, tidal and nuclear power. New methods are required for converting this clean energy into power, such as hydrogen fuel cells for cars. The bottom line is that carbon dioxide emissions must be reduced by 70–80% in order to stabilise atmospheric carbon dioxide concentrations and to stop the increase in global temperatures.

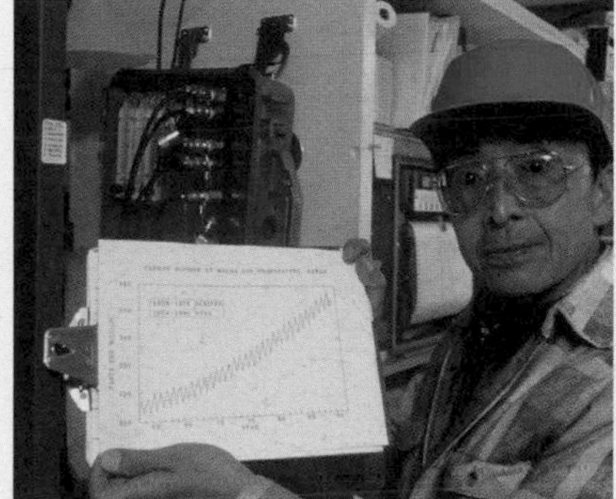

Figure 3 Atmospheric carbon study. A graph showing the concentration of atmospheric carbon dioxide (CO_2) in the years 1958–1989. There is an obvious upward trend, the annual peak concentration rising from 318 parts per million (p.p.m.) in 1958 to 356 p.p.m. in 1989.

EU strategy for the future

In 2007, the EU (European Union) agreed a long-term strategy on energy policy. This was an attempt to lead the world in the fight against global warming.

The deal sets the following binding targets:

- At least 20% of energy used in the EU will come from *renewable* sources by 2020.
- At least 10% of the fuels used in transport will be *biofuels* by 2020.
- EU greenhouse gas emissions will be reduced to 20% below 1990 levels by 2020.

It was also agreed that the EU would further reduce greenhouse gas emissions to 30%, if nations including the US, Russia, China and India followed this bold lead by the EU.

In 2007, the UK published a draft Climate Change Bill which aims to establish a framework to achieve a 60% cut in the UK's carbon emissions by 2050, compared with 1990 levels. There would also be an intermediate target of between 26% and 32% reduction by 2020. The UK is likely to become the first country to set such a long-range and significant carbon reduction target into law.

Cleaner cars

Cars produced in Europe are set to become cleaner after 2012. The EU has agreed new stringent carbon dioxide emission standards.

From 2012, new cars must emit no more than 130 g of carbon dioxide for every kilometre travelled. This is a substantial reduction, as average EU car emissions were 162 g per kilometre in 2005. The big question is whether this will lead to an overall reduction in greenhouse emissions. Car usage continues to increase, with everyone seemingly wanting someone else to cut back. So this reduction in emissions per car could easily be offset by the increased emissions from more cars!

Governments listen to the scientists

We seem to have now reached the stage at which politicians believe what the majority of scientists have been saying for years – 'Human activity is contributing to global warming.'

With the case for global warming now widely accepted, politicians are now looking for measures to reduce carbon dioxide emissions. Measures such as the Kyoto Protocol and the EU initiatives indicate that governments are listening.

Questions

1. List three consequences of global warming for the Earth.
2. Petrol contains a mixture of hydrocarbons, including isomers of C_8H_{18}.
   - (a) Write a balanced equation for the complete combustion of C_8H_{18}. Assume that petrol contains just C_8H_{18}.
   - (b) Look at the section headed in this spread: 'Cleaner cars'. Calculate the reduction in CO_2 emission from a car per kilometre in 2012 compared with 2005 by volume.

Solutions to the greenhouse effect

By the end of this spread, you should be able to . . .

- **Outline how chemists investigate solutions to environmental problems, such as carbon capture and storage (CCS).**
- **Understand the removal of waste carbon dioxide as a liquid injected deep in the oceans.**
- **Outline carbon storage in deep geological formations, by reaction with metal oxides to form stable carbonate minerals.**

Alternative fuels

We are familiar with many of the solutions for reducing greenhouse gas production, including:

- wind turbines
- tidal power
- solar panels
- nuclear plants.

All these technologies can contribute to our energy requirements, although each has its own drawbacks, such as wind turbine noise or radioactive waste.

Carbon capture and storage

Carbon capture and storage (CCS), or sequestration, captures carbon dioxide from power stations and stores it away safely, instead of it being released into the atmosphere. CSS is seen as an *immediate* strategy to get rid of waste carbon dioxide gas.

The Earth has many safe, stable areas in which to store gases we don't want. Underground porous rock can act as a sponge to store carbon dioxide gas and keep it from leaking away. Old oil and gas fields are some of the best natural containers for carbon dioxide gas. The UK has many of these in the North Sea.

In Scotland, BP is developing the world's first industrial project to create *decarbonised fuels*. The project plans to generate electricity using hydrogen manufactured from natural gas. The reduction in carbon dioxide emissions is about 90%.

At present, a power station burns methane from natural gas as its fuel. The methane is obtained from several gas fields, offshore from Scotland under the North Sea. The reaction is shown below.

$$CH_4(g) + 2O_2(g) \longrightarrow CO_2(g) + 2H_2O(l)$$

The CO_2 produced is emitted into the atmosphere, contributing to the greenhouse effect.

A decarbonised fuel will be produced by reforming natural gas into a mixture of H_2 and CO_2. This process uses well-established technology already used by the oil industry.

$$CH_4(g) + 2H_2O(g) \longrightarrow CO_2(g) + 4H_2(g)$$

The CO_2 will then be separated and piped offshore to an oilfield which is nearing the end of its productive life.

It is extremely difficult to extract the final 30% of the oil from an oilfield. However, the CO_2 pumped in will enable the remaining oil to be extracted. This oil bonus will partly contribute to the cost of the decarbonised fuel process.

The new plant could be operating before 2009, when it would become the first system of its type in the world.

Figure 1 North Sea oil platform

The diagram shown in Figure 2 illustrates this process.

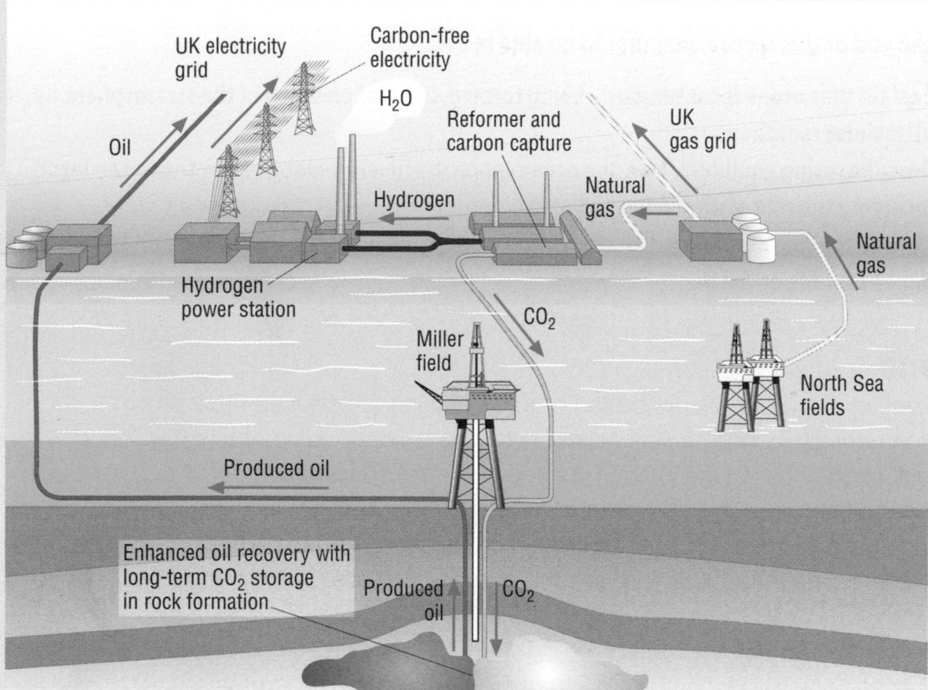

Figure 2 Capture and storage of CO_2 beneath the North Sea

If successful, this new gas separation technology could be fitted to existing power stations and petrochemical plants. The hydrogen could be burnt as a truly clean fuel, and the carbon dioxide could be captured and stored.

Using this technology, hydrogen would be burnt to produce energy, with water as the only emission. Hydrogen could also be converted into electricity in fuel cells.

How much CO_2 can the UK store?
The short answer to this question is a *lot*!

The UK has many old oil and gas fields, which are nearing the end of their productive life. It has been estimated that there is space for about 10 years worth of total UK CO_2 emissions in oilfields, and a further 30 years worth in gas fields. However, we need to start now, while the oil and gas platforms are still intact. Even when these fields have been filled with CO_2, it has been estimated that the UK has other porous rocks capable of storing up to 500 years worth of CO_2 production.

Storage as carbonates

Mineral storage aims to store carbon in stable minerals. CO_2 would be trapped by converting it into a carbonate rock. In this process, CO_2 is reacted with metal oxides to produce stable carbonates.

$$CaO(s) + CO_2(g) \longrightarrow CaCO_3(s)$$

$$MgO(s) + CO_2(g) \longrightarrow MgCO_3(s)$$

This process occurs naturally and has produced much of our limestone rock as calcium carbonate. Unfortunately, the natural reaction is very slow and efforts to increase the rate are very energy intensive. A power station using mineral storage would need 60–180% more energy than a power plant without CCS. More research is required if mineral storage is to become a viable form of CCS.

Questions

1 Explain how CO_2 could be removed as an emission when natural gas is used as a fuel.

2 What is mineral storage and what are the current problems with its attempted use?

2.4 (4) The ozone layer

By the end of this spread, you should be able to . . .

* **Explain that ozone is continuously being formed and broken down in the stratosphere by ultraviolet radiation.**
* **Describe, using equilibria, how the ozone concentration is maintained in the ozone layer, including the role of ultraviolet radiation.**
* **Outline the role of ozone in the absorption of harmful ultraviolet radiation and the resulting benefit for life on Earth.**

Figure 1 Ozone, O_3

Key definitions

The **troposphere** is the lowest layer of the Earth's atmosphere, extending from the Earth's surface up to about 7 km (above the poles) and to about 20 km (above the tropics).

The **stratosphere** is the second layer of the Earth's atmosphere, containing the 'ozone layer', about 10 km to 50 km above the Earth's surface.

Too little there ... many popular consumer products such as air conditioners and refrigerators have involved CFCs or halons during either manufacture or use. Over time, these chemicals damage the Earth's protective ozone layer.

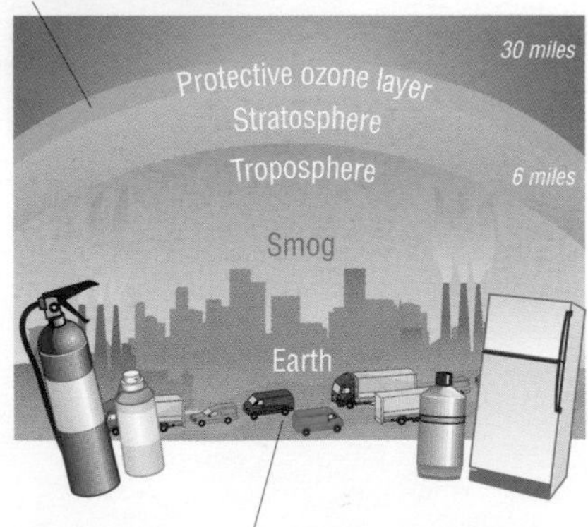

Too much here ... cars, trucks, power plants and factories all emit air pollution that forms ground-level ozone, a primary component of smog.

Figure 2 Ozone in the atmosphere

Ozone

Ozone, O_3, is a molecule consisting of three oxygen atoms (see Figure 1). Environmentally, ozone can be either *bad* or *good*, depending where it is found.

- *Bad*: ozone near to the Earth's surface in the **troposphere** is an air pollutant with *harmful* effects on the respiratory systems of animals.
- *Good*: ozone in the upper atmosphere in the **stratosphere** *protects* living organisms by preventing harmful ultraviolet light from reaching the Earth's surface.

The ozone layer

The ozone layer is found in the region of the atmosphere called the *stratosphere*, about 10–50 km above the Earth's surface. Ultraviolet radiation from the Sun has wavelengths in the range 270–400 nm. The ozone layer filters out the shorter wavelengths, i.e. less than 320 nm. These shorter wavelengths would be very damaging to life. The ozone converts this ultraviolet radiation into heat and consequently, the ozone layer is at a higher temperature than other parts of the upper atmosphere.

Much of our knowledge of the ozone layer is a result of work by the British meteorologist G.M.B. Dobson. He devoted much of his life to the observation and study of atmospheric ozone and developed a simple spectrophotometer to measure stratospheric ozone from the ground.

Between 1928 and 1958, Dobson set up a worldwide network of ozone monitoring stations which still operate today. Ozone concentration is measured in units called *Dobsons* in his honour.

The ozone–oxygen cycle

Ultraviolet (UV) radiation

Ozone is continuously being formed and broken down in the stratosphere by the action of UV radiation. There are *three* different types of UV radiation, based on the wavelength of the radiation:

- UV-a (320–400 nm) – reaches the Earth's surface. It has less energy than the shorter wavelengths and is not as damaging. UV-a radiation doesn't cause us much concern.
- UV-b (280–320 nm) radiation – can cause sunburn and sometimes genetic damage, which can result in skin cancer, if exposure to UV-b is prolonged.
- UV-c (200–280 nm) – is entirely screened out by the ozone layer.

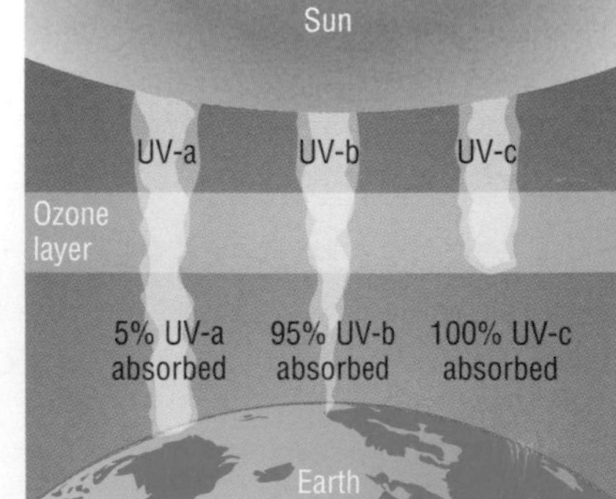

Figure 3 Absorption of UV radiation by the ozone layer

Although ozone screens out *most* UV-b, some does reach the Earth's surface. Any decrease in the ozone layer would allow more UV-b radiation to reach the surface, causing increased genetic damage to living organisms.

UV protection

Scientists have developed sunscreens to protect our skin from damaging UV radiation. If you look on the label of a sunscreen, you will usually see that it protects against both UV-a and UV-b (UV-c is absorbed in the mid-stratosphere). High levels of exposure to UV-b increase the risk of skin cancer.

Figure 4 Sun Protection Factors (SPF). SPF is measured by timing how long it takes skin covered in sunscreen to burn. SPF 15 means that it takes 15 times longer to burn than when wearing no sunscreen. SPF only considers burning times, so it only applies to UV-b.

Formation of ozone

The first step in the formation of ozone is the absorption of UV radiation with a wavelength less than 240 nm by O_2 molecules. This is high-energy radiation, capable of breaking an O_2 molecule into two oxygen atoms.

$$O_2 + (\text{radiation} < 240\ \text{nm}) \rightarrow 2O$$

The O atoms then react with O_2 molecules to form *ozone* molecules, O_3. This process generates heat.

$$O_2 + O \rightarrow O_3 + \text{heat}$$

The heat is absorbed by air molecules in the stratosphere, raising its temperature. Ozone is formed mainly in the upper reaches of the stratosphere.

How the ozone layer works

The ozone molecules formed in the above reaction then absorb UV with wavelengths between 240 and 310 nm. In this process, O_3 molecules are converted back to O_2 molecules and O atoms. Chemically, this is the *reverse* of the reaction that forms ozone.

$$O_3 + (\text{radiation} < 310\ \text{nm}) \rightarrow O_2 + O$$

The atomic oxygen, O, produced immediately reacts with other O_2 molecules to reform ozone:

$$O_2 + O \rightarrow O_3 + \text{heat}$$

and so the cycle continues.

- In this way, the *chemical* energy released when O and O_2 combine is converted into *kinetic* energy of molecular motion (heat).
- The overall effect is to convert penetrating UV radiation into heat, without any net loss of ozone.

This cycle keeps the ozone layer in a stable balance. A natural steady state is reached in which ozone is being formed at the *same* rate as it is being broken down:

$$O_2 + O \rightleftharpoons O_3$$

It is this process that protects living organisms from the harmful effects of high-energy UV radiation.

Removal of ozone

When an oxygen atom and an ozone molecule combine, they form two O_2 molecules:

$$O_3 + O \rightarrow 2O_2$$

Luckily for us, the removal rate is slow, since the concentration of O atoms is very low. However, human activity can affect this balance and this is discussed in spread 2.4.5.

Questions

1 Outline the filtering of ultraviolet radiation by the ozone layer.

2 Outline, including equations, how the concentration of ozone is maintained in the ozone layer.

2.4 (5) Ozone depletion

By the end of this spread, you should be able to . . .

- **Understand that radicals from CFCs and NOx may catalyse the breakdown of ozone.**
- **Explain that apparent benefits may be offset by unexpected and detrimental side effects.**

Beware the radicals

The Ozone Depletion Potential (ODP) of a substance is the *relative* amount of breakdown to the ozone layer caused by the substance. The ODP of a substance is compared with that for trichlorofluoromethane, $CFCl_3$, which has an ODP of 1.0.

In spread 2.4.4, you saw that ozone is in a stable balance between photochemical production and loss. However, the introduction of new compounds into the environment can lead to ozone loss and a disruption of this natural equilibrium. It is now recognised that most chlorine radicals in the stratosphere are generated by human activity. This has upset the natural ozone–oxygen balance, leading to problems in maintaining the protective ozone layer.

Chlorine radicals in the stratosphere mainly come from chlorofluorocarbons, CFCs (see spread 2.2.7). The stability of CFC molecules means that CFCs can only be broken down by the extremely energetic UV radiation found above most of the ozone layer.

CFCs take a long time to reach the ozone layer, perhaps decades! Once there, UV radiation provides the energy required for the process that damages the ozone layer to begin.

When UV radiation strikes a CFC molecule in the stratosphere, a C–Cl bond breaks, producing a chlorine radical, Cl•.

e.g. $CFCl_3 \longrightarrow Cl\cdot + \cdot CFCl_2$ *initiation*

Radicals are extremely reactive. Once a chlorine radical has been generated, it can react with an O_3 molecule, breaking it apart and destroying the ozone.

The breakdown of ozone takes place in two propagation steps:

- $Cl\cdot + O_3 \longrightarrow ClO\cdot + O_2$ *propagation step 1*
- $ClO\cdot + O \longrightarrow Cl\cdot + O_2$ *propagation step 2*

overall: $O_3 + O \longrightarrow 2O_2$

Another chlorine radical is produced and is free to attack another ozone molecule. The two propagation steps above repeat in a cycle many thousands of times. A single CFC molecule can destroy 100 000 ozone molecules. This process is very similar to the radical substitution of alkanes by chlorine, discussed in spread 2.1.12.

Another radical that destroys ozone is nitrogen oxide, •NO, from lightning or aircraft engines.

As with chlorine radicals, the breakdown of ozone proceeds in two propagation steps:

- $\cdot NO + O_3 \longrightarrow \cdot NO_2 + O_2$ *propagation step 1*
- $\cdot NO_2 + O \longrightarrow \cdot NO + O_2$ *propagation step 2*

overall: $O_3 + O \longrightarrow 2O_2$

Ozone levels over the northern hemisphere have been dropping by 4% per decade. Around the north and south poles, much larger and seasonal declines have been seen. These are the *ozone holes* (see Figure 1).

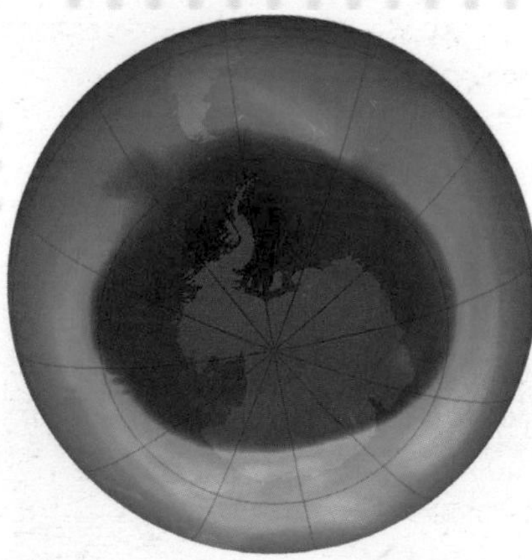

Figure 1 Antarctic ozone hole 2005. Coloured satellite image of low atmospheric ozone levels over Antarctica on 11th September 2005. The ozone hole (dark blue) is 27 million square kilometres in size. The largest hole was 29.2 million square kilometres in 2000

Regulation

In 1978, Sweden became the first nation to ban CFC-containing aerosols. A few other countries, including the USA, Canada and Norway, followed suit later that year, but the EU rejected a similar proposal. CFCs continued to be used in other applications, such as refrigeration and industrial cleaning. Then came the discovery of the Antarctic ozone hole in 1985. This led to international agreement that something had to be done urgently and the Montreal Protocol was initiated.

Montreal Protocol

The Montreal Protocol is viewed as one of the most successful global environmental agreements ever to be signed. Nations recognised scientific evidence that worldwide emissions of CFCs could deplete the ozone layer, resulting in damage to both human health and the environment.

The protocol introduced steps to control global emissions of CFCs. It also set a time frame for their complete removal in all but a limited number of products (those where no suitable alternative had been found). CFCs are still used in some inhalers, but they have been replaced in virtually all other uses. In the EU, CFCs have been completely banned since 1995. The first protocol came into force in 1989 and has since been through several amendments. In total, 30 countries signed up to the first Montreal Protocol. By 2006, this had increased to 197 nations.

Important restrictions are:

- CFCs – zero production by 2000.
- Tetrachloromethane (used as a solvent) – zero production by 2000.
- Production of halons (used in fire extinguishers) – phased out by 2000 (with some exceptions if no known alternatives).
- 1,1,1-trichloroethane (used as a solvent) – zero by 2005.
- HCFCs and HFCs – to replace CFCs in about 15% of applications.

Unfortunately, hydrochlorofluorocarbons (HCFCs) and hydrofluorocarbons (HFCs) are now also thought to contribute to global warming. The Montreal Protocol currently calls for a complete phase-out of HCFCs by 2030, but does not place any restriction on HFCs. Another downside is that each HCFC molecule is up to 10 000 times more potent as a greenhouse gas than carbon dioxide. At least there may be enough time in which to find a better alternative to CFCs.

On the mend?

On August 2nd 2003, scientists announced that the depletion of the ozone layer may be slowing down following the international ban on CFCs. Satellites and ground stations have confirmed that the stratospheric ozone depletion rate has slowed down significantly during the past decade. CFCs have extremely long atmospheric lifetimes, ranging from 50 to over 100 years, so the final recovery of the ozone layer is expected to take several lifetimes.

The fridge legacy

CFCs were first introduced for good scientific reasons. Their non-toxic and non-flammable properties made them ideal as *safe* refrigerants and aerosol propellants. You can read more about their development in spread 2.2.7. Since refrigeration helps to control and prevent germs and disease, the use of CFCs in fridges has doubtless saved many lives.

If CFCs had been used solely as aerosol propellants, then a ban on their use would have swiftly prevented their release into the environment. However, what about the countless old CFC-containing refrigerators still being used worldwide?

Since 2002, the EU has required all member states to remove CFCs and HCFCs from refrigeration equipment *before* the appliances are scrapped. It is estimated that up to 3 million domestic refrigeration and freezer units are disposed of each year in the UK. A further half million commercial units are also replaced annually.

Figure 2 Refrigerators and freezers stacked ready for recycling at a refuse disposal site in Sussex, UK. The machines have their compressors, coolant (refrigerant) and polyurethane foam removed before they are crushed and sold for scrap. The coolants and foam contain CFCs, harmful gases that deplete the ozone layer

Questions

1 Outline, including equations, how a single molecule of a CFC can remove many ozone molecules from the ozone layer.

2 List three restrictions in the Montreal Protocol aimed at preserving the ozone layer.

3 Why are HCFCs and HFCs not suitable as long-term replacements for CFCs?

2.4 (6) Controlling air pollution

By the end of this spread, you should be able to . . .

- **Explain the formation of carbon monoxide, oxides of nitrogen and unburnt hydrocarbons from the internal combustion engine.**
- **State environmental concerns relating to the toxicity of these molecules and their contribution to low-level ozone and photochemical smog.**
- **Outline how a catalytic converter decreases toxic emissions via adsorption, chemical reaction and desorption.**
- **Outline the use of infrared spectroscopy in monitoring air pollution.**

The internal combustion engine

Traffic emissions are one of the major threats to clean air. The internal combustion engine in a modern car emits various atmospheric pollutants. These are mainly:

- carbon monoxide
- oxides of nitrogen
- unburnt hydrocarbons.

Carbon monoxide (CO)

This is a poisonous gas, emitted into the atmosphere from the incomplete combustion of hydrocarbons and other organic compounds. In the UK, the majority of atmospheric carbon monoxide comes from traffic pollution, especially in urban areas.

Carbon monoxide molecules can exist for about one month in the environment before being oxidised to carbon dioxide.

For humans, carbon monoxide pollution has serious health implications. The haemoglobin in our blood carries oxygen to tissues and organs. Carbon monoxide can bind strongly to haemoglobin, reducing the amount of oxygen supplied to tissues and organs. The heart and the brain are severely affected; people with cardiovascular disease are especially susceptible to this type of poisoning. Symptoms of carbon monoxide poisoning include reduced manual dexterity, disturbed vision, tiredness and an inability to perform complex tasks.

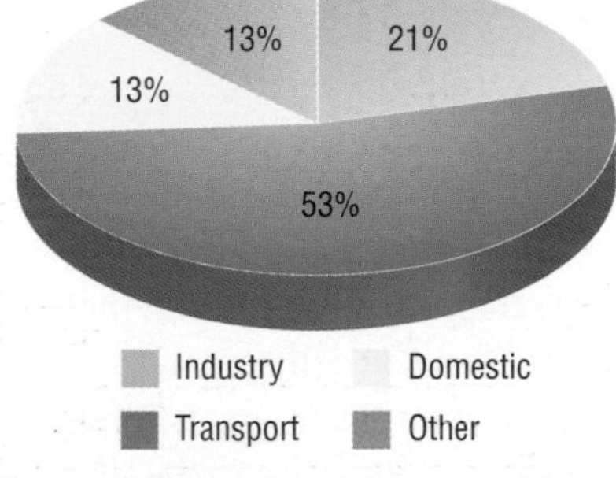

Figure 1 Emissions of nitrogen oxides

Oxides of nitrogen (NO_x)

During the burning of fuels in the internal combustion engine, air is drawn into the cylinder along with the fuel. The fuel is burned in the presence of oxygen, generating energy. Nitrogen oxides are also produced during this high-temperature process, as some of the nitrogen from the air is oxidised by the oxygen. Two oxides of nitrogen are produced: nitrogen monoxide (NO) and nitrogen dioxide (NO_2). The concentration of these gases is higher in urban areas, where traffic levels are high.

Nitrogen dioxide has a variety of environmental impacts, including the formation of low-level ozone (discussed below). Much of the nitrogen dioxide released into the atmosphere is converted into nitric acid, a contributor to acid rain. Nitrogen oxides are respiratory irritants and even low levels affect asthmatics.

Unburnt hydrocarbons

Volatile organic compounds (VOCs) are released in vehicle exhaust gases, usually from unburnt fuels. Two compounds are of particular concern, benzene and buta-1,3-diene. These compounds are both found in petrol in small quantities (2%) and are known to be human carcinogens.

Once released into the atmosphere, the unburnt hydrocarbons and nitrogen dioxide react together to form low-level ozone. The energy for this reaction is provided by sunlight. The mechanism leading to ozone formation involves radicals. Low-level ozone is a serious pollutant, causing breathing difficulties and increasing susceptibility to infections.

Figure 2 Photochemical air pollution in Boston, Massachusetts, USA

Concentrations of low-level ozone build up on humid sunny days, when the still air contains large quantities of hydrocarbons and oxides of nitrogen. Urban areas are very prone to low-level ozone formation. Infrared spectroscopy is currently being developed to monitor environmental pollution.

The catalytic converter

A catalytic converter has been fitted to all petrol cars sold in the EU since January 1993. A typical catalytic converter is made from platinum, rhodium and palladium supported on a honeycomb mesh. The honeycomb arrangement is used to provide a large surface area (see Figure 3).

Figure 3 A typical three-way catalytic converter has a surface area as big as two football pitches

The hot exhaust gases pass over the catalytic surface and the harmful gases are converted into less harmful products, which are then released into the atmosphere. There are two types of catalytic converters commonly in use.

Oxidation catalysts are used on diesel engines to decrease emissions of carbon monoxide and unburnt hydrocarbons. Combined with complex filter systems, they also remove particulate matter and nitrogen oxides.

The following reactions take place:

$$2CO(g) + O_2(g) \longrightarrow 2CO_2(g)$$

$$C_{12}H_{26}(l) + 18\frac{1}{2}O_2(g) \longrightarrow 12CO_2(g) + 13H_2O(l)$$

	Emissions/grams per kilometre		
	CO	Unburnt hydrocarbons	NO_x
No catalyst	5.99	1.67	1.04
With catalyst	0.61	0.07	0.04

Table 1 How a catalytic converter reduces emissions from a 1.8 litre petrol engine

The second type of catalytic system, known as the *three-way* catalyst, is fitted to petrol engines. In this system, nitrogen monoxide reacts with carbon monoxide to form the non-toxic gases nitrogen and carbon dioxide.

$$2NO(g) + 2CO(g) \longrightarrow N_2(g) + 2CO_2(g)$$

How the catalyst functions

The catalyst provides a surface on which the reaction takes place.

- The CO and NO gas molecules diffuse over the catalytic surface of the metal. Some of the molecules are held on to the metal surface by **adsorption**.
- Temporary bonds are formed between the catalytic surface and the gas molecules.
- These bonds hold the gas molecules in the correct position on the metal surface, where they *react* together.
- After the reaction, the CO_2 and N_2 products are *desorbed* from the surface and diffuse away from the catalytic surface.

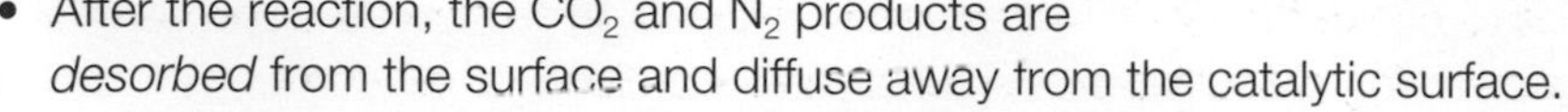

N_2, H_2O, CO_2 — Three-way catalyst — HC, CO, NOx

Figure 4 A three–way catalytic converter

Key definition

Adsorption is the process that occurs when a gas, liquid or solute is held to the surface of a solid or, more rarely, a liquid.

Questions

1 How are harmful NO_x and CO formed in a car engine?

2 Write equations to show how a catalytic converter removes NO_x, unburnt hydrocarbons (e.g. C_8H_{18}) and CO.

3 Outline how a catalytic converter functions.

2.4 (7) Green chemistry

By the end of this spread, you should be able to . . .

* **Describe principles and discuss issues of chemical sustainability.**
* **Understand the importance of establishing international cooperation to promote the reduction of pollution levels.**

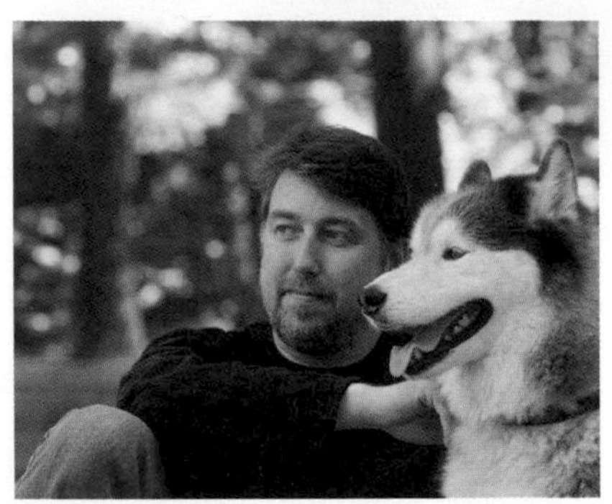

Figure 1 Paul Anastas, the 'father of green chemistry'. Paul is an organic chemist and the man widely credited with starting the grassroots movement to change the way chemists work. Paul Anastas summarised his philosophy in his '*12 principles of green chemistry*'. Since 1996, an annual Presidential Green Chemistry Challenge has been held where prizes are awarded to companies that have developed greener processes, reaction conditions or products. You can read more about this on the Internet

Sustainability and the green chemist

For the chemist, *sustainability* really means the development of processes that prevent the depletion of natural resources.

Chemicals have sometimes been perceived by the public as pollutants and harmful to the planet. This is now changing with a new movement known as *green chemistry*. The philosophy behind green chemistry is to improve our quality of life, make a profit and save the planet, all at the same time.

A fundamental idea behind green chemistry is that the chemist is *responsible* for the effect on the planet of a newly introduced chemical. Cleaning up chemical mess is costly. By cutting down on waste and developing more efficient processes, the cost of waste and environmental disposal can be slashed.

Green chemistry poses the questions:

- Why are we using hazardous substances at all? Can't we use non-hazardous alternatives?
- Can we use reactants or make products that are non-hazardous?
- Why are we producing waste? Why are we throwing things away that we can use elsewhere?
- Can we carry out processes using less energy?

University courses are now being offered that develop strategies to look after our resources. This goes beyond just recycling. This is more about looking after the Earth's atoms, making the most of available energy, and looking after the Earth itself.

Some of the philosophies of green chemistry are discussed below, together with some examples of how companies have responded.

Using renewable resources

- Using resources such as plant-based substances or solar energy instead of using finite resources, such as fossil fuels, that will eventually run out.

Green bags

The chemical company Proctor & Gamble recently announced a novel way to make plastic bags from renewable resources such as agricultural waste. It uses bacteria to ferment vegetable oils and sugars in the waste and converts this into plastics. This is a triumph for green chemistry, but perfecting the technique took nearly 15 years.

Saving money

- Companies will save money from not having to treat hazardous waste, or by using fewer chemicals or less energy.

Novel catalysts

Many new processes are being developed using catalysts to drastically decrease the amount of water and energy required.

The 12 principles

1 Prevention
2 Atom economy
3 Less hazardous chemical synthesis
4 Designing safer chemicals
5 Safer solvents and auxiliaries
6 Design for energy efficiency
7 Use of renewable feedstocks
8 Reduce derivatives
9 Catalysis
10 Design for degradation
11 Real-time analysis for pollution prevention
12 Inherently safer chemistry for accident prevention

Preventing waste

- Preventing any waste in the first place, so time and money is not spent on cleaning up later.

Green ibuprofen

Ibuprofen is a commonly used painkiller. Originally, it was synthesised using *six* steps with less than 40% atom economy. The company BHC has developed a new method for this synthesis that uses fewer materials, has only *three* catalytic steps and a 99% atom economy. Any waste that is produced is either recycled or is sold on as a useful by-product.

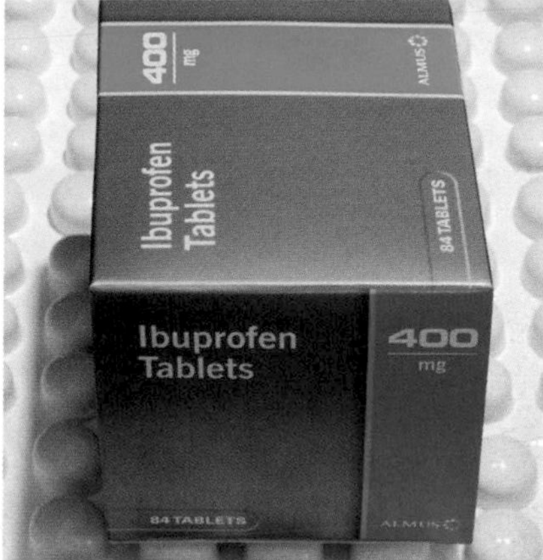

Figure 2 Ibuprofen is both a painkilling drug (analgesic) and an anti-inflammatory drug. It is widely marketed under many brand names, such as Advil, Motrin, Nuprin and Proflex

Maximising atom economy

- Designing synthetic methods to maximize all materials used in the process into the final product.

100% atom economy?

Crucial to green chemistry is the redesign of existing processes to improve the *atom economy*. Barry Trost was recognised for his work in advancing the concept of **atom economy**.

Trost wanted to produce less chemical waste when carrying out organic reactions for making new products.

Trost suggested that for every ton of finished product between 20 and 100 tons of starting materials are sometimes used! The rest is waste. The ideal situation would be that 100% of the starting materials would be incorporated into the desired product and also that no undesirable by-products would be made.

Figure 3 Barry M. Trost advanced the concept of atom economy

Recycling and biodegradability

- At the end of their use, materials should be recycled, or easily broken down in the environment into harmless substances.

Sea-Nine2 and PYROCOOL

Organotin compounds are used to prevent barnacles and marine plants from growing on ship's hulls. The World-Wide Fund for Nature (WWF) has been campaigning to phase out these compounds because of their high toxicity. A new alternative called Sea-Nine2 has been developed, which is non-toxic to marine life and is broken down quickly in the environment.

Traditional fire extinguishers use organic halogen compounds, which are ozone-depleting chemicals. The product PYROCOOL® FEF has been developed as a new fire-extinguishing and cooling agent. It is just as effective in extinguishing fires as traditional halogenated chemicals, but is non-toxic, does not deplete the ozone layer or persist in the environment.

International control

Hazardous chemicals do not recognise national boundaries. Ocean currents or winds can carry them from one nation to another. Many countries have now signed up to international treaties to reduce the use of hazardous chemicals.

These include:

- the Montreal Protocol to protect the ozone layer.
- the Stockholm Convention on Persistent Organic Pollutants (POPs), to encourage governments to eliminate or reduce the release of toxic POPs.
- the Rio Declaration on Environment and Development, to guide future sustainable development.

Questions

1 Butan-2-ol can be prepared from but-2-ene and from 2-bromobutane.

Preparation 1 $CH_3CH{=}CHCH_3 + H_2O \longrightarrow CH_3CH_2CHOHCH_3$

Preparation 2 $CH_3CH_2CHBrCH_3 + NaOH \longrightarrow CH_3CH_2CHOHCH_3 + NaBr$

(a) Show by calculation that the atom economy of Preparation 1 is greater than that of Preparation 2.

(b) What could be done to increase the atom economy of Preparation 2?

2.4  CO_2 – villain to saviour

By the end of this spread, you should be able to . . .

* **Discuss some of the ways of reducing CO_2 emissions into the environment.**
* **Outline some of the positive benefits of using CO_2 as a substitute for more harmful materials.**

Using CO_2

Carbon dioxide has had an unfair press coverage. It is the bad guy, constantly associated with global warming. So, here is the good news! There *are* some beneficial uses for the CO_2 that we produce. An example is for the CO_2 formed as a by-product of fermentation to make ethanol.

$$C_6H_{12}O_6 \longrightarrow 2C_2H_5OH + 2CO_2$$

Rather than allowing the CO_2 gas to escape into the environment, it can be collected and used.

Figure 1 Expanded polystyrene chips each in the shape of an 'S'. These are widely used as a thermally insulating and protective packaging material. Here the expanded polystyrene is made porous and is expanded with CO_2 trapped from other processes

CO_2 in foam

Chlorofluorocarbons, CFCs, were once used as the blowing agents to manufacture expanded polystyrene foam for packaging (see Figure 1). Because of environmental concerns associated with ozone depletion, CFCs are no longer used. CFCs are also greenhouse gases, contributing to global warming.

In 1996, the Dow Chemical Company developed a process that uses CO_2 as the blowing agent instead of CFCs. The company is using CO_2 waste from industrial and natural sources, ensuring that there is no net increase in global CO_2. So, environmentally harmful CFCs have been replaced by CO_2 that has already been produced.

CO_2 as a solvent

Solvents are used to dissolve chemical reactants, and to extract or purify products. Solvents are usually recycled, but many *organic* solvents are volatile, flammable or toxic. If traces of these types of solvents escape, risks are posed to both health and the environment.

We are now turning to CO_2 to eliminate these risks. It seems strange that CO_2 is being considered as a solvent since it is a gas at room temperature and pressure. However, by altering temperature and pressure CO_2 can be converted into an unusual state called a *supercritical fluid*. The resulting solvent is known as supercritical carbon dioxide or $scCO_2$.

The critical temperature of $scCO_2$ is 31°C, which is remarkably low. This means that $scCO_2$ can be used as a solvent with materials that would usually decompose at higher temperatures. The properties of $scCO_2$ can be modified by controlling the pressure.

CO_2 for decaffeinated coffee

Some people like the taste of coffee, but not the stimulating effects of the caffeine. The trick is to extract the caffeine without removing the flavour or introducing toxins through the decaffeination process.

Until recently, conventional liquid solvents were used.

- *Dichloromethane, CH_2Cl_2* – removes caffeine, but very little flavouring. However, CH_2Cl_2 has been identified as a possible carcinogen and is no longer used.
- *Ethyl ethanoate, $CH_3COOCH_2CH_3$* – removes caffeine but also some flavouring, and is slightly toxic.
- *Water* – is non-toxic, but removes the flavouring. Keeping the flavour involves an over-complicated process.

Figure 2 Cup of coffee with roasted coffee beans – the caffeine stimulant in coffee can be removed using $scCO_2$

Supercritical carbon dioxide is now widely used for decaffeination, removing 97–99% of the caffeine.

- The pressure of the $scCO_2$ is adjusted so that only caffeine is removed.
- The flavouring remains and there is no problem with toxicity.

Caffeine is extracted from green coffee beans, before roasting. The process is summarised in Figure 3. The extracted caffeine is sold on for other uses, such as in pharmaceuticals and soft drinks.

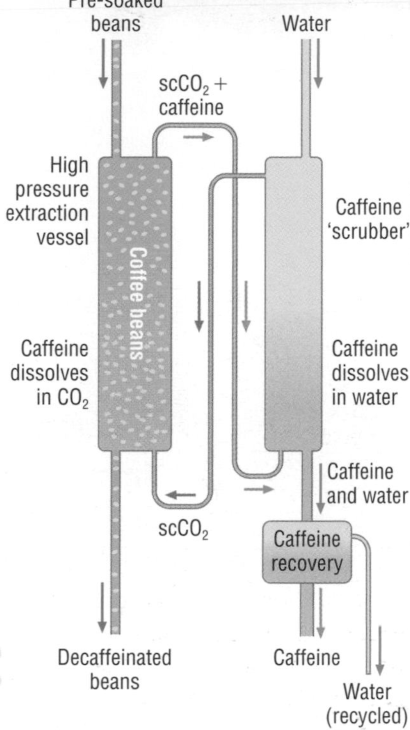

Figure 3 Decaffeinating coffee with $scCO_2$

CO_2 for beer

CO_2 produces the natural fizz in beer, but it can also be used to extract the characteristic flavour of beer from hops. As with coffee, water and dichloromethane could be used in theory, but these are the same risks and problems as with decaffeinating coffee (see above). $scCO_2$ is a much better alternative and is now widely used.

Figure 4 Dried hop flowers – hops grow throughout Europe and are extensively cultivated in northern Europe for the production of beer. Preparations of the flowers are also used in herbal medicine to aid digestion and to relieve fever

CO_2 for dry cleaning

For many years, chlorinated hydrocarbons have been used for dry cleaning. Although having extremely good solvent properties – exploited for removing grease and oil deposits from clothing – these solvents are toxic. Tetrachloroethene, C_2Cl_4 (*perc*), is the solvent most widely used. Perc is suspected of causing cancer and in addition its disposal can contaminate ground water. Prior to this, the extremely toxic tetrachloromethane, CCl_4, was used, but CCl_4 is now known to be carcinogenic and its use has been banned.

Once again, $scCO_2$ comes to the rescue. Used together with a *wetting agent*, it is a safer solvent that still has the required grease-dissolving properties. This method is now superseding the use of perc by some commercial dry cleaners.

CO_2 for toxic waste treatment

Many organic compounds dissolve in $scCO_2$. This property can be used to remove toxic materials from waste. The solvent's properties are adjusted using a much higher temperature and pressure than those used for decaffeination. This allows toxins to be removed from toxic waste mixtures.

CO_2 for chemical synthesis

There are many possible reactions that can be carried out with $scCO_2$. It can be used either as a solvent or as one of the reagents. The ability to control its solvent properties means that the desired product is obtained with fewer co-products. This in turn leads to simpler product separation.

Questions

1 (a) Outline how waste CO_2 has reduced the use of CFCs.
  (b) How is supercritical carbon dioxide made?

2 (a) What health problems are caused by the use of chlorinated hydrocarbons?
  (b) Outline two uses of supercritical CO_2 that have allowed less use of chlorinated hydrocarbons.

2.4 Resources summary

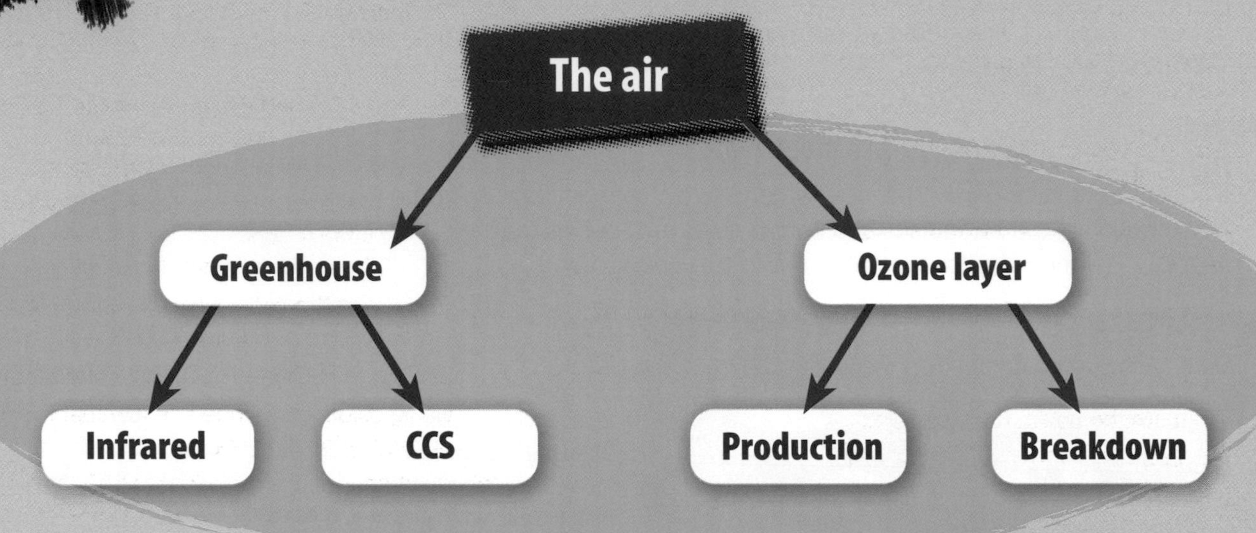
The air
Greenhouse
Ozone layer
Infrared
CCS
Production
Breakdown

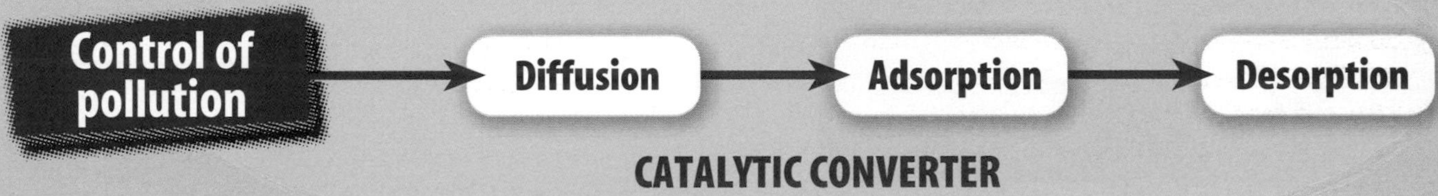
Control of pollution
Diffusion
Adsorption
Desorption
CATALYTIC CONVERTER

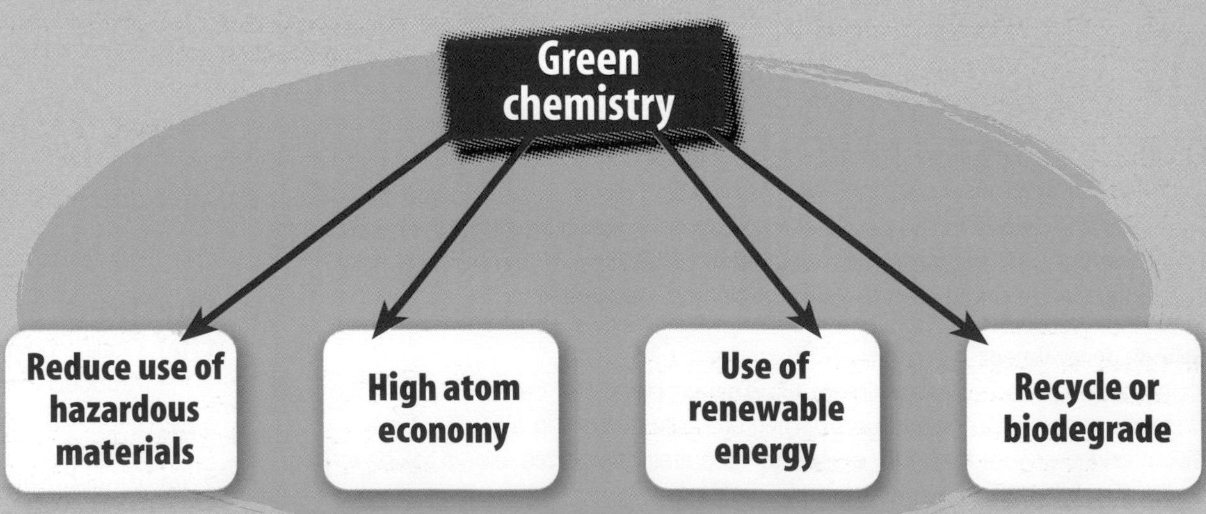
Green chemistry
Reduce use of hazardous materials
High atom economy
Use of renewable energy
Recycle or biodegrade

Practice questions

1. Name three greenhouse gases. For each gas, state the bonds that absorb infrared radiation.

2. Outline solutions to greenhouse gas emission involving carbon capture and storage (CCS).

3. Explain how the ozone content in the stratosphere is maintained.

4. Use equations to show how nitrogen oxides break down ozone.

5. How are NO and CO formed as pollutants in a car engine?

6. Summarise how a catalytic converter removes CO and NO.

7. What are the health hazards from:
   (a) CO; **(b)** NO;
   (c) O_3; **(d)** ultraviolet radiation?

8. How does green chemistry enable chemicals and resources to be preserved?

9. A deodorant contains a mixture of butane, 2-methylpropane and propane as the propellant.
   The alkanes are highly flammable.
   (a) What is the molecular formula of 2-methylpropane?
   (b) Suggest a property of these alkanes that makes them suitable as a propellant in an aerosol.
   (c) Write a balanced equation for the complete combustion of butane.
   (d) Why are alkanes now being used as propellants in aerosol sprays?

10. **(a)** Trifluorochloromethane, CF_3Cl, is an example of a CFC.
    (i) Draw a diagram to show the shape of a CF_3Cl molecule.
    (ii) Predict an approximate value for the bond angles in a CF_3Cl molecule.
    (b) When CFCs are exposed to strong ultraviolet radiation in the upper atmosphere, homolytic fission takes place to produce radicals.
    (i) Explain what is meant by the term homolytic fission.
    (ii) Write an equation to show the homolytic fission of CF_3Cl by UV radiation.
    (c) Cl radicals are one of the factors responsible for depleting the ozone layer in the stratosphere.
    (i) The equations below represent two possible steps that occur during this process.
    Complete these equations and construct an overall equation for the reaction.
    $Cl + O_3 \longrightarrow ClO + \ldots$
    $\ldots + O \longrightarrow Cl + O_2$
    (ii) Use the equations above to identify a catalyst in the reaction scheme. Explain your answer.

11. Nitrogen oxides, commonly referred to as NO_x, are emitted as pollutant gases from car exhausts. NO_x consists of a mixture of nitrogen monoxide, NO, and nitrogen dioxide, NO_2.
    (a) **(i)** NO is formed when air is heated to the high temperature inside a car engine.
    Write an equation for this reaction.
    (ii) NO_2 is formed when NO reacts with oxygen as the emissions cool.
    Write an equation for this reaction.
    (b) On leaving the car's exhaust system, NO_2 dissolves in water to form a mixture of acids:
    $2NO_2(g) + H_2O(l) \rightarrow HNO_2(aq) + HNO_3(aq)$
    Using oxidation numbers, show that this is a disproportionation reaction.
    (c) In urban traffic, a car releases 150 cm^3 of NO_x per kilometre, measured at room temperature and pressure.
    (i) Calculate the amount, in moles, of NO_x emitted by the car each kilometre.
    (ii) Calculate the number of molecules of NO_x emitted by the car each kilometre.
    (d) **(i)** The mass of NO_x released by the car is 0.250 g per kilometre.
    Using your answer to **(c)(i)**, calculate the relative molecular mass of the NO_x.
    (ii) Deduce whether there are more NO_2 molecules than NO molecules in this sample of NO_x. Explain your reasoning.
    (e) What measures are taken to cut down on NO_x emissions from car exhausts?

12. In the stratosphere, ozone forms when oxygen radicals react with oxygen molecules.
    $O_2 + O \longrightarrow O_3$
    (a) **(i)** Where does the energy come from to form ozone in the stratosphere?
    (ii) Draw a dot-and-cross diagram of an oxygen radical.
    (iii) Suggest why oxygen radicals are really 'diradicals'.
    (iv) Draw a dot-and-cross diagram of an ozone molecule. Show outer electrons only.
    (b) Chlorine radicals formed from CFCs deplete the ozone layer in a chain reaction.
    It has been estimated that 1 g of chlorine radicals destroys 150 kg of ozone during the one- to two-year residence time of the chlorine radicals.
    Calculate how many ozone molecules are destroyed in this chain reaction by a single chlorine radical before the radical is destroyed.

2.4 Examination questions

1 This question is about the greenhouse effect and carbon dioxide. The table shows the greenhouse factors (relative effects) of various gases in the Earth's atmosphere.

gas	greenhouse factor
CO_2	1
CH_4	30
N_2O	150
O_3	20 000
CFCs	10 000–25 000

(a) (i) Explain what is meant by the term *greenhouse gas*. [2]

(ii) Name the compound, not included in the table, which makes the largest contribution to the overall greenhouse effect. [1]

(iii) Suggest a reason why nitrogen is **not** a greenhouse gas. [1]

(iv) In addition to the greenhouse factor, what else affects the contribution of a particular substance to the overall greenhouse effect? [2]

[Total: 6]

(Jun 03 2815/3)

2 Under unpolluted conditions, the concentration of ozone in the stratosphere remains fairly constant. Ozone is constantly being formed and broken down, but the two processes occur at the same rate.
The formation of ozone in the stratosphere occurs by the two-step process shown below.

$O_2 \rightarrow O + O$

$O_2 + O \rightarrow O_3$

(a) (i) What is the source of energy for this process? [1]

(ii) What is the importance of the ozone layer to life on Earth? [1]

(b) Explain how the release of CFCs into the lower atmosphere affects the concentration of ozone in the stratosphere. [4]

(c) HFCs, such as CF_3CH_2F, can be used as alternatives to CFCs.
Outline the chemical properties of HFCs which can make them less damaging than CFCs to the ozone layer. [2]

[Total: 8]

(Jun 03 2815/3)

3 Arcton 133 is a CFC with the molecular formula $C_2H_2ClF_3$.

(a) When Arcton 133 is released into the atmosphere, its molecules can absorb energy. The C–Cl bond breaks forming free radicals.

$C_2H_2ClF_3 \rightarrow C_2H_2F_3\cdot + Cl\cdot$

(i) What source of energy is required for this reaction to take place? [1]

(ii) Chlorine free radicals catalyse the breakdown of ozone, O_3.
Write **two** equations to show how this happens. [2]

(iii) Write an equation for the overall reaction in **(a) (ii)**. [1]

(iv) What **type** of catalysis is shown here? Explain your answer. [2]

(b) In some applications, CFCs are being replaced by hydrocarbons such as alkanes.

(i) What is the M_r of Arcton 133, $C_2H_2ClF_3$? [1]

(ii) The formulae of some alkanes are shown below.

C_5H_{12} C_6H_{14} C_7H_{16} C_8H_{18} C_9H_{20}

Draw a circle around the molecular formula of the alkane whose M_r is most similar to that of Arcton 133. [1]

(iii) Suggest why hydrocarbons are replacing CFCs. [1]

(iv) Apart from cost, suggest **one** possible **disadvantage** of using a hydrocarbon instead of a CFC. [1]

[Total: 10]

(Jan 01 2813/1)

4 Ozone is a harmful constituent of photochemical smog, which can form in the troposphere. Ozone is produced photochemically from nitrogen dioxide and oxygen.

(a) What is meant by the term *photochemically*? [1]

(b) The following equations show the formation of ozone from nitrogen dioxide and oxygen in the troposphere.

$NO_2 \rightarrow NO + O$

$O + O_2 \rightarrow O_3$

(i) Which of these reactions requires light energy? Explain your answer. [1]

(ii) Draw a dot-and-cross diagram of nitrogen monoxide, NO, showing outer electrons only. Explain why it can be called a *free radical*. [2]

(iii) One source of nitrogen oxides in the troposphere is the emissions from car exhausts.
Describe how catalytic converters in car exhaust systems are effective in decreasing the emission of nitrogen oxides. [3]

(c) State **one** way in which ozone is damaging to human health. [1]

(d) Ozone reacts with rubber, a naturally occurring unsaturated polymer.

(i) Suggest how ozone reacts with rubber. An equation is **not** required. [2]

(ii) Suggest how the appearance of rubber is changed by this reaction. [1]

[Total: 11]

(Jan 03 2815/3)

Answers to examination questions will be found on the Exam Café CD.

5 This question looks at some aspects of the use of petrol as a fuel for cars.
Petrol contains octane, C_8H_{18}. Two of the stages that occur when petrol, containing octane, is used in a car engine are shown below.

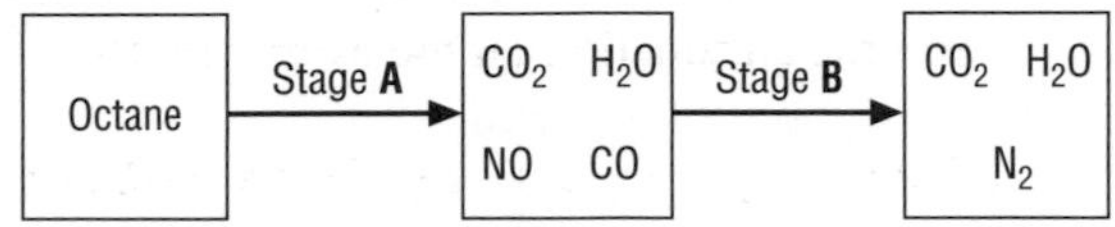

(a) Stage **A** includes the complete combustion of octane.
(i) Write the equation for this reaction. [2]
(ii) Suggest how NO is produced. [1]
(b) Stage **B** requires a catalyst.
(i) Name **two** metals generally present in the catalyst. [1]
(ii) The catalyst is a heterogeneous catalyst. Describe how it works. [3]
(iii) Using the substances shown above, write the equation for the reaction that occurs in stage **B**. [2]
(c) If stage **B** does not happen, further reactions occur and pollution levels rise.
Suggest **one** pollutant whose level in the atmosphere would rise. [1]
[Total: 10]
(Jan 05 2813/1)

6 Ethanol, C_2H_5OH, is manufactured on a large scale for a wide range of uses such as alcoholic drinks, as an industrial solvent and as a raw material for the synthesis of many organic compounds.
(a) Ethanol, C_2H_5OH, is manufactured on a large scale by two methods:
- Fermentation, using yeast, of sugars, such as glucose, $C_6H_{12}O_6$.
$C_6H_{12}O_6(aq) \rightarrow 2C_2H_5OH(aq) + 2CO_2(g)$
The ethanol is then distilled off.
- Hydration of ethane, C_2H_4, with steam in the presence of an acid catalyst.
$C_2H_4(g) + H_2O(g) \rightarrow C_2H_5OH(g)$

Compare the sustainability of these methods of manufacturing ethanol. In your answer, you should make clear how the points you make relate to:
- availability of starting material;
- atom economy and waste products.

✎ You will be awarded marks for the quality of written communication. [6]
[Total: 6]
(Jun 07 Specimen F322 paper)

7 This question is about pollutants in the atmosphere.
(a) CFCs, such as CCl_2F_2, affect ozone concentrations in the stratosphere.
(i) In this question, one mark is available for the quality of use and organisation of scientific terms. Explain how CFCs in the stratosphere alter the balance of oxygen and ozone. You should include equations in your answer. [6]
Quality of Written Communication [1]
(ii) Hydrofluorocarbons, such as $CHClF_2$, may be used instead of CFCs because they do not reach the stratosphere in significant amounts. Explain why. [2]
(b) Another pollutant in the atmosphere is nitrogen monoxide. It is readily converted to nitrogen dioxide. The mechanism for the oxidation of NO to NO_2 in polluted air can be summarised in two equations.
$2NO \rightarrow N_2O_2$ $\quad$ $N_2O_2 + O_2 \rightarrow 2NO_2$
(i) State one way in which NO is formed in the troposphere. [1]
(ii) Write an overall equation for the two reactions shown above. [1]
(iii) Deduce the oxidation state of nitrogen in N_2O_2. [1]
(c) State two environmental consequences of the presence of **nitrogen monoxide** in the atmosphere, other than acid rain. You should specify the region of the atmosphere for each of your examples. [2]
[Total: 14]
(Jun 06 2815/3)

8 The oxygen-only reactions involved in the formation of ozone in the stratosphere may be summarised as follows.
$O_2 \xrightarrow{hf} O + O \quad \Delta H^{\ominus} = +497\ kJ\ mol^{-1}$
$O + O_2 \rightarrow O_3 \quad \Delta H^{\ominus} = -390\ kJ\ mol^{-1}$
(a) **(i)** Explain why the standard conditions, implied by $\Delta H^{\ominus}$, do not really describe the reaction conditions in the stratosphere. [2]
(ii) How many electrons are there in an O radical? [1]
(b) The presence of CFCs in the stratosphere can lead to ozone depletion through a sequence of reactions such as that given below.
1 $O_3 \rightarrow O_2 + O$
2 $Cl + O_3 \rightarrow O_2 + ClO$
3 $ClO + O \rightarrow Cl + O_2$
4 $ClO + O_3 \rightarrow ClO_2 + O_2$
(i) Where does the energy come from for **step 1** of the sequence? [1]
(ii) One chlorine atom can destroy about 5000 molecules of ozone by a chain reaction. Which **two** steps in the above sequence propagate this chain reaction? Explain your answer. [2]
(iii) Complete a dot-and-cross diagram for the radical ClO_2. Show outer electron shells only. [1]
(c) **(i)** What role does ozone have in the stratosphere which is beneficial to people? [1]
(ii) State **one** damaging result of ozone pollution in the *troposphere*. [1]
(iii) Emissions of CFCs have been significantly reduced in recent years. Suggest why this has not yet resulted in restoration of the ozone layer. [2]
(iv) Explain why hydrofluoroalkanes, such as CH_2FCF_3, may be used as environmentally safer alternatives to CFCs. [2]
[Total: 13]
(Jun 02 2815/03)

Answers

Spread answers

1.1.1

1 Atoms are tiny particles that make up elements.
Atoms are indivisible.
All atoms of a given element are the same.
Atoms of one element are different from those of every other element.

2 An atom is nuclear.
The positive charge of an atom and most of its mass are concentrated in a nucleus at the centre.
Negative electrons orbit the nucleus, just like the planets orbit around the Sun.
Most of the volume of an atom would be in the space between the tiny nucleus and the orbiting electrons.
The overall positive and negative charges must balance.

1.1.2

1 (a) 3p, 4n, 3e; (b) 11p, 13n, 11e; (c) 9p, 10n, 9e; (d) 26p, 29n, 26e; (e) 19p, 20n, 18e; (f) 9p, 10n, 10e; (g) 20p, 19n, 18e; (h) 8p, 9n, 10e.

2 (a) $^{27}_{13}Al$; (b) $^{34}_{16}S^{2-}$.

1.1.3

1 (a) 10.80; (b) 28.11; (c) 52.06.

2 (a) 36.5; (b) 44.0; (c) 34.1; (d) 17.0; (e) 98.1.

3 (a) 159.6; (b) 62.0; (c) 331.2; (d) 132.1; (e) 310.3.

1.1.4

1 (a) 180.3 g; (b) 79.8 g; (c) 26.5 g.

2 (a) 0.170 mol; (b) 0.150 mol; (c) 0.0300 mol.

1.1.5

1 (a) ZnO; (b) Al_2O_3; (c) Ag_2SO_4.

2 (a) N_2O_4; (b) $C_4H_8O_2$; (c) $C_6H_{12}O_6$.

1.1.6

1 (a) (i) 1.5 mol; (ii) 45 mol; (iii) 0.167 mol.
(b) (i) 144 dm^3; (ii) 6 dm^3; (iii) 10.8 dm^3.

2 (a) 0.7 g; (b) 3.52 g; (c) 0.154 g.

3 (a) 0.48 dm^3; (b) 1.32 dm^3; (c) 0.72 dm^3.

1.1.7

1 (a) 8 mol; (b) 3.75×10^{-3} mol; (c) 3.04×10^{-3} mol.

2 (a) 3 mol dm^{-3}; (b) 2.00 mol dm^{-3}; (c) 0.175 mol dm^{-3}.

3 (a) 10.584 g dm^{-3}; (b) 4.5625 g dm^{-3}; (c) 13.969 g dm^{-3}.

1.1.8

1 (a) $4Li(s) + O_2(g) \longrightarrow 2Li_2O(s)$.
(b) $2Al(s) + 3Cl_2(g) \longrightarrow Al_2Cl_6(s)$.
(c) $2Al(s) + 3H_2SO_4(aq) \longrightarrow Al_2(SO_4)_3(aq) + 3H_2(g)$.
(d) $C_3H_8(g) + 5O_2(g) \longrightarrow 3CO_2(g) + 4H_2O(l)$.
(e) $Zn(s) + 4HNO_3(aq) \longrightarrow Zn(NO_3)_2(aq) + 2H_2O(l) + 2NO_2(g)$.
(f) $3Cu(s) + 8HNO_3(aq) \longrightarrow 3Cu(NO_3)_2(aq) + 4H_2O(l) + 2NO(g)$.

2 (a) $2Ca(s) + O_2(g) \longrightarrow 2CaO(s)$
(b) $Mg(s) + 2AgNO_3(aq) \longrightarrow Mg(NO_3)_2(aq) + 2Ag(s)$
(c) $2Pb(NO_3)_2(s) \longrightarrow 2PbO(s) + 4NO_2(g) + O_2(g)$
(d) $Cu(s) + 2H_2SO_4(l) \longrightarrow 2H_2O(l) + CuSO_4(aq) + SO_2(g)$

1.1.9

1 (a) $2NaHCO_3(s) \longrightarrow Na_2CO_3(s) + CO_2(g) + H_2O(l)$.
(b) 0.72 dm^3.

2 (a) $2Pb(NO_3)_2(s) \longrightarrow 2PbO(s) + 4NO_2(g) + O_2(g)$.
(b) 384 cm^3.

3 (a) $MgCO_3(s) + 2HNO_3(aq) \longrightarrow Mg(NO_3)_2(aq) + H_2O(l) + CO_2(g)$.
(b) (i) 0.72 dm^3; (ii) 0.600 mol dm^{-3}.

1.1.10

1 (a) H_2SO_4; (b) HNO_3; (c) CH_3COOH; (d) KOH; (e) $Ca(OH)_2$; (e) NH_3.

2 (a) proton donor; (b) proton acceptor;
(c) a soluble base, releasing $OH^-(aq)$ ions.

1.1.11

1 (a) $HCl + KOH \longrightarrow KCl + H_2O$.
(b) $2HNO_3 + Ca(OH)_2 \longrightarrow Ca(NO_3)_2 + 2H_2O$.
(c) $H_2SO_4 + 2NaOH \longrightarrow Na_2SO_4 + 2H_2O$.
(d) $2HNO_3 + MgCO_3 \longrightarrow Mg(NO_3)_2 + CO_2 + H_2O$.
(e) $2H_3PO_4 + 3Na_2CO_3 \longrightarrow 2Na_3PO_4 + 3CO_2 + 3H_2O$.

2 (a) (i) $NH_3 + HCl \longrightarrow NH_4Cl$.
(ii) $2NH_3 + H_2SO_4 \longrightarrow (NH_4)_2SO_4$.
(iii) $3NH_3 + H_3PO_4 \longrightarrow (NH_4)_3PO_4$.
(b) NH_4Cl: 26.2%; $(NH_4)_2SO_4$: 21.2%; $(NH_4)_3PO_4$: 28.2%.

1.1.12

1 (a) $BaCl_2 \cdot 2H_2O$; (b) $ZnSO_4 \cdot 7H_2O$;
(c) $Fe(NO_3)_3 \cdot 6H_2O$.

2 $CaCl_2 \cdot 6H_2O$.

1.1.13

1 $0.137\ mol\ dm^{-3}$.
2 $118\ g\ mol^{-1}$.

1.1.14

1 **(a)** K: 0; **(b)** Br: 0; **(c)** N: –3, H: +1;
(d) Ca: +2, F: –1; **(e)** Al: +3, O: –2; **(f)** N: +5, O: 2–;
(g) P: +5, O: –2.
2 **(a)** +1; **(b)** +5; **(c)** –2;
(d) +5; **(e)** +3.
3 **(a)** Cl: +5; **(b)** S: +4; **(c)** Mn: +7;
(d) Mn: +6; **(e)** Cr: +6.
4 **(a)** ClO_2^-; **(b)** ClO_4^-; **(c)** PO_3^{3-};
(d) CrO_4^{2-}.

1.1.15

1 **(a) (i)** $Fe + 2HCl \longrightarrow FeCl_2 + H_2$.
(ii) $2Al + 3H_2SO_4 \longrightarrow Al_2(SO_4)_3 + 3H_2$.
(b) (i) Fe has been oxidised from 0 to +2; H has been reduced from +1 to 0.
(ii) Al has been oxidised from 0 to +3; H has been reduced from +1 to 0.
2 **(a)** Br has been oxidised from –1 to 0; Cl has been reduced from 0 to –1.
(b) S has been oxidised from +4 to +6; O has been reduced from 0 to –2.
(c) Br has been oxidised from –1 to 0; S has been reduced from +6 to +4.

1.2.1

1 $Cl^{3+}(g) \longrightarrow Cl^{4+}(g) + e^-$.
2

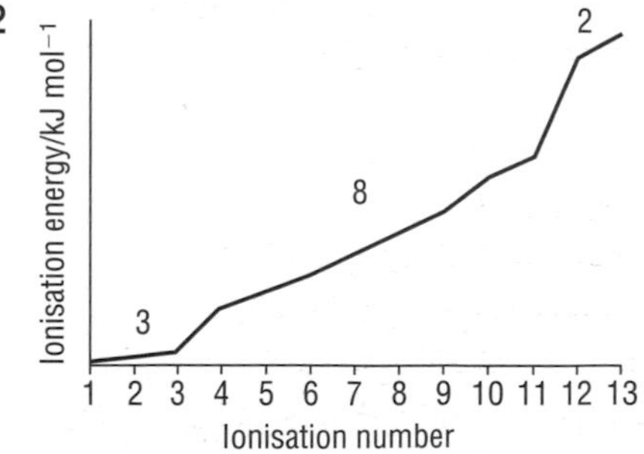

3 Si. There is a large increase between the 4th and 5th ionisation energies, showing that the 5th electron is removed from a different shell closer to the nucleus. The element is in Group 4 and must be Si.

1.2.2

1 $n = 1$: 2; $n = 2$: 8; $n = 3$: 18; $n = 4$: 32; $n = 5$: 50.
2 **(a)** A region in space that can hold up to two electrons.
(b) 2.
(c) s: 1; p: 3; d: 5; f: 7.

1.2.3

1

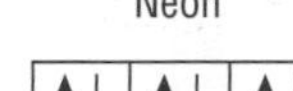

Fluorine: $1s^22s^22p^5$ Neon: $1s^22s^22p^6$

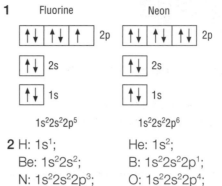

2 H: $1s^1$; He: $1s^2$; Li: $1s^22s^1$;
Be: $1s^22s^2$; B: $1s^22s^22p^1$; C: $1s^22s^22p^2$;
N: $1s^22s^22p^3$; O: $1s^22s^22p^4$; F: $1s^22s^22p^5$;
Ne: $1s^22s^22p^6$.

1.2.4

1 **(a)** s-block; **(b)** p-block;
(c) d-block; **(d)** f-block.
2 **(a)** $1s^22s^22p^63s^23p^1$; **(b)** $1s^22s^22p^63s^23p^2$;
(c) $1s^22s^22p^63s^23p^4$; **(d)** $1s^22s^22p^63s^23p^64s^23d^1$;
(e) $1s^22s^22p^63s^23p^64s^23d^7$; **(f)** $1s^22s^22p^63s^23p^64s^23d^{10}$.
3 **(a)** $1s^22s^22p^6$; **(b)** $1s^22s^22p^6$;
(c) $1s^22s^22p^6$; **(d)** $1s^22s^22p^63s^23p^63d^5$;
(e) $1s^22s^22p^63s^23p^6$;
(f) $1s^22s^22p^63s^23p^64s^23d^{10}4p^6$.
4 **(a)** $[Ne]3s^2$; **(b)** $[Xe]6s^2$; **(c)** $[Ne]3s^23p^5$.

1.2.5

1 **(a)** ionic; **(b)** ionic; **(c)** covalent;
(d) ionic; **(e)** covalent; **(f)** covalent.

1.2.6

1 **(a)**

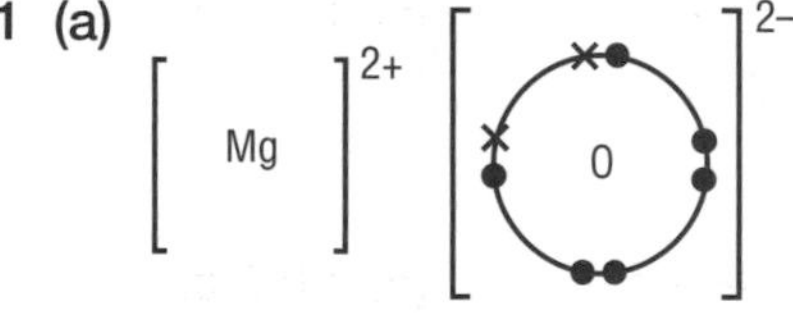

(b)

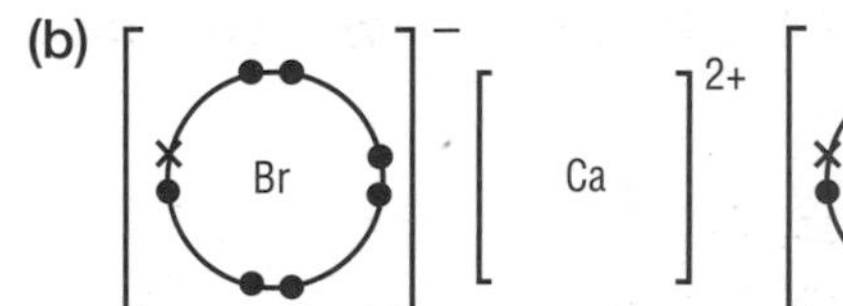

(c)

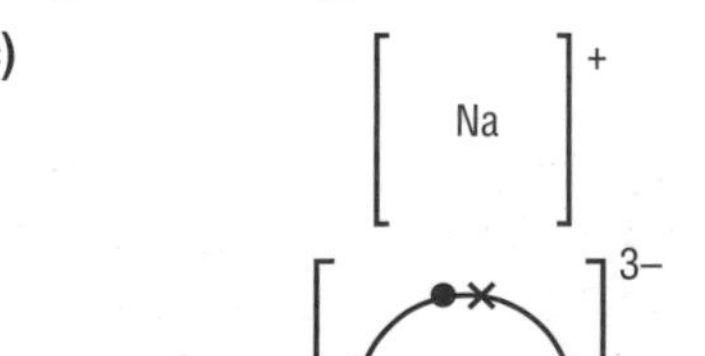

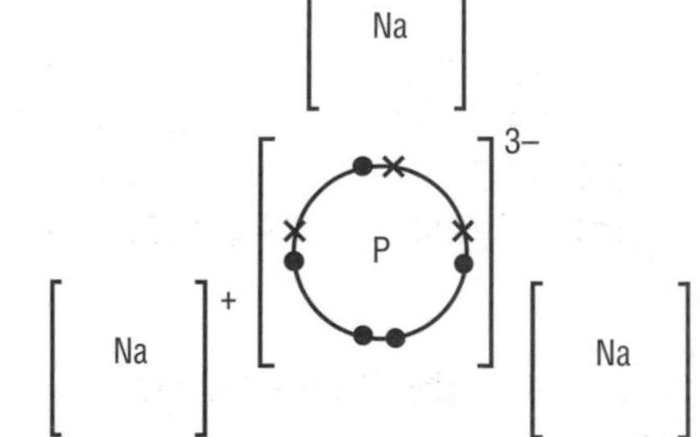

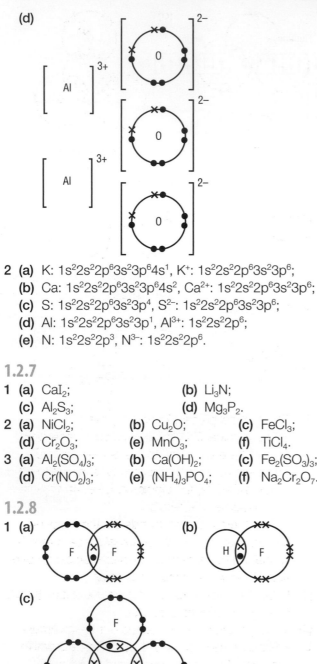

(d)

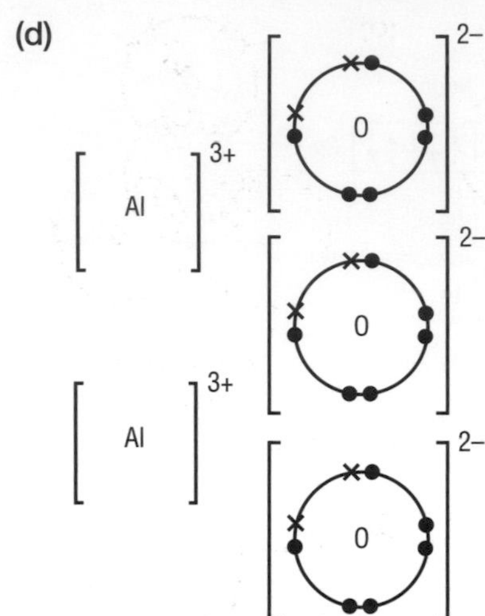

2 (a) K: $1s^22s^22p^63s^23p^64s^1$, K^+: $1s^22s^22p^63s^23p^6$;
(b) Ca: $1s^22s^22p^63s^23p^64s^2$, Ca^{2+}: $1s^22s^22p^63s^23p^6$;
(c) S: $1s^22s^22p^63s^23p^4$, S^{2-}: $1s^22s^22p^63s^23p^6$;
(d) Al: $1s^22s^22p^63s^23p^1$, Al^{3+}: $1s^22s^22p^6$;
(e) N: $1s^22s^22p^3$, N^{3-}: $1s^22s^22p^6$.

1.2.7

1 (a) CaI_2; (b) Li_3N;
(c) Al_2S_3; (d) Mg_3P_2.

2 (a) $NiCl_2$; (b) Cu_2O; (c) $FeCl_3$;
(d) Cr_2O_3; (e) MnO_3; (f) $TiCl_4$.

3 (a) $Al_2(SO_4)_3$; (b) $Ca(OH)_2$; (c) $Fe_2(SO_3)_3$;
(d) $Cr(NO_2)_3$; (e) $(NH_4)_3PO_4$; (f) $Na_2Cr_2O_7$.

1.2.8

1 (a)

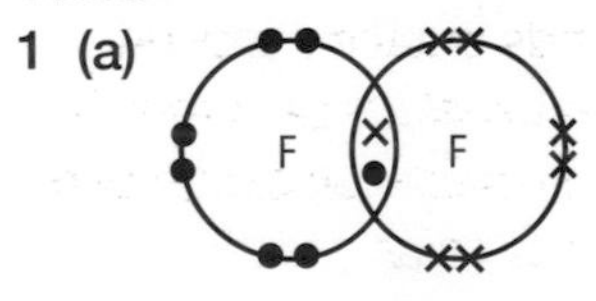

(b)

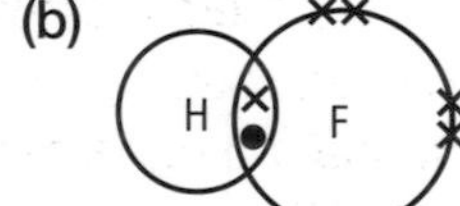

(c)

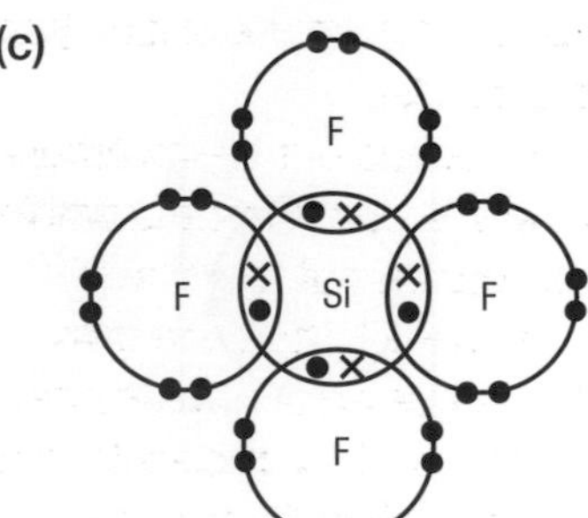

(d)

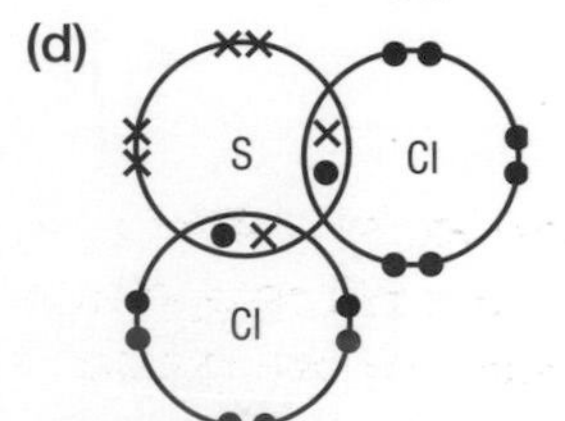

2 (a)

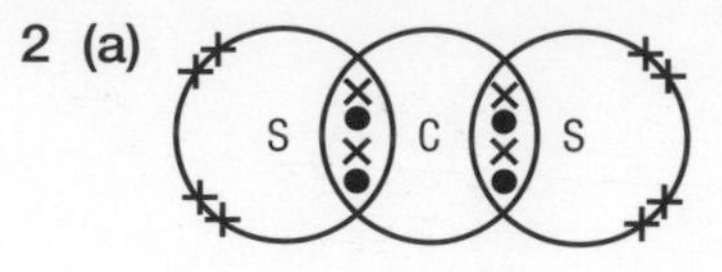

(b)

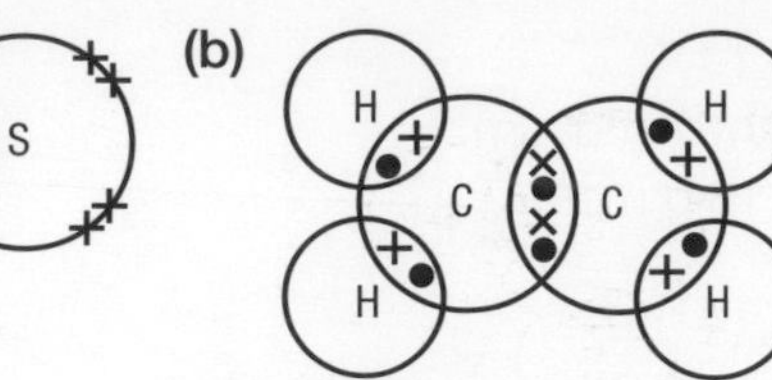

(c)

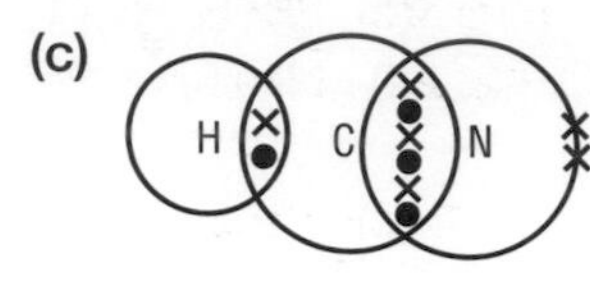

(d)

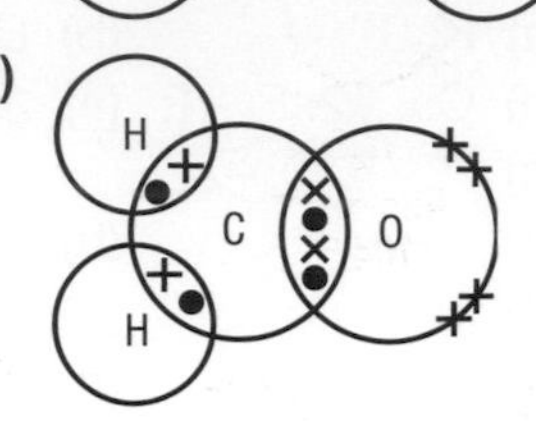

1.2.9

1 (a)

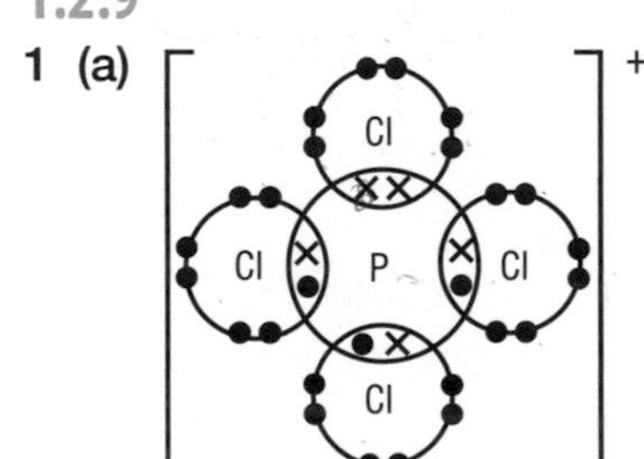

(b)

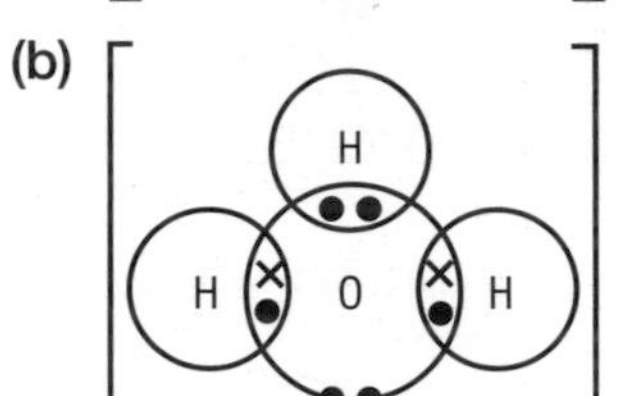

(c)

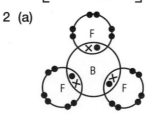

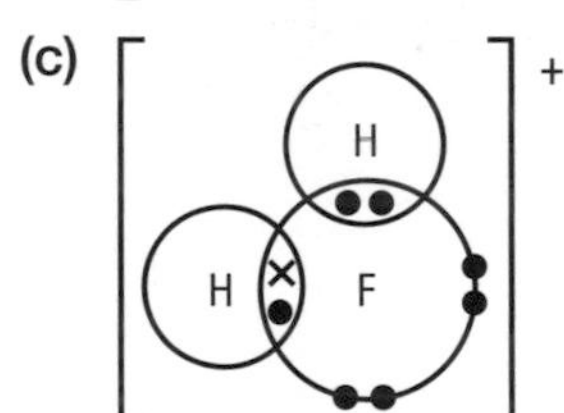

2 (a)

(b)

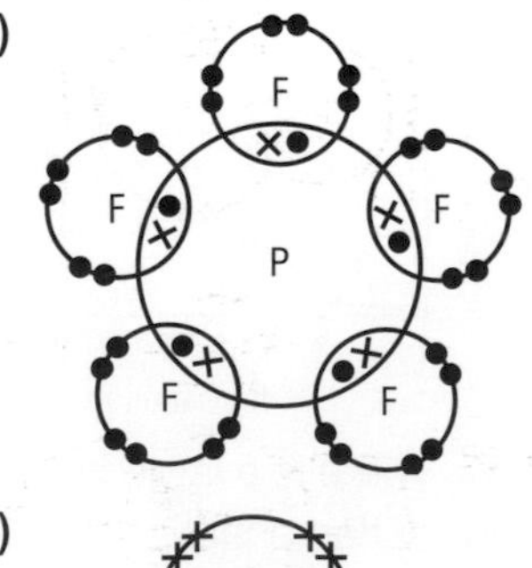

(c)

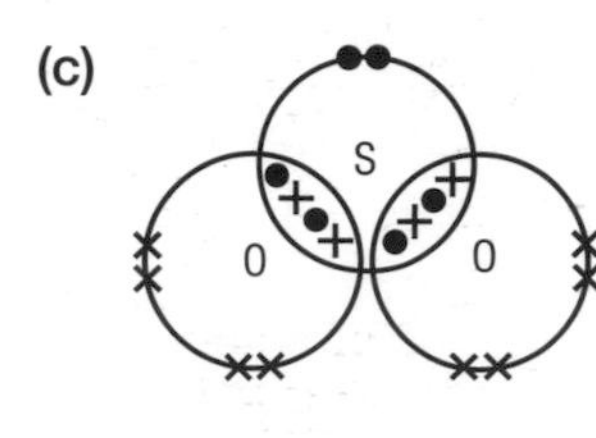

(d)

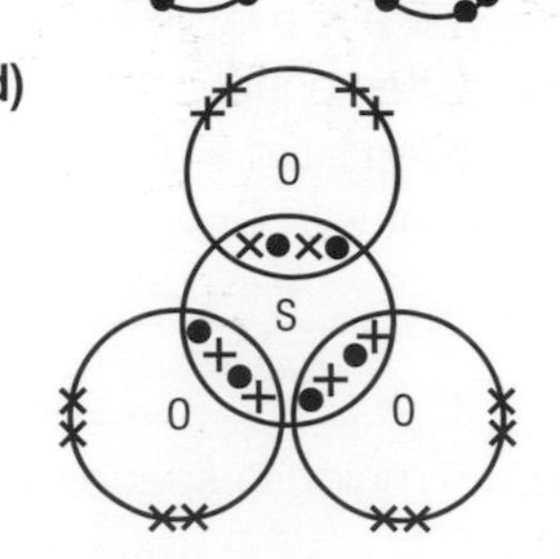

1.2.10

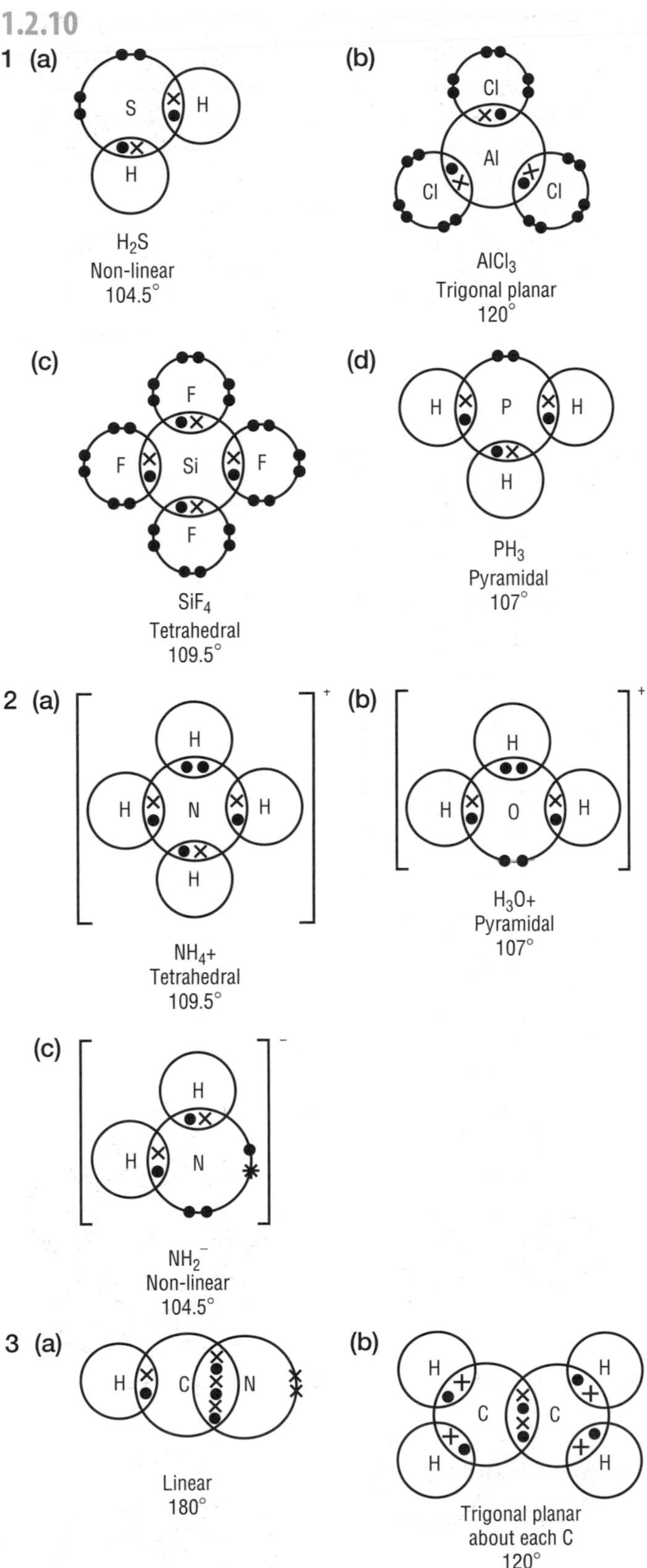

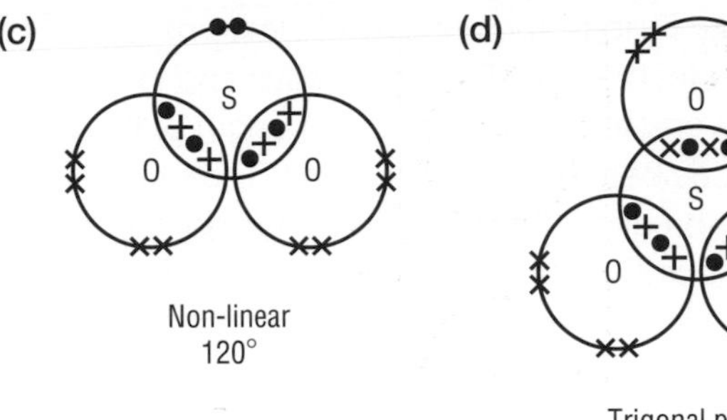

1.2.11

1 (a)

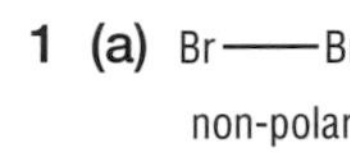

non-polar

(b)

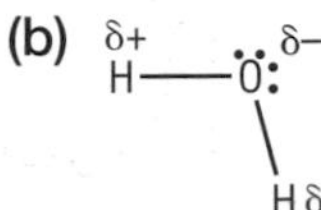

(c) O═O

non-polar

(d)

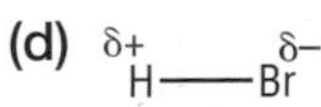

(e) δ− N, δ+ H, δ+ H, δ+ H

(f)

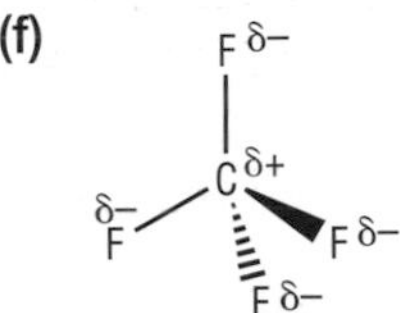

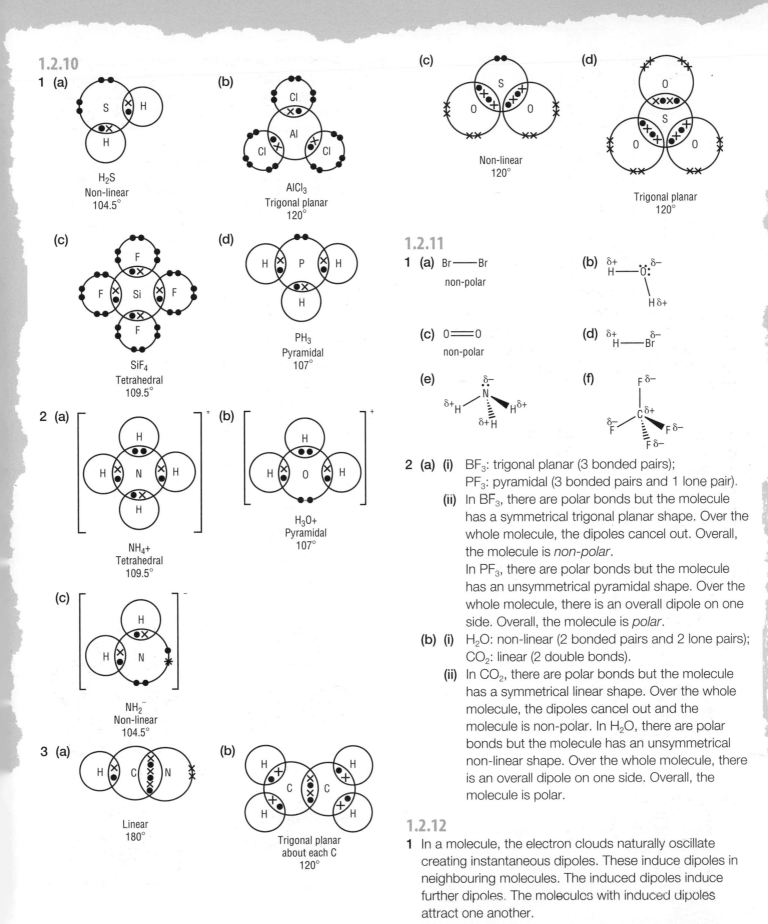

2 (a) (i) BF_3: trigonal planar (3 bonded pairs);
PF_3: pyramidal (3 bonded pairs and 1 lone pair).

(ii) In BF_3, there are polar bonds but the molecule has a symmetrical trigonal planar shape. Over the whole molecule, the dipoles cancel out. Overall, the molecule is *non-polar*.
In PF_3, there are polar bonds but the molecule has an unsymmetrical pyramidal shape. Over the whole molecule, there is an overall dipole on one side. Overall, the molecule is *polar*.

(b) (i) H_2O: non-linear (2 bonded pairs and 2 lone pairs);
CO_2: linear (2 double bonds).

(ii) In CO_2, there are polar bonds but the molecule has a symmetrical linear shape. Over the whole molecule, the dipoles cancel out and the molecule is non-polar. In H_2O, there are polar bonds but the molecule has an unsymmetrical non-linear shape. Over the whole molecule, there is an overall dipole on one side. Overall, the molecule is polar.

1.2.12

1 In a molecule, the electron clouds naturally oscillate creating instantaneous dipoles. These induce dipoles in neighbouring molecules. The induced dipoles induce further dipoles. The molecules with induced dipoles attract one another.

2 Moving from F_2 to I_2, the number of electrons increases. With more electrons, any instantaneous and induced dipoles are greater, *increasing* the strength of the van der Waals' forces between the molecules. The more energy required to break the van der Waals' forces, the higher the boiling point. As the strength of the van der Waals' forces increases in the order F_2<Cl_2<Br_2<I_2, the boiling points also increase.

1.2.13

1 CH_3OH and CH_3NH_2.

2 **(b)** 2 molecules of H_2O

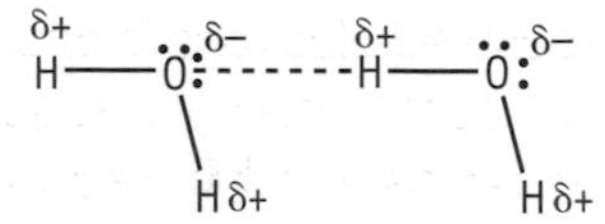

δ+ H—Ö:δ− ------ δ+ H—Ö:δ−
| H δ+ | H δ+

(b) 2 molecules of NH_3

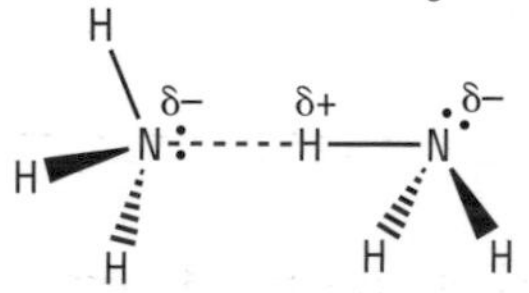

H, H, H —N:δ− ----- δ+ H—N·δ− H, H

(c) 2 molecules of C_2H_5OH

CH_3CH_2—Ö:δ− ------ δ+ H—Ö:δ−
| H δ+ | CH_2CH_3

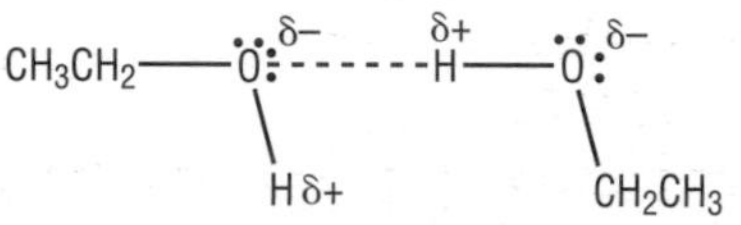

(d) 1 molecule of H_2O and 1 molecule of CH_3OH

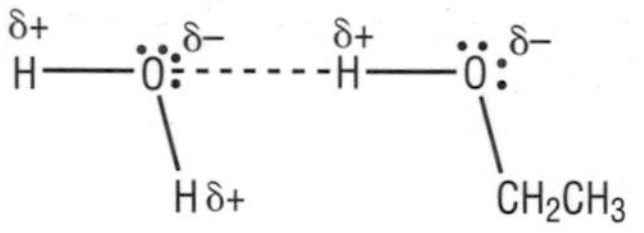

δ+ H—Ö:δ− ------ δ+ H—Ö:δ−
| H δ+ | CH_2CH_3

1.2.14

1 Metallic bonding is the attraction between positive ions and negatively charged delocalised electrons.

2 The delocalised electrons can move freely within the metallic lattice, allowing the metal to conduct electricity.

3 In metallic bonding, the electrons are delocalised. It is impossible to see which electrons *belong* to which positive ions.
In a covalent bond, the electrons are paired and shared between two atoms. The bonded pair of electrons is *localised*.

1.2.15

1 To melt or boil an ionic compound, strong ionic bonds need to be broken. This requires a large amount of energy.

2 When solid, the ions have fixed positions in a giant ionic lattice. As the ions are unable to move, the solid cannot conduct electricity. When molten or aqueous, the ions have broken free from the lattice and are able to move and conduct electricity.

3 Water has *polar* molecules which attract the ions in an ionic lattice. This attraction weakens the ionic bonds and the ions are pulled free from the lattice. Eventually, the ions become surrounded by water molecules.

1.2.16

1 **(a)** **(i)** NaCl: giant ionic lattice;
(ii) SiO_2: giant covalent lattice;
(iii) Br_2: simple molecular lattice;
(iv) C_2H_5OH: simple molecular lattice.
(b) **(i)** NaCl: high;
(ii) SiO_2: high;
(iii) Br_2: low;
(iv) C_2H_5OH: low, but higher than hydrocarbons of similar molecular mass.
(c) **(i)** NaCl: poor as solid – conducts when molten or in solution;
(ii) SiO_2: poor;
(iii) Br_2: poor;
(iv) C_2H_5OH: poor.
(d) **(i)** NaCl: good in polar solvents;
(ii) SiO_2: insoluble in polar and non-polar solvents;
(iii) Br_2: good in non-polar solvents;
(iv) C_2H_5OH: good in polar and non-polar solvents.

2 Structure. Diamond has a tetrahedral structure, bonded by covalent bonds throughout. This makes diamond very hard. Graphite has a hexagonal layer structure. Each layer is bonded together strongly by covalent bonds but the layers are weakly bonded together. This allows the layers to slide over one another and graphite is soft and slippery.
Conductivity. In diamond all the electrons are used in covalent bonds. There are no mobile electrons and diamond does not conduct. In graphite there are delocalised electrons between the layers. These can move, conducting electricity across an applied potential difference.
Solubility. The covalent bonds in the diamond and graphite structures are too strong to be broken by either polar or non-polar bonds. Diamond and graphite are non-polar so there will be no interaction with polar solvents such as water.

1.3.1

1 **(a)** Average A_r from Ca (40.1) and Ba (137.3) = 88.7, compared with actual A_r value for Sr of 87.6.
(b) Average A_r from Cl (35.5) and I (126.9) = 81.2, compared with actual A_r value for Br of 79.9.
(c) Average A_r from Li (6.9) and K (39.1) = 23.0, compared with actual A_r value for Na of 23.0.

2 (a) Zn(25) and Y(24); Ce/La(33) and Zr(32); U(40) and Sn(39); Te(43) and I(42); Tl 53, Pb 54 Th 56, Hg 52, Bi 55 and Os 51.
(b) Atomic number.
(c) Gl, Be: beryllium; Bo, B: boron.
(d) Rows: groups; columns: periods.

1.3.2

1 Gallium, scandium and germanium.
2 (a) Ar (39.9) before K (39.1);
Co (58.9) before Ni (58.7);
Te (127.6) before I (126.9).
(b) The number of protons in the nucleus.

1.3.3

1 Across the Periodic Table, for each successive element, there is one more electron in the outer shell.
2 (a) Each Group 2 element has two electrons in its outer shell: ns^2.
(b) Each Group 7 element has seven electrons in its outer shell: ns^2np^5.

1.3.4

1 Ionisation energy shows a general *increase* across each period.
Across each period, the number of protons increases, so there is more nuclear attraction acting on the electrons.
Electrons are added to the same shell, so the nuclear attraction draws the outer shell inwards slightly.
There is the same number of inner shells and so electron shielding will hardly change.
The increased nuclear charge is the significant factor.
2 Ionisation energy shows a general *decrease* down each group.
The number of shells increases. The distance of the outer electrons from nucleus increases, increasing the atomic radius. There is a weaker force of attraction on the outer electrons.
There are more inner shells, so the electrons are more effectively shielded from the nuclear charge. Again, there is less attraction.
The number of protons in the nucleus also increases, but the resulting increased attraction is far outweighed by the increase in distance and shielding.

1.3.5

1 Carbon has a giant covalent structure with strong covalent bonds between the atoms in this structure. So, a *large* amount of energy is required to break these strong forces, resulting in a high boiling point.
Nitrogen has a simple molecular structure with weak van der Waals' forces between the N_2 molecules in this structure. Thus, a *small* amount of energy is required to break these weak forces, resulting in a low melting point.
2 The metallic structure of aluminium is made from Al^{3+} ions and delocalised electrons. The metallic structure of magnesium is made from Mg^{2+} ions and delocalised electrons. Al^{3+} ions are smaller than Mg^{2+} ions. Each Al atom has three electrons in its outer shell. Each Mg atom has two electrons in its outer shell. The metallic bonding in Al is *stronger* than that in Mg. In Al, the attraction between the smaller, more positively charged ions and the greater number of delocalised electrons is *stronger* than in Mg. In Al, a larger amount of energy is required to break these strong metallic bonds than Mg, resulting in a higher melting point.

1.3.6

1 (a) Mg has changed from 0 to +2; H has changed from +1 to 0.
(b) Mg has been oxidised; H has been reduced.
(c) HCl is the oxidising agent; Mg is the reducing agent.
2 (a) (i) $Ba + 2H_2O \longrightarrow Ba(OH)_2 + H_2$.
(ii) $2Sr + O_2 \longrightarrow 2SrO$.
(b) (i) Ba has been oxidised from 0 to +2; H has been reduced from +1 to 0.
(ii) Sr has been oxidised from 0 to +2; O has been reduced from 0 to –2.

1.3.7

1 (a) $BaO + 2HCl \longrightarrow BaCl_2 + H_2O$.
(b) $RaCO_3 + 2HNO_3 \longrightarrow Ra(NO_3)_2 + CO_2 + H_2O$.
(c) $Sr + 2H_2O \longrightarrow Sr(OH)_2 + H_2$.
(d) $MgCO_3 \longrightarrow MgO + CO_2$.
(e) $CaO + 2HNO_3 \longrightarrow Ca(NO_3)_2 + H_2O$.

1.3.8

1 Br_2: 0 to Br^-: –1; I^-: –1 to I_2: 0.
2 (a) In Cl_2: one Cl atom has been oxidised from 0 to +1 in HClO;
one Cl atom has been reduced from 0 to –1 in HCl.
Cl_2 has been both oxidised and reduced (disproportionation).
(b) In Cl_2: one Cl atom has been oxidised from 0 to +1 in NaClO;
one Cl atom has been reduced from 0 to –1 in NaCl.
Cl_2 has been both oxidised and reduced (disproportionation).

1.3.9

1 A Group 7 element occurs as diatomic molecules, X_2, with a single covalent bond between the two atoms. They have comparatively low melting and boiling points, as they have a simple molecular structure with weak van der Waals' forces between the molecules The melting and boiling points increase down the group. They are oxidising agents, reacting by gaining electrons to form halide ions. The reactivity decreases down the group.

2 A halogen exists as a diatomic molecule. This has no charge (i.e.uncharged) with two halogen atoms covalently bonded together, e.g. Cl_2. A halide ion is a halogen atom that has gained an electron. A halide ion has a 1– charge, e.g. Cl^-.

3 An aqueous solution of silver nitrate, $AgNO_3$(aq), is added to the solutions.

With NaCl: a white precipitate of AgCl forms which is soluble in dilute NH_3(aq).

With NaBr: a cream precipitate of AgBr forms which is soluble in concentrated NH_3(aq).

With NaI: a yellow precipitate of AgI forms which is insoluble in concentrated NH_3(aq).

2.1.1

1 Group 4.

2 Four.

3 Carbon has the ability to form chains by bonding to other carbons. It can form long chains and rings of compounds. It can form single, double or triple bonds with another carbon atom. A carbon atom can also bond with atoms of other elements such as oxygen, hydrogen, nitrogen, phosphorus and the halogens.

2.1.2

1 (a)

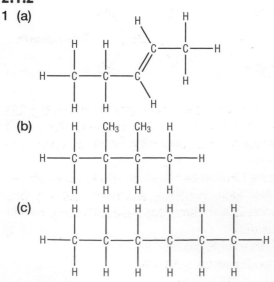

```
                   H     H
                    \    |
     H   H           C———C———H
     |   |          //   |
 H———C———C———C          H
     |   |    \
     H   H     H
```

(b)

```
     H   CH3  CH3  H
     |    |    |   |
 H———C————C————C———C———H
     |    |    |   |
     H    H    H   H
```

(c)

```
     H   H   H   H   H   H
     |   |   |   |   |   |
 H———C———C———C———C———C———C———H
     |   |   |   |   |   |
     H   H   H   H   H   H
```

(d)

```
     H   CH3  CH3  CH3  H   H
     |    |    |    |   |   |
 H———C————C————C————C———C———C———H
     |    |    |    |   |   |
     H    H    H    H   H   H
```

(e)

```
     H       CH3
     |      /
 H———C———C        H
     |    \\      |
     H      C—————C———H
           /      |
          H       H
```

2 (a) 2,3,3-trimethylpentane;

(b) 2,3-dimethylpentane;

(c) 3-methylpent-2-ene;

(d) 2,3-dimethylbut-2-ene.

2.1.3

1 (a) 2-chloropropan-1-ol;

(b) 2-bromo-1-chloro-3-iodo-2-methylbutane;

(c) 2-methylpropanal acid;

(d) ethane-1,2-diol.

2.1.4

1 (a) Empirical: CH_2Br;

Molecular: $C_2H_4Br_2$.

(b)

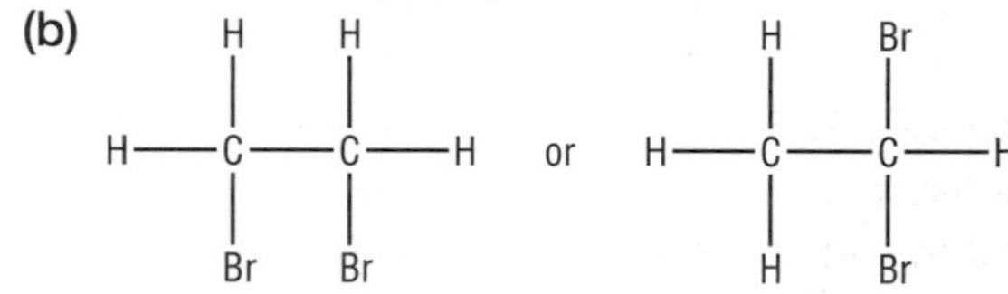

2 (a) C_5H_{10}, C_6H_{12}, C_7H_{14}, C_8H_{16}.

(b) $C_5H_{11}OH$, $C_6H_{13}OH$, $C_7H_{15}OH$, $C_8H_{17}OH$.

3 (a) $C_6H_{11}NO_3S$

(b) Mass of empirical formula unit = 177.1 which is the same as M_r value.

2.1.5

1 (a)

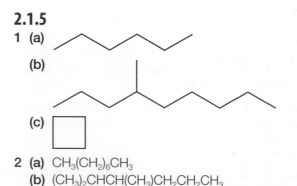

(b)

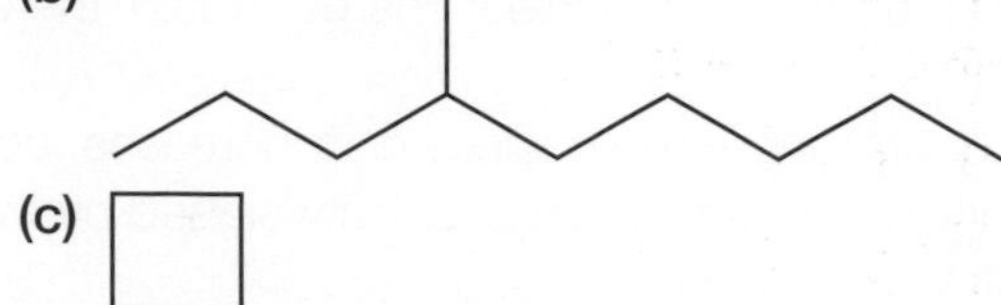

(c)

2 (a) $CH_3(CH_2)_6CH_3$

(b) $(CH_3)_2CHCH(CH_3)CH_2CH_2CH_3$

3 (a) 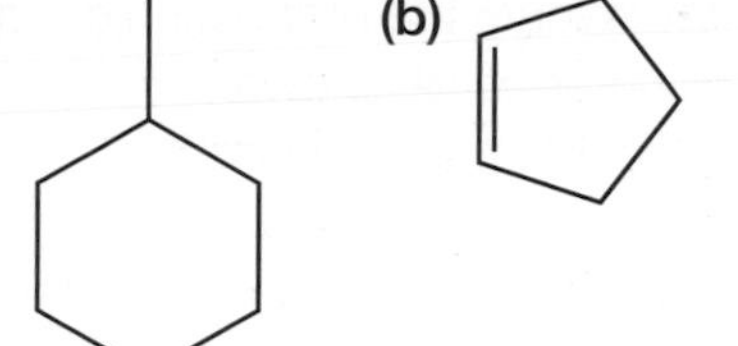 **(b)**  **(c)** (methylbenzene structure)

2.1.6

1 HO–$CH_2CH_2CH_2CH_2CH_2CH_3$ (skeletal formula)

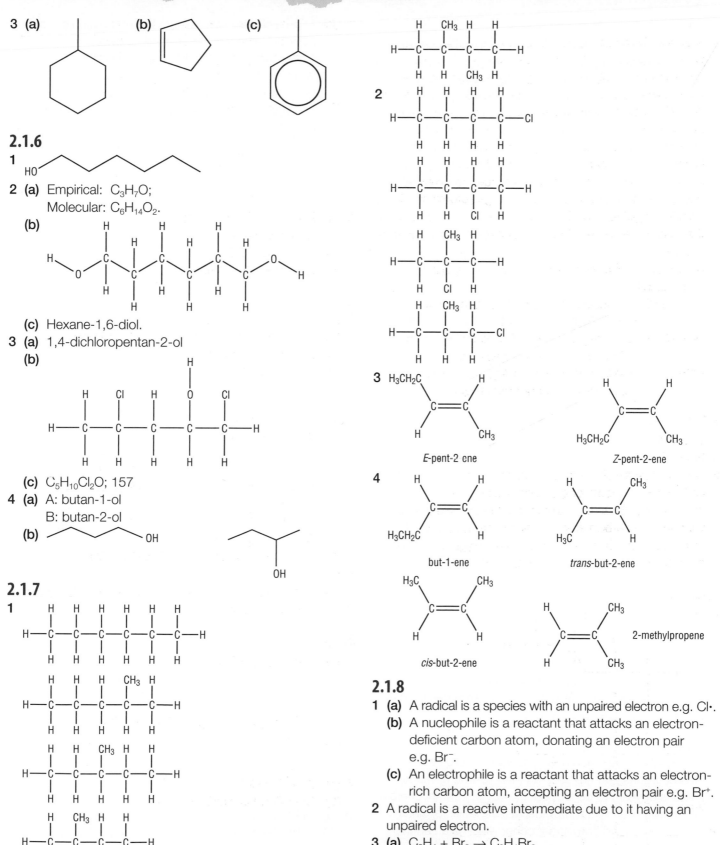

2 (a) Empirical: C_3H_7O;
Molecular: $C_6H_{14}O_2$.

(b)

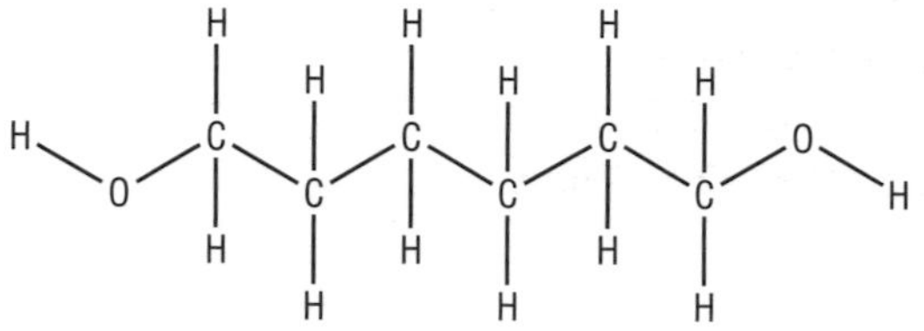

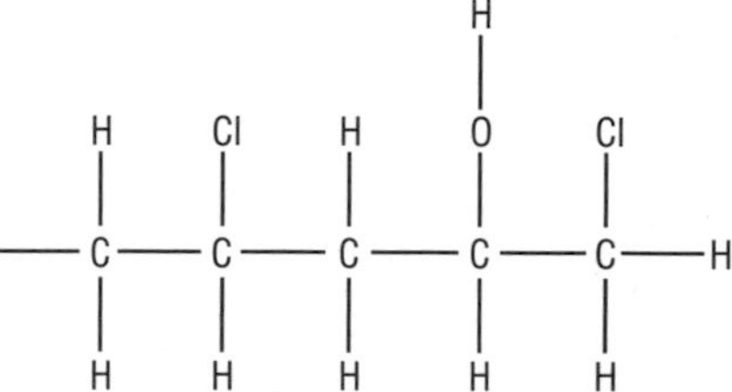

(c) Hexane-1,6-diol.

3 (a) 1,4-dichloropentan-2-ol

(b) $CH_2Cl–CH(OH)–CH_2–CHCl–CH_3$ (displayed formula)

(c) $C_5H_{10}Cl_2O$; 157

4 (a) A: butan-1-ol
B: butan-2-ol

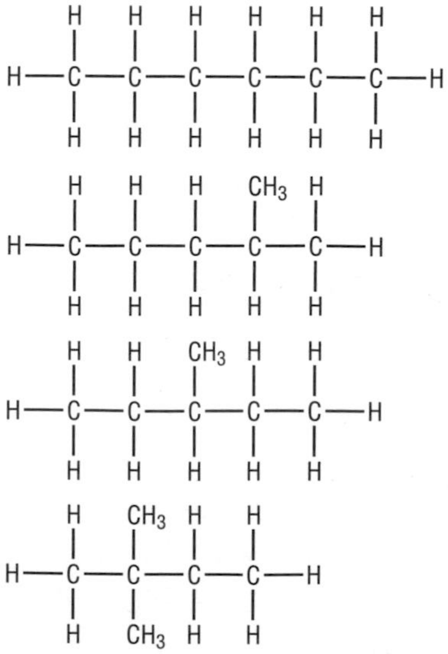

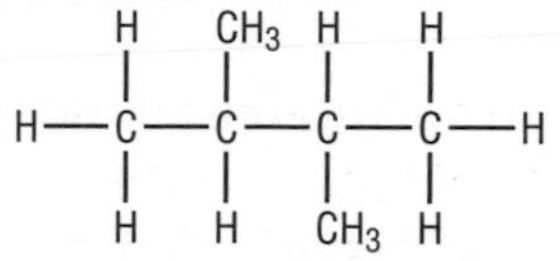

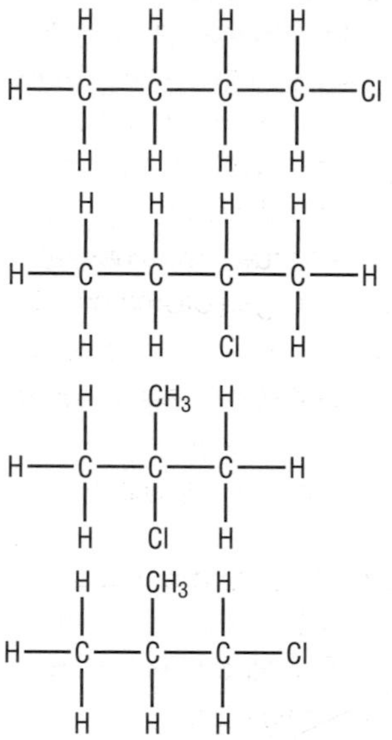

(b)  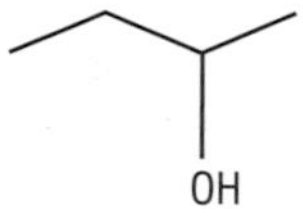 $CH_3CH_2CH_2CH_2OH$ $CH_3CH_2CH(OH)CH_3$ (skeletal formulae)

2.1.7

1 $CH_3CH_2CH_2CH_2CH_2CH_3$; $CH_3CH_2CH_2CH(CH_3)CH_3$; $CH_3CH_2CH(CH_3)CH_2CH_3$; $CH_3C(CH_3)_2CH_2CH_3$; $CH_3CH(CH_3)CH(CH_3)CH_3$ (displayed formulae)

2 $CH_3CH_2CH_2CH_2Cl$; $CH_3CH_2CHClCH_3$; $CH_3CCl(CH_3)CH_3$; $CH_3CH(CH_3)CH_2Cl$ (displayed formulae)

3 H_3CH_2C(H)C=C(H)CH_3 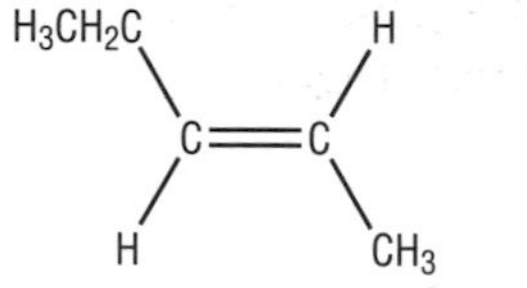 *E*-pent-2-ene; H_3CH_2C(H)C=C(H)CH_3 *Z*-pent-2-ene

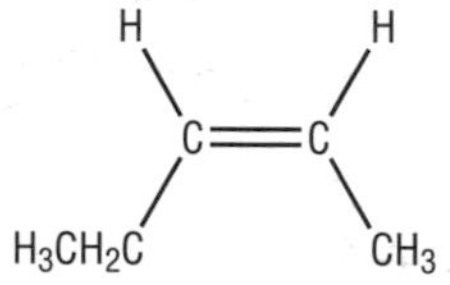

4 but-1-ene; 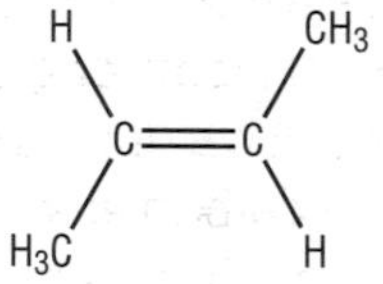 *trans*-but-2-ene; 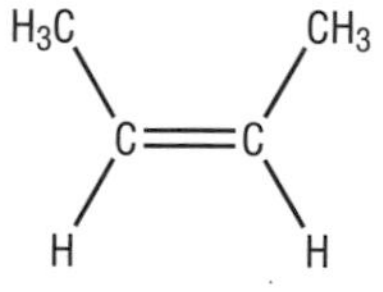 *cis*-but-2-ene; 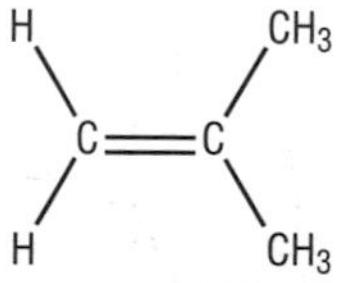 2-methylpropene

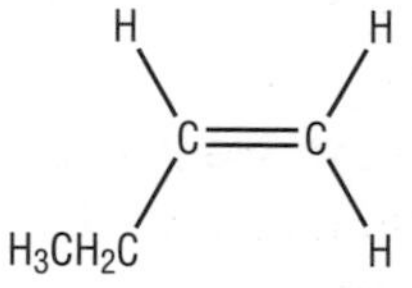

2.1.8

1 (a) A radical is a species with an unpaired electron e.g. Cl•.

(b) A nucleophile is a reactant that attacks an electron-deficient carbon atom, donating an electron pair e.g. Br^-.

(c) An electrophile is a reactant that attacks an electron-rich carbon atom, accepting an electron pair e.g. Br^+.

2 A radical is a reactive intermediate due to it having an unpaired electron.

3 (a) $C_2H_4 + Br_2 \rightarrow C_2H_4Br_2$

(b) $C_2H_5Br + OH^- \rightarrow C_2H_5OH + Br^-$

(c) $C_2H_5OH \rightarrow C_2H_4 + H_2O$

2.1.9

1 Fractional distillation separates the fractions in a mixture of liquids on the basis of their different boiling points.

2 **(a)** The *longer* the carbon chain, the *higher* the boiling point. This is explained by the increasing van der Waals' forces between the molecules, as the chain lengthens.

(b) The *more* branched the alkanes, the *lower* the boiling points. Branched alkanes have fewer points of contact and thus reduced van der Waals' forces of attraction.

Branched alkanes cannot be packed as close to each other as unbranched alkanes of similar molecular mass, which also reduces the intermolecular forces.

2.1.10

1 Plentiful supply: $C_3H_8(g) + 5O_2(g) \longrightarrow 3CO_2(g) + 4H_2O(l)$

Limited supply: $C_3H_8(g) + 3\frac{1}{2}O_2(g) \longrightarrow 3CO(g) + 4H_2O(l)$

Volume of oxygen: $= 5 \times 24\ dm^3$

$= 120\ dm^3$.

2 $C_{10}H_{22} \longrightarrow C_2H_4 + C_8H_{18}$.

3

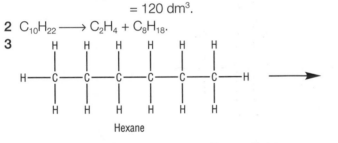

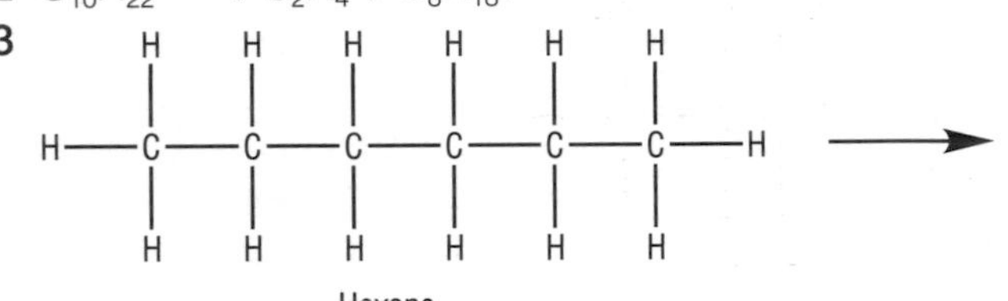

Hexane

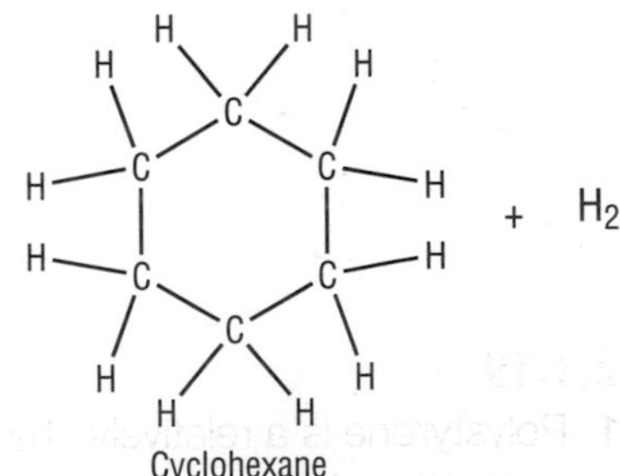

Cyclohexane + H_2

4 Cracking leads to the formation of shorter-chained alkanes, which are better fuels with higher octane numbers. Alkenes are also produced which can be used in the manufacture of polymers.

2.1.11

1 Crude oil is a finite and non-renewable source of hydrocarbons that are used as fuels. Crude oil takes millions of years to be formed. However, ethanol offers an alternative source of motor fuel. It is *renewable*, since it can be produced from sugar cane, which is an annual crop.

2 Brazil has a climate that favours the production of sugar cane. Brazil has few oil reserves of its own.

3 Fossil fuels are hydrocarbons, primarily coal and petroleum (fuel oil or natural gas), formed from the fossilised remains of dead organisms. Ethanol obtained from the hydration of ethane is non-renewable, as ethane is derived from crude oil. However, when ethanol is made via fermentation, its raw material is now sugar which is renewable.

2.1.12

1 **(a)** $Br_2 \longrightarrow 2Br\cdot$

(b) $C_3H_8 + Br\cdot \longrightarrow C_3H_7\cdot + HBr$

$C_3H_7\cdot + Br_2 \longrightarrow C_3H_7Br + Br\cdot$

2 $C_2H_5\cdot + C_2H_5\cdot \longrightarrow C_4H_{10}$

2.1.13

1 C_4H_8.

2 **(a)**

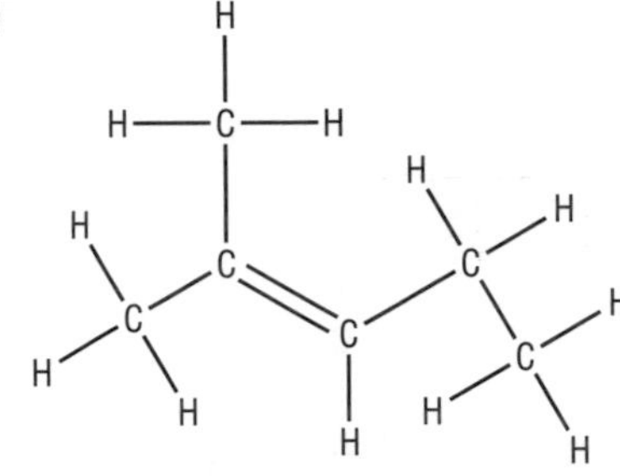

2-methylpent-2-ene

(b)

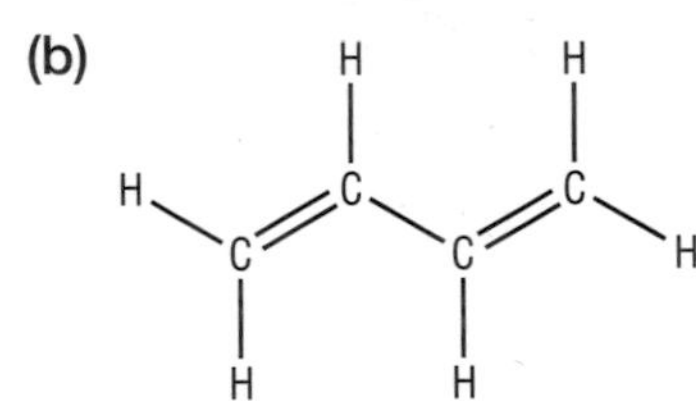

Buta-1,3-diene

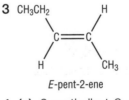

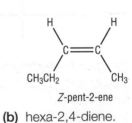

3 *E*-pent-2-ene (CH₃CH₂ and H on one carbon, H and CH₃ on the other, CH₃CH₂ and CH₃ on opposite sides); *Z*-pent-2-ene (H and H on the same side, CH₃CH₂ and CH₃ on the same side)

4 **(a)** 2-methylbut-2-ene; **(b)** hexa-2,4-diene.

2.1.14

1 The π-bond is weaker than the σ-bond and so the π-bond breaks more easily.

2 **(a)** $CH_3CH{=}CHCH_3 + H_2 \longrightarrow CH_3CH_2CH_2CH_3$ butane

(b) $CH_3CH{=}CHCH_3 + Br_2 \longrightarrow CH_3CHBrCHBrCH_3$ 2,3-dibromobutane

2.1.15

1 (a)

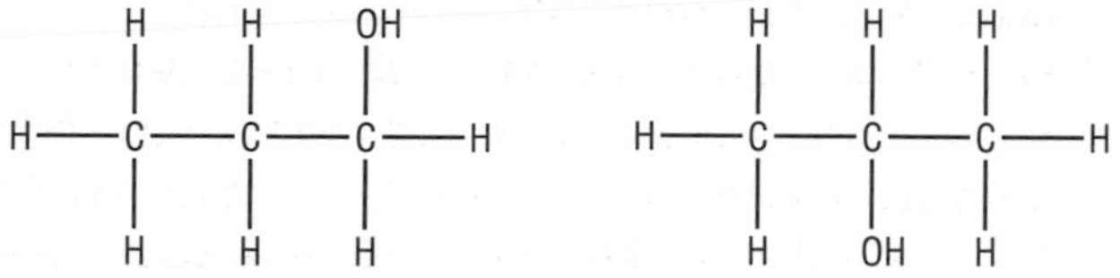

(b) H_2O

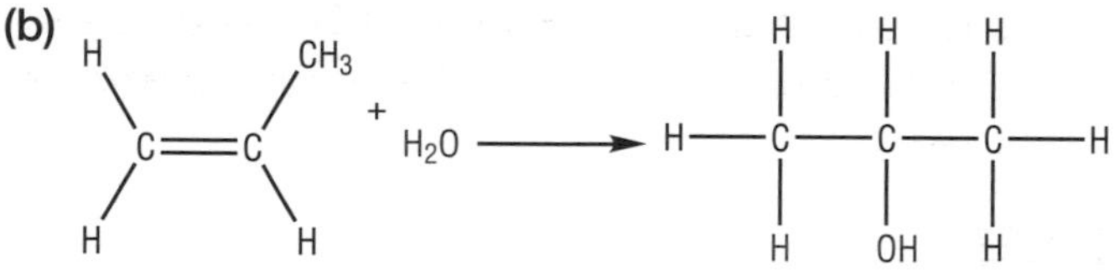

2

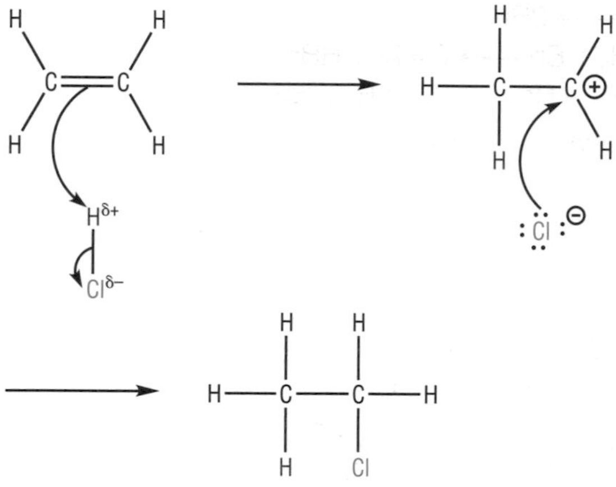

2.1.16

1

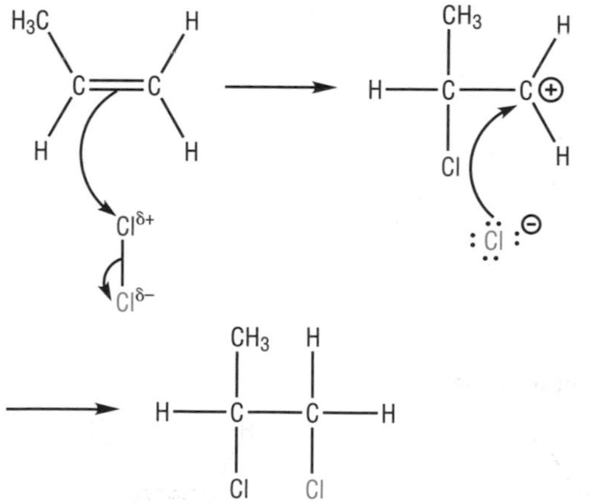

2 (a) Reaction A – hydrogen and nickel catalyst.
Reaction B – H_2O, phosphoric acid catalyst, high temperature.

(b) 2-chlorobutane.

3 (a)

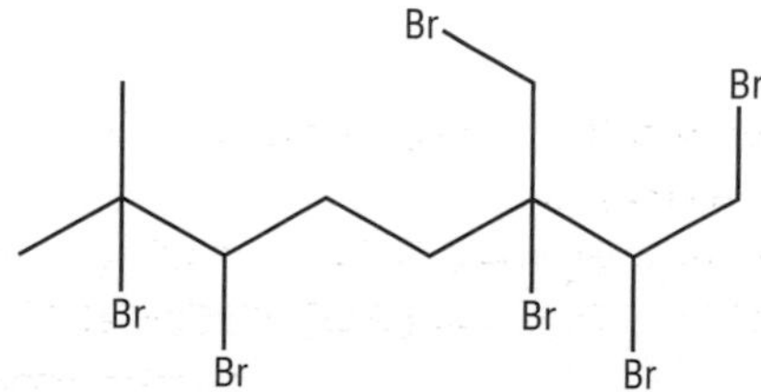

(b) (i) $C_{10}H_{16}Br_6$; (ii) $C_5H_8Br_3$

2.1.17

1 There are many possibilities including: 1,2-dichloroethane; ethane-1,2-diol; ethanoic acid; ethanol; and poly(ethene).

2 Greater control over the product formed leads to more straight-chained polymers.

3

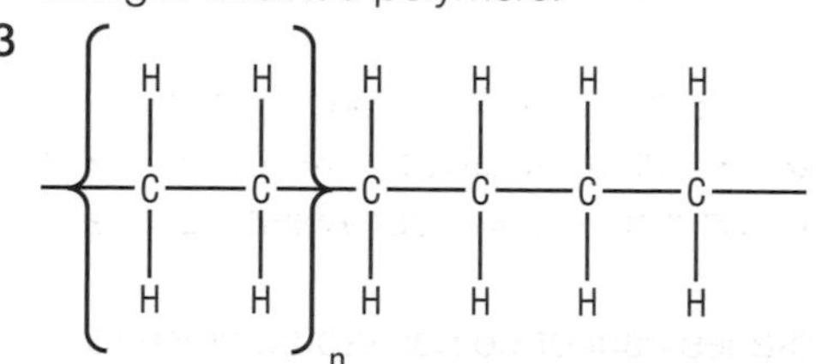

2.1.18

1

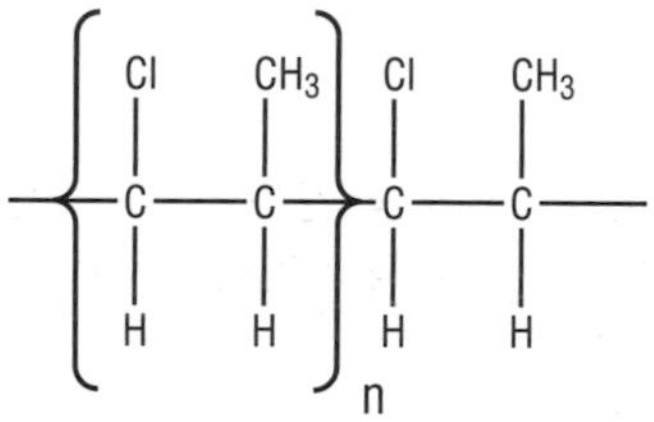

2

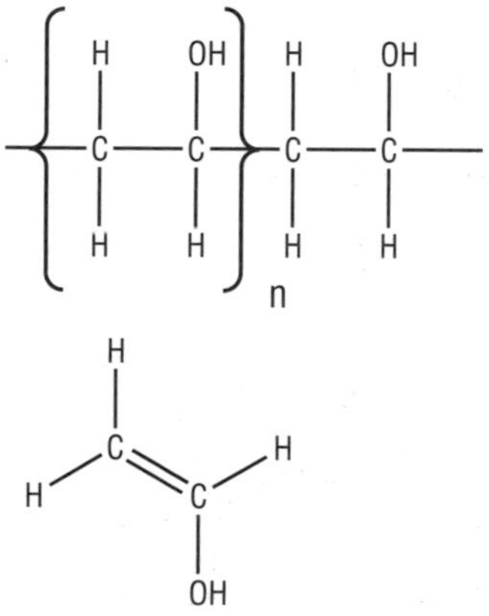

2.1.19

1 Polystyrene is a relatively cheap plastic used in foam packaging, insulation, model-making and the food retail trade.

2 A non-biodegradable material does not break down naturally or safely in the environment by biological means.

3 The non-biodegradability of polystyrene means that waste polystyrene ends up in landfill sites, which are rapidly running out of space.

4 HCl and CO.

2.1.20

1 (a) 16.4 kg.

(b) $2.97 \times 10^8\ dm^3$.

2.2.1

1 The disadvantages of the hydration of ethene include higher temperatures and pressures for this process, resulting in higher energy costs. Furthermore, plants that are required to withstand high pressures are costly to build. The process also relies on crude oil, which is non-renewable.
2 Fractional distillation followed by cracking.
3 Ethanol is a *renewable* energy source which is made from the fermentation of sugar. Sugar cane is an annual crop.
4 Brazil has no natural deposits of crude oil, but does have a climate which favours the growth of sugar cane.
5 $CH_3OH(l) + 1\frac{1}{2}O_2(g) \longrightarrow CO_2(g) + 2H_2O(l)$.

2.2.2

1 **(a)** propan-1-ol; primary.
(b) butan-2-ol; secondary.
(c) 3-ethylpentan-3-ol; tertiary.
(d) pentan-2-ol; secondary.

2

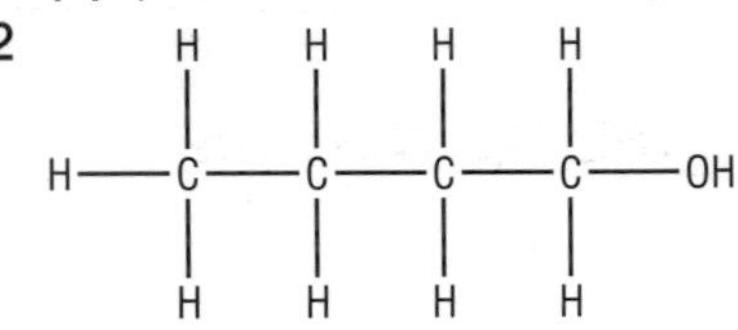

Butan-1-ol
Primary

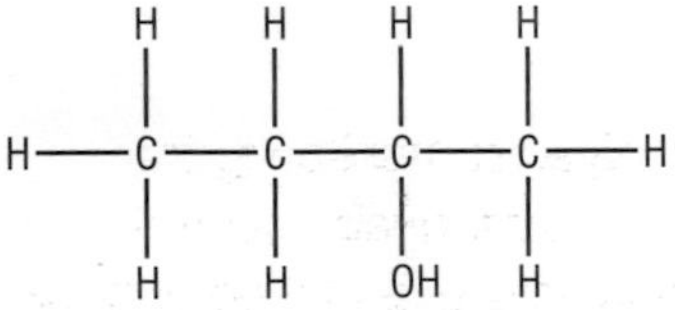

Butan-2-ol
Secondary

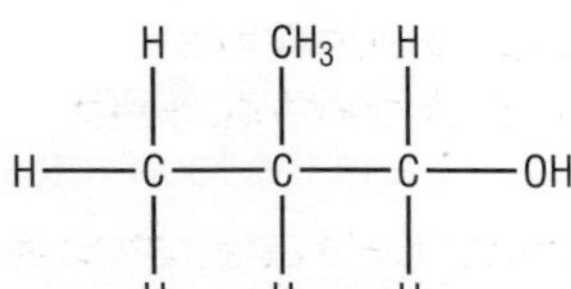

2-methylpropan-1-ol
Primary

2-methylpropan-2-ol
Tertiary

3

Hydrogen bond

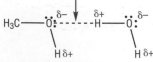

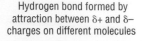

Hydrogen bond formed by attraction between δ+ and δ– charges on different molecules

Solubility of alcohols *decreases* as the chain length *increases*. Alcohols are soluble in water due to their ability to form hydrogen bonds with the water molecules. However, the hydrogen bonds only occur between the OH group and the water molecule. As the carbon chain length increases the alcohols become less polar and thus less soluble. So, since methanol has a shorter carbon chain than pentan-1-ol, it will be more soluble in water.

2.2.3

1 $CH_3CH_2CH_2CH_2CH_2OH(l) + 7\frac{1}{2}O_2(g) \longrightarrow 5CO_2(g) + 6H_2O(l)$

$CH_3CH_2CH_2CH_2CH(CH_3)CH_2OH(l) + 10\frac{1}{2}O_2(g) \longrightarrow 7CO_2(g) + 8H_2O(l)$

2

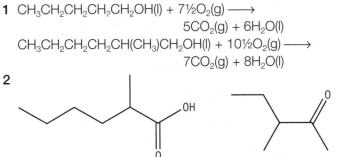

2-methylhexanoic acid

3-methylpentan-2-one

3 Continual boiling and condensation of a solvent.

2.2.4

1 $CH_3CH_2CH_2CH_2COOH(l) + CH_3CH_2OH(l) \longrightarrow CH_3CH_2CH_2CH_2COOCH_2CH_3(l) + H_2O(l)$

Organic product: ethyl pentanoate.

2

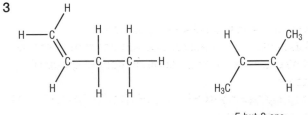

3

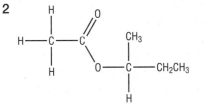

But-1-ene

E-but-2-ene

Z-but-2-ene

2.2.5

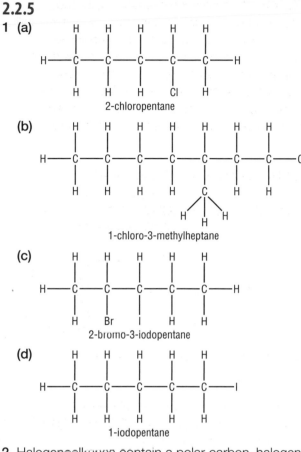

1 (a) 2-chloropentane

(b) 1-chloro-3-methylheptane

(c) 2-bromo-3-iodopentane

(d) 1-iodopentane

2 Halogenoalkanes contain a polar carbon–halogen bond, which arises because a halogen atom is more electronegative than a carbon atom.

3 The halogenoalkanes contain a slightly positive carbon atom, part of the carbon–halogen bond. This carbon atom can be attacked by an electron-rich nucleophile.

2.2.6

1 (a) A nucleophile is an electron pair donor.

(b) Substitution is a reaction in which an atom or group of atoms is replaced by another atom or group of atoms.

2 $CH_3CH_2CH_2CH_2Br + KOH \longrightarrow CH_3CH_2CH_2CH_2OH + KBr$
Alcohol: butan-1-ol.

3 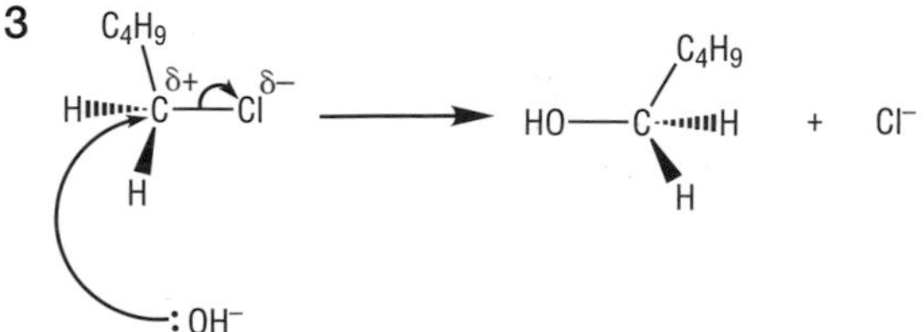

2.2.7

1 Carbon–fluorine bonds are very strong, which makes poly(tetrafluoroethene) (PTFE or Teflon) inert and resistant to chemical attack. PTFE's chemical inertness along with its heat resistance, electrical insulating properties and non-stick surface makes it an excellent choice for non-stick pans.

2 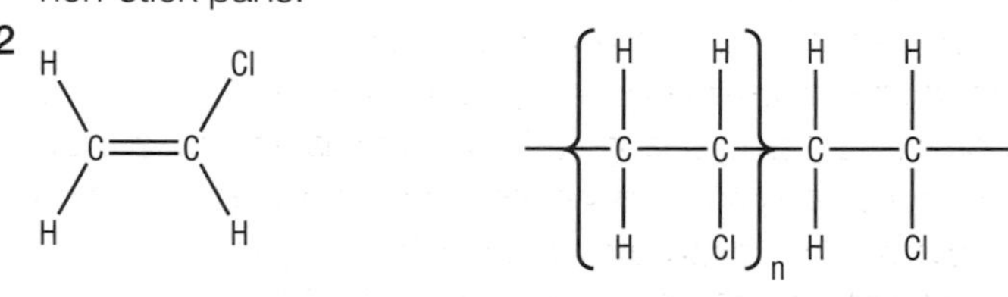

3 Uses include: electrical insulation in wiring; plastic window frames; and guttering and drain pipes.

2.2.8

1 Percentage yield: 98.5%.

2 Percentage yield: 83.3%.

2.2.9

1 $CH_3CH(I)CH_3 + NaOH \longrightarrow CH_3CH(OH)CH_3 + NaI$
Atom economy: 28.6%.

2 Atom economy: 75.7%.

2.2.10

1 The absorption of infrared radiation causes the bonds to vibrate.

2 Stretching and bending.

3 Applications include: breathalysers; monitoring the degree of unsaturation in polymers; quality control applications in perfume manufacture; and drug analysis.

2.2.11

1 Carbonyl, C=O.

2 (a) pentan-2-ol: strong broad O–H absorption at 3200–3500 cm^{-1}, indicating an alcohol functional group.

(b) 2-hydroxypentanal: strong sharp C=O absorption at 1640–1750 cm^{-1}, indicating a carbonyl group and a strong broad absorption at 3200–3550 cm^{-1}, indicating an alcohol functional group.

(c) butanoic acid: strong sharp C=O absorption at 1640–1750 cm^{-1}, indicating a carbonyl group and a strong broad O–H absorption at 2500–3300 cm^{-1}, indicating an OH group in a carboxylic acid.

2.2.12

1 6.93.

2 87.71.

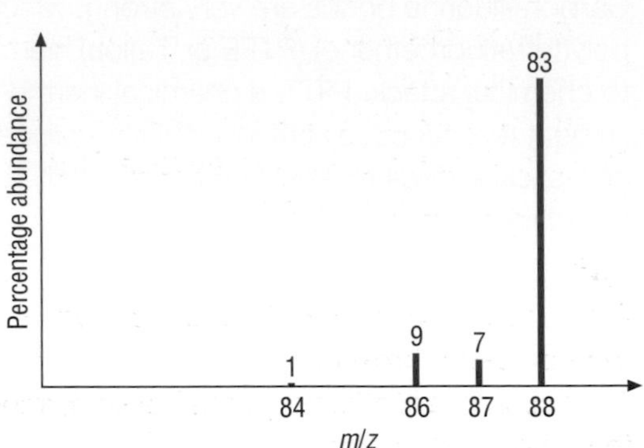

2.2.13

1 (a) $CH_3CH_2CH_2OH + e^- \longrightarrow CH_3CH_2CH_2OH^+ + 2e^-$
molecular ion peak: $m/z = 60$.

(b) $CH_3CH_2CH_2CH_3 + e^- \longrightarrow CH_3CH_2CH_2CH_3^+ + 2e^-$
molecular ion peak: $m/z = 58$.

(c) $CH_3CH_2CH_2CH_2CH_2CH_2CH_2CH_3 + e^- \longrightarrow CH_3CH_2CH_2CH_2CH_2CH_2CH_2CH_3^+ + 2e^-$
molecular ion peak: $m/z = 114$.

2 **A:** molecular ion peak, $m/z = 70$.
B: molecular ion peak, $m/z = 100$.

2.2.14

1 **A**: $m/z = 70$, $H_2C{=}CHCH_2CH_2CH_3^+$.
B: $m/z = 55$, $H_2C{=}CHCH_2CH_2^+$.
C: $m/z = 27$, $H_2C{=}CH^+$.

2 (a) molecular ion peak is the furthest to the right at $m/z = 86$
molecular mass = 86
molecular formula = C_6H_{14}

(b)

(c)

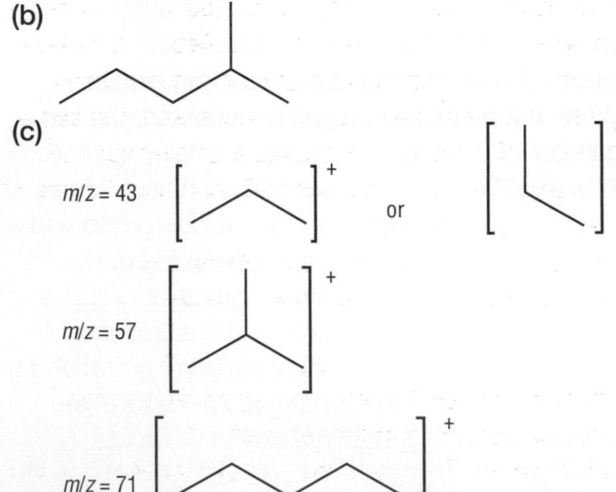

2.2.15

1 (a) Ethane and chlorine in ultraviolet radiation.

Initiation $Cl_2 \longrightarrow 2Cl\cdot$

Propagation $Cl\cdot + C_2H_6 \longrightarrow C_2H_5\cdot + HCl$
$C_2H_5\cdot + Cl_2 \longrightarrow C_2H_5Cl + Cl\cdot$

Termination $C_2H_5\cdot + Cl\cdot \longrightarrow C_2H_5Cl$
$Cl\cdot + Cl\cdot \longrightarrow Cl_2$
$C_2H_5\cdot + C_2H_5\cdot \longrightarrow C_4H_{10}$

(b) C_3H_7, H, H, $\delta+$ C–I $\delta-$, $:OH^-$ → HO—C(C_3H_7)(H)(H) + I^-

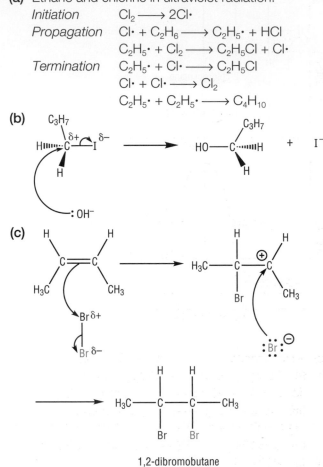

(c)

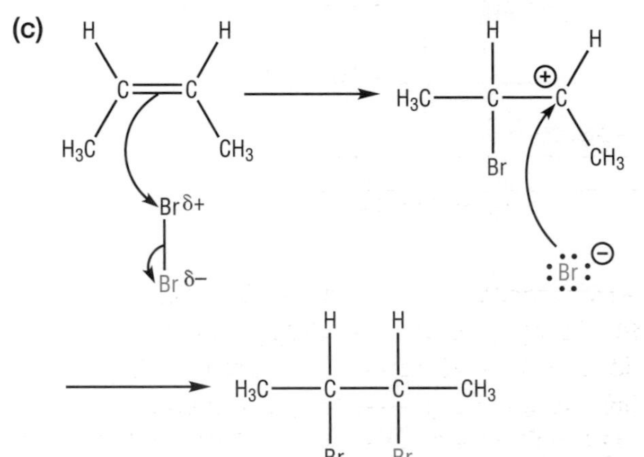

1,2-dibromobutane

2.3.1

1 Enthalpy, *H*, is the heat content that is stored in a chemical system.

2 A reaction is *exothermic* if the total enthalpy of the products is *less* than the total enthalpy of the reactants. A reaction is *endothermic* if the total enthalpy of the products is *greater* than the total enthalpy of the reactants.

2.3.2

1 4450 kJ of energy is released.

2 Photosynthesis is the reverse reaction to respiration.

3 Light from the Sun.

2.3.3

1

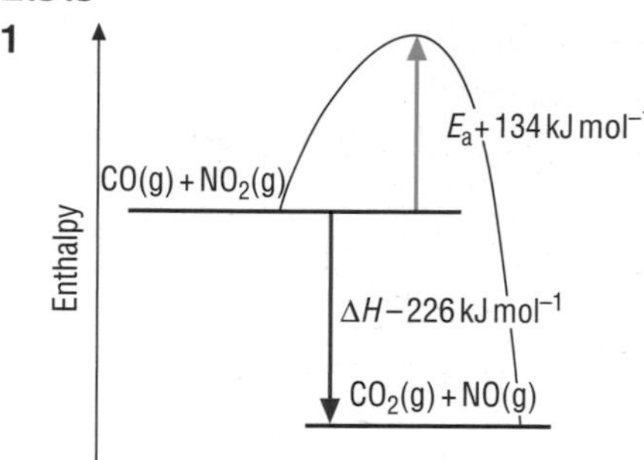

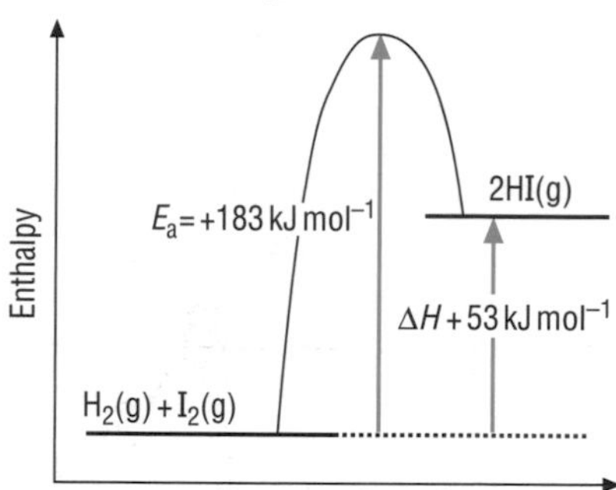

2.3.4

1 **(a)** −46 kJ mol^{-1}. **(b)** +29 kJ mol^{-1}.

2 **(a)** **(i)** $CH_4(g) + 2O_2(g) \longrightarrow CO_2(g) + 2H_2O(l)$.
(ii) $C_3H_8(g) + 5O_2(g) \longrightarrow 3CO_2(g) + 4H_2O(l)$.
(iii) $CS_2(l) + 3O_2(g) \longrightarrow CO_2(g) + 2SO_2(g)$.
(iv) $CH_3OH(l) + 1\frac{1}{2}O_2(g) \longrightarrow CO_2(g) + 2H_2O(l)$.
(v) $C_2H_5OH(l) + 3O_2(g) \longrightarrow 2CO_2(g) + 3H_2O(l)$.

3 **(a)** **(i)** $2C(s) + 2H_2(g) \longrightarrow C_2H_4(g)$.
(ii) $2C(s) + 3H_2(g) \longrightarrow C_2H_6(g)$.
(iii) $2C(s) + 3H_2(g) + \frac{1}{2}O_2(g) \longrightarrow C_2H_5OH(l)$.
(iv) $Ca(s) + \frac{1}{2}O_2(g) \longrightarrow CaO(s)$.
(v) $2Al(s) + 1\frac{1}{2}O_2(g) \longrightarrow Al_2O_3(s)$.

2.3.5

1 Amount of Zn reacted = 0.327/65.4 = 5.00×10^{-3} mol.
Energy released = 55 × 4.18 × 9.5 = 2184 J = 2.184 kJ.
$\Delta H_r = -2.184/5.00 \times 10^{-3} = -436.8$ kJ mol^{-1}.

2 Amount of HCl and NaOH reacted = 2.00 × 25/1000 = 5.00×10^{-2} mol.
Energy released = (25+25) × 4.18 × 12.0 = 2508 J = 2.508 kJ.
$\Delta H_r = -2.508/5.00 \times 10^{-2} = -50.16$ kJ mol^{-1}.

2.3.6

1 **(a)** Amount of $CH_3CH_2CH_2CH_2OH$ reacted = 1.51/74.0 = 2.04×10^{-2} mol.
Energy released = 300 × 4.18 × 42 = 52 668 J = 52.668 kJ.
$\Delta H = -52.668/2.04 \times 10^{-2} = -2580$ kJ mol^{-1}.

(b) Amount of decane, $C_{10}H_{22}$ reacted = 0.826/142.0 = 5.82×10^{-3} mol.
Energy released = 180 × 4.18 × 49.0 = 36 868 J = 36.868 kJ.
$\Delta H = -36.868/5.82 \times 10^{-3} = -6330$ kJ mol^{-1}.

2 **(i)** Some heat is lost to the surroundings.
(ii) Incomplete combustion.

2.3.7

1 A reaction is *exothermic* if the bonds formed are *stronger* than the bond broken.
A reaction is *endothermic* if the bonds formed are *weaker* than the bond broken.

2 **(a)** [612 + 436] − [(2 × 413) +347] = −125 kJ mol^{-1}
(b) [(2 × 347) + (8 × 413) + (5 × 497)] − [(6 × 805) + (8 × 463)] = −2051 kJ mol^{-1}
(c) [945 + (3 × 436)] − [(6 × 391)] = −93 kJ mol^{-1}

2.3.8

1 **(a)** [(4 × −394) + (5 × −286)] − [−2877] = −129 kJ mol^{-1}
(b) [(2 −394) + (3 × −286)] − [−1367] = −279 kJ mol^{-1}

2.3.9

1 **(a)** (2 × 33) − (2 × 90) = −114 kJ mol^{-1}
(b) [(4 × 90) + (6 × −286)] − [(4 × −46)] = −1172 kJ mol^{-1}
(c) [(4 × −174)] − [(2 × −286) − (4 × 33)] = −256 kJ mol^{-1}

2 **(a)** **(i)** $5C(s) + 6H_2(g) \longrightarrow C_5H_{12}(l)$
(ii) $C_6H_{14}(l) + 9\frac{1}{2}O_2(g) \longrightarrow 6CO_2(g) + 7H_2O(l)$
(b) **(i)** (5 × −394 + 6 × −286) − (−173) = −3513 kJ mol^{-1}
(ii) (6 × −394 + 7 × −286) − (−4163) = −203 kJ mol^{-1}

2.3.10

1 **(a)** When the concentrations of the reactants are increased, there are *more* molecules in the *same* volume. More collisions can take place in a certain length of time and the rate of reaction increases.
(b) When the pressure of a *gas* is increased, the *same* number of molecules occupies a *smaller* volume. There will be more frequent collisions and the rate of reaction will increase.

2 The activation energy for a chemical reaction is the energy required to start a reaction by breaking bonds.

2.3.11

1 A catalyst increases the *rate* of a chemical reaction without being used up in the process.

2 A catalyst lowers the activation energy by providing an alternative route for the reaction to follow.

3 The Haber process for ammonia production:

$$N_2(g) + 3H_2(g) \rightleftharpoons 2NH_3(g)$$

iron catalyst

4

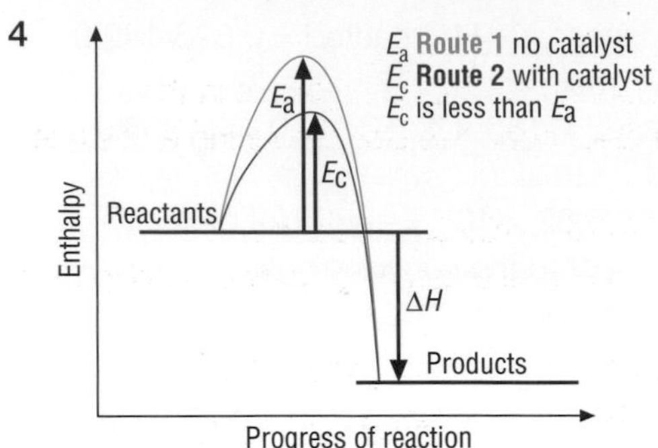

2.3.12

1 Making fertilisers, explosives, manufacture of nitric acid.
2 The minimum energy required to start a reaction by breaking the bonds in the reactant molecules.
3 An enzyme can only catalyse a particular reactant molecule (substrate).
4 Shorter washing times, reduced energy consumption, reduced water consumption.

2.3.13

1 The activation energy is the energy required to start a reaction by breaking bonds. Only the molecules with energy *greater* than the activation energy, E_a, are able to react.
2 An increase in temperature increases the reaction rate. The molecules move faster with more kinetic energy, resulting in more collisions in a certain length of time. As the molecules have more energy, more of the collisions exceed the activation energy and lead to a chemical reaction.
3 **(a), (b)**

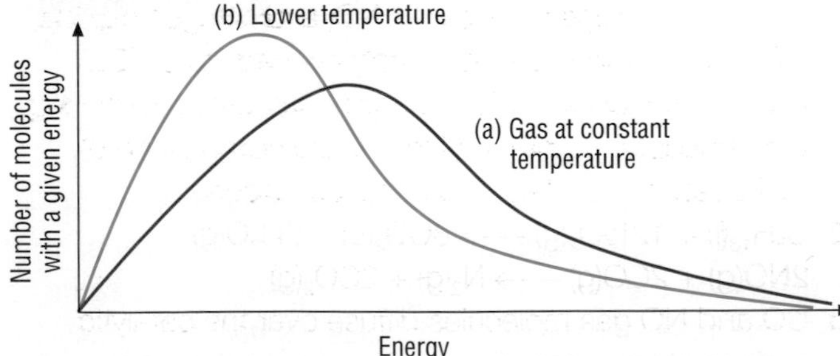

2.3.14

1 Increasing the concentration of chloride ions causes the position of equilibrium to move in the direction that decreases this increased concentration. The system opposes the change by *decreasing* the concentration of chloride ions, by removing them. The position of equilibrium moves to the *right-hand* side, forming more products. You would observe the pale blue colour of $[Cu(H_2O)_6]^{2+}$ disappearing, replaced by the yellow colour of $[CuCl_4]^{2-}$. In practice, the equilibrium colour will be green (blue and yellow), and the colour will turn from blue/green to green/yellow.

2 **(a)** Increasing the total pressure shifts the position of equilibrium to the side with *fewer* gas molecules, as this will *decrease* the pressure. In this equilibrium, there are nine moles of gas on the left-hand side and 10 moles of gas on the right-hand side. The equilibrium position will move towards the reactants on the *left-hand* side, $NH_3(g)$ and $O_2(g)$.
(b) Removing NO(g) reduces its concentration and causes the position of equilibrium to move in the direction that replaces the NO(g) concentration. The position of equilibrium moves to the *right-hand* side forming more products, NO(g) and $H_2O(g)$.

2.3.15

1 Increasing the temperature of the system causes the position of equilibrium to oppose the change and move in the direction that *decreases* the temperature, the endothermic (ΔH: +ve) direction. In this reaction, the equilibrium moves to the *right* ($\Delta H = +132\ kJ\ mol^{-1}$). The equilibrium mixture will contain more CO(g) and $H_2(g)$ and less C(s) and $H_2O(g)$.
2 **(a)** The equilibrium will oppose the change by *increasing* the pressure. This will be achieved by moving to the side with a *greater* number of moles. As there are three moles on the left-hand side for every two moles on the right-hand side, the equilibrium moves to the *left*, towards $SO_2(g)$ and $O_2(g)$.
(b) High pressure and low temperature.
(c) A catalyst does *not* affect the equilibrium position. A catalyst speeds up *both* forward and reverse reactions, so equilibrium is achieved more quickly.

2.4.1

1 Water vapour, carbon dioxide and methane.
2 The molecules of greenhouse gases can absorb infrared radiation, causing the molecule to vibrate. The vibrating molecule then emits some of this energy in the form of radiation. This radiation may then be absorbed by other greenhouse gas molecules or at the Earth's surface. This absorption–emission process enables the heat to be kept close to the planet's surface.
3 A gas's contribution to global warming is related to: the gas's lifetime in the atmosphere; the ability of the gas to absorb infrared radiation; and its atmospheric concentration.

2.4.2

1 Any three from the following:
In the Antarctic, the ice sheet is melting into the oceans.
Tropical areas are experiencing more frequent and destructive storms and floods.

In Europe, glaciers are disappearing and there have been extreme heat waves.
In the Arctic, ice and permafrost are disappearing.

2 (a) $C_8H_{18} + 12\frac{1}{2}O_2 \longrightarrow 8CO_2 + 9H_2O$

(b) Reduction of greenhouse gases = 32 g.

Amount of $CO_2 = \frac{32}{44}$ mol;

volume of $CO_2 = \frac{32}{44} \times 24.0 = 17.45\ dm^3$.

2.4.3

1 Natural gas is first reformed into hydrogen and carbon dioxide:

$CH_4(g) + 2H_2O(g) \longrightarrow CO_2(g) + 4H_2(g)$

The CO_2 produced is captured and stored in porous rock in old oilfields. The hydrogen is burnt to generate power, with water vapour as the product.

$2H_2(g) + O_2(g) \longrightarrow 2H_2O(l)$

2 In mineral storage, CO_2 is reacted with metal oxides to form stable carbonate rock, for example:

$CaO(s) + CO_2(g) \longrightarrow CaCO_3(s)$

Unfortunately, the natural reaction for this process is very slow and efforts to increase the rate are extremely energy intensive.

2.4.4

1 Ultraviolet (UV) radiation from the Sun has wavelengths in the range 270 to 400 nm. The ozone layer filters out the shorter wavelengths, less than 320 nm. These shorter wavelengths would be very damaging to life on Earth. The ozone converts this UV radiation into heat and so the ozone layer has a higher temperature that other parts of the upper atmosphere.

2 O_2 molecules absorb high energy UV radiation with a wavelength less than 240 nm, forming atomic oxygen:

$O_2 + (\text{radiation} < 240\ nm) \rightarrow 2O$

The O atoms react with O_2 molecules to form ozone molecules, O_3. This process generates heat.

$O_2 + O \rightarrow O_3 + \text{heat}$

The O_3 molecules formed absorb UV (wavelengths between 240 and 310 nm), converting O_3 molecules back to O_2 molecules and O atoms.

$O_3 + (\text{radiation} < 310\ nm) \rightarrow O_2 + O$

The atomic oxygen produced immediately reacts with other O_2 molecules to reform ozone:

$O_2 + O \rightarrow O_3 + \text{heat}$

. . . and so the cycle continues.

The overall effect is to convert penetrating UV radiation into heat, without any net loss of ozone:

$O_2 + O \rightleftharpoons O_3$

2.4.5

1 When UV radiation strikes a CFC molecule in the stratosphere, a C–Cl bond breaks, producing a chlorine radical, Cl•.

e.g. $CFCl_3 \longrightarrow Cl\bullet + \bullet CFCl_2$ *initiation*

The breakdown of ozone takes place in two propagation steps:

$Cl\bullet + O_3 \rightarrow ClO\bullet + O_2$ *propagation step 1*

$ClO\bullet + O \rightarrow Cl\bullet + O_2$ *propagation step 2*

The two propagation steps above repeat in a cycle, many thousands of times, with each cycle removing another ozone molecule.

2 Any three from:

Production of CFCs stopped by 2000.

Production of tetrachloromethane (used as a solvent) stopped by 2000.

Production of halons (used in fire extinguishers) phased out by 2000.

Production of 1,1,1-trichloroethane (used as a solvent) phased out by 2005.

The use of HCFCs to replace CFCs in about 15% of applications.

3 HCFCs and HFCs are now thought to contribute to global warming.

2.4.6

1 Nitrogen oxides are produced during the high temperature combustion process as some of the nitrogen from the air is oxidised by the oxygen. Two oxides of nitrogen are produced: nitrogen monoxide, NO; and nitrogen dioxide, NO_2. So the mixture is commonly referred to as NO_x.
Carbon monoxide, CO, is a poisonous gas, emitted into the atmosphere from the incomplete combustion of hydrocarbons and other organic compounds.

2 $C_8H_{18}(l) + 12\frac{1}{2}O_2(g) \longrightarrow 8CO_2(g) + 9H_2O(g)$

$2NO(g) + 2CO(g) \longrightarrow N_2(g) + 2CO_2(g)$

3 CO and NO gas molecules diffuse over the catalytic surface of the metal. Some of the molecules are adsorbed on to the metal surface, forming temporary bonds. These bonds hold the molecules in the correct position to react together on the metal surface. After the reaction, the CO_2 and N_2 products are desorbed from the surface and diffuse away from the catalytic surface.

2.4.7

1 (a) atom economy =

$$\frac{\text{molecular mass of the desired product}}{\text{sum of molecular masses of all products}} \times 100$$

Preparation 1 is an addition reaction so the atom economy must be 100%.

(c)

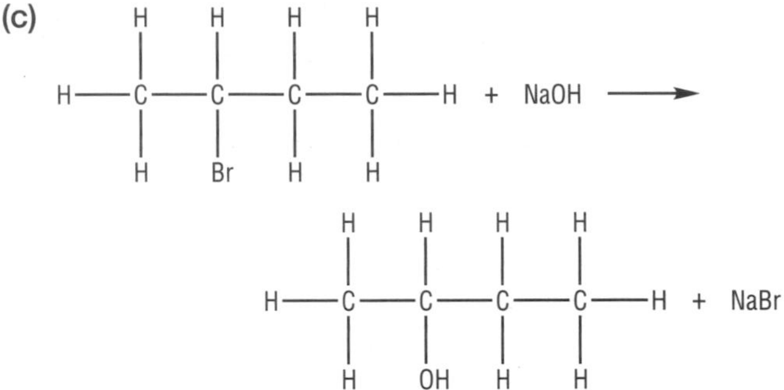

(d) It acts as a catalyst.

(e) Addition of bromine water to the alkene causes the bromine water to be decolourised.

(f) The rate of reaction would decrease as the C–Cl bond in the chloroalkane is stronger than the C–Br bond in the bromoalkane.

14 **(a)** $C_4H_{10}O$.

(b) There is a C=O group present.

(c) $m/z = 15$: CH_3^+ fragment and $m/z = 29$: $C_2H_5^+$ fragment.

(d) $C_4H_{10}O$.

(e)

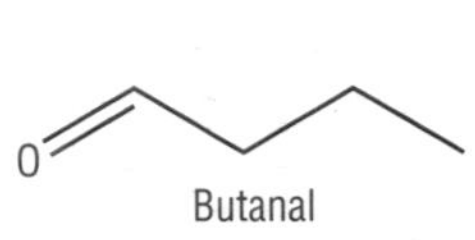

Butanal

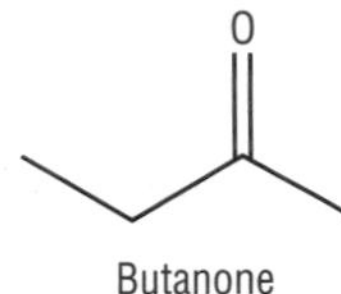

Butanone

2.3 Energy practice

1 An exothermic reaction is one that gives out heat energy to the surroundings.

$CH_4(g) + 2O_2(g) \longrightarrow CO_2(g) + 2H_2O(l)$ (*any* combustion reaction is a good example)

2 Standard enthalpy change of combustion is the enthalpy change that takes place when *one* mole of a substance reacts *completely* with oxygen under standard conditions, 1 atm pressure and 25 °C (298 K), all reactants and products being in their standard states.

3 **(a)** $4C(s) + 5H_2(g) \longrightarrow C_4H_{10}(g)$

(b) $Na(s) + ½H_2(g) + C(s) + 1½O_2(g) \longrightarrow NaHCO_3(s)$

4 **(a)** Amount of $C_5H_{11}OH$ reacted = 1.24/88.0
= 0.0141 mol.

Energy released = 500 × 4.18 × 19 = 39710 J
= 39710 kJ.

$\Delta H = -39.710/0.0141 = -2816$ kJ mol^{-1}.

5

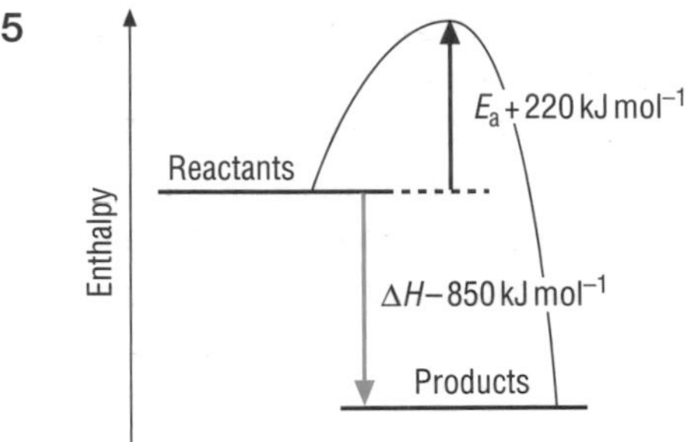

6 **(a)** Average bond enthalpy is the *average* enthalpy change that occurs when breaking, by homolytic fission, *one* mol of a given type of bond in the molecules of a gaseous species.

(i) (612 + 463) – (413 +347 + 358) = –43 kJ mol^{-1}

(ii) [(9 × 413) + (3 × 347) + 358 + 463 + (6 × 497)] – [(8 × 805) + (10 × 463)] = –2509 kJ mol^{-1}

7 [(4 × –394) + (2 × –286)] – [–2597] = +449 kJ mol^{-1}

8 **(a)** When the concentration of a reactant is increased, there are *more* molecules in the *same* volume. More collisions take place in a certain length of time and the rate of reaction *increases*.

(b) An increase in temperature *increases* the reaction rate. The molecules move faster with more kinetic energy, resulting in *more* collisions in a certain length of time. As the molecules have more energy, more of the collisions *exceed* the activation energy, leading to a chemical reaction.

9

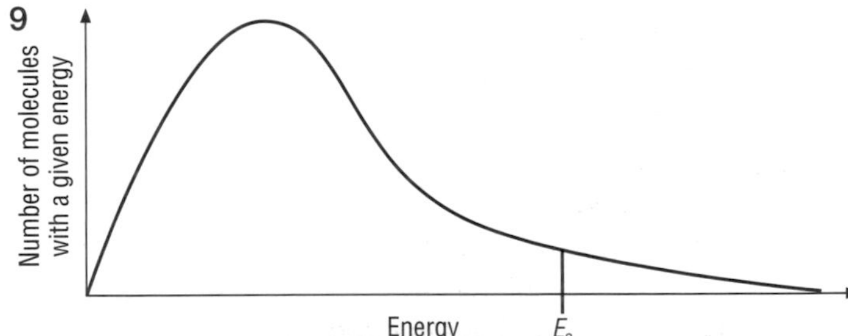

10 le Chatelier's principle states that when a system in dynamic equilibrium is subjected to a change, the system readjusts itself to minimise the effect of the change and to reach a new equilibrium.

11 For maximum equilibrium yield, *high* pressure and *low* temperature would be used.

The production of $NH_3(g)$ from $N_2(g)$ and $H_2(g)$ produces fewer gas molecules (4 molecules → 2 molecules). *Increasing* the total pressure causes the position of equilibrium to move to the side with *fewer* gas molecules, as this will decrease the pressure. Here it moves to the *right* side, so high pressure favours ammonia production.

The forward reaction is exothermic (ΔH –ve), favoured by using low temperature. The production of $NH_3(g)$ from $N_2(g)$ and $H_2(g)$ is *exothermic* and releases heat energy (ΔH: –ve), increasing the temperature of the surroundings. *Decreasing* the temperature causes the position of equilibrium to move to the exothermic side, releasing energy and increasing the temperature of the surroundings. Since the equilibrium has shifted to the *right*, low temperatures favours ammonia production. A catalyst would also be used in this process, as it would allow the ammonia production to take place *faster* and at lower temperatures. Lower temperatures will increase the equilibrium yield.

12 (a) (i) When a system in equilibrium is subjected to a change, the system readjusts itself to minimise the effect of the change.

(b) (i) The system appears to become browner. The position of equilibrium moves to the left to minimise the change that has been made. It does this by shifting in the endothermic direction which is to the left.

(ii) An increase in pressure causes the equilibrium to move to the side with fewer gas molecules. In this case, the right-hand side has fewer gas molecules, so the system becomes lighter in colour as the equilibrium shifts to the right.

(iii) No change – a catalyst increases the rate of both the forward and reverse reactions equally.

13 (a) The enthalpy change when one mole of substance in its standard state is completely burned in an excess of oxygen.

(b) (i) $C_3H_7OH(l) + 4\frac{1}{2}O_2(g) \longrightarrow 3CO_2(g) + 4H_2O(l)$

(ii) 2.82 kJ

(iii) amount = 0.100/60.0 = 1.67×10^{-3} mol

$\Delta H_c = -1689$ kJ mol^{-1}

(iv) Heat lost to the atmosphere.

Experiment carried out under non-standard conditions.

Incomplete combustion.

2.4 Resources practice

1 Carbon dioxide: C=O bonds absorb infrared radiation.
Water vapour: O–H bonds absorb infrared radiation.
Methane: C–H bonds absorb infrared radiation.

2 Carbon capture and storage (CCS), or sequestration, captures CO_2 from power stations and stores it away safely, instead of being released into the atmosphere.

3 O_2 molecules absorb high-energy UV radiation with a wavelength less than 240 nm, forming atomic oxygen:

O_2 + (radiation < 240 nm) $\rightarrow$ 2O

The O atoms react with O_2 molecules to form ozone molecules, O_3. This process generates heat.

$O_2 + O \rightarrow O_3$ + heat

The O_3 molecules formed absorb UV (wavelengths between 240 and 310 nm), converting O_3 molecules back to O_2 molecules and O atoms.

O_3 + (radiation < 310 nm) $\rightarrow O_2 + O$

The atomic oxygen produced immediately reacts with other O_2 molecules to reform ozone:

$O_2 + O \rightarrow O_3$ + heat

. . . and so the cycle continues.

The overall effect is to convert penetrating UV radiation into heat, without any net loss of ozone:

$O_2 + O \rightleftharpoons O_3$

4 Nitrogen oxides, such as NO and NO_2 are both radicals. The breakdown of ozone proceeds in two propagation steps.

$\cdot NO + O_3 \rightarrow NO_2 + O_2$ *propagation step 1*

$NO_2 + O \rightarrow \cdot NO + O_2$ *propagation step 2*

This process repeats many thousand of times in a chain reaction. Each propagation cycle removes an ozone molecule.

5 Nitrogen oxides are produced during the high temperature combustion process as some of the nitrogen from the air is oxidised by the oxygen. Two oxides of nitrogen are produced: nitrogen monoxide, NO; and nitrogen dioxide, NO_2. So the mixture is commonly referred to as NO_x.

Carbon monoxide, CO, is a poisonous gas, emitted into the atmosphere from the incomplete combustion of hydrocarbons and other organic compounds.

6 CO and NO gas molecules diffuse over the catalytic surface of the metal. Some of the molecules are adsorbed onto the metal surface, forming temporary bonds. These bonds hold the molecules in the correct position to react together on the surface of the metal:

$2NO(g) + 2CO(g) \longrightarrow N_2(g) + 2CO_2(g)$

After the reaction, the CO_2 and N_2 products are desorbed from the surface and diffuse away from the catalytic surface.

7 (a) The haemoglobin in our blood carries oxygen to tissues and organs. Carbon monoxide can also bind strongly to haemoglobin, reducing the amount of oxygen supplied to tissues and organs. The heart and the brain are severely affected and people with cardiovascular disease are especially susceptible to this type of poisoning.

(b) Nitrogen oxides are respiratory irritants and low levels of nitrogen oxides affect asthmatics.

(c) Ozone is an air pollutant with harmful effects on the respiratory systems of animals.

(d) There are three types of UV based on the wavelength of the radiation: UV-a; UV-b; and UV-c. Exposure to UV-b can cause sunburn and sometimes genetic damage, which can result in skin cancer.

8 The principles of green chemistry aim to generate less waste, use processes with improved atom economy, use less energy, recycle materials and use renewable resources.

9 (a) C_4H_{10}

(b) Liquid under pressure but vaporise when pressure is released.

(c) $C_4H_{10}(g) + 6\frac{1}{2}O_2(g) \longrightarrow 4CO_2(g) + 5H_2O(l)$

(d) Alkanes have replaced CFCs which cause damage to the ozone layer.

10 (a) (i)

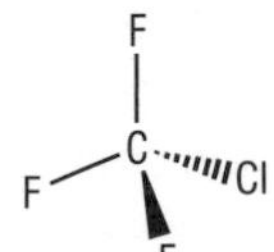

(ii) 109.5°

(b) (i) The breaking of a covalent bond with one of the bonded electrons going to each atom, forming two radicals.

(ii) $CF_3Cl \longrightarrow Cl\cdot + \cdot CF_3$

(c) (i) $Cl + O_3 \longrightarrow ClO + O_2$

$ClO + O \longrightarrow Cl + O_2$

(ii) Cl· is the catalyst. Cl· is used up in the first step but is regenerated in the second step.

11 (a) (i) $N_2(g) + O_2(g) \longrightarrow 2NO(g)$

(ii) $2NO(g) + O_2(g) \longrightarrow 2NO_2(g)$

(b) N in NO_2 has oxidation state of +4

In forming HNO_2, oxidation state changes from +4 to +3: reduction.

In forming HNO_3, oxidation state changes from +4 to +5: oxidation.

(c) (i) amount of NO_x = 150/24 000 = 6.25×10^{-3} mol

(ii) number of molecules of $NO_x = 6.25 \times 10^{-3} \times 6.02 \times 10^{23}$

$= 3.76 \times 10^{21}$ molecules

(d) (i) Relative molecular mass of NO_x = $0.250/6.25 \times 10^{-3} = 40$

(ii) $M_r(NO) = 30$; $M_r(NO_2) = 46$

40 is closer to 46 than 30, so there are more molecules of NO_2 in the NO_x sample.

(e) Catalytic converters convert NO_x into non-toxic products.

12 (a) (i) From UV radiation

(ii)

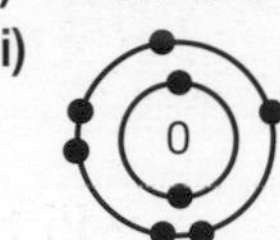

(iii) O is a diradical as there are two unpaired electrons present.

(iv)

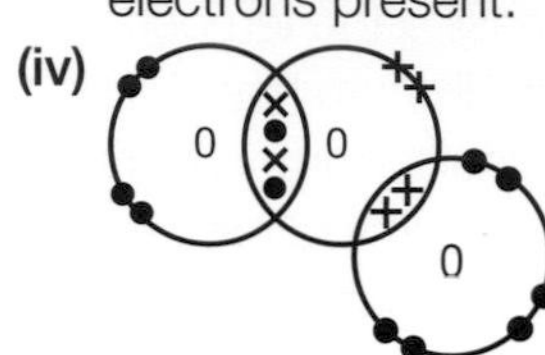

(b) Amount of O_3 in 150 kg = $150 \times 10^3/48.0 = 3.125 \times 10^6$ mol

amount of Cl radicals in 1 g = 1 /35.5 = 2.82×10^{-2} mol

1 mol Cl destroys $3.125 \times 10^3/2.82 \times 10^{-2} = 1.11 \times 10^5$ mol O_3

1 Cl radical destroys 1.11×10^5 O_3 molecules

Examination answers

Answers to examination questions will be found on the Exam Café CD.

Glossary

acid A species that is a proton donor.

activation energy The minimum energy required to start a reaction by the breaking of bonds.

addition polymer A very long molecular chain, formed by repeated addition reactions of many unsaturated alkene molecules (monomers).

addition polymerisation The process in which unsaturated alkene molecules (monomers) add on to a growing polymer chain one at a time to form a very long saturated molecular chain (the addition polymer).

addition reaction A reaction in which a reactant is added to an unsaturated molecule to make a saturated molecule.

adsorption The process that occurs when a gas, liquid or solute is held to the surface of a solid or, more rarely, a liquid.

alicyclic hydrocarbon A hydrocarbon with carbon atoms joined together in a ring structure.

aliphatic hydrocarbon A hydrocarbon with carbon atoms joined together in straight or branched chains.

alkali A type of base that dissolves in water forming hydroxide ions, $OH^-(aq)$ ions.

alkanes The homologous series with the general formula: C_nH_{2n+2}.

alkyl group An alkane with a hydrogen atom removed, e.g. CH_3, C_2H_5; alkyl groups are often shown as 'R'.

amount of substance The quantity whose unit is the mole. Chemists use 'amount of substance' as a means of counting atoms.

anhydrous A substance containing no water molecules.

anion A negatively charged ion.

atom economy

$$\text{atom economy} = \frac{\text{molecular mass of the desired product}}{\text{sum of molecular masses of all products}} \times 100$$

atomic orbital A region within an atom that can hold up to two electrons, with opposite spins.

atomic (proton) number The number of protons in the nucleus of an atom.

average bond enthalpy The average enthalpy change that takes place when breaking by homolytic fission 1 mol of a given type of bond in the molecules of a gaseous species.

Avogadro constant, N_A The number of atoms per mole of the carbon-12 isotope (6.02×10^{23} mol^{-1}).

base A species that is a proton acceptor.

biodegradable material A substance that is broken down naturally in the environment by living organisms.

Boltzmann distribution The distribution of energies of molecules at a particular temperature, usually shown as a graph.

bond enthalpy The enthalpy change that takes place when breaking by homolytic fission 1 mol of a given bond in the molecules of a gaseous species.

carbanion An organic ion in which a carbon atom has a negative charge.

carbocation An organic ion in which a carbon atom has a positive charge.

catalyst A substance that increases the rate of a chemical reaction without being used up in the process.

cation A positively charged ion.

***cis–trans* isomerism** A special type of *E/Z* isomerism in which there is a non-hydrogen group and hydrogen on each C of a C=C double bond: the *cis* isomer (*Z* isomer) has the H atoms on each carbon on the same side; the *trans* isomer (*E* isomer) has the H atoms on each carbon on different sides.

compound A substance formed from two or more chemically bonded elements in a fixed ratio, usually shown by a chemical formula.

concentration The amount of solute, in mol, per 1 dm^3 (1000 cm^3) of solution.

coordinate bond A shared pair of electrons which has been provided by one of the bonding atoms only; also called a dative covalent bond.

covalent bond A bond formed by a shared pair of electrons.

cracking The breaking down of long-chained saturated hydrocarbons to form a mixture of shorter-chained alkanes and alkenes.

curly arrow A symbol used in reaction mechanisms to show the movement of an electron pair in the breaking or formation of a covalent bond.

dative covalent bond A shared pair of electrons which has been provided by one of the bonding atoms only; also called a coordinate bond.

dehydration An elimination reaction in which water is removed from a saturated molecule to make an unsaturated molecule.

delocalised electrons Electrons that are shared between more than two atoms.

displacement reaction A reaction in which a more reactive element displaces a less reactive element from an aqueous solution of the latter's ions.

displayed formula A formula showing the relative positioning of all the atoms in a molecule and the bonds between them.

disproportionation The oxidation and reduction of the same element in a redox reaction.

dynamic equilibrium The equilibrium that exists in a closed system when the rate of the forward reaction is equal to the rate of the reverse reaction.

***E/Z* isomerism** A type of stereoisomerism in which different groups attached to each carbon of a C=C double bond may be arranged differently in space because of the restricted rotation of the C=C bond.

electron configuration The arrangement of electrons in an atom.

electronegativity A measure of the attraction of a bonded atom for the pair of electrons in a covalent bond.

electron shielding The repulsion between electrons in different inner shells. Shielding reduces the net attractive force from the positive nucleus on the outer-shell electrons.

electrophile An atom (or group of atoms) that is attracted to an electron-rich centre or atom, where it accepts a pair of electrons to form a new covalent bond.

electrophilic addition A type of addition reaction in which an electrophile is attracted to an electron-rich centre or atom, where it accepts a pair of electrons to form a new covalent bond.

elimination reaction The removal of a molecule from a saturated molecule to make an unsaturated molecule.

empirical formula The simplest whole-number ratio of atoms of each element present in a compound.

endothermic reaction A reaction in which the enthalpy of the products is greater than the enthalpy of the reactants, resulting in heat being taken in from the surroundings (ΔH +ve).

enthalpy, H The heat content that is stored in a chemical system.

(standard) enthalpy change of combustion, $\Delta H_c^{\ominus}$ The enthalpy change that takes place when one mole of a substance reacts completely with oxygen under standard conditions, all reactants and products being in their standard states.

(standard) enthalpy change of formation, $\Delta H_f^{\ominus}$ The enthalpy change that takes place when one mole of a compound in its standard state is formed from its constituent elements in their standard states under standard conditions.

(standard) enthalpy change of reaction, $\Delta H_r^{\ominus}$ The enthalpy change that accompanies a reaction in the molar quantities expressed in a chemical equation under standard conditions, all reactants and products being in their standard states.

enthalpy cycle A diagram showing alternative routes between reactants and products that allows the indirect determination of an enthalpy change from other known enthalpy changes using Hess' law.

enthalpy profile diagram A diagram for a reaction to compare the enthalpy of the reactants with the enthalpy of the products.

esterification The reaction of an alcohol with a carboxylic acid to produce an ester and water.

exothermic reaction A reaction in which the enthalpy of the products is smaller than the enthalpy of the reactants, resulting in heat loss to the surroundings (ΔH –ve).

fractional distillation The separation of the components in a liquid mixture into fractions which differ in boiling point (and hence chemical composition) by means of distillation, typically using a fractionating column.

fragmentation The process in mass spectrometry that causes a positive ion to split into pieces, one of which is a positive fragment ion.

functional group The part of the organic molecule responsible for its chemical reactions.

general formula The simplest algebraic formula of a member of a homologous series. For example, the general formula of the alkanes is C_nH_{2n+2}.

giant covalent lattice A three-dimensional structure of atoms, bonded together by strong covalent bonds.

giant ionic lattice A three-dimensional structure of oppositely charged ions, bonded together by strong ionic bonds.

giant metallic lattice A three-dimensional structure of positive ions and delocalised electrons, bonded together by strong metallic bonds.

greenhouse effect The process in which the absorption and subsequent emission of infrared radiation by atmospheric gases warms the lower atmosphere and the planet's surface.

group A vertical column in the Periodic Table. Elements in a group have similar chemical properties and their atoms have the same number of outer-shell electrons.

Hess' law If a reaction can take place by more than one route and the initial and final conditions are the same, the total enthalpy change is the same for each route.

heterogeneous catalysis A reaction in which the catalyst has a different physical state from the reactants; frequently, reactants are gases whilst the catalyst is a solid.

heterolytic fission The breaking of a covalent bond with both of the bonded electrons going to one of the atoms, forming a cation (+ ion) and an anion (– ion).

homogeneous catalysis A reaction in which the catalyst and reactants are in the same physical state, which is most frequently the aqueous or gaseous state.

homologous series A series of organic compounds with the same functional group, but with each successive member differing by CH_2.

homolytic fission The breaking of a covalent bond with one of the bonded electrons going to each atom, forming two radicals.

hydrated Crystalline and containing water molecules.

hydrocarbon An organic compound of hydrogen and carbon only.

hydrogen bond A strong dipole–dipole attraction between an electron-deficient hydrogen atom ($O–H^{\delta+}$ or $N–H^{\delta+}$) on one molecule and a lone pair of electrons on a highly electronegative atom (H–**O:**$^{\delta-}$ or H–**N:**$^{\delta-}$) on a different molecule.

hydrolysis A reaction with water or aqueous hydroxide ions that breaks a chemical compound into two compounds.

initiation The first step in a radical substitution in which the free radicals are generated by ultraviolet radiation.

intermolecular force An attractive force between neighbouring molecules. Intermolecular forces can be van der Waals' forces (induced dipole–dipole forces), permanent dipole–dipole forces or hydrogen bonds.

ion A positively or negatively charged atom or (covalently bonded) group of atoms (a molecular ion).

ionic bond The electrostatic attraction between oppositely charged ions.

(first) ionisation energy The energy required to remove one electron from each atom in one mole of gaseous atoms to form one mole of gaseous 1+ ions.

(second) ionisation energy The energy required to remove one electron from each ion in one mole of gaseous 1+ ions to form one mole of gaseous 2+ ions.

(successive) ionisation energy A measure of the energy required to remove each electron in turn, e.g. the second ionisation energy is the energy required to remove one electron from each ion in one mole of gaseous 1+ ions to form one mole of gaseous 2+ ions.

isotopes Atoms of the same element with different numbers of neutrons and different masses.

le Chatelier's principle When a system in dynamic equilibrium is subjected to a change, the position of equilibrium will shift to minimise the change.

limiting reagent The substance in a chemical reaction that runs out first.

lone pair An outer shell pair of electrons that is not involved in chemical bonding.

mass (nucleon) number The number of particles (protons and neutrons) in the nucleus.

mechanism A sequence of steps showing the path taken by electrons in a reaction.

metallic bond The electrostatic attraction between positive metal ions and delocalised electrons.

molar mass, *M* The mass per mole of a substance. The units of molar mass are g mol^{-1}.

molar volume The volume per mole of a gas. The units of molar volume are $dm^3\,mol^{-1}$. At room temperature and pressure the molar volume is approximately 24.0 $dm^3\,mol^{-1}$.

mole The amount of any substance containing as many particles as there are carbon atoms in exactly 12 g of the carbon-12 isotope.

molecular formula The number of atoms of each element in a molecule.

molecular ion, M^+ The positive ion formed in mass spectrometry when a molecule loses an electron.

molecule A small group of atoms held together by covalent bonds.

monomer A small molecule that combines with many other monomers to form a polymer.

nomenclature A system of naming compounds.

nucleophile An atom (or group of atoms) that is attracted to an electron-deficient centre or atom, where it donates a pair of electrons to form a new covalent bond.

nucleophilic substitution A type of substitution reaction in which a nucleophile is attracted to an electron-deficient centre or atom, where it donates a pair of electrons to form a new covalent bond.

oxidation Loss of electrons or an increase in oxidation number.

oxidation number A measure of the number of electrons that an atom uses to bond with atoms of another element. Oxidation numbers are derived from a set of rules.

oxidising agent A reagent that oxidises (takes electrons from) another species.

percentage yield

$$\%\ \text{yield} = \frac{\text{actual amount, in mol, of product}}{\text{theoretical amount, in mol, of product}} \times 100$$

period A horizontal row of elements in the Periodic Table. Elements show trends in properties across a period.

periodicity A regular periodic variation of properties of elements with atomic number and position in the Periodic Table.

permanent dipole A small charge difference across a bond resulting from a difference in electronegativities of the bonded atoms.

permanant dipole–dipole force An attractive force between permanent dipoles in neighbouring polar molecules.

pi-bond (π-bond) The reactive part of a double bond formed above and below the plane of the bonded atoms by sideways overlap of p-orbitals.

polar covalent bond A bond with a permanent dipole.

polar molecule A molecule with an overall dipole, taking into account any dipoles across bonds.

polymer A long molecular chain built up from monomer units.

precipitation reaction The formation of a solid from a solution during a chemical reaction. Precipitates are often formed when two aqueous solutions are mixed together.

principal quantum number, *n* A number representing the relative overall energy of each orbital, which increases with distance from the nucleus. The sets of orbitals with the same *n* value are referred to as electron shells or energy levels.

propagation The two repeated steps in radical substitution that build up the products in a chain reaction.

radical A species with an unpaired electron.

radical substitution A type of substitution reaction in which a radical replaces a different atom or group of atoms.

rate of reaction The change in concentration of a reactant or a product in a given time.

redox reaction A reaction in which both reduction and oxidation take place.

reducing agent A reagent that reduces (adds electrons to) another species.

reduction Gain of electrons or a decrease in oxidation number.

reflux The continual boiling and condensing of a reaction mixture to ensure that the reaction takes place without the contents of the flask boiling dry.

relative atomic mass, A_r The weighted mean mass of an atom of an element compared with one-twelfth of the mass of an atom of carbon-12.

relative formula mass The weighted mean mass of a formula unit compared with one-twelfth of the mass of an atom of carbon-12.

relative isotopic mass The mass of an atom of an isotope compared with one-twelfth of the mass of an atom of carbon-12.

relative molecular mass, M_r The weighted mean mass of a molecule compared with one-twelfth of the mass of an atom of carbon-12.

repeat unit A specific arrangement of atoms that occurs in the structure over and over again. Repeat units are included in brackets, outside of which is the symbol n.

salt A chemical compound formed from an acid, when a H^+ ion from the acid has been replaced by a metal ion or another positive ion, such as the ammonium ion, NH_4^+.

saturated hydrocarbon A hydrocarbon with single bonds only.

shell A group of atomic orbitals with the same principal quantum number, *n*. Also known as a main energy level.

simple molecular lattice A three-dimensional structure of molecules, bonded together by weak intermolecular forces.

skeletal formula A simplified organic formula, with hydrogen atoms removed from alkyl chains, leaving just a carbon skeleton and associated functional groups.

species Any type of particle that takes part in a chemical reaction.

specific heat capacity, *c* The energy required to raise the temperature of 1 g of a substance by 1 °C.

spectator ions Ions that are present but take no part in a chemical reaction.

standard conditions A pressure of 100 kPa (1 atmosphere), a stated temperature, usually 298 K (25 °C), and a concentration of 1 mol dm^{-3} (for reactions with aqueous solutions).

standard enthalpies see enthalpy.

standard solution A solution of known concentration. Standard solutions are normally used in titrations to determine unknown information about another substance.

standard state The physical state of a substance under the standard conditions of 100 kPa (1 atmosphere) and 298 K (25 °C).

stereoisomers Compounds with the same structural formula but with a different arrangement of the atoms in space.

stoichiometry The molar relationship between the relative quantities of substances taking part in a reaction.

stratosphere The second layer of the Earth's atmosphere, containing the 'ozone layer', about 10 km to 50 km above the Earth's surface.

structural formula A formula showing the minimal detail for the arrangement of atoms in a molecule.

structural isomers Molecules with the same molecular formula but with different structural arrangements of atoms.

sub-shell A group of the same type of atomic orbitals (s, p, d or f) within a shell.

substitution reaction A reaction in which an atom or group of atoms is replaced with a different atom or group of atoms.

termination The step at the end of a radical substitution when two radicals combine to form a molecule.

thermal decomposition The breaking up of a chemical substance with heat into at least two chemical substances.

troposphere The lowest layer of the Earth's atmosphere, extending from the Earth's surface up to about 7 km (above the poles) and to about 20 km (above the tropics).

unsaturated hydrocarbon A hydrocarbon containing carbon-to-carbon multiple bonds.

van der Waals' forces Very weak attractive forces between induced dipoles in neighbouring molecules.

volatility The ease with which a liquid turns into a gas. Volatility increases as boiling point decreases.

water of crystallisation Water molecules that form an essential part of the crystalline structure of a compound.

The Periodic Table of the Elements

Key

Relative atomic mass
Atomic symbol
Name
Atomic (proton number)

1	2											3	4	5	6	7	0
							1.0 **H** Hydrogen 1										4.0 **He** Helium 2
6.9 **Li** Lithium 3	9.0 **Be** Beryllium 4											10.8 **B** Boron 5	12.0 **C** Carbon 6	14.0 **N** Nitrogen 7	16.0 **O** Oxygen 8	19.0 **F** Fluorine 9	20.2 **Ne** Neon 10
23.0 **Na** Sodium 11	24.3 **Mg** Magnesium 12											27.0 **Al** Aluminium 13	28.1 **Si** Silicon 14	31.0 **P** Phosphorus 15	32.1 **S** Sulfur 16	35.5 **Cl** Chlorine 17	39.9 **Ar** Argon 18
39.1 **K** Potassium 19	40.1 **Ca** Calcium 20	45.0 **Sc** Scandium 21	47.9 **Ti** Titanium 22	50.9 **V** Vanadium 23	52.0 **Cr** Chromium 24	54.9 **Mn** Manganese 25	55.8 **Fe** Iron 26	58.9 **Co** Cobalt 27	58.7 **Ni** Nickel 28	63.5 **Cu** Copper 29	65.4 **Zn** Zinc 30	69.7 **Ga** Gallium 31	72.6 **Ge** Germanium 32	74.9 **As** Arsenic 33	79.0 **Se** Selenium 34	79.9 **Br** Bromine 35	83.8 **Kr** Krypton 36
85.5 **Rb** Rubidium 37	87.6 **Sr** Strontium 38	88.9 **Y** Yttrium 39	91.2 **Zr** Zirconium 40	92.9 **Nb** Niobium 41	95.9 **Mo** Molybdenum 42	(98) **Tc** Technetium 43	101.1 **Ru** Ruthenium 44	102.9 **Rh** Rhodium 45	106.4 **Pd** Palladium 46	107.9 **Ag** Silver 47	112.4 **Cd** Cadmium 48	114.8 **In** Indium 49	118.7 **Sn** Tin 50	121.8 **Sb** Antimony 51	127.6 **Te** Tellurium 52	126.9 **I** Iodine 53	131.3 **Xe** Xenon 54
132.9 **Cs** Caesium 55	137.3 **Ba** Barium 56	138.9 **La*** Lanthanum 57	178.5 **Hf** Hafnium 72	180.9 **Ta** Tantalum 73	183.8 **W** Tungsten 74	186.2 **Re** Rhenium 75	190.2 **Os** Osmium 76	192.2 **Ir** Iridium 77	195.1 **Pt** Platinum 78	197.0 **Au** Gold 79	200.6 **Hg** Mercury 80	204.4 **Ti** Thallium 81	207.2 **Pb** Lead 82	209.0 **Bi** Bismuth 83	(209) **Po** Polonium 84	(210) **At** Astatine 85	(222) **Rn** Radon 86
(223) **Fr** Francium 87	(226) **Ra** Radium 88	(227) **Ac*** Actinium 89	(261) **Rf** Rutherfordium 104	(262) **Db** Dubnium 105	(266) **Sg** Seaborgium 106	(264) **Bh** Bohrium 107	(277) **Hs** Hassium 108	(268) **Mt** Meitnerium 109	(271) **Ds** Darmstadtium 110	(272) **Rg** Roentgenium 111	Elements with atomic numbers 112–116 have been reported but not fully authenticated						

140.1 **Ce** Cerium 58	140.9 **Pr** Praseodynium 59	144.2 **Nd** Neodymium 60	144.9 **Pm** Promethium 61	150.4 **Sm** Samarium 62	152.0 **Eu** Europium 63	157.2 **Gd** Gadolinium 64	158.9 **Tb** Terbium 65	162.5 **Dy** Dysprosium 66	164.9 **Ho** Holmium 67	167.3 **Er** Erbium 68	168.9 **Tm** Thulium 69	173.0 **Yb** Ytterbium 70	175.0 **Lu** Lutetium 71
232.0 **Th** Thorium 90	(231) **Pa** Protactinium 91	238.1 **U** Uranium 92	(237) **Np** Neptunium 93	(242) **Pu** Plutonium 94	(243) **Am** Americium 95	(247) **Cm** Curium 96	(245) **Bk** Berkelium 97	(251) **Cf** Californium 98	(254) **Es** Einsteinium 99	(253) **Fm** Fermium 100	(256) **Md** Mendeleevium 101	(254) **No** Nobelium 102	(257) **Lr** Lawrencium 103

Data Sheet

General information

- 1 mol of gas molecules occupies 24.0 dm^3 at room temperature and pressure, RTP.
- Avogadro constant, $N_A = 6.02 \times 10^{23}$ mol^{-1}.

Characteristic infrared absorptions in organic molecules

Bond	Location	Wavenumber/cm^{-1}
C–O	alcohols, esters, carboxylic acids	1000–1300
C=O	aldehydes, ketones, carboxylic acids, esters, amides	1640–1750
C–H	organic compound with a C–H bond	2850–3100
O–H	carboxylic acids	2500–3300 (very broad)
N–H	amines, amides	3200–3500
O–H	alcohols, phenols	3200–3550 (broad)

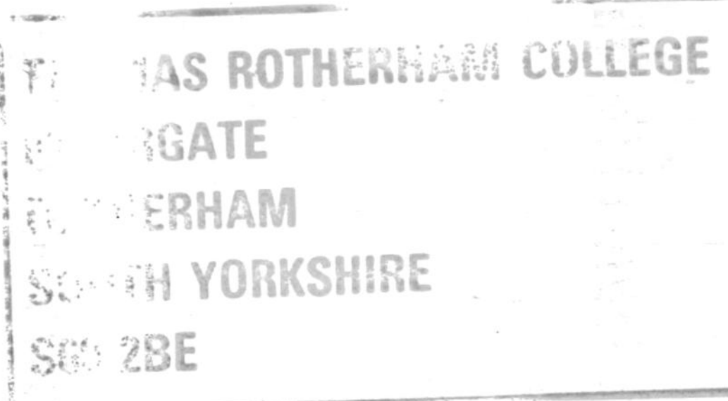

Index

Your Exam Café CD-Rom

In the back of this book you will find an Exam Café CD-ROM. This CD contains advice on study skills, interactive questions to test your learning, a link to our unique partnership with New Scientist, and many more useful features. Load it onto your computer to take a closer look.

Amongst the files on the CD are PDF files, for which you will need the Adobe Reader program, and editable Microsoft Word documents for you to alter and print off if you wish.

Minimum system requirements:

- Windows 2000, XP Pro or Vista
- Internet Explorer 6 or Firefox 2.0
- Flash Player 8 or higher plug-in
- Pentium III 900 MHz with 256 Mb RAM

To run your Exam Café CD, insert it into the CD drive of your computer. It should start automatically; if not, please go to My Computer (Computer on Vista), click on the CD drive and double-click on 'start.html'.

If you have difficulties running the CD, or if your copy is not there, please contact the helpdesk number given below.

Software support

For further software support between the hours of 8.30–5.00 (Mon-Fri), please contact:

Tel: 01865 888108

Fax: 01865 314091

Email: software.enquiries@pearson.com